The Reproductive Neuroendocrinology of Aging and Drug Abuse

Edited by

Dipak K. Sarkar, Ph.D., D. Phil.
Associate Professor of Physiology
Department of Veterinary and Comparative Anatomy, Pharmacology and Physiology
College of Veterinary Medicine
Washington State University
Pullman, Washington

Charles D. Barnes, Ph.D.
Professor and Chairman
Department of Veterinary and Comparative Anatomy, Pharmacology and Physiology
College of Veterinary Medicine
Washington State University
Pullman, Washington

CRC Press
Boca Raton Ann Arbor London Tokyo

Library of Congress Cataloging-in-Publication Data

The reproductive neuroendocrinology of aging and drug abuse / edited by Dipak K. Sarkar and Charles D. Barnes.
p. cm.
Includes bibliographical references and index.
ISBN 0-8493-2451-3
1. Reproduction—Physiological aspects. 2. Reproduction—Endocrine aspects. 3. Reproduction—Effect of drugs on. 4. Neuroendocrinology. 5. Drug abuse—Endocrine aspects. 6. Aging. I. Sarkar, Dipak K. II. Barnes, Charles D., 1935–.
[DNLM: 1. Reproduction—physiology. 2. Endocrine Glands—physiology. 3. Nervous System—physiology. 4. Aging—metabolism. 5. Substance Abuse—metabolism. WQ 205 R42877 1994]
QP251.R444545 1994
612.6—dc20
DNLM/DLC
for Library of Congress 94-34781
CIP

Direct all inquiries to CRC Press, Inc., 2000 Corporate Blvd., N.W., Boca Raton, Florida 33431.

International Standard Book Number 0-8493-2451-3
Library of Congress Card Number 94-34781
Printed in the United States of America 1 2 3 4 5 6 7 8 9 0
Printed on acid-free paper

CONTENTS

ABOUT THE EDITORS

Dipak Kumar Sarkar, Ph.D., D.Phil., is an Associate Professor of Physiology in the Department of Veterinary and Comparative Anatomy, Pharmacology and Physiology at Washington State University.

Dr. Sarkar received his B.Sc. with honors degree in 1970 and subsequently a Ph.D. degree in Physiology in 1975. His Ph.D. work on ovarian steroidogenesis was performed under the direction of Professor C. Deb. He then relocated to Oxford, UK and in 1979 completed a D.Phil. degree in Physiology at the University of Oxford. His D.Phil. work on luteinizing hormone-releasing hormone was performed under the direction of Professor George Fink. He did postdoctoral training for three years at Oxford University, Yale University and Michigan State University. At Michigan State University, Dr. Sarkar conducted research on prolactin and aging under the guidance of Dr. Joseph Meites. He moved to San Diego in 1983 and served as an Assistant Professor of Reproductive Medicine in the University of California at San Diego. He has held his present position at Washington State University since 1988.

Dr. Sarkar is a member of the American Society for Cancer Research, British Endocrine Society, Endocrine Society, International Brain Research Organization, International Society of Neuroendocrinology, Society for Neuroscience, and Research Society for Alcoholism. He has received several awards, including Merit Scholarships from the Indian Government (1971–1978), Best Thesis Award from the British Society for Endocrinology (1979), Melon Foundation Faculty Scholar Award (1983–1985) and Basil O'Connor Award from the March of Dimes (1985–1987). He has been recipient of research grants from the National Institutes of Health, March of Dimes, Melon Foundation, Wellcome Trust and Gynex Corporation. Dr. Sarkar serves on the Neuroscience and Behavior Review Committee of NIAA and was on the editorial boards of *Endocrinology* and *Neuroendocrinology*. He has published more than 100 research reports and reviews.

Dr. Sarkar's research interests concern the topic of this book. He is studying the mechanisms by which the hypothalamus and pituitary control the reproductive cycle and how this regulatory process alters during alcohol abuse, aging and pituitary tumorigenesis.

Charles D. Barnes, Ph.D., is Professor and Chairman of the Department of Veterinary and Comparative Anatomy, Pharmacology and Physiology in the College of Veterinary Medicine at Washington State University.

Dr. Barnes received his B.S. degree from Montana State University in 1958 with double majors in biology and physics. In 1961 he received an M.S. degree in physiology and biophysics from the University of Washington, and in 1962 he earned his Ph.D. in physiology from the University of Iowa.

After two years as a postdoctoral fellow in the Department of Pharmacology at the University of California at San Francisco, he became Assistant Professor of Anatomy and Physiology at Indiana University in 1964. He advanced to Associate Professor in 1968, and in 1971 became Professor of Life Sciences at Indiana State University. In 1975 he was named Chairman of the Department of Physiology at Texas Tech University College of Medicine, where he remained until taking his present position in 1983.

Dr. Barnes is a fellow of the American Association for the Advancement of Science and a member of the American Association of Anatomists, American Institute of Biological Sciences, American Physiological Society, American Association of Veterinary Anatomists, American Society of Pharmacology and Experimental Therapeutics, American Society of Veterinary Physiologists and Pharmacologists, International Brain Research Organization, Society for Experimental Biology and Medicine, Society for Neuroscience, Society of General Physiologists, and the Western Pharmacological Society. He has been the recipient of many research grants from the National Institutes of Health and the National Science Foundation.

Dr. Barnes is the author of more than 160 papers and has been the author or editor of 20 books. His current research interests relate to the modulation of nervous system output by centers in the brainstem.

CONTRIBUTORS

(IN ALPHABETICAL ORDER)

KATHLEEN C. CHAMBERS Ph.D.
Department of Psychology and
Program in Neural, Informational and Behavioral Sciences
University of Southern California
Los Angeles, California 99089-1061

WILLIAM LES DEES Ph.D.
Department of Veterinary Anatomy and Public Health
Texas A&M University
College Station, Texas 77843-4458

MARY ANN EMANUELE M.D.
Departments of Medicine, Molecular and Cellular Biochemistry
Loyola University of Chicago
Stritch School of Medicine
Maywood, Illinois 60153

NICHOLAS V. EMANUELE M.D.
Department of Medicine and
Molecular Biology Program
Loyola University of Chicago
Stritch School of Medicine
Maywood, Illinois 60153

GREGORY F. ERICKSON Ph.D.
Department of Reproductive Medicine
UCSD School of Medicine
University of California, San Diego
La Jolla, California 92093-0625

CALEB E. FINCH Ph.D.
Neurogerontology Division, Andrus Gerontology Center and
Department of Biological Sciences
University of Southern California
Los Angeles, California 90089-0191

JUDITH S. GAVALER Ph.D.
Women's Research
Baptist Medical Center and Oklahoma Medical Research Foundation
Oklahoma City, Oklahoma 73112-4481

MICHAEL D. GRISWOLD Ph.D.
Department of Biochemistry and Biophysics
Washington State University
Pullman, Washington 99164-4660

MARGARET M. HALLORAN Ph.D.
Department of Molecular and Cellular Biochemistry
Loyola University of Chicago
Stritch School of Medicine
Maywood, Illinois 60153

LAWRENCE E. HEISLER Ph.D.
Department of Physiology
Queens University
Kingston, Ontario, Canada K7L3N6

JILL K. HINEY M.S.
Department of Veterinary Anatomy and Public Health
Texas A&M University
College Station, Texas 77843-4458

MARK R. KELLEY Ph.D.
Department of Pediatrics
Indiana University Medical Center
Indianapolis, Indiana 46202

PHILIP S. LAPOLT Ph.D.
Department of OB/GYN
UCLA School of Medicine
Los Angeles, California 90024-1740

ANN MILLER LAWRENCE M.D., Ph.D.
Department of Medicine and
Department of Molecular and Cellular Biochemistry
Loyola University of Chicago
Stritch School of Medicine
Maywood, Illinois 60153 and
Research and Medical Services
Department of Veteran Affairs Hospital
Hines, Illinois 60141

JOHN K.H. LU Ph.D.
Department of OB/GYN
UCLA School of Medicine
Los Angeles, California 90024-1740

YING JUN MA Ph.D.
Oregon Regional Primate Research Center
Beaverton, Oregon 97006

ROBERT F. MCGIVERN Ph.D.
Department of Psychology
San Diego State University
San Diego, California 92110

MICHAEL P. MCGUINNESS Ph.D.
Department of Biochemistry and Biophysics
Washington State University
Pullman, Washington 99164-4660

JOSEPH MEITES Ph.D.
Department of Physiology
Michigan State University
East Lansing, Michigan 48864

NANCY K. MELLO Ph.D.
Alcohol and Drug Abuse Research Center
Harvard Medical School-McLean Hospital
Belmont, Massachusetts 02178

JACK H. MENDELSON M.D.
Alcohol and Drug Abuse Research Center
Harvard Medical School-McLean Hospital
Belmont, Massachusetts 02178

JAMES F. NELSON Ph.D.
Department of Physiology
University of Texas Health Science Center at San Antonio
San Antonio, Texas 78284-7756

CHRISTOPHER L. NYBERG M.S.
Department of Veterinary Anatomy and Public Health
Texas A&M University
College Station, Texas 77843-4458

SERGIO R. OJEDA D.V.M.
Oregon Regional Primate Research Center
Beaverton, Oregon 97006

DENNIS D. RASMUSSEN Ph.D.
Department of Medicine
University of Washington and
American Lake VA Medical Center
Tacoma, Washington 98498

WILLIAM J. RAUM M.D., Ph.D.
Department of Medicine, Division of Endocrinology
UCLA and Harbor/UCLA Medical Center
Torrance, California 90509

EVA REDEI Ph.D.
Departments of Pharmacology and Psychiatry
University of Pennsylvania
Philadelphia, Pennsylvania 19104-6141

CATHERINE RIVIER Ph.D.
The Salk Institute
La Jolla, California 92037

DIPAK K. SARKAR Ph.D., D.Phil.
Department of Veterinary and Comparative Anatomy,
Pharmacology and Physiology (VCAPP)
Washington State University
Pullman, Washington 99164-6520

JOHN J. TENTLER Ph.D.
Department of Molecular and Cellular Biochemistry
Loyola University of Chicago
Stritch School of Medicine
Maywood, Illinois 60153

SIEW KOON TEOH M.D.
Alcohol and Drug Abuse Research Center
Harvard Medical School-McLean Hospital
Belmont, Massachusetts 02178

SHAHAB UDDIN Ph.D.
Research and Medical Services
Department of Veteran Affairs Hospital
Hines, Illinois 60141

DEAN A. VAN VUGT Ph.D.
Division of Reproductive Endocrinology
Department of Obstetrics and Gynecology and
Department of Physiology
Queens University
Kingston, Ontario, Canada K7L3N6

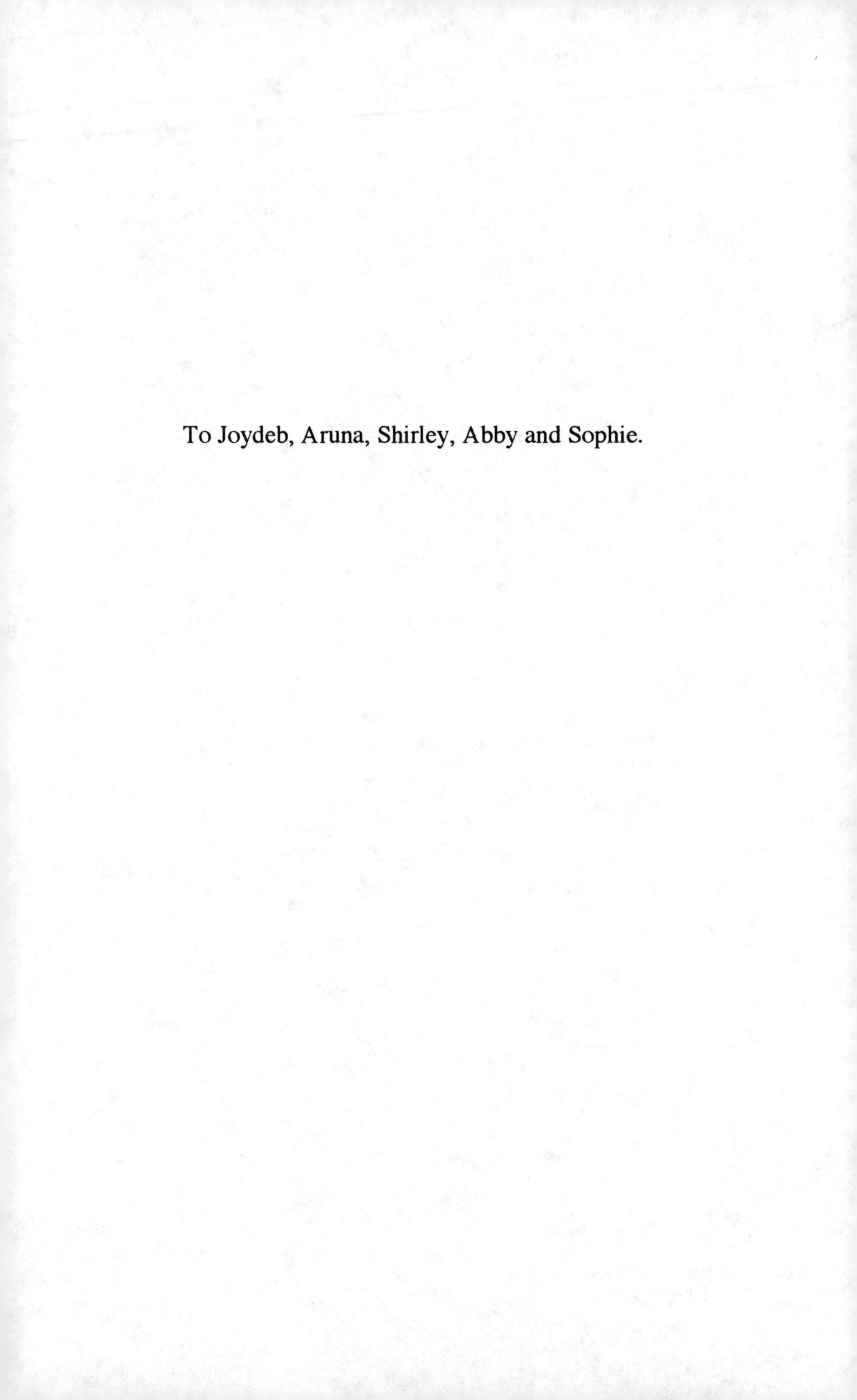

To Joydeb, Aruna, Shirley, Abby and Sophie.

PREFACE

The physiology of reproduction has been the subject of extensive research for the past few decades. This research has shown that the interaction between the nervous system and the endocrine system is essential for maintaining reproduction in mammalian species. The organs that participate in the neuroendocrine interaction are the brain (mainly the hypothalamus), pituitary and gonads (ovaries and testes). These organs communicate via secreting hormones and work in an integrated fashion to maintain reproduction.

A considerable number of recent works, mostly done in laboratory animals (rats and mice), and a few in human and non-human primates, suggest that the neuroendocrine control of reproduction is altered during aging and following drug abuse, resulting in a variety of reproductive problems, including acyclicity, hypogonadism and impotence.

The purpose of this volume is to discuss the neuroendocrine regulation of reproduction and the changes that occur during aging and drug abuse. Most of the leading investigators working in this field are represented here.

This reference introduces topics related to the neuroendocrine control mechanisms governing the normal reproductive function in both males and females, and it discusses the importance of the differentiation of the neuroendocrine brain during the developmental period for the normal reproductive process. It describes genotypic influences in reproductive aging and the changes in the reproductive neuroendocrine axis during aging. The volume covers information regarding the influences of stress and immune components in reproduction; it summarizes the effects of drug abuse in the neuroendocrine brain differentiation and how this alteration induces abnormalities in the reproductive process. Finally, this comprehensive volume compares the effects of various drugs, such as alcohol, cocaine, morphine, and nicotine, on the maintenance of reproduction.

The reader may find it interesting that many changes seen in the neuroendocrine axis following drug abuse have similarity to those seen during the aging process. These similarities are intriguing to us and led us to compare the two apparently different aspects of human living in one volume. We hope that this fascinating area of reproductive neuroendocrinology will impress the reader as it has us, and that some readers will be provoked to uncover new information that has yet to be discovered.

Dipak K. Sarkar
Editor

Charles D. Barnes
Editor

ACKNOWLEDGMENTS

Our work as editors has been greatly simplified by the submission of outstanding chapters and by the cooperation of the contributing authors.

As with all books, we benefited from endeavors of those individuals who assisted with our project. This volume would not have been completed without the hard work of our departmental editor, Jeanne Jensen, who formatted the entire book, took care of all correspondence with authors, and assisted in obtaining copyright permissions from many publishers. We greatly appreciate her dedication, and that of her assistant, Melissa Lee, who also helped check and proofread many of the chapters.

During the early stages of work for this book, we obtained invaluable support from the late Dr. Thomas Akesson, who assisted in organizing a series of seminars given at Washington State University's Pullman campus by several authors whose work appears in this volume.

REPRODUCTIVE NEUROENDOCRINOLOGY OF AGING AND DRUG ABUSE

Chapter 1

NEUROENDOCRINE CONTROL OF GONADOTROPIN RELEASE

D.K. Sarkar

TABLE OF CONTENTS

0-8493-2451-3/95/$0.00+.50

ABBREVIATIONS

6-OHDA–6-hydroxy-dopamine
ACTH–adrenocorticotropin hormone
CRF–corticotropin-releasing factor
CSF–cerebral spinal fluid
FSH–follicle-stimulating hormone
GAP– GnRH-associated peptide
GnRH–gonadotropin-releasing hormone
hCG–human chorionic gonadotropin
IHDA–incertohypothalamic dopaminergic
i.p.–intraperitoneal
IR–immunoreactive
LH–luteinizing hormone
LHRH–luteinizing hormone-releasing hormone
LI–like
mRNA–messenger ribonucleic acid
NAL–naloxone
NPY–neuropeptide Y
NRS–normal rabbit serum
OVLT–organum vasculosum of the lamina terminalis
PMSG–pregnant mare serum gonadotropin
POMC–proopiomelanocortin
TIDA–tuberoinfundibular dopamine
VO–vaginal opening

INTRODUCTION

Reproduction is essential for the survival of every species. In mammals, reproductive function is maintained by endocrine interaction between the hypothalamus, pituitary and gonads. Although several dozens of hormones are involved in the endocrine interaction within the hypothalamic-pituitary-gonadal axis, there appear to be a selective number of hormones that play the primary role in conveying the signal from one component of the axis to another. For example, luteinizing hormone-releasing hormone (LHRH), which is also known as gonadotropin-releasing hormone (GnRH), is the primary transducer of the hypothalamic impulse to the pituitary. Luteinizing hormone (LH) and follicle-stimulating hormone (FSH) secreted from the gonadotropes of the anterior pituitary are the primary hormones that convey the pituitary signals to testes in the male and to ovaries in the female. Testosterone from the testes and estradiol and progesterone from the ovaries are the primary hormones that convey the gonadal input to the hypothalamus and the pituitary. This chapter discusses the LHRH-regulated secretion of pituitary gonadotropes and the influences of the gonadal hormone on this regulatory process. The pituitary control of gonadal function will be discussed in chapters 3 and 4 of this volume. The changes of LHRH-regulated gonadotropin release during aging, drug abuse and stress are discussed in several chapters of this book.

MODE OF GONADOTROPIN RELEASE

Gonadotropes of the anterior pituitary gland secrete two hormones, LH and FSH, that are essential for maintaining gonadal function and reproduction. These gonadotropins have a common hypophysiotropic neurohormone (LHRH) and sometimes a common cell of origin in the anterior pituitary.[13,49,160,211] Both LH and FSH are glycoproteins and are composed of two polypeptide subunits (alpha and beta) that are bound together in a noncovalent association of high affinity (for reviews see Chin[33] and Pierce[162]). Within a species, the alpha subunit of LH and FSH has essentially the same amino acid sequence, but the beta subunits of these two gonadotropins differ and confer the biological specificity on the dimeric molecule. Hormones that are related to LH and FSH are human chorionic gonadotropin (hCG), secreted by the human chorion, and pregnant mare serum gonadotropin (PMSG), produced by specialization of the uterine endometrium of the pregnant mare. There are multiple isoforms of gonadotropins with variations in biopotency in the pituitary; some of these isoforms are secreted into the blood. The primary action of FSH on the ovary and testis is to control processes involved in germ cell development. LH controls steroid production from steroidogenic cells in the ovary and testis, and it participates in the process leading to ovulation of the mature follicle (for reviews see Catt and Pierce[28] and chapters 3 and 4 of this volume).

Cyclic Gonadotropin Release

The secretion of LH appears to be pulsatile during the peripubertal period in the majority of the mammals studied.[56,151,215] The secretion of FSH is less pulsatile than LH secretion during this period. The pulses are characterized by a rapid increase over a short period of time (a few minutes) followed by a fall in the basal secretion for a relatively longer duration (several minutes). In general, FSH pulses are longer than LH pulses, perhaps reflecting the different clearance rates of the two gonadotropins.[7,28,156] The pulsatile secretion of gonadotropin is maintained throughout the reproductive life in males (see below). However, in females of species showing estrous cycles (e.g., rats, ewes) and menstrual cycles (women, rhesus monkeys), following the onset of puberty, the intermittent secretory patterns of gonadotropin are interrupted by a large increase in LH and FSH secretion (gonadotropin surge) preceding ovulation and lasting for several hours.[187] The induction of this preovulatory gonadotropin surge occurs in a periodic manner until terminated temporarily during pregnancy (all species) and during some seasons (seasonal breeder). In other mammalian species (e.g., rabbit, vole, cat), copulation induces the LH surge and, subsequently, ovulation and the reproductive cycle. Using the rat as an animal model, the primary focus of this chapter will be the discussion of LHRH-regulated LH secretion during the reproductive cycle.

Acyclic Gonadotropin Release

Unlike the cyclic profiles of circulating LH and FSH in females, males release LH only in a pulsatile pattern but release FSH in a relatively steady fashion. The LH pulses are much higher than the FSH pulses.[28] Circadian variation of LH discharges is evident in pubertal boys during sleep.[22,74] The nocturnal increase in LH release is believed to represent the early CNS-mediated determinant of puberty. In both adult and pubertal males, an LH pulse is often followed by an increase in circulating testosterone.[23,88] In some species there is evidence for increased LH pulse amplitude during sexual stimuli and ejaculation.[72,122] Thus, there are clear differences between male and female profiles of LH and FSH. The mechanism of the sex differences between male and female patterns of gonadotropin secretion has been well studied, and each species seems to fall into one of two categories[50,78,96,224] (see also chapter 11 of this volume). In laboratory rats, sheep, cattle and horses, the absence of an LH surge may be due to masculinization of the neuroendocrine brain responsible for cyclic LHRH release. In these species, testosterone from the fetal and neonatal testis appears to cause anatomical and functional changes in the brain. This, in turn, leads to the prevention of a cyclic center responsive to estradiol and is responsible for a surge amount of LHRH release. In these species, unlike females, males fail to secrete a surge of LH following gonadectomy and injection of a bolus amount of estradiol. In man, rhesus monkey, pig and goat, the lack of cyclic LH release is due to the presence of testes and not to the masculinization of the neuroendocrine brain. Castration and estradiol administration cause a surge release of LH in these species.

PHYSIOLOGICAL MECHANISMS OF CYCLIC GONADOTROPIN RELEASE

The two major components of the mechanism that regulates the secretion of gonadotropin from the anterior pituitary gland in females are the neural stimuli arising within the central nervous system and sex steroid hormones. The role of these two components in the mechanism of gonadotropin release has been studied extensively in rats and has been described in detail in several recent reviews;[58,91,221] therefore, only the main points and data related to LHRH regulated LH secretion will be considered in this chapter.

Generation of the Neural Signal

LHRH Neurons

As mentioned before, LHRH is the main hormone that conveys the hypothalamic signal to control gonadotropin release from the pituitary in species showing rhythmic ovulatory cycles. Immunoneutralization of endogenous LHRH using LHRH antibody or pharmacological manipulations of LHRH receptors using LHRH antagonists blocks preovulatory LH release and reduces castration-induced LH release.[17,109,132,171,177] Although there was some concern as to whether or not LHRH plays an active role in generation of the preovulatory LH surge in the rhesus monkey,[103] recent evidence suggests that the secretion of LHRH into hypophysial portal blood is elevated prior to the LH surge in monkeys,[117,147,230] as it is in rats,[34,116,187] sheep[2,38,140] and rabbits.[97,212] LHRH levels in portal blood and the median eminence release of LHRH are temporarily correlated with the LH release during the different days of the reproductive cycle in these species, providing further evidence that LHRH actively participates in conveying the hypothalamic signal to the gonadotropes.

LHRH is a decapeptide hormone.[25,131] It is synthesized as part of a larger precursor protein (pro-LHRH) in a group of neurons in the hypothalamus.[1,48,100,129,136,198] The pro-LHRH is extended at the C and, possibly, the N terminals. LHRH was found to be preceded by a 23-amino-acid signal peptide sequence and followed by a 56-amino-acid peptide GnRH (LHRH)-associated peptide (GAP). An enzymatic processing site separates the LHRH and GAP moieties. The processing of pro-LHRH to LHRH may involve cleavage by endopeptidase and carboxypeptidase E and alpha-amidating monooxygenase.[222]

LHRH-containing neurons are localized and scattered within the hypothalamus. The number of LHRH neurons is relatively lower (1000–5000 in most mammals) than most of the other hypophysiotropic hormones present in the hypothalamus.[82,202] The fact that there is diversity in the distribution of LHRH neurons in different species has been recently reviewed.[82,202] In the rat, LHRH perikarya are most abundant in the medial preoptic area in close proximity to the organum vasculosum of the lamina terminalis (OVLT). LHRH-containing fiber terminals are concentrated in the median eminence and the OVLT. It is generally believed that LHRH neuron terminals in the median eminence release LHRH into the capillaries of the pituitary portal vessel. LHRH is transported via the blood of these vessels to the gonadotropes, where it acts to cause the release of gonadotropin.

Several studies have now demonstrated that LHRH neurons undergo substantial prenatal and postnatal development.[173,195,228] LHRH neurons are first identified in the olfactory placode, and then LHRH cells appear to migrate through the nasal septum into the forebrain to establish the adult-like LHRH distribution. This unusual developmental pattern of LHRH neurons may explain species differences in the LHRH position in the forebrain that may result from differential migration along the olfactory bulb-hypothalamus-median eminence. A disorder in the LHRH migration during the developmental process has been identified in various hypogonadal syndromes, including Kallmann's disorder.[196] In the rat, most LHRH neurons are small (10–12 μm) and smooth at the time of birth, but there is a shift from smooth to spiny LHRH neurons during postnatal life that plateaus at the time of puberty.[227] The physiological significance of this morphological differentiation of the LHRH is not apparent but may relate to the functional status of the neurons. The spiny LHRH neurons have been shown to contain more Golgi complexes and mitochondria, suggesting an active metabolic state.[226] LHRH neurons receive synaptic or dendritic processes[82,113,128,155,202] from other LHRH neurons. These synaptic connections between LHRH neurons may be physiologically relevant in synchronizing LHRH neuronal activity to cause pulsatile or preovulatory surges of LHRH. Another interesting anatomical aspect of LHRH neurons is that the dendrites and perikarya of these neurons often lie close to small blood vessels. This interesting anatomical feature may facilitate the delivery of blood-borne factors to LHRH neurons[82] in order to have a rapid cross-talk between each component of the hypothalamic-pituitary-ovarian axis.

LHRH neurons are also innervated by the synapses of other transmitter systems. Electron microscopy studies have revealed synaptic innervation of LHRH neurons by β-endorphin,[45,115] gamma amino butyric acid (GABA),[112] catecholamines,[107,114] 5-hydroxytryptamine (5-HT),[101] neuropeptide Y (NPY),[214] corticotropin-releasing factor (CRF),[126] substance P,[213] and vasopressin.[232] This morphological evidence suggests that various transmitter systems may participate in the communication within the hypothalamic-pituitary-gonadal axis.

As stated earlier, LHRH neurons are scattered throughout the hypothalamus. The question arises whether all of these or a group of these LHRH neurons work in synchrony to generate the gonadotropin surge. Recently this issue has been addressed by identifying LHRH neurons expressing the protooncogene product, c-fos, which is expressed during the active phase of neurotransmission and is not expressed during the inactive phase of some neurons.[2,82,110,176] In rats, using the protooncogene expression as a method of assessing neurotransmission, activation of a distinct subpopulation of LHRH neurons during the preovulatory LH surge was revealed. The location of this population of LHRH cells was in the vicinity of the OVLT below the anterior commissure and extended into the preoptic area and anterior hypothalamus.

LHRH Surge-Generating Center

It is interesting to note that the location of the active LHRH cells in the hypothalamus around the time of the LH surge is in the area known to be crucial for the generation of the neural signal for LHRH. Studies employing hypothalamic deafferentation in which the hypothalamus was systematically separated by knife cuts, electrochemical lesions of hypothalamic nuclei, or electrical stimulation of different areas of the hypothalamus revealed that the region of the preoptic area and suprachiasmatic nucleus is capable of generating an LH surge.[19,58,76,79,90,91,105,206] The functional importance of the preoptic area as a surge center in rats is also emphasized by the sexual differences in preoptic morphology (formation of a larger sexual dimorphic nucleus in male than in female) and by the ability of the preoptic area to cause an LH and LHRH surge release in response to electrical stimulation.[52,60,78] It should be noted that the mediobasal hypothalamus may be more involved in generating the LH surge in primate species, but LHRH neurons are also more concentrated in the mediobasal hypothalamus of primate species.[42,127] In many species, other than rats, electrical stimulation of the preoptic area caused a small increase in LH release that was less than the surge release of LH.[127] Perhaps the surge-generating center is more widely spread in the hypothalami of these species. Although preoptic and suprachiasmatic areas may be the areas actively involved in generating the LH surge, the maintenance of the frequency and amplitude of the LH surge depends on the integrity of the neural inputs from the midbrain and forebrain limbic system.[58,91,127,146,167]

The precise role of CNS areas that project to the hypothalamus is not apparent. The limbic system, which is involved in the generation of emotion and in conveying olfactory signals, has been implicated in controlling reproductive behavior. However, evidence for a direct influence of the forebrain limbic system components in the generation of the LHRH surge is lacking.[32] The midbrain and hindbrain appear to control the LHRH surge generator via ascending noradrenergic projections to the hypothalamus.[12,58,91,177] The neocortex of the brain may send visual and auditory signals to the hypothalamus via multisynaptic pathways.[87] In addition, visual signals can be directly transmitted to the hypothalamus via the retino-suprachiasmatic tract or via multisynaptic pathways involving the retina, suprachiasmatic nuclei, superior cervical ganglia, and the pineal gland. Visual signals have a significant impact in controlling cyclicity in animal species (such as sheep) that are dependent on photoperiods.[170,217]

Developmental Maturation of the LHRH Surge-Generating Center

Generation of the LH surge and activation of the surge generator are not initiated until puberty. Morphological evidence of maturational changes in LHRH-immunopositive cells is documented during late fetal and neonatal

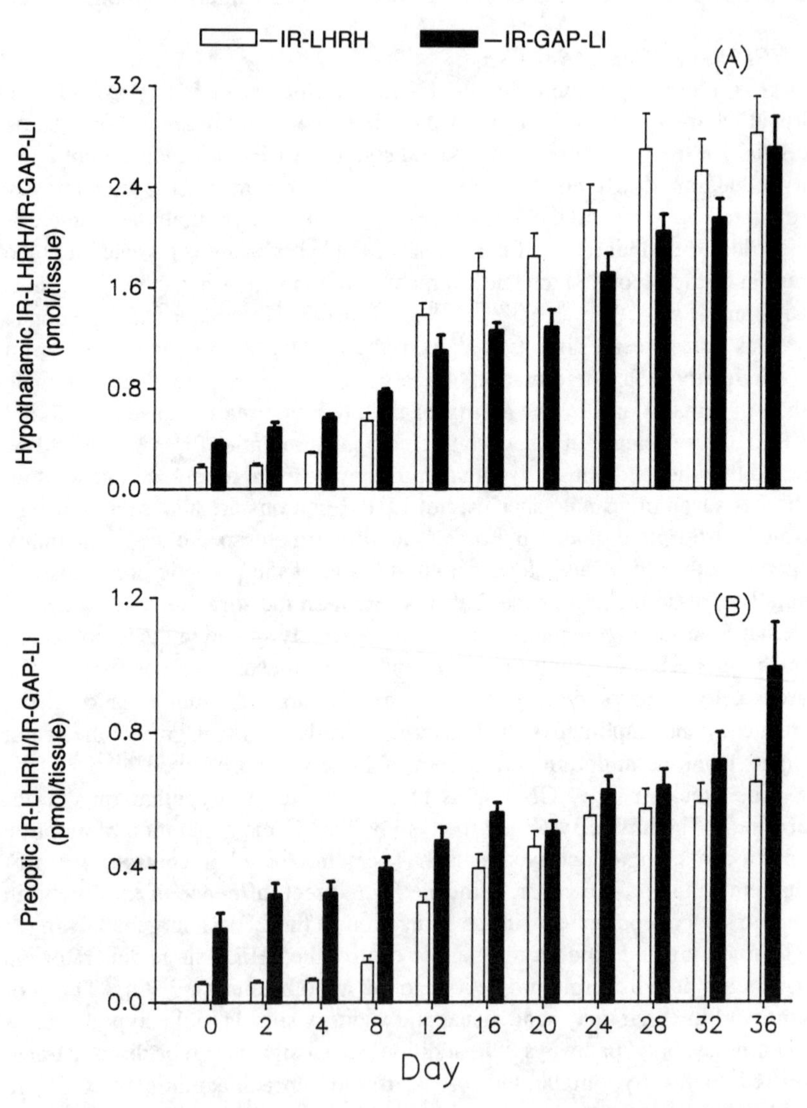

FIGURE 1. Mean + SEM content of IR-LHRH (□) and IR-GAP-LI (■) in acetic acid extracts of hypothalamus (*A*) and preoptic area (*B*) in developing female rats from birth to day 30 of postnatal life. The rise in content was correlated with advancing development ($p < 0.0001$). $N = 4\text{–}10$.

life.[31,229] Biochemical evidence also shows developmental changes of functional LHRH neurons in the preoptic and mediobasal hypothalamus[184] (see Figure 1) during postnatal and juvenile periods. Accompanying these changes in LHRH neuronal activity is an increase in catecholamines, NPY, opioid peptide and various other neurotransmitter nerve endings apposed to LHRH cell bodies.[36,95,144,169,229] A large number of reports now suggest that perhaps establishment of the neural circuitry is important in the activation of the LHRH surge-generating center (see References 75, 151, 152 and 164 for reviews). The control mechanism for the organization of this circuitry is poorly understood. A substantial number of reports suggest that estradiol, neurotransmitters and growth factors play critical roles in the organizational process.[89,130,143,151,153,208-210] Recently it was shown that the synaptic inputs to LHRH neurons are sexually dimorphic,[30] indicating that steroid-induced synaptic plasticity underlies the mechanism leading to cyclic and acyclic gonadotropin release in the rodent and other species. Estradiol has been shown to enhance both neurite outgrowth and synaptogenesis.[130,153,208-210] Estradiol enhances synaptogenesis not only in the fetal hypothalamus, but also in the juvenile as well as adult hypothalami.[41,130] Although few LHRH neurons contain estradiol receptors,[201] many estradiol-sensitive neurons synapse with LHRH neurons.[108] Therefore, estradiol may affect those estradiol-sensitive neurons that synapse with LHRH neurons. In addition to estradiol, there appear to be varieties of polypeptide growth factors that act as neurotropic factors to mature the hypothalamus.[152] The postulated mechanisms by which these growth factors and estradiol affect development of the reproductive neuroendocrine axis are described in chapter 2 of this book.

LHRH Secretion in the Blood of Pituitary Portal Vessels

It is now well established that the pituitary portal vessels are the vascular pathway that enables neurotransmission between the hypothalamus and the anterior pituitary gland.[59,79,148,183] In support of this concept are the observations that i) section of the pituitary stalk produces atrophy of the gonad,[80] ii) transplantation of the pituitary gland under the kidney capsule following hypophysectomy results in gonadal atrophy in rats,[148] iii) pituitary transplantation under the hypothalamus in hypophysectomized female rats reactivates appreciable reproductive functions,[148] iv) LHRH neuron terminals are in the external zone of the median eminence in the immediate vicinity of primary capillary plexus of the pituitary portal vascular system, v) the major flow of blood in the portal vessels is directed from the hypothalamus toward the pituitary gland,[73,225] and vi) pituitary portal blood contains a severalfold higher concentration of LHRH than does peripheral or pharyngeal blood.[60,187]

Earlier attempts to measure LHRH activity in the blood of pituitary portal vessels and to correlate the activity with varying secretion rates of gonadotropes in rats were unsuccessful due to the anesthetic used for collection of portal blood.[200] However, we have found a surge release of LHRH into the blood of pituitary portal vessels just prior to the preovulatory LH surge in female rats anesthetized with a steroid anesthetic known as althesin and Saffan[187] (see also Figures 2B and 3). Although Saffan is comprised of C21 steroids alphaxolone (3-α-hydroxy-5-α-pregnane-11,20-dione) and alphadolone acetate (21-acetoxy-2-α-hydroxy-5-α-pregnane-11,20-dione), it does not block

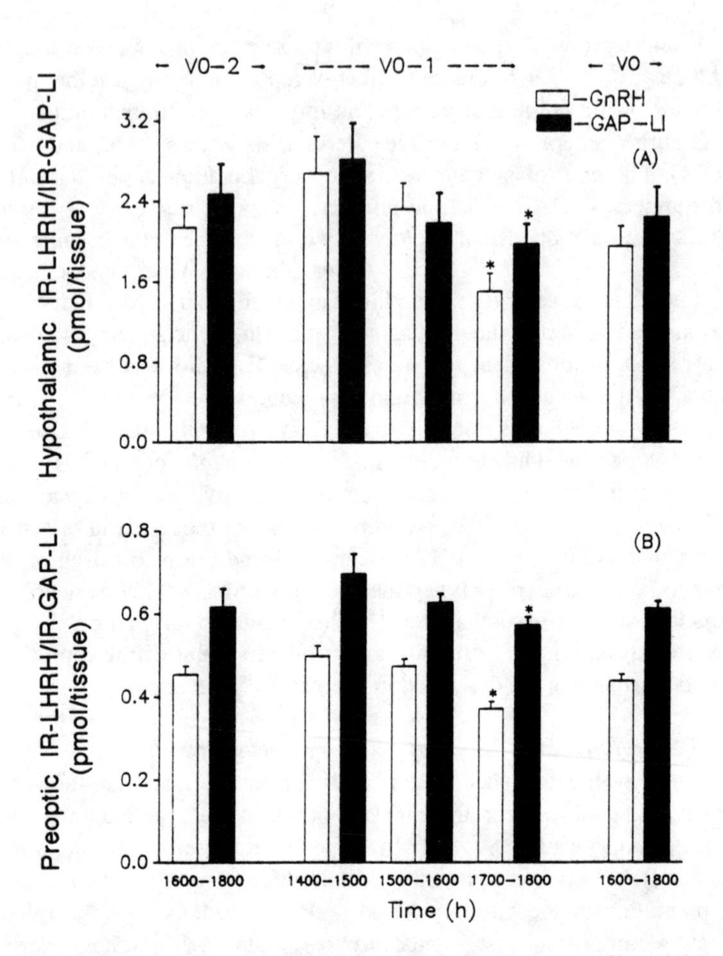

FIGURE 2A. Mean ± SEM content of IR-LHRH and IR-GAP-LI in extracts of hypothalamus (*A*) and preoptic area (*B*) in pubertal female rats at the time of vaginal opening (VO) and either 1 (VO-1) or 2 (VO-2) days before vaginal opening. Times of tissue sample collection are shown on the x axis. $N=15$. (Taken from Sarkar and Mitsugi[184] with permission from *Neuroendocrinology*.)

the LH surge or ovulation.[39,187] This anesthetic does not change the electrophysiological activity of the diencephalon,[207] and its analgesic property differs from other commonly used anesthetics with regard to its mechanism of action.[63]

Changes of LHRH Secretion in the Blood of Pituitary Portal Vessels During the Reproductive Cycle in Rats

Estimation of LHRH levels in the portal blood collected under althesin anesthesia showed a clear-cut increase on the afternoon of the day preceding vaginal opening[180,184] (Figure 2B). Comparison of the concentration of LHRH in plasma of the pituitary portal vessels with concentration of LH in peripheral

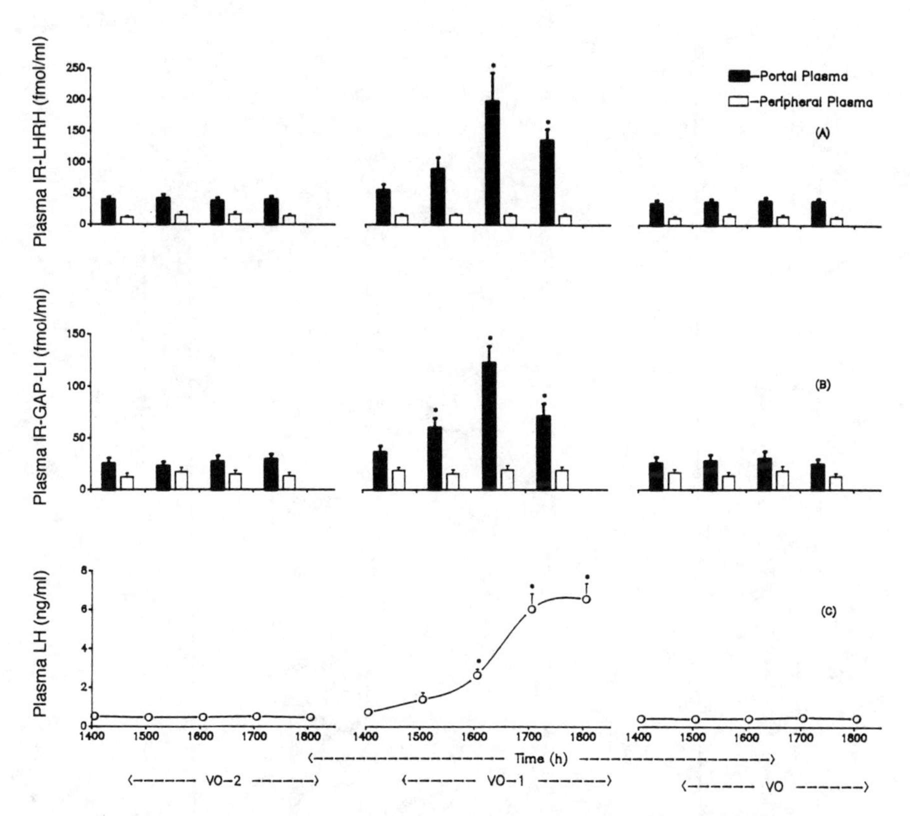

FIGURE 2B. Cosecretion of IR-LHRH (*A*) and IR-GAP-LI (*B*) into pituitary portal plasma at the time of vaginal opening (VO) and either 1 (VO-1) or 2 (VO-2) days before vaginal opening in pubertal female rats. Peripheral plasma LH levels in these rats are shown for comparison (*C*). $*p < 0.05$, from the rest of the groups. Data are mean ± SEM of 5–14 rats. (Taken from Sarkar and Mitsugi[184] with permission from *Neuroendocrinology*.)

plasma of rats suggests that the LH surge begins shortly after the LHRH surge. The levels of portal LHRH and peripheral gonadotropins appear to correspond well with the reduction of LHRH content in the preoptic area and hypothalamus and of gonadotropin content in the pituitary gland[180,184] (Figure 2A). These results indicate that the neural control of gonadotropin release is manifested by the increased secretion of LHRH from the preoptic and hypothalamic areas.

The preovulatory surge of gonadotropin on the day of proestrus also coincides with a surge of LHRH in the blood of the pituitary portal vessels of adult female rats, suggesting that, as in pubertal rats, LHRH mediates neural control of gonadotropin release in adult rats[187] (Figure 3). The changes of LHRH concentration in portal blood at the time of the LH surge in the rat are similar to those found in the portal blood of the anesthetized monkey and conscious ewe and rabbits.[34,38,117,139,140,147,180,187,212,230]

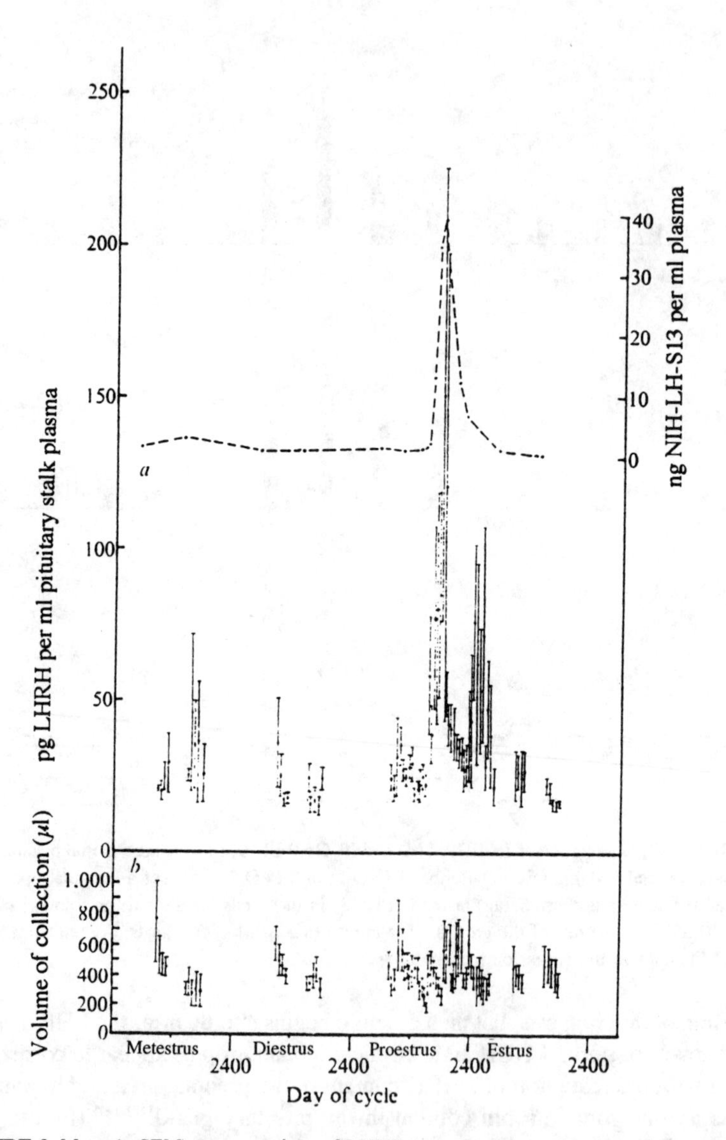

FIGURE 3. Mean (± SEM) concentrations of LHRH (***a***) and volumes (***b***) of 30-minute collections of pituitary stalk (portal) blood. Most means are based on 5–15 samples, a few on 3–4 samples. The mean concentrations of plasma LH (---) are shown for comparison. (Data taken from Sarkar *et al.*[187] with permission from *Nature*, ©1976, Macmillan Magazines Limited.)

It should be noted that a large variation in portal blood LHRH concentrations has been detected by us during the preovulatory LHRH release,[200] suggesting the possibility of a pulsatile secretion of LHRH release from the median eminence into the portal blood. Indeed, our recent studies involving frequent sampling of portal blood revealed that LHRH levels are pulsatile throughout the estrous cycle.[191] The increased LHRH levels in portal plasma during the afternoon of proestrus in rats is in agreement with the studies where LHRH neuronal activity was measured by determining LHRH levels in push-pull perfusate from the median eminence of proestrous rats.[116] The LHRH levels in the push-pull perfusate also showed pulsatility during the estrous cycle. Similarly, LHRH levels in pituitary portal blood, third ventricular CSF or in the push-pull perfusate of the median eminence have been shown to increase at the time of the LH and FSH surge in rhesus monkeys and ewes.[2,38,117,140,147,230] Infusion studies in monkeys also suggest that the intermittent mode of LHRH may be more effective than the constant or a single mode of LHRH in eliciting gonadotropin release from the anterior pituitary gland.[14] Hence, it could be concluded that, like LH, LHRH is secreted both in pulsatile and surging manners during the reproductive cycle. It is believed that the pulsatile nature of endogenous LHRH secretion is involved in activating the priming effect of LHRH,[57,58,161] leading to increased pituitary gonadotropin responsiveness to LHRH[4,62,180] and optimal release of gonadotropin.

In addition to pulsatile and preovulatory release of LHRH, our recent studies in rats suggest that there is a daily afternoon increase in LHRH secretion in pituitary portal blood in each day of the estrous cycle (Figure 4). The physiological significance of this diurnal rhythm of LHRH secretion in the control of LH regulation is not apparent, since a diurnal pattern of LH release during the rat estrous cycle has not been documented. However, a daily afternoon increase in LH release has been documented during the prepubertal periods in female rats,[215] in adult cyclic females receiving barbiturate sedation[51] and in estradiol-primed ovariectomized rats.[181] Furthermore, diurnal changes in pulsatile LH secretion have been reported during the menstrual cycle in both women and female monkeys.[56,85,174]

Role of Gonadal Steroids

Estradiol

The data discussed earlier suggest that the neural signal for LH release is mediated by an enhanced release of LHRH into the pituitary portal blood. Evidence from studies in which estradiol has been administered to adult ovariectomized rats indicated that the development of the neural signal depended on maintenance of an increased estradiol concentration in plasma[26,57,58,91,159,177,180] This view is further supported by the observation that the first (pubertal) surge of LHRH is preceded by an increase in plasma estradiol concentration and not by any major changes of the other hormones secreted from the hypothalamic-pituitary-gonadal axis,[134] although the increase in plasma estradiol may depend on increased pulsatile LH release.[124,151,215] Interestingly, the hypothalamic-pituitary axis is able to respond positively to estradiol several days prior to the onset of puberty,[124,151,180] and therefore, the generation of the neuronal signal for pubertal gonadotropin surge appears to depend primarily on the ovarian capacity to secrete estradiol to greatly elevate the amount of the steroid in plasma.[151]

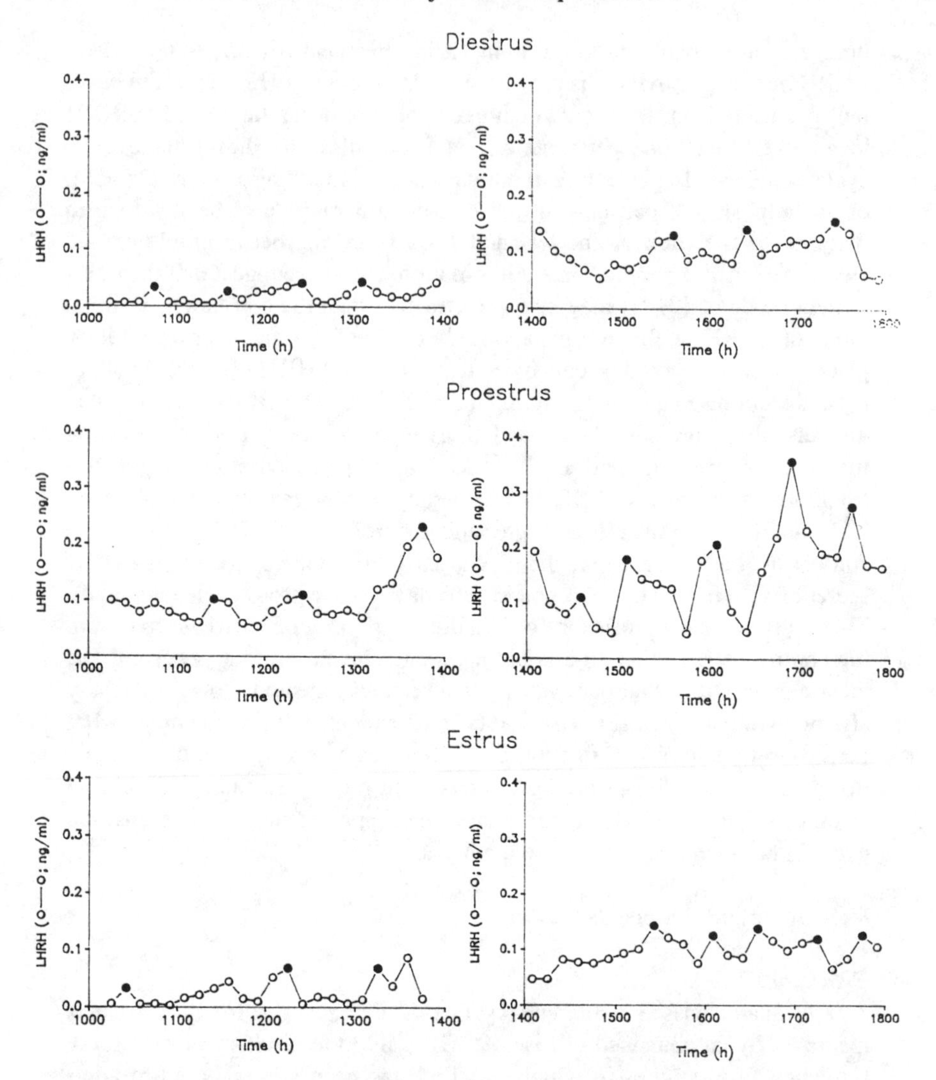

FIGURE 4. Diurnal rhythm in the pulsatile release of LHRH during the estrus cycle. Graphs indicate individual pituitary portal plasma LHRH concentrations in representative female rats during 1000–1800 hours of diestrus, proestrus and estrus. The times of collection are shown on the x-axis. The peaks of the pulses are indicated by filled symbols.

Like the pubertal surge of LHRH, the preovulatory surge of LHRH during the estrous cycle in adult animals also depends on the increased estradiol concentration in plasma. This is evidenced by the observation in rats that, i) a rise in plasma levels of estradiol precedes the preovulatory surge of LH and LHRH on the day of proestrus, ii) the blockade of preovulatory estradiol release by ovariectomy on diestrus reduces the release of LHRH into the portal blood during the afternoon of proestrus,[93,187,199] and iii) exogenous estradiol in the ovariectomized rats reinstates the LHRH surge on the afternoon of proestrus.[179] The stimulatory effect of estradiol on LHRH secretion also has been documented in the ewe and rhesus monkey.[2,38,62,139,230] In addition to the positive feedback effect on LHRH release, estradiol has been reported to act directly on the pituitary to enhance pituitary LH synthesis and secretion[3,57,58] and to regulate LHRH receptor expression.[40] Thus, the stimulatory effect of estradiol is exerted at the level of both the brain and anterior pituitary.

Progesterone

In rats, the level of progesterone in plasma rises before the LH surge on the day of proestrus because of the elevation of adrenal secretion of this steroid.[93,199] Ovarian secretion of progesterone is only elevated after the onset of the LH surge and peaks after the LH surge. The increase in early progesterone from the adrenal gland does not appear to be essential for the generation of the preovulatory LHRH release, since a preovulatory level of LHRH release can be reinstated in the estradiol-primed ovariectomized rat and the ovariectomized and adrenalectomized rat in the absence of progesterone.[179] However, adrenal progesterone may magnify estradiol action on LHRH secretion. Progesterone administration enhances LHRH secretion from the median eminence *in vitro*[22] and increases the number of LHRH neurons displaying c-fos activity.[111] Although we were unable to detect an increase in LHRH release in portal plasma following progesterone administration in estradiol-primed ovariectomized rats, a moderate increase in LHRH release was observed following treatment with progesterone in estradiol-primed ovariectomized and adrenalectomized rats[179] (Figure 5). It should be mentioned that progesterone produced significant stimulatory effects on LHRH release in portal blood in rats exposed to constant light.[179] These rats were switched from spontaneous to reflex ovulators, and exogenous progesterone stimulated a surge of LH release in these rats.[24] Adrenal steroids may also be responsible for the daily afternoon increase in LHRH release during the estrous cycle,[191] since there is a daily rhythm in adrenal corticoid and progesterone secretion.[94] In addition to its action on the brain, progesterone has been shown to magnify the estradiol stimulatory action on the pituitary gland's responsiveness to LHRH.[3,57,58]

As described above, the secretion of ovarian progesterone reaches a peak after the LH surge. It appears that this marked increase in progesterone secretion is involved in extinguishing the estradiol-induced neural signal for preovulatory LH release. Administration of progesterone at noon of the day of expected proestrus in estradiol-primed ovariectomized rats inhibits LHRH secretion.[179,181] Progesterone also extinguishes the daily neural signal for LH release in estradiol-primed ovariectomized rats.[66,81] The inhibitory action of progesterone in the presence of estradiol is also evident in sheep[81] and humans.[216] It appears that, in the presence of high levels of progesterone, estradiol inhibits LH release (for reviews see Fink[57,58]). Progesterone by itself has little effect, perhaps because the estradiol is prerequisite for the expression

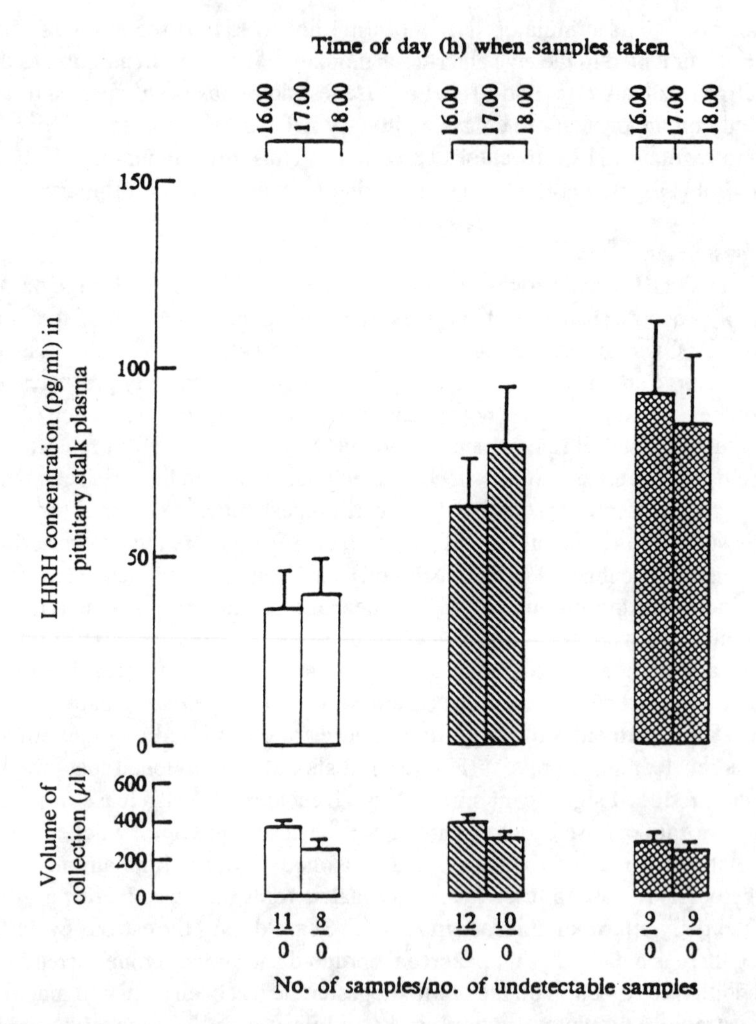

FIGURE 5. Mean (± SEM) concentrations of luteinizing hormone-releasing hormone (LHRH) in pituitary stalk (portal) plasma, and volumes of stalk blood collected at various times (indicated at top) on the expected day of proestrus. The experiment was carried out in long-term adrenalectomized animals. They were treated with oil plus oil (open bars), 10 μg estradiol benzoate (OB) plus oil (hatched bars), or 10 μg OB plus 2–5 mg progesterone (cross-hatched bars). Values below bars refer to total number of samples at each time/number of samples in which LHRH was not detectable. (Taken from Sarkar and Fink[179]; reproduced with the permission of *Journal of Endocrinology*, Ltd.)

of progesterone receptors in the hypothalamus and the pituitary gland.[125] Thus, the negative and positive feedback actions of steroids also appear to depend on the timing and the level of exposure to the steroids (for reviews see Fink[57,58] and Fink *et al.*[61]).

Role of Catecholaminergic and Peptidergic Neurons

As described above, the majority of the available data in the literature indicate that very few LHRH neurons contain steroid receptors,[201] but a large number of steroid-sensitive neurons synapse with LHRH neurons.[108] Therefore, steroid actions on LHRH neurons are believed to be mediated by various neurotransmitters. Three classes of neurotransmitters (aminergic, peptidergic and amino acid) have been shown to alter steroid-activated LH release and have been described in several recent reviews[10,58,91,95,221] (see also chapters 2 and 10 in this volume). The following section discusses only a selected group of catecholaminergic and peptidergic systems in the hypothalamus that have been shown to make synaptic contact with the LHRH neurons and affect LHRH release into the pituitary portal plasma of rats.

Catecholamines

The role of catecholamines in generating the LH surge has been studied extensively, and abundant evidence now indicates that these neurotransmitters play an important function in controlling pituitary LH secretion.[12,59,73,95,169,193] At physiological concentrations, the catecholamines do not appear to affect gonadotropin secretion by a direct action on the pituitary. Hence, the action of this central monoaminergic system is probably brought about by influencing LHRH neurons. Anatomical studies revealed that a group of noradrenergic and adrenergic neuron terminals in the hypothalamus and the preoptic area originate from the ascending projections from nuclei in the midbrain and hindbrain.[12,82] Many of these noradrenergic and adrenergic neuronal terminals are in close proximity to LHRH neurons in the preoptic and anterior hypothalamus and in the internal layer of the median eminence.[107,114] The dopamine system of the preoptic area and hypothalamus is derived from two major sources: one group of dopamine neurons is the tuberoinfundibular group (TIDA), which originates in the arcuate and ventromedial nuclei and terminates in close proximity with LHRH axon terminals in the median eminence;[12,61,82,229] the second group of dopamine neurons is the incertohypothalamic dopaminergic (IHDA) system, which originates from cell groups in zona incerta and is distributed in the preoptic area and hypothalamus. Some of these nerve terminals are close to LHRH neuronal perikarya.[114]

We have studied the role of catecholamines in the control of preovulatory LHRH release employing pharmacological manipulations of catecholaminergic activity in prepubertal rats given PMSG to induce LH and LHRH surges.[182,188] Treatment with PMSG increases the production of estradiol from the ovary and induces preovulatory surges of LH and LHRH and ovulation in prepubertal rats.[134,180] However, the LHRH/LH secretion in these animals is reduced when the central noradrenergic system has been pharmacologically manipulated by various catecholamine synthesis blockers and precursors or has been chemically lesioned using intraventricular administration of a neurotoxin 6-hydroxy-dopamine (6-OHDA). These results support the notion that normal norepinephrine synthesis is important for the occurrence of preovulatory surge release of LHRH and LH.

The role of the dopaminergic system is more complex. Stimulation of LHRH release was evident after preferential lesion of TIDA neurons using a combined intraventricular 6-OHDA and intraperitoneal (i.p.) desipramine treatment, suggesting an inhibitory effect of dopamine. In agreement with this conclusion is the finding that the dopaminergic antagonist, domperidone, stimulates LHRH release.[182] This dopamine antagonist does not cross the blood-brain barrier, and therefore, when administered peripherally, it blocks dopaminergic transmission from the TIDA neurons to LHRH neurons at the level of the median eminence. These data suggest that the TIDA system may be inhibitory to the LHRH system. In contrast to the TIDA neurons, the evidence for a stimulatory dopaminergic system came from the study in which dopaminergic neurons in the dorsal nuclei of the hypothalamus were destroyed by intraventricular administration of 6-OHDA, together with ip desipramine. The treatment resulted in a reduced release of both LHRH and LH. The dorsal nuclei of the hypothalamus receives fibers of IHDA, and therefore, it is possible that this dopaminergic system is stimulatory to LHRH neurons. This is also consistent with the finding that the dopamine antagonist, haloperidol, which can block both TIDA and IHDA neuronal activity, can also inhibit PMSG-induced LHRH and LH release. The haloperidol effect was in contrast to the finding that the LHRH release was increased after domperidone treatment. Several subtypes of dopamine receptors are now identified in the hypothalamus.[98] Thus, these data raise the possibility of the presence of different functional and pharmacological subtypes of dopamine receptors that regulate LHRH secretion. Clearly, more direct evidence is needed to resolve this issue.

Neuropeptide Y

Immunohistochemical mapping studies have described two distinct sources of hypothalamic NPY. An intrinsic hypothalamic projection originates within the arcuate nuclei and provides extensive intra- and extra-hypothalamic terminal fields.[8] The other is the ascending catecholaminergic projections from the brainstem with colocalized NPY.[53,144,192] Changes in activity and catecholamine turnover in this pathway frequently accompany functional changes in the reproductive system.[12,59,95,169,182,188] This association led to the question of whether catecholamines and NPY interact to regulate LHRH secretion. We have shown that there is a striking similarity in the developmental changes of postnatal NPY and LHRH content in the hypothalamus and preoptic area, and that a preovulatory surge of LHRH is accompanied by a surge of NPY in pituitary portal blood prior to the LH surge in pubertal female rats.[204] In addition, administration of NPY antisera reduced pubertal LH and LHRH[137] (Figure 6), whereas, injection of exogenous NPY in prepubertal rats accelerated the onset of puberty.[137,138] These findings, together with the reported

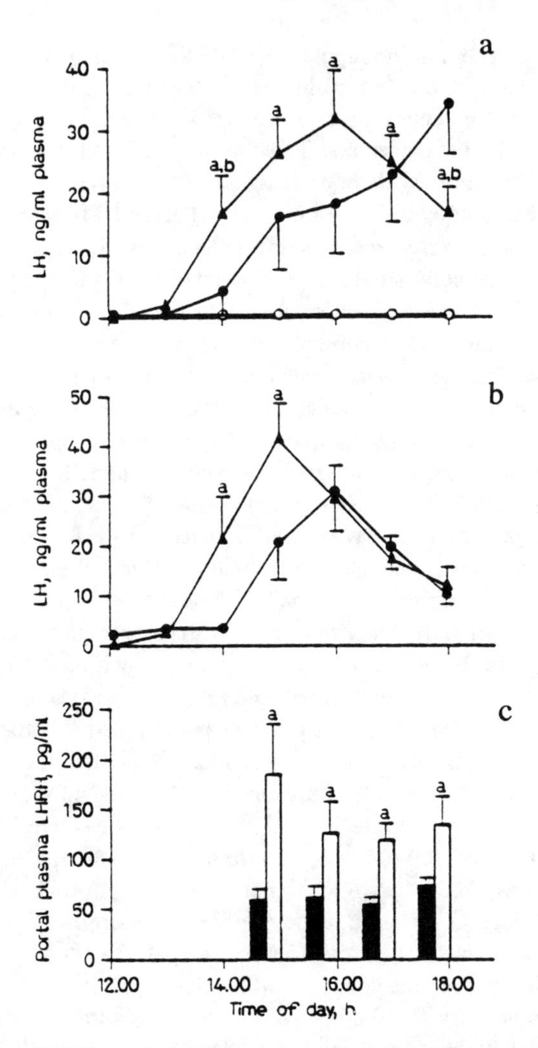

FIGURE 6. Effect of NPY immunoneutralization on LH and LHRH secretion during the first proestrus in rats. (*a*) Peripheral plasma LH profiles in rats infused with 30 μl/15 minutes of NRS (▲; $N = 10$) or rabbit anti-NPY serum (○, ●; $N = 12$) into the third ventricle (intracerebroventricularly) on the day of first proestrus. Animals treated with anti-NPY showed either a delayed (significant) LH surge (●; $N = 6$) or no significant increase in plasma LH (○; $N = 6$). $^{a}p < 0.05$ compared with anti-NPY (○) at respective times; $^{b}p < 0.05$ compared with anti-NPY (●) at respective times. (*b*) Peripheral plasma LH profiles in rats that received intravenous injection of 300 μl of NRS (▲; $N = 7$) or anti-NPY (●; $N = 9$). $^{a}p < 0.05$ vs. anti-NPY-treated rats at respective times. (*c*) Pituitary portal plasma LHRH concentration in rats treated with intracerebroventricular infusion of NRS (□; $N = 8$) or anti-NPY (■; $N = 6$). $^{a}p < 0.05$ vs. anti-NPY-treated rats at respective times. (Taken from Minami and Sarkar[137] with the permission of *Neuroendocrinology*.)

alteration in catecholamine inputs,[12,59,95,169,182,188] along with NPY innervation of LHRH-positive neural elements,[214] clearly suggest an involvement of central NPY in the generation of an LHRH and the onset of puberty.

Several studies have been conducted in the past several years to determine the role of NPY in preovulatory LH release in adult female rats. It has been shown that exogenous NPY stimulates LHRH and LH release only in the presence of ovarian steroids and inhibits LHRH and LH release in the absence of steroids in ovariectomized rats (for reviews see Kalra and Crowley[92] and Sutton *et al.*[205]). These responses resemble the documented effects of the adrenergic transmitters, norepinephrine and epinephrine, on LH and LHRH release. These findings, together with the observation that adrenergic transmitters colocalize with NPY in several regions of the rat brain, suggest an interaction between catecholamines and NPY in the regulation of LHRH release. However, innervation to the region containing LHRH cells and terminals is also provided by non-CA-containing NPY neurons in the hypothalamus.[36,144] Gonadal steroids have been shown to alter the hypothalamic content of NPY in gonadectomized male and female rats[92] and alter secretion of NPY from the hypothalamic slices *in vitro*.[6,77,118] Hence, NPY may regulate LHRH release either independently or in concert with catecholamines. Experimental studies using exogenous hormones have verified such possibilities.[6,77,118]

High amounts of NPY in the median eminence have prompted speculation that NPY may be released into the pituitary portal blood. Anatomically, NPY-containing fibers and terminals have been identified within the subependymal, internal and external zones of the median eminence, the infundibular stalk and adjacent to the third ventricle lumen,[6,53,133,197] all sites from which released NPY might be expected to reach the pituitary portal blood. Indeed, we and others have found a severalfold higher concentration of NPY in pituitary portal blood than in systemic blood.[133,204,205] The origin of pituitary portal blood NPY appears to be from the arcuate NPY system, since bilateral electrolytic lesions reduce portal blood levels of NPY.[205]

A role for pituitary portal blood NPY in the regulation of LH secretion has been suggested by the observation that the level of portal blood NPY increased in parallel with the level of LHRH prior to the preovulatory surge of LH in adult cyclic rats and pubertal female rats.[204,205] An intravenous or intraventricular injection of NPY antiserum has been shown to prevent the LH surge in steroid-primed ovariectomized rats and pubertal rats.[77,118,137,205] In studies where systemic immunoneutralization of NPY was carried out, the inhibition of LH release was not accompanied by any change in portal blood levels of LHRH.[205] These data suggest an effect of portal NPY, presumably at the level of the anterior pituitary, which is not modulated via a change in hypothalamic LHRH release. However, these data do not rule out effects of NPY at the central level (for review see Kalra and Crowley[92]). The action of NPY at the anterior pituitary is further supported by the data demonstrating high densities of NPY receptors in this gland. Recent studies using both *in vivo* and *in vitro* pituitary culture systems revealed that NPY increases the pituitary LH responsiveness to LHRH. The data presented above clearly provide evidence for a mediatory role of hypothalamic NPY in the generation of the signal for LHRH release.

β-Endorphin

As in the case of catecholamine and NPY neurons, β-endorphin-containing neurons also have synaptic contact both at the perikarya and at the terminals of the LHRH neurons. A majority of perikarya of the β-endorphin neurons are located in the arcuate nucleus region and supplied to the median eminence, preoptic area and various areas of the hypothalamus.[20,64] This opioid peptide is derived from a precursor polypeptide proopiomelanocortin (POMC) which yields β-endorphin as well as nonopioid peptides [α-melanocyte-stimulating hormone (α-MSH) and ACTH].[203] Considerable evidence now suggests a physiological role for hypothalamic β-endorphin in the regulation of LH secretion during the ovulatory cycle.[16,37,54,95,135,142] Not only do opiate agonists block ovulation and inhibit the preovulatory surge of LH and LHRH in rats,[11,35,44,46,83,106,154] but the secretion of hypothalamic β-endorphin and LHRH into pituitary portal blood during the estrous cycle is inversely associated,[185] and NAL prematurely evokes the LH surge in steroid-primed or cyclic rats.[5,67,84,123]

Since administration of opiate antagonists affects LH and LHRH secretion differentially depending on the day of the estrous cycle[158,163] or on the steroidal environment,[15,149,157,175] a role for steroids in endogenous opioid regulation or in mediation of LHRH secretion has also been suggested. Immunohistochemical studies have shown that a substantial number of hypothalamic β-endorphin neurons accumulate estradiol.[86,141] In the monkey, estradiol moderately increases the concentration of β-endorphin in pituitary portal blood, suggesting a possible stimulatory role of estradiol on β-endorphin secretion.[54] Indirect evidence similarly suggests that estradiol may stimulate β-endorphin secretion, since long-term ovariectomy lowers β-endorphin in pituitary portal blood in the rat[185] and monkey.[54] However, replacement with estradiol in long-term ovariectomized rats reduces POMC mRNA levels[172] and β-endorphin concentrations in the hypothalamus,[104,218] which suggests a decreased synthesis and possibly a release of β-endorphin.

Because a large amount of β-endorphin is secreted from the median eminence into the pituitary portal blood of female rats,[149,185,186] the question arises whether portal β-endorphin regulates pituitary gonadotropin activity directly or reflects only the activity at the median eminence. The data currently available do not support a direct action of this opioid peptide on the pituitary gland.[16,37,54,135,142] We have shown that the portal levels of β-endorphin decrease prior to the preovulatory surge of LHRH.[185] An inverse relationship between the pattern of median eminence β-endorphin and LHRH secretion prior to the preovulatory LH release has recently been confirmed in conscious ewes[2,45,46] and has provided direct evidence for the notion that the concentration of portal blood β-endorphin reflects the activity of β-endorphin neurons at the level of the median eminence.

Recently, we have further characterized the interaction between LHRH and β-endorphin at the median eminence during different days of the estrous cycle by determining the levels of LHRH and β-endorphin in portal plasma samples collected at 10-minute intervals between 1000–1800 hours on the day of diestrus, proestrus and estrus in cyclic female rats.[191] This study revealed that both LHRH and β-endorphin are secreted into pituitary portal blood in a pulsatile fashion during the estrous cycle in female rats. In addition, cyclic animals showed a daily increase in β-endorphin and LHRH release (Figures 4 and 7). The increase in β-endorphin release was observed each day in the late morning and early afternoon and was followed by an increase in LHRH release. The pulsatile nature of the portal blood LHRH and β-endorphin levels is consistent with the data obtained from ewe portal blood samples,[38,66] the rat and ewe median eminence push-pull perfusates,[2,45,44,166] and from the *in vitro* perfusion of rat hypothalamic tissues.[99] The early morning increase in β-endorphin release prior to the preovulatory LHRH release on the day of proestrus has been previously documented in rats.[185] Also, in the ewe, an increase in β-endorphin levels in the push-pull perfusate from the median eminence has been observed prior to the preovulatory release of LHRH and LH during the follicular phase.[2]

The data showing the diurnal rhythm in the secretion of hypothalamic β-endorphin into portal blood is also consistent with the reports showing an increase in hypothalamic β-endorphin protein and POMC mRNA levels on each day of the estrous cycle.[71,194] A comparison of diurnal patterns of β-endorphin and LHRH release during the cycle revealed a close negative interrelationship (Figures 4 and 7). The observed close negative coupling between β-endorphin and LHRH release led us to propose that the hypothalamic β-endorphin system is inhibitory to LHRH secretion and that β-endorphin neuronal activity may play a critical role in the circadian underpinning of LHRH release throughout the estrous cycle.

The physiological significance of the diurnal rhythmicity in LHRH release during the estrous cycle is not yet clear. The increase in LHRH release on the afternoon of proestrus is large and is sufficient to induce preovulatory release of gonadotropin required for ovulation. However, the afternoon LHRH release on the days of estrus and diestrus is relatively smaller in amplitude than on the day of proestrus. Also, the pituitary responsiveness to LHRH is lower on these days than that of proestrus because of the negative feedback effect of steroids, particularly of progesterone.[57] Therefore, it remains to be demonstrated whether the moderate increase in LHRH release on the afternoons of diestrus and estrus will be reflected by the secretion of LH from the pituitary gland and hormones from the ovarian tissues.

Is the diurnal rhythm in LHRH secretion essential for the maintenance of reproductive cyclicity? Several recent reports have shown an aging-associated loss of the diurnal rhythms in the hypothalamic content of neurotransmitters

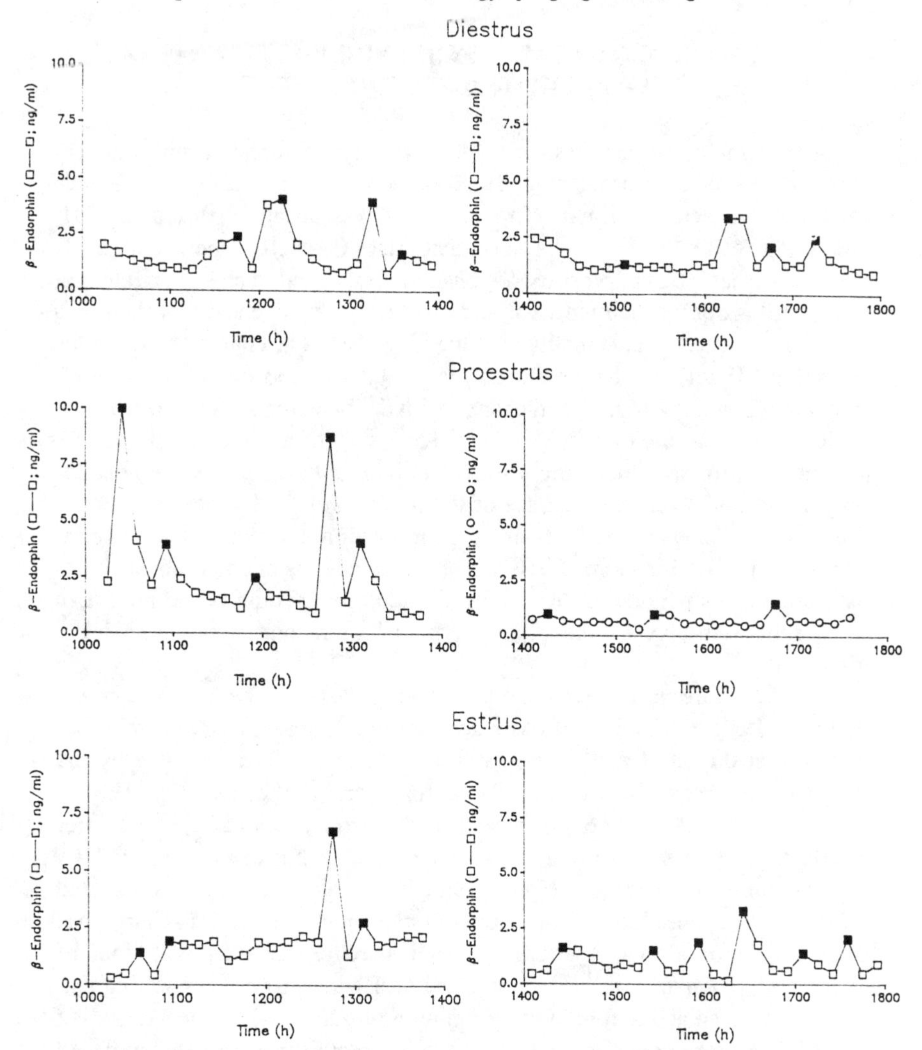

FIGURE 7. Diurnal rhythm in the pulsatile release of β-endorphin during the estrus cycle. Graphs indicate individual pituitary portal plasma β-endorphin concentrations in representative female rats during 1000–1800 hours of diestrus, proestrus and estrus. The times of collection are shown on the x-axis. The peaks of the pulses are indicated by filled symbols.

and neuropeptides (including β-endorphin) known to regulate the secretion of LHRH.[43,55,150,219,220] The secretion of estradiol from the ovary gradually declines during aging.[121] Hence, it will be important to test whether the aging-associated decrease in ovarian estradiol secretion results in a loss of the daily rhythm in β-endorphin secretion (and/or other LHRH regulatory hormones) and leads to altered diurnal rhythmicity in LHRH release. Also, there is a need to demonstrate whether or not an abnormality in LHRH diurnal rhythm is associated with lost or reduced amplitude of LH surges in aging rats.

PHYSIOLOGICAL MECHANISMS OF BASAL GONADOTROPIN RELEASE

In both male and female species, plasma levels of gonadotropins are very low for most of the reproductive life. Basal gonadotropin release in both male and female species is primarily characterized by a pulsatile (particularly LH) secretion.[9,47,56,65,68,231] The question that arises is whether the pulsatile LH secretion is an inherent property of gonadotropes. Detailed characterization of LH profiles during the menstrual and estrous cycles revealed that there is a changing frequency and amplitude of the LH pulses. The characteristics of the pulsatile LH release change during the pubertal period in various species, including humans and rats, and change with the breeding seasons in seasonal breeders, such as the ewe.[74,75,120,151,164] Thus, the pulsatile nature of gonadotropin secretion does not simply reflect an inherent property of the gonadotropes but rather reflects changes of the function of the hypothalamic-pituitary-gonadal axis in order to maintain reproduction. Indeed, pulsatile administration of LHRH is more effective than a single bolus or continuous administration of this peptide in stimulating gonadotropin synthesis and release in rats, rhesus monkeys, sheep, hypogonadal mice and hypogonadal patients.[29,47,102]

Several studies have been conducted during the past decade to understand the regulatory mechanism of basal gonadotropin release. It is generally believed that the basal and the pulsatile release of LH is maintained by the negative feedback effect of gonadal steroids, and in the case of FSH, it is maintained by the inhibitory effects of inhibin from the gonad.[58,102,119] After inhibitory feedback from the ovaries or testes has been removed by gonadectomy, an increase in high amplitude LH pulses has been documented in many species including rats, monkeys and sheep. The LH discharge has been shown to follow the pattern of LHRH secretion into the portal blood of both ovariectomized rats[51,179] (Figure 8A and B) and sheep,[38] suggesting that LHRH plays an active role in the generation of pulsatile LH release. Unlike the LHRH surge generator, which is absent in male rats and sheep due to sexual differentiation of the neuroendocrine brain (see above and chapter 11 in this volume), the LHRH pulse generator appears to be active in both the male and female of these species.[58,127]

Studies involving implantation of steroids in the pituitary gland of rats and electrochemical lesioning of the hypothalamic nuclei in rhesus monkeys have indicated that the inhibitory effect of steroids is at the pituitary level.[18,21,145,165] However, in the rat[181,190] and possibly other species,[27,147] steroids also inhibit the LHRH pulse generator. Steroid removal by ovariectomy or orchidectomy increases LHRH secretion as compared to basal LHRH release in these species. Although the effect of gonadal steroids on the pulse generator has not been well defined, the data currently available suggest that estradiol, proges-

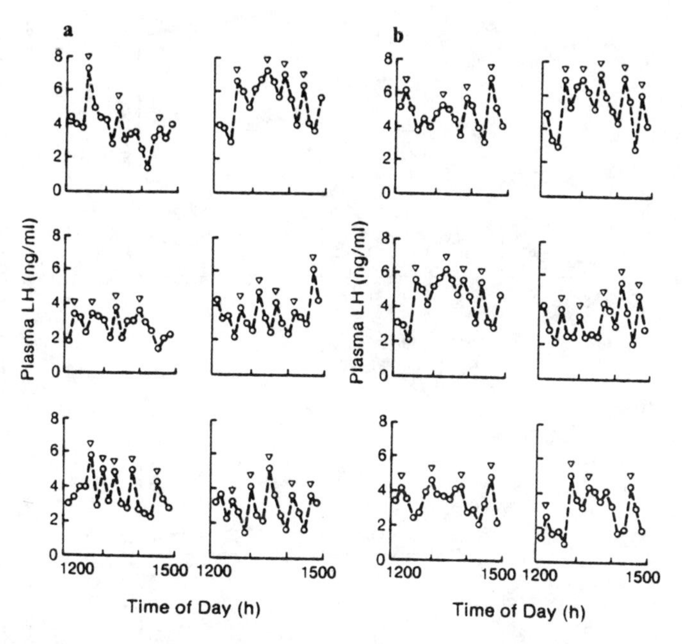

FIGURE 8A. Individual systemic plasma LH in unanesthetized ovariectomized rats (*a*) and ovariectomized rats anesthetized with Saffan (0.4–0.5 ml/100 g; (*b*). Plasma LH concentrations are pulsatile in nature, and the peaks of the pulses are indicated by triangles. (Taken from Sarkar[178] with kind permission of S. Karger AG, Basel.)

terone and testosterone can negatively influence castration-induced LHRH release in species such as rats and sheep.[119]

The circuitry involved in the generation of LHRH pulses appears to be located in the mediobasal part of the hypothalamus, since deafferentation of this part of the hypothalamus does not affect pulsatile LH release but blocks the LH surge.[18] The neurotransmitters, shown to affect surge LHRH release, also affect pulsatile LHRH release, but appear, in some instances, to produce an opposite effect.[58,69,70,91,150,189] For example, noradrenergic and NPY systems are excitatory to the LHRH/LH surge, but inhibitory to LHRH/LH pulses. Opioid peptides are inhibitory to both the surge and pulsatile LHRH release. It should be noted that LHRH neurons also receive synaptic terminals from neighboring and within LHRH neurons. Thus, there may also be autocrine and paracrine regulation of LHRH release by LHRH. Indeed, data shown in Figure 8 indicate that exogenous LHRH significantly inhibits the pulsatile secretion of LHRH in pituitary portal blood of ovariectomized female rats,[178] providing evidence for LHRH autoregulation. In accordance with this is the evidence for an intrinsic pulsatile activity in immortalized LHRH secreting neurons.[223]

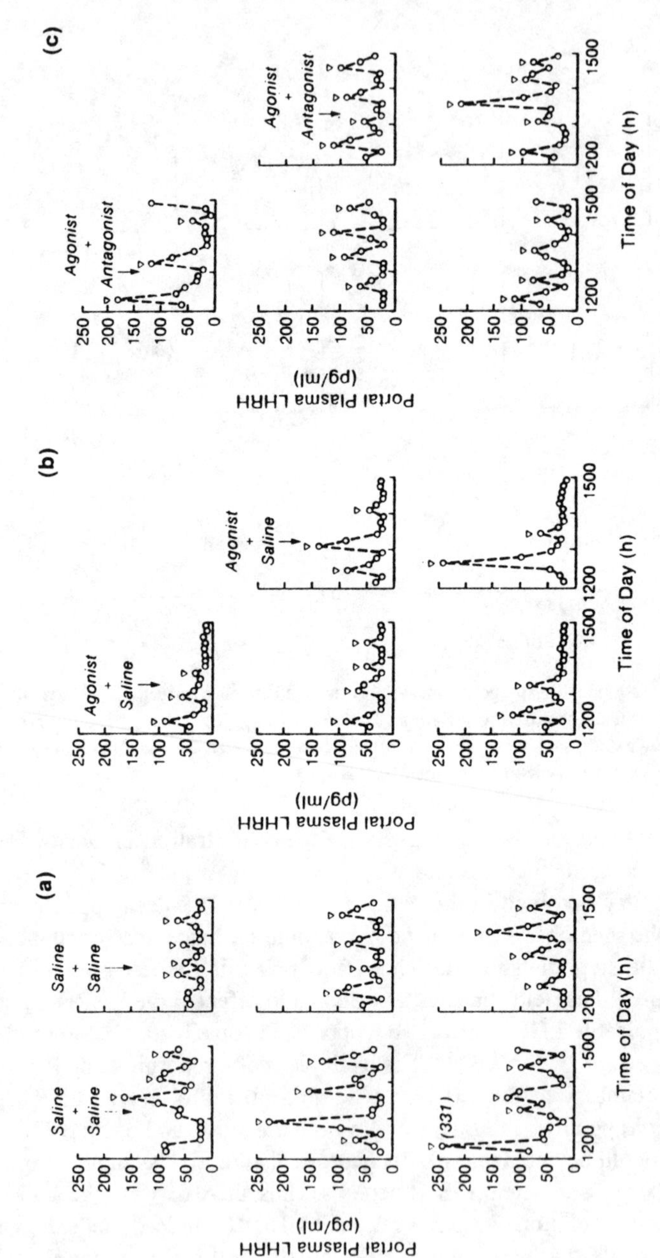

FIGURE 8B. Individual portal plasma LHRH in ovariectomized and hypophysectomized rats following injection of saline plus saline (*a*), LHRH agonist (0.1 ng/100g) plus saline (*b*) or LHRH agonist (0.1 ng/100 g) plus LHRH antagonist (10 ng/100 g) (*c*). Arrows indicate time of injection. Plasma LHRH concentrations are pulsatile in nature, and the peaks of the pulses are indicated by triangles. (Taken from Sarkar[178] with kind permission of S. Karger AG, Basel.)

SUMMARY

The events that lead to preovulatory gonadotropin release in rats can be summarized as follows. Between late diestrus and early proestrus in cyclic female rats, and between 1 and 2 days before vaginal opening (puberty) in pubertal female rats, estradiol secretion from the ovary increases. Increased secretion of estradiol may prevent the TIDA and β-endorphin inhibitory neurotransmission and activate the stimulatory noradrenergic, IHDA and NPY neurotransmission on the LHRH neuronal system. Adrenal progesterone synergizes with the estradiol action on LHRH. The increase in the LHRH output during the afternoon of the day preceding ovulation further enhances the responsiveness to LHRH (gonadotropes are already sensitized by the increased ovarian estradiol and the adrenal progesterone), probably by activating the priming effect of LHRH. The peak LHRH output and peak pituitary responsiveness to LHRH coincide with the late afternoon of proestrus and on the day preceding vaginal opening and, as a result, release massive amounts of gonadotropins. High levels of LH in plasma, in turn, release a threshold amount of ovarian progesterone, which probably activates β-endorphin and TIDA systems and inhibits further release of LHRH. The release of FSH before ovulation promotes Graffian follicle maturation in the ovary and causes the secretion of a threshold amount of estradiol, which serves to turn on the neural signal for gonadotropin release during the next proestrus. The basal gonadotropin release is probably maintained by the LHRH autofeedback regulation and the negative feedback effect of steroids, which may depend on the inhibitory influences of β-endorphin, catecholamines and NPY.

ACKNOWLEDGMENTS

I am indebted to Dr. George Fink, who introduced me to neuroendocrine research and guided me at Oxford during my initial research on LHRH regulation. I also acknowledge Drs. Joseph Meites and Samuel S.C. Yen for their important contributions over the years. Drs. Naoto Mitsugi, Shiro Minami, Shally Frautschy, Susan Friedman, Juan P. Advis and Paul Plotsky are acknowledged for their collaborative efforts.

REFERENCES

1. **Adelman, J.P., Mason, A.J., Hayflick, J.S., and Seeburg, P.H.,** Isolation of the gene and hypothalamic cDNA for the common precursor of gonadotropin releasing hormone and prolactin release-inhibiting factor in human and rat, *Proc. Natl. Acad. Sci. USA,* 83, 179, 1986.
2. **Advis, J.P., Hoffman, G.E., Sarkar, D.K., and Conover, C.D.,** Expression of *c-fos* antigens in hypothalamic LHRH and β-endorphin neurons, and *in vivo* median eminence release of LHRH and β-endorphin estimated by push-pull cannula sampling, before, during and after the ewe preovulatory LH surge, *Prog. 72nd Ann. Meet. Endocr. Soc.*, Washington DC, 461, 1721, 1991.
3. **Aiyer, M.S., and Fink, G.,** The role of sex steroid hormones in modulating the responsiveness of the anterior pituitary gland to luteinizing hormone-releasing factor in the female rat, *J. Endocrinol.*, 62, 553, 1974.
4. **Aiyer, M.S., Fink, G., and Greig, F.,** Changes in the sensitivity of the pituitary gland to luteinizing hormone releasing factor during the oestrous cycle of the rat, *J. Endocrinol.*, 60, 47, 1974.
5. **Allen, L.G., Hahn, E., Caton, D., and Kalra, S.P.,** Evidence that a decrease in opioid tone on proestrus changes the episodic pattern of luteinizing hormone (LH) secretion: Implications in the preovulatory LH hypersecretion, *Endocrinology*, 122, 1004, 1988.
6. **Allen, L.G., Kalra, P.S., Crowley, W.R., and Kalra, S.P.,** Comparison of the effects of neuropeptide Y and adrenergic transmitters on LH release and food intake in male rats, *Life Sci.*, 37, 617, 1985.
7. **Amin, H.K., and Hunter, W.M.,** Human pituitary follicle-stimulating hormone: Distribution, plasma clearance and urinary excretion as determined by radioimmunoassay, *J. Endocrinol.*, 48, 307, 1970.
8. **Bai, F.L., Yamano, M., Shiotani, Y., Emson, P.C., Smith, A.D., Powell, J.F., and Tohyama, M.,** An arcuato-paraventricular and -dorsomedial hypothalamic neuropeptide Y-containing system which lacks noradrenaline in the rat, *Brain Res.*, 331, 172, 1985.
9. **Baird, D.T., Swantson, I., and Scramuzzi, R.J.,** Pulsatile release of LH and secretion of ovarian steroids in sheep during the luteal phase of estrous cycle, *Endocrinology,* 98, 1490, 1976.
10. **Barraclough, C.A.,** Neural control of the synthesis and release of luteinizing hormone-release hormone, in *Functional Anatomy of the Neuroendocrine Hypothalamus,* Ciba Foundation Symposium, Wiley, Chichester, 168, 233, 1992.
11. **Barraclough, C.A., and Sawyer, C.H.,** Inhibition of the release of pituitary ovulatory hormone in the rat by morphine, *Endocrinology,* 57, 329, 1955.
12. **Barraclough, C.A., and Wise, P.M.,** The role of catecholamines in the regulation of pituitary luteinizing hormone and follicle-stimulating hormone secretion, *Endocr. Rev.*, 91, 1982.
13. **Batten, T.F.C., and Hopkins, C.R.,** Discrimination of LH, FSH, TSH and ACTH in dissociated porcine anterior pituitary cells by light and electron microscope immunocytochemistry, *Cell Tiss. Res.*, 192, 107, 1978.
14. **Belchetz, P.E., Plant, T.M., Nakai, Y., Keogh, E.J., and Knobil, E.,** Hypophysial responses to continuous and intermittent delivery of hypothalamic gonadotropin-releasing hormone, *Science*, 202, 631, 1978.
15. **Bhanot, R., and Wilkinson, M.,** The inhibitory effect of opiates on gonadotrophin secretion is dependent upon gonadal steroids, *J. Endocrinol.,* 102, 133, 1984.
16. **Bicknell, R.J.,** Endogenous opioid peptides and hypothalamic neuroendocrine neurons, *J. Endocrinol.*, 107, 437, 1985.
17. **Blake, C.A., and Kelch, R.,** Administration of antiluteinizing hormone-releasing hormone serum to rats: Effects on preovulatory secretion of luteinizing hormone and follicle-stimulating hormone, *Endocrinology*, 109, 2175, 1981.

18. **Blake, C.A., and Sawyer, C.H.,** Effects of hypothalamic deafferentation on the pulsatile rhythm in plasma concentration of luteinizing hormone in ovariectomized rats, *Endocrinology*, 94, 730, 1974.
19. **Blake, C.A., Weiner, R.I., Gorski, R.A., and Sawyer, C.H.,** Secretion of pituitary luteinizing hormone and follicle stimulating hormone in female rats made persistently estrous or diestrous by hypothalamic deafferentation, *Endocrinology*, 90, 855, 1972.
20. **Bloom, F., Battenberg, E., Rossier, J., Ling, N., and Guillemin, R.,** Neurons containing β-endorphin in rat brain exist separately from those containing enkephalin: Immunocytochemical studies, *Proc. Natl. Acad. Sci. USA*, 75, 1591, 1978.
21. **Bogdanove, E.M., Nolin, J.M., and Campbell, G.T.,** Qualitative and quantitative gonad-pituitary feedback, *Recent. Prog. Horm. Res.*, 31, 567, 1975.
22. **Boyar, R., Finkelstein, J., Roffwarg, H., Kapen, S., Weitzman, E., and Hellman, L.,** Synchronization of augmented luteinizing hormone secretion with sleep during puberty, *N. Eng. J. Med.*, 287, 582, 1972.
23. **Boyar, R.M., Rosenfeld, R.S., Finkelstein, J.W., Kapen, S., Roffwarg, H., and Weitzman, E.D.,** Ontogeny of luteinizing hormone and testosterone secretion, *J. Steroid Biochem.*, 6, 803, 1975.
24. **Brown-Grant, K., Davidson, J.M., and Greig, F.,** Induced ovulation in albino rats exposed to constant light, *J. Endocrinol.*, 57, 7, 1973.
25. **Burgus, R., Butcher, M., Amoss, M., Ling, N., Monahan, M., Rivier, J., Fellows, R., Blackwell, R., Vale, W., and Guillemin, R.,** Primary structure of the ovine hypothalamic luteinizing hormone-releasing factor (LRF), *Proc. Natl. Acad. Sci. USA*, 69, 278, 1972.
26. **Caligaris, L., Astrada, J.J., and Taleisnik, S.,** Release of luteinizing hormone induced by estrogen injection into ovariectomized rats, *Endocrinology*, 88, 810, 1971.
27. **Caraty, A., Locatelli, A., and Martin, G.B.,** Biphasic response in the secretion of gonadotropin-releasing hormone in ovariectomized ewes injected with oestradiol, *J. Endocrinol.*, 123, 375, 1989.
28. **Catt, K.J., and Pierce, J.G.,** Gonadotropic hormones of the adenohypophysis, in *Reproductive Endocrinology*, Yen, S.S.C., and Jaffe, R.B., Eds., W.B. Saunders Co., Philadelphia, 1986, 75.
29. **Charlton, H.M., Halpin, D.M.G., Iddon, C., Rosie, R., Levy, G., McDowell, I.F.W., Megson, A., Morris, J.F., Bramwell, A., Speight, A., Ward, B.J., Broadhead, J., Davey-Smith, G., and Fink, G.,** The effects of daily administration of single and multiple injections of gonadotropin-releasing hormone on pituitary and gonadal function in the hypogonadal (hpg) mouse, *Endocrinology*, 113, 535, 1983.
30. **Chen, W.-P., Witkin, J.W., and Silverman, A.-J.,** Sexual dimorphism in the synaptic input to gonadotropin-releasing hormone neurons, *Endocrinology*, 126, 695, 1990.
31. **Chiappa, S., and Fink, G.,** Releasing factor and hormonal changes in the hypothalamic-pituitary-gonadotropin and -adrenocorticotrophin systems before and after birth and puberty in male, female and androgenized female rats, *J. Endocrinol.*, 72, 211, 1977.
32. **Chiappa, S.A., Fink, G., and Sherwood, N.M.,** Immunoreactive luteinizing hormone releasing factor (LRF) in pituitary stalk plasma from female rats: Effects of stimulating diencephalon, hippocampus and amygdala, *J. Physiol.*, 267, 625, 1977.
33. **Chin, W.W.,** Organization and expression of glycoprotein hormone genes, in *The Pituitary Gland*, Imura, H., Ed., Raven Press, New York, 1985, 103.
34. **Ching, M.,** Correlative surges of LHRH and LHRH, LH and FSH in pituitary stalk plasma and systemic plasma of rat during proestrus, *Neuroendocrinology*, 34, 279, 1982.
35. **Ching, M.,** Morphine suppresses the proestrous surge of GnRH in pituitary portal plasma of rats, *Endocrinology*, 112, 2209, 1983.
36. **Chronwall, B.M., DiMaggio, D.A., Massar, V.J., Pickel, V.M., Ruggiero, D.A., and O'Donohue, T.L.,** The anatomy of neuropeptide Y-containing neurons in rat brain, *Neuroscience*, 15, 1159, 1985.

37. **Cicero, T.J.,** Effect of exogenous and endogenous opiates on the hypothalamic-pituitary-gonadal axis in the male, *Fed. Proc.*, 39, 2551, 1980.
38. **Clarke, I.J., Cumins, J.T., Jenkin, M., and Phillips, D.J.,** The estrogen-induced surge of LH requires a signal pattern of gonadotropin-releasing hormone input to the pituitary gland in the ewe, *J. Endocrinol.*, 122, 127, 1989.
39. **Clarke, W.P., and Gala, R.R.,** The influence of anesthetics on the estrogen-induced afternoon prolactin surge. Althesin does not block the surge, *Life Sci.*, 29, 277, 1981.
40. **Clayton, R.N.,** Gonadotropin-releasing hormone: Its actions and receptors, *J. Endocrinol.*, 120, 11, 1989.
41. **Clough, R.W., Rodriguez-Sierra, J.F.,** Synaptic changes in the hypothalamus of the prepubertal female rat administered estrogen, *Am. J. Anat.*, 167, 205, 1983.
42. **Cogen, P.H., Antunes, J.L., Louis, K.M., Dyrenfurth, I., and Ferin, M.,** The effects of anterior hypothalamic disconnection on gonadotropin secretion in the female rhesus monkey, *Endocrinology*, 107, 677, 1980.
43. **Cohen, I.R., and Wise, P.M.,** Age-related changes in the diurnal rhythm of serotonin turnover in microdissected brain areas of estradiol-treated ovariectomized rats, *Endocrinology*, 122, 2626, 1988.
44. **Conover, C.D., Gore, A.C., Roberts, J.L., Sarkar, D.K., Rabi, J., and Advis, J.P.,** Median eminence *in vivo* release, infundibular/arcuate peptide content, and arcuate steady-state mRNA of β-endorphin, before, during and after a preovulatory LH surge in ewes, *Prog. 22nd Ann. Meet. Soc. Neurosci.*, Anaheim, CA, Abstr. 52.13, 1992.
45. **Conover, C.D., Kuligs, R.O., Sarkar, D.K., and Advis, J.P.,** β-Endorphin regulation of LHRH release in ewes: Immunocytochemistry, tissue content and *in vivo* release analysis, *Prog. 20th Ann. Meet. Soc. Neurosci.*, St. Louis, Abstr. 167.14, 1990.
46. **Contijoch, A.M., Gonzales, C.D, Malamed, S., Sarkar, D.K., Troncoso, S., and Advis, J.P.,** β-endorphin regulation of LHRH release at the median eminence level: Immunocytochemical and physiological evidence in hens, *Neuroendocrinology*, 57, 365, 1993.
47. **Crowley Jr., W.F., Filcari, M., Spratt, D.I., and McNeilly, A.S.,** The physiology of gonadotropin-releasing hormone (GnRH) secretion in men and women, *Recent Prog. Horm. Res.*, 41, 473, 1985.
48. **Curtis, A., and Fink, G.,** A high molecular weight precursor of luteinizing hormone-releasing hormone from rat hypothalamus, *Endocrinology*, 112, 390, 1983.
49. **Dada, M.O., Campbell, G.T., and Blake, C.A.,** A quantitative immunocytochemical study of the luteinizing hormone and follicle-stimulating hormone cells in the adenohypophysis of adult male rats and adult female rats throughout the estrous cycle, *Endocrinology*, 113, 970, 1983.
50. **Dial, G.D., Wiseman, B.S., Ott, R.S., Smith, A.L., and Hixon, J.E.,** Absence of sexual dimorphism in the goat: Induction of luteinizing hormone discharge in the castrated male and female and in the intersex with estradiol benzoate, *Theriogenology*, 23, 351, 1985.
51. **Everett, J.W., and Sawyer, C.H.,** A 24-hour periodicity in the "LH-release apparatus" of female rats, disclosed by barbiturates sedation, *Endocrinology*, 47, 198, 1950.
52. **Everett, J.W., Krey, L.C., and Tyrey, L.,** The quantitative relationship between electrochemical preoptic stimulation and LH release in proestrous versus late-diestrous rats, *Endocrinology*, 93, 947, 1973.
53. **Everitt, B.J., Hökfelt, T., Terenius, L., Tatemoto, K., Mutt, V., and Goldstein, M.,** Differential co-existence of neuropeptide Y (NPY)-like immunoreactivity with catecholamines in the central nervous system of the rat, *Neuroscience*, 11, 2, 1984.
54. **Ferin, M., Van Vugt, D., and Wardlaw, S.,** The hypothalamic control of the menstrual cycle and the role of endogenous opioid peptides, *Recent Prog. Horm. Res.*, 40, 441, 1984.
55. **Field, E., and Kuhn, C.M.,** Opioid inhibition of luteinizing hormone release declines with age and acyclicity in female rats, *Endocrinology*, 123, 2626, 1988.
56. **Filicori, M., Santoro, N., Merriam, G.R., and Crowley Jr., W.F.,** Characterization of the physiological pattern of episodic gonadotropin secretion throughout the human menstrual cycle, *J. Clin. Endocrinol. Metab.*, 62, 113, 1986.

57. **Fink, G.**, Feedback actions of target hormones on hypothalamus and pituitary with special reference to gonadal steroids, *Ann. Rev. Physiol.*, 41, 571, 1979.
58. **Fink, G.**, Gonadotropin secretion and its control, in *The Physiology of Reproduction*, Knobil, E., and Neill, J.D., Eds., Raven Press, New York, 1988, 1349.
59. **Fink, G., and Geffen, L.B.**, The hypothalamo-hypophysial system: Model for central peptidergic and monoaminergic transmission, in *International Review of Physiology, Neurophysiology III, Vol. 17*, Porter, R., Ed., University Park Press, Baltimore, 1978, 1.
60. **Fink, G., and Jamieson, M.G.**, Immunoreactive luteinizing hormone releasing factor in rat stalk blood: Effects of electrical stimulation of the medial preoptic area, *J. Endocrinol.*, 68, 71, 1976.
61. **Fink, G., Aiyer, M., Chiappa, S., Henderson, S., Jamieson, M., Levy-Perez, V., Pickering, A., Sarkar, D., Sherwood, N., Speight, A., and Watts, A.**, Gonadotropin-release hormone, in *Hormonally Active Brain Peptide*, McKerns, E., and Pantic, G., Eds., Plenum Publishing, New York, 1982, 397.
62. **Fink, G., Chippa, S.A., Pickering, A., and Sarkar, D.**, Control of rat oestrous cycle, in *Endocrinology, Vol. 1*, James, V.H.T., Ed., Excerpta Medica, Amsterdam, 1976, 186.
63. **Fink, G., Sarkar, D.K., Dow, R.C., Dick, H., Borthwick, N., Malnick, S., and Twine, M.**, Sex difference in response to alphaxalone anesthesia may be oestrogen dependent, *Nature (Lond.)*, 298, 270, 1982.
64. **Finley, J., Lindstrom, P., and Petrusz, P.**, Immunocytochemical localization of β-endorphin-containing neurons in the hypothalamus, *Neuroendocrinology*, 33, 28, 1981.
65. **Fox, S.R., and Smith, S.M.**, Changes in the pulsatility pattern of luteinizing hormone secretion during the rat estrous cycle, *Endocrinology*, 116, 1485, 1985.
66. **Freeman, M.C., Dupke, K.C., and Croteau, C.M.**, Extinction of the estrogen-induced daily signal for LH release in the rat: A role for the proestrous surge of progesterone, *Endocrinology*, 99, 223, 1976.
67. **Gabriel, S.M., Berglund, L.A., and Simpkins, J.W.**, A decline in endogenous opioid influence during the steroid-induced hypersecretion of luteinizing hormone in the rat, *Endocrinology*, 118, 558, 1986.
68. **Gallo, R.V.**, Pulsatile LH release during the periods of low level LH secretion in the rat estrous cycle, *Biol. Reprod.*, 24, 771, 1981.
69. **Gallo, R.V.**, Further studies on norepinephrine-induced suppression of pulsatile luteinizing hormone release in ovariectomized rats, *Neuroendocrinology*, 39, 120, 1984.
70. **Gallo, R.V., and Drouva, S.V.**, Effects of intraventricular infusion of catecholamines on luteinizing hormone release in ovariectomized rats and ovariectomized steroid-primed rats, *Neuroendocrinology*, 29, 149, 1979.
71. **Genazzani, A.R., Trentini, G.P., Petraglia, F., De Gaetani, C.F., Criscuolo, M., Ficarra, G., De Ramundo, B.M., and Cleva, M.**, Estrogen modulates the circadian rhythm of hypothalamic beta-endorphin contents in female rats, *Neuroendocrinology*, 52, 221, 1990.
72. **Graham, J.M., and Desjardins, C.**, Classical conditioning: Induction of luteinizing hormone and testosterone secretion in anticipation of sexual activity, *Science*, 210, 1039, 1980.
73. **Green, J.D., and Harris, G.H.**, The neurovascular link between the neurohypophysis and adenohypophysis, *J. Endocrinol.*, 5, 136, 1947.
74. **Grumbach, M.M., Roth, J.C., Kaplan, S.L., and Kelch, R.P.**, Hypothalamic-pituitary regulation of puberty in man: Evidence and concepts derived from clinical research, in *The Control of the Onset of Puberty*, Grumbach, M., Grave, M.G.D., and Mayer, F.E., Eds., John Wiley and Sons, New York, 1974, 115.
75. **Grumbach, M.M., Sizonenko, P.C., and Aubert, M.L.**, *Control of the Onset of Puberty*, Williams and Wilkins, Baltimore, 1990, 710.
76. **Halasz, B., and Gorski, R.A.**, Gonadotropic hormone secretion in female rats after partial or total interruption of neural afferents to the medial basal hypothalamus, *Endocrinology*, 80, 608, 1967.

77. **Harfstrand, A., Fuxe, K., Agnate, L.F., Eneroth, P., Zini, I., Zoli, M., Anderson, K., Von Euler, L., Terenius, L., Mutt, V., and Goldstein, M.,** Studies on neuropeptide Y-catecholamine interaction in the hypothalamus and in the forebrain of the male rat. Relationship to neuroendocrine function, *Neurochem. Int.*, 8, 355, 1986.
78. **Harlan, R.E., Gordon, J.H., and Gorski, R.A.,** Sexual differentiation of the brain: Implication for neuroscience, in *Reviews of Neuroscience, Vol. 4*, Schneider, D.M., Ed., Raven Press, New York, 1979, 13.
79. **Harris, G.W.,** Ed., *Neural Control of the Pituitary Gland*, Edward Arnold, London, 1955.
80. **Harris, G.W., and Jackobsohn, D.,** Functional grafts of the anterior pituitary gland, *Proc. R. Soc. London Ser. B*, 139, 263, 1952.
81. **Hauger, R.L., Karsch, F.J., and Foster, D.L.,** A new concept for control of the ewe based on the temporal relationship between luteinizing hormone, estradiol and progesterone in peripheral serum and evidence that progesterone inhibits tonic LH secretion, *Endocrinology*, 101, 807, 1977.
82. **Hoffman, G.E., Lee, W.-S., and Wray, S.,** Gonadotropin releasing hormone (GnRH), in *Neuroendocrinology*, Nemeroff, C.B., Ed., CRC Press, Boca Raton, 1992, 185.
83. **Hulse, G.K., and Coleman, B.J.,** The effects of morphine sulfate on ovulation in the immature rat treated with PMSG, *Pharmacol. Biochem. Behav.*, 19, 269, 1982.
84. **Ieiri, T., Chen, H.T., Campbell, G.A., and Meites, J.,** Effects of naloxone and morphine on the proestrous surge of prolactin and gonadotropins in the rat, *Endocrinology* 105, 1568, 1980.
85. **Ji, W.-Z., Kaynard, A.H., Pau, K.Y.F., Hess, D.L., Baughman, W.L., and Spies, H.G.,** Endogenous opiates regulate the nocturnal reduction in luteinizing hormone pulse frequency during the luteal phase of the macaque menstrual cycle, *Biol. Reprod.*, 41, 1024, 1989.
86. **Jirikowski, G.F., Merchenthaler, I.L., Rueger, G.E., and Stumpf, W.E.,** Estradiol target sites immunoreactive for β-endorphin in the arcuate nucleus of the rat and mouse hypothalamus, *Neurosci. Lett.*, 65, 121, 1986.
87. **Jones, E.G., and Powell, T.P.S.,** An anatomical study of converging sensory pathways within the cerebral cortex of the monkey, *Brain*, 93, 793, 1970.
88. **Judd, H.L., Parker, D.C., Rakoff, J.S., Hopper, E.R., and Yen, S.S.C.,** Elucidation of mechanism(s) of the nocturnal rise of testosterone in men, *J. Clin. Endocrinol. Metab.*, 38, 134, 1974.
89. **Junier, M.-P., Ma, Y.J., Costa, M.E., Hoffman, G., Hill, D.E., and Ojeda, S.R.,** Transforming growth factor alpha contributes to the mechanism by which hypothalamic injury induces precocious puberty, *Proc. Natl. Acad. Sci. USA*, 88, 9743, 1991.
90. **Kaasjager, W.A., Woodbury, D.M., Van Dieten, J.A.M.J., and Van Rees, G.P.,** The role played by the preoptic region and the hypothalamus in spontaneous ovulation and ovulation induced by progesterone, *Neuroendocrinology*, 7, 54, 1971.
91. **Kalra, S.P.,** Mandatory neuropeptide-steroid signaling for the preovulatory luteinizing hormone-releasing hormone discharge, *Endocr. Rev.*, 14, 507, 1993.
92. **Kalra, S.P., and Crowley, W.R.,** Neuropeptide Y: A novel neuroendocrine peptide in the control of pituitary hormone secretion, and its relation to luteinizing hormone, *Frontiers in Neuroendocrinology*, 13, 1, 1992.
93. **Kalra, S.P., and Kalra, P.S.,** Temporal interrelationship between the circulating levels of estradiol, progesterone and LH during the rat estrous cycle: Effects of exogenous progesterone, *Endocrinology*, 95, 1711, 1974.
94. **Kalra, P.S., and Kalra, S.P.,** Circadian periodicities of serum androgens, progesterone, gonadotropins and LHRH in male rats: The effect of hypothalamic deafferentation, castration or adrenalectomy, *Endocrinology*, 101, 1821, 1977.
95. **Kalra, S.P., and Kalra, P.S.,** Neural regulation of luteinizing hormone secretion in the rat, *Endocr. Rev.*, 4, 311, 1983.
96. **Karsch, F.J., Dierschke, D.J., and Knobil, E.,** Sexual differentiation of pituitary function: Apparent difference between primates and rodents, *Science*, 179, 484, 1973.

97. **Kaynard, A.H., Pau, F.K.-Y., Hess, D.L., and Spies, H.G.,** Gonadotropin-releasing hormone and norepinephrine release from the rabbit mediobasal and anterior hypothalamus during the mating-induced luteinizing hormone surge, *Endocrinology*, 127, 1176, 1990.
98. **Kebabian, J.W.,** Brain dopamine receptors: 20 years of progress, *Neurochem. Res.*, 18, 101, 1993.
99. **Kim, K., and Ramirez, V.D.,** *In vitro* LHRH release from superfused hypothalamus as a function of the rat estrous cycle: Effects of progesterone, *Neuroendocrinology*, 42, 392, 1986.
100. **King, J.C., and Anthony, E.L.P.,** Biosynthesis of LHRH: Inference from immunocytochemical studies, *Peptides*, 4, 963, 1983.
101. **Kiss, J., and Halasz, B.,** Demonstration of serotoninergic axons terminating on luteinizing hormone-releasing hormone neurons in the preoptic area of the rat using a combination of immunocytochemistry and high resolution autoradiography, *Neuroscience*, 14, 69, 1985.
102. **Knobil, E., and Hotchkiss, J.,** The menstrual cycle and its neuroendocrine control, in *The Physiology of Reproduction*, Knobil, E., and Neill, J.D., Eds., Raven Press, New York, 1988, 1971.
103. **Knobil, E., Plant, T.M., Wildt, L., Belchetz, P.E., and Marshall, G.,** Control of the rhesus monkey menstrual cycle: Permissive role of hypothalamic gonadotropin releasing hormone, *Science*, 207, 1371, 1980.
104. **Knuth, U.A., Sikand, G.S., Casanueva, F.F., Havlicek, V., and Friesen, H.G.,** Changes in β-endorphin content in discrete areas of the hypothalamus throughout proestrus and diestrus of the rat, *Life Sci.*, 33, 1443, 1983.
105. **Koves, K., and Halasz, B.,** Location of the neural structures triggering ovulation in the rat, *Neuroendocrinology*, 6, 180, 193, 1970.
106. **Koves, K., Marton, J., Molnar, J., and Halasz, B.,** (D-Met^2Pro^5)-enkephalinamide-induced blockage of ovulation and its reversal by naloxone in the rat, *Neuroendocrinology*, 32, 82, 1981.
107. **Kuljis, R.O., and Advis, J.P.,** Immunocytochemical and physiological evidence of a synapse between dopamine and luteinizing hormone releasing hormone-containing neurons in the ewe median eminence, *Endocrinology*, 124, 1579, 1989.
108. **Langub Jr., M.C., Maley, B.E., and Watson Jr., R.E.,** Ultrastructural evidence for luteinizing hormone-releasing hormone neuronal control of estrogen responsive neurons in the preoptic area, *Endocrinology*, 128, 27, 1991.
109. **Lee, K.H., and Yen, S.S.C.,** The effect of an antiserum to luteinizing hormone-releasing hormone on the progesterone-induced luteinizing hormone surge in ovariectomized, estrogen primed rats, *Endocrinology*, 106, 867, 1980.
110. **Lee, W.-S., Smith, M.S., and Hoffman, G.E.** Luteinizing hormone-releasing hormone (LHRH) neurons express c-Fos during the proestrous LH surge, *Proc. Natl. Acad. Sci. USA*, 1990.
111. **Lee, W.-S., Smith, S.M., and Hoffman, G.E.,** Progesterone enhances the surge of luteinizing hormone by increasing the activation of luteinizing hormone-releasing neurons, *Endocrinology*, 127, 2604, 1990.
112. **Leranth, C., MacLusky, N.J., Sakamoto, H., Shanabrough, M., and Naftolin, F.,** Glutamic acid decarboxylase-containing axons synapse on LHRH neurons in the rat medial preoptic area, *Neuroendocrinology*, 40, 536, 1985.
113. **Leranth, C., Seguraum, L.M.G., Palkovits, M., MacLusky, N.J., Shanabrough, M., and Naftolin, F.,** The LH-RH containing neuronal network in the preoptic area of the rat: Demonstration of LR-RH containing nerve terminals in synaptic contact with LHRH neurons, *Brain Res.*, 345, 332, 1985.
114. **Leranth, C., MacLusky, N.J., Shanabrough, M., and Naftolin, F.,** Catecholaminergic innervation of LHRH and GAD immunopositive neurons in the rat medial preoptic area: An electron microscopic double immunostaining and degeneration study, *Neuroendocrinology*, 48, 581, 1988.

115. **Leranth, C., MacLusky, N.J., Shanabrough, M., and Naftolin, F.,** Immunohistochemical evidence for synaptic connections between proopiomelanocortin-immunoreactive axons and LHRH neurons by the preoptic area of the rat, *Brain Res.*, 449, 167, 1988.
116. **Levine, J.E., and Ramirez, V.D.,** Luteinizing hormone-releasing hormone release during the rat estrous cycle and after ovariectomy as estimated with push-pull cannulae, *Endocrinology*, 111, 1439, 1982.
117. **Levin, J.E., Norman, R.L., Gliessman, P.M., Oyama, T.T., Bangsberg, D.R., and Spies, H.G.,** *In vivo* gonadotropin-releasing hormone release and serum luteinizing hormone measurements in ovariectomized estrogen-treated rhesus macaques, *Endocrinology*, 117, 711, 1985.
118. **Li, S., and Pelletier, G.,** The role of dopamine in the control of neuropeptide Y neurons in the rat arcuate nucleus, *Neurosci. Lett.*, 69, 74, 1986.
119. **Lincoln, D.W., Fraser, H.M., Lincoln, G.A., Martin, G.B., and McNeilly, A.S.,** Hypothalamic pulse generators, *Recent Prog. Horm. Res.*, 41, 369, 1985.
120. **Lincoln, G.A., and Short, R.V.,** Seasonal breeding: Nature's contraceptive, *Recent Prog. Horm. Res.*, 36, 1, 1980.
121. **Lu, J.K.H., Matt, D.W., and LaPolt, P.S.,** Modulatory effects of estrogen and progestin on female reproductive organ, in *Ovarian Secretion and Cardiovascular and Neurological Function*, Naftolin, F., and De Cherney, H., Eds., Raven Press, New York, 1990, 297.
122. **Lunstra, D.D., Boyd, G.W., and Corah, L.R.,** Effects of natural mating stimuli on serum luteinizing hormone, testosterone and estradiol-17β in yearling beef bulls, *J. Animal Sci.*, 67, 3277, 1989.
123. **Lustig, R.H., Pfaff, D.W., and Fishman, F.,** Opioidergic modulation of the oestradiol-induced LH surge in the rat: Roles of ovarian steroids, *J. Endocrinol.*, 116, 55, 1988.
124. **MacKinnon, P.C.B., Puig-Duran, E., and Laynes, R.,** Reflection of the attainment of puberty in the rat: Have circadian signals a role to play in its onset? *J. Reprod. Fertil.*, 52, 401, 1978.
125. **MacLusky, N.J., and McEwen, B.S.,** Oestrogen modulates progestin receptor concentrations in some rat brain regions but not in others, *Nature (Lond)*, 274, 276, 1978.
126. **MacLusky, N.J., Naftolin, F., and Leranth, C.,** Immunocytochemical evidence for direct synaptic connections between corticotropin-releasing factor (CRF) and gonadotropin releasing hormone (GnRH)-containing neurons in the preoptic area of the rat, *Brain Res.*, 439, 391, 1988.
127. **Malven, P.V.,** Ed., *Mammalian Neuroendocrinology*, CRC Press, Boca Raton, 1993, 181.
128. **Marshall, P.E., and Goldsmith, P.C.,** Neuroregulatory and neuroendocrine GnRH pathways in the hypothalamus and forebrain of the baboon, *Brain Res.*, 193, 353, 1980.
129. **Mason, A.J., Hayflick, J.S., Zoeller, R.T., Young III, W.S., Phillips, H.S., and Seeburg, P.H.,** A deletion truncating the gonadotropin-releasing hormone gene is responsible for hypogonadism in the hpg mouse, *Science*, 234, 1366, 1986.
130. **Matsumoto, A., and Arai, Y.,** Sexual dimorphism in "wiring pattern" in the hypothalamic arcuate nucleus and its modification by neonatal hormonal environment, *Brain Res.*, 190, 238, 1980.
131. **Matsuo, H., Baba, Y., Nair, R.M.G., Arimura, A., and Schally, A.V.,** Structure of the porcine LH- and FSH-releasing hormone. I. The proposed amino acid sequence, *Biochem. Biophys. Res. Commun.*, 43, 1334, 1971.
132. **McCormack, J.T., Plant, T.M., Hess, D.S., and Knobil, E.,** The effect of luteinizing hormone releasing hormone (LHRH) antiserum administration on gonadotropin secretion in the rhesus monkey, *Endocrinology*, 100, 663, 1977.
133. **McDonald, J.K., Koenig, J.I., Gibbs, D.M., Collins, P., and Noe, B.D.,** High concentrations of neuropeptide Y in pituitary portal blood of rats, *Neuroendocrinology*, 23, 538, 1987.
134. **Meijs-Roelofs, H.M.A., Uilenbroek, J.T.J., deGreef, W.J., deJong, F.H., and Kramer, P.,** Gonadotrophin and steroid levels around the time of first ovulation in the rat, *J. Endocrinol.*, 67, 275, 1975.

135. **Meites, J.**, Introduction to effects of endogenous opioid peptides on the neuroendocrine system, *Adv. Physiol. Sci.*, 14, 269, 1981.
136. **Millar, R.P., Wegener, I., and Schally, A.V.**, Putative prohormonal luteinizing hormone-releasing hormone, in *Neuropeptides: Biochemical and Pharmacological Studies*, Millar, R.P., Ed., Churchill Livingstone, Edinburgh, 1981, 111.
137. **Minami, S., and Sarkar, D.K.**, Central administration of neuropeptide Y induces precocious puberty in female rats, *Neuroendocrinology*, 56, 930, 1992.
138. **Minami, S., Frautschy, S.A., Plotsky, P.M., Sutton, S.H., and Sarkar, D.K.**, Facilitatory role of neuropeptide Y (NPY) on the onset of puberty: The effect of immunoneutralization of NPY on LH and LHRH release, *Neuroendocrinology*, 52, 112, 1990.
139. **Moenter, S.M., Caraty, A., and Karsch, F.J.**, The estradiol-induced surge of gonadotropin-releasing hormone in the ewe, *Endocrinology*, 127, 1375, 1990.
140. **Moenter, S.M., Brand, R.C., and Karsch, F.J.**, Dynamics of gonadotropin-releasing hormone (GnRH) secretion during the GnRH surge: Insight into the mechanism of GnRH surge induction, *Endocrinology*, 130, 2978, 1992.
141. **Morrell, J.I., McGinty, J.F., and Pfaff, D.W.**, A subset of β-endorphin- or dynorphin-containing neurons in the medial basal hypothalamus accumulates estradiol, *Neuroendocrinology*, 41, 417, 1985.
142. **Motta, M., and Martini, L.**, Effect of opioid peptides on gonadotrophin secretion, *Acta Endocrinol.*, 99, 321, 1982.
143. **Naftolin, F., Garcia-Segura, L.M., Keefe, D., Leranth, C., MacLusky, N.J., and Brawer, J.R.**, Estrogen effects on the synaptology and neural membranes of the rat hypothalamic arcuate nucleus, *Biol. Reprod.*, 42, 21, 1990.
144. **Nakagawa, Y., Shiosaka, S., Emson, P.C., and Tohyama, M.**, Distribution of neuropeptide Y in the forebrain and diencephalon: An immunohistochemical analysis, *Brain Res.*, 361, 52, 1985.
145. **Nakai, Y., Plant, T.M., Hess, D.L., Keogh, E.J., and Knobil, E.**, On the sites of the negative and positive feedback action of estradiol in the control of gonadotropin release in the rhesus monkey, *Endocrinology*, 102, 1008, 1978.
146. **Nauta, W.J.H.**, Central nervous organization and the endocrine motor system, in *Advances in Neuroendocrinology*, Nalbandov, A.V., Ed., University of Illinois Press, Urbana, 1963, 5.
147. **Neill, J.D., Patton, J.M., Dailey, R.A., Tsou, R.C., and Tindall, G.T**, Luteinizing hormone-releasing hormone in pituitary stalk blood of rhesus monkey: Relationship to level of LH release, *Endocrinology*, 101, 430, 1978.
148. **Nikitovitch-Winer, M.B., and Everett, J.W.**, Functional restitution of pituitary graft retransplanted from kidney to median eminence, *Endocrinology*, 63, 916, 1958.
149. **Nikolarakis, K.E., Pfeiffer, D.G., Almeida, O.F.X., and Herz, A.**, Opioid modulation of LHRH release *in vitro* depends upon levels of testosterone *in vivo*, *Neuroendocrinology*, 44, 314, 1986.
150. **Nishihara, M., Hiruma, H., and Kimura, F.**, Interaction between the noradrenergic and opioid peptidergic systems in controlling the electrical activity of luteinizing hormone-releasing hormone pulse generator in ovariectomized rats, *Neuroendocrinology*, 54, 321, 1991.
151. **Ojeda, S.R., and Urbanski, H.F.**, Puberty in rats, in *The Physiology of Reproduction*, Knobil, E., and Neill, J.D., Eds., Raven Press, New York, 1988, 1699.
152. **Ojeda, S.R., Dissen, G.A., and Junier, M.-P.**, Neurotrophic factors and female sexual development, *Frontiers in Neuroendocrinology*, 13, 120, 1992.
153. **Olmos, G., Aguilera, P., Tranque, P., Naftolin, F., and Garcia-Segura, L.M.**, Estrogen-induced synaptic remodeling in adult brain is accompanied by the reorganization of neuronal membranes, *Brain Res.*, 425, 57, 1987.
154. **Packman, P.M., and Rothchild, J.A.**, Morphine inhibition of ovulation: Reversal by naloxone, *Endocrinology*, 99, 7, 1976.

155. **Pelletier, G.,** Demonstration of contacts between neurons staining for LHRH in the preoptic area of the rat brain, *Neuroendocrinology*, 46, 457, 1987.
156. **Pepperell, R.J., Kretser, D.M., and Burger, H.G.,** Studies on the metabolic clearance rate and production rate of human luteinizing hormone and on the initial half-time of its subunit in man, *J. Clin. Invest.*, 56, 118, 1975.
157. **Petraglia, F., Locatelli, V., Penalra, A., Cocchi, D., Genazzani, A.R., and Muler, E.E.,** Gonadal steroid modulation of naloxone-induced LH secretion in the rat, *J. Endocrinol.*, 101, 33, 1984.
158. **Petraglia, F., Locatelli, V., Facchinetti, F., Bergamaschi, M., Genazzani, A.R., and Cocchi, D.,** Oestrous cycle-related LH responsiveness to naloxone: Effect of high oestrogen levels on the activity of opioid receptors, *J. Endocrinol.*, 108, 89, 1986.
159. **Pfaff, D.W.,** Impact of estrogen on hypothalamic nerve cells: Ultrastructural, chemical and electrical effects, *Recent Prog. Horm. Res.*, 39, 127, 1983.
160. **Phifer, R.F., Midgley, A.R., and Spicer, S.S.,** Immunohistologic and histologic evidence that follicle-stimulating hormone and luteinizing hormone are present within the same cell type in the human pars distalis, *J. Clin. Endocrinol. Metab.*, 36, 125, 1973.
161. **Pickering, A., and Fink, G.,** Priming effect of luteinizing hormone-releasing factor *in vitro*: Role of protein synthesis, contractile elements, Ca^{2+} and cyclic AMP, *J. Endocrinol.*, 81, 223, 1979.
162. **Pierce, J.G.,** Gonadotropin: Chemistry and biosynthesis, in *The Physiology of Reproduction*, Knobil, E., and Neill, J.D., Eds., Raven Press, New York, 1988, 1335.
163. **Piva, F., Maggi, P., Limonta, M., Motta, M., and Martini, L.,** Effect of naloxone on luteinizing hormone, follicle-stimulating hormone, and prolactin secretion in the different phases of the estrous cycle, *Endocrinology*, 117, 766, 1985.
164. **Plant, T.M.,** Puberty in primates, in *The Physiology of Reproduction*, Knobil, E., and Neill, J.D., Eds., Raven Press, New York, 1988, 1763.
165. **Plant, T.M., Nakai, Y., Belchetz, P., Keogh, E.J., and Knobil, E.,** The sites of action of estradiol and phentolamine in the inhibition of the pulsatile, circhoral discharges of LH in rhesus monkey (*Macaca mulatta*), *Endocrinology*, 102, 1015, 1978.
166. **Prasad, B.M., Conover, C.D., Sarkar, D.K., Rabii, J., and Advis, J.P.,** Feed restriction in prepubertal lambs: Effect on puberty onset and *in vivo* release of luteinizing-hormone-releasing hormone, neuropeptide Y and beta-endorphin from the median eminence, *Neuroendocrinology*, 57, 1171, 1993.
167. **Raisman, G., and Brown-Grant, K.,** The 'suprachiasmatic syndrome': Endocrine and behavioral abnormalities following lesions of the suprachiasmatic nuclei in the female rat, *Proc. R. Soc. Lond. Ser. B*, 198, 297, 1977.
168. **Ramirez, V.D., and Dluzen, D.E.,** Is progesterone a pre-hormone in the CNS? *J. Steroid Biochem.*, 27, 589, 1987.
169. **Ramirez, V.D., Feder, H.H., and Sawyer, C.H.,** The role of brain catecholamines in the regulation of LH secretion: A critical inquiry, in *Frontiers in Neuroendocrinology, Vol. 8*, Martini, L., and Ganong, W.F., Eds., Raven Press, New York, 1984, 27.
170. **Reiter, R.J.,** The mammalian pineal gland: Structure and function, *Am. J. Anat.*, 62, 287, 1981.
171. **Rivier, C., Rivier, J., and Vale, W.,** Anti-reproductive effects of a potent GnRH antagonist in the male rats, *Science*, 210, 93, 1980.
172. **Roberts, J.L., Wilcox, J.N., and Blum, M.,** The regulation of proopiomelanocortic gene expression by estrogen in the rat hypothalamus, in *Neuroendocrine Molecular Biology*, Fink, A., Harmer, A.J., and McKerns, K.W., Eds., Plenum Press, New York, 1986, 261.
173. **Ronnekleiv, O.K., and Resko, J.A.,** Ontogeny of gonadotropin-releasing hormone-containing neurons in early fetal development in rhesus monkeys, *Endocrinology*, 126, 498, 1990.
174. **Rossmanith, W.G., and Yen, S.S.C.,** Sleep-associated decrease in luteinizing hormone pulse frequency during the early follicular phase of the menstrual cycle: Evidence for an opioidergic mechanism, *J. Clin. Endocrinol. Metab.*, 65, 715, 1987.

175. **Rossmanith, W.G., Mortola, J.F., and Yen, S.S.C.,** Role of endogenous opioid peptides in the initiation of the midcycle luteinizing hormone surge in normal cycling women, *J. Clin. Endocrinol. Metab.*, 67, 695, 1988.
176. **Sagar, S.M., Sharp, F.R., and Curran, T.,** Expression of c-fos protein in brain: Metabolic mapping at the cellular level, *Science*, 240, 1328, 1988.
177. **Sarkar, D.K.,** Does LHRH meet the criteria for a hypothalamic releasing factor? *Psychoneuroendocrinology*, 3, 259, 1983.
178. **Sarkar, D.K.,** *In vivo* secretion of LHRH in ovariectomized rats is regulated by a possible autofeedback mechanism, *Neuroendocrinology*, 45, 510, 1987.
179. **Sarkar, D.K., and Fink, G.,** Effects of gonadal steroids on output of luteinizing hormone-releasing factor into pituitary stalk blood in the female rat, *J. Endocrinol.*, 80, 303, 1979.
180. **Sarkar, D.K., and Fink, G.,** Mechanism of the first spontaneous gonadotrophin surge and that induced by pregnant mare serum, and effects of neonatal androgen in rats, *J. Endocrinol.*, 83, 339, 1979.
181. **Sarkar, D.K., and Fink, G.,** Luteinizing hormone-releasing factor in pituitary stalk plasma from long-term ovariectomized rats: Effects of steroids, *J. Endocrinol.*, 86, 511, 1980.
182. **Sarkar, D.K., and Fink, G.,** Gonadotropin-releasing hormone surge: Possible modulation through post-synaptic alpha-adrenoreceptors and two pharmacologically distinct dopamine receptors, *Endocrinology*, 108, 862, 1981.
183. **Sarkar, D.K., and Minami, S.,** Pituitary portal blood collection in rats: A powerful technique for studying the hypothalamic hormone secretion, in *Neuroendocrine Research Methods*, Brunstein, B.D., Ed., Howard Academic Publishers, London, 1991, 235.
184. **Sarkar, D.K., and Mitsugi, N.,** Correlative changes of the gonadotropin-releasing hormone (GnRH) and GnRH associated peptide (GAP) immunoreactivities in the hypothalamus, preoptic area and pituitary portal plasma during neonatal and pubertal periods of female rats, *Neuroendocrinology*, 52, 15, 1990.
185. **Sarkar, D.K., and Yen, S.S.C.,** Changes in β-endorphin-like immunoreactivity in pituitary portal blood during the estrous cycle and after ovariectomy in rats, *Endocrinology*, 116, 2075, 1985.
186. **Sarkar, D.K., and Yen, S.S.C.,** Hyperprolactinemia decreases the luteinizing hormone releasing hormone concentration in pituitary portal plasma: A possible role for β-endorphin as a mediator, *Endocrinology*, 116, 2080, 1985.
187. **Sarkar, D.K., Chiappa, S.A., and Fink, G.,** Gonadotropin-releasing hormone surge in pro-oestrous rats, *Nature (Lond.)*, 264, 461, 1976.
188. **Sarkar, D.K., Smith, G.C., and Fink, G.,** Effect of manipulating central catecholamines on puberty and the surge of luteinizing hormone and gonadotropin-releasing hormone induced by pregnant mare serum gonadotropin in female rats, *Brain Res.*, 213, 335, 1981.
189. **Sarkar, D.K., Mitsugi, N., and Mitchell, M.D.,** The mechanism of action of prolactin on gonadotropin release, in *Neuroendocrine Control of Hypothalamic-Pituitary System*, H. Imura, Ed., Japan Scientific Press, Tokyo, 1988, 47.
190. **Sarkar, D.K., Friedman, S.J., Yen, S.S.C., and Frautschy, S.A.,** Chronic inhibition of hypothalamic-pituitary-ovarian axis and weight gain by a brain-targeted delivery of estradiol, *Neuroendocrinology*, 50, 204, 1989.
191. **Sarkar, D.K., Minami, S., Xie, Q.-W., Advis, J., and Prasad, B.M.,** Pulsatile secretion of LHRH and β-endorphin into pituitary portal blood during different days of the estrous cycle, *Prog. 22nd Ann. Meet. Soc. Neurosci.*, Anaheim, CA, Abstr. 52.9, 1992.
192. **Sawchenko, P.E., Swanson, L.W., Grzanna, R., Howe, P.R.C., Bloom, S.R., and Polak, J.M.,** Colocalization of neuropeptide Y immunoreactivity in brainstem catecholaminergic neurons that project to the paraventricular nucleus of the hypothalamus, *J. Comp. Neurol.*, 241, 138, 1985.
193. **Sawyer, C.H.,** Stimulation of ovulation in the rabbit by the intraventricular injection of epinephrine or norepinephrine, *Anat. Rec.*, 112, 385, 1952.

194. **Scarbrough, K.M., Jakuboski, M., Levin, N., Wise, P.M., and Roberts, J.L.,** The effect of time of day on levels of hypothalamic POMC primary transcript, processing intermediate and messenger RNA in proestrous and estrous rats, *Endocrinology*, 134, 555, 1994.
195. **Schwanzel-Fukuda, M., and Pfaff, D.,** Origin of luteinizing hormone-releasing hormone neurons, *Nature (London)*, 338, 161, 1989.
196. **Schwanzel-Fukuda, M., Bick, D., and Pfaff, D.W.,** Luteinizing hormone releasing hormone (LHRH)-expressing cells do not migrate normally in an inherited hypogonadal (Kallmann) syndrome, *Mol. Brain Res.*, 6, 311, 1989.
197. **Sebatino, F.D., Murnane, J.M., Hoffman, R.A, and McDonald, J.K.,** Distribution of neuropeptide Y-Like immunoreactivity in the hypothalamus of the adult golden hamster, *J. Comp. Neurol.*, 257, 93, 1987.
198. **Seeburg, P.H., and Adelman, J.P.,** Characterization of cDNA for the precursor of luteinizing hormone-releasing hormone, *Nature (Lond.)*, 311, 666, 1984.
199. **Shaikh, A.A., and Shaikh, S.A.,** Adrenal and ovarian steroid secretion in the rat estrous cycle temporally related to gonadotropins and steroid levels found in peripheral plasma, *Endocrinology*, 96, 37, 1975.
200. **Sherwood, N.M., Chiappa, S.A., Sarkar, D.K., and Fink, G.,** Gonadotropin-releasing hormone (GnRH) in pituitary stalk blood from proestrous rats: Effects of anesthetics and relationship between stored and released GnRH and luteinizing hormone, *Endocrinology*, 107, 1410, 1980.
201. **Shivers, B.D., Harlan, R.E., Morell, J.I., and Pfaff, D.W.,** Absence of oestradiol concentration in cell nuclei of LHRH-immunoreactive neurons, *Nature (Lond.)*, 304, 345, 1983.
202. **Silverman, A.,** The gonadotropin-releasing hormone (GnRH) neuronal systems: Immunocytochemistry, in *The Physiology of Reproduction*, Knobil, E., and Neill, J.D., Eds., Raven Press, New York, 1988, 1283.
203. **Smith, A.I., and Funder, W.,** Proopiomelanocortin processing in the pituitary, central nervous system, and peripheral tissues, *Endocr. Rev.*, 135, 1988.
204. **Sutton, S.W., Mitsugi, N., Plotsky, P.M., and Sarkar, D.K.,** Neuropeptide Y (NPY): A possible role in the initiation of puberty, *Endocrinology*, 123, 2152, 1988.
205. **Sutton, S.W., Toyama, T.T., Otto, S., and Plotsky, P.M.,** Evidence that neuropeptide Y (NPY) released into the hypophysial-portal circulation participates in priming gonadotropes to the effects of gonadotropin releasing hormone (GnRH), *Endocrinology*, 123, 1208, 1988.
206. **Terasawa, E., Watson, S.J., and Bridson, W.E.,** A role for medial preoptic nucleus on afternoon of proestrus in female rats, *Am. J. Physiol.*, 238, E533, 1980.
207. **Timms, R.J.,** The use of the anesthetic steroids aphaxalone-alphadolone in studies of the forebrain in the cat, *J. Physiol.*, 256, 7, 1976.
208. **Toran-Allerand, C.D, Ellis, L., and Pfenninger, K.H.,** Estrogen and insulin synergism in neurite growth enhancement *in vitro*: Mediation of steroid effects by interactions with growth factors? *Dev. Brain Res.*, 41, 87, 1988.
209. **Toran-Allerand, D.,** Sex steroids and the development of the new-born mouse hypothalamus and preoptic area *in vitro*: New implications for sexual differentiation, *Brain Res.*, 106, 407, 1976.
210. **Torres-Aleman, I., Naftolin, F., and Robbins, R.J.,** Trophic effects of basic fibroblast growth factor and fetal rat hypothalamic cells: Interaction with insulin-like growth factor I, *Dev. Brain Res.*, 52, 253, 1990.
211. **Tougard, C., and Tixier-Vidal, A.,** Lactotropes and gonadotropes, in *The Physiology of Reproduction*, Knobil, E., Neill, J.D., Ewing, L.L., Greenwald, G.S., Market, C.L., and Pfaff, D.W., Eds., Raven Press, New York, 1988, 1305.
212. **Tsou, R.C., Dailey, R.A., McLanahan, C.S., Parent, A.D., Tindall, G.T., and Neill, J.D.,** Luteinizing hormone releasing hormone (LHRH) levels in pituitary stalk plasma during the preovulatory gonadotropin surge of rabbits, *Endocrinology* 101, 534, 1977.

213. **Tsuruo, Y., Kawano, H., Hisano, S., Kagotani, Y., Daikoku, S., Zhang, T., and Yanaihara, N.**, Synaptic regulation of LHRH-containing neurons by substance P in rats, *Neurosci. Lett.*, 110, 261, 1990.
214. **Tsuruo, Y., Kawano, H., Kagotani, Y., Daikoku, S., Zhang, T., and Yanaihara, N.**, Morphological evidence for neuronal regulation of luteinizing hormone-releasing hormone-containing neurons by neuropeptide Y in the rat preoptic area, *Neurosci. Lett.*, 110, 261, 1990.
215. **Urbanski, H.F., and Ojeda, S.R.**, The juvenile-pubertal transition period in the female rat: Establishment of a diurnal pattern of pulsatile luteinizing hormone secretion, *Endocrinology*, 117, 644, 1985.
216. **Van Look, P.F.A.**, Failure of positive feedback, *Clin. Obstet. Gynecol.*, 3, 555, 1976.
217. **Versi, E., Chiappa, S.A., Fink, G., and Charlton, H.M.**, Pineal influences hypothalamic GnRH content in the vole, *Mictrotus agrestis*, *J. Reprod. Fertil.*, 67, 365, 1983.
218. **Wardlaw, S.L., Wang, P.J., and Frantz, A.G.**, Regulation of β-EP and ACTH in brain by estradiol, *Life Sci.*, 33, 1941, 1983.
219. **Weiland, N.G., and Wise, P.M.**, Age-associated alterations in catecholaminergic concentrations, neuronal activity, and alpha-1 receptor densities in female rats, *Neurobiol. Aging*, 10, 323, 1989.
220. **Weiland, N.G., Scarbrough, K., and Wise, P.M.**, Aging abolishes the estradiol-induced suppression and diurnal rhythm of proopiomelanocortin gene expression in the arcuate nucleus, *Endocrinology*, 131, 2959, 1992.
221. **Weiner, R.I., Findell, P.R., and Kordon, C.**, Role of classic and peptide neuromediators in the neuroendocrine regulation of LH and prolactin, in *The Physiology of Reproduction*, Knobil, E., and Neill, J.D., Eds., Raven Press, New York, 1988, 1235.
222. **Weiner, R.I., Wetsel, W., Goldsmith, P., Martinez de la Escalera, G., Windle, J., Padula, C., Choi, A., Negro-Vilar, A., and Mellon, P.**, Gonadotropin-releasing hormone neuronal cell lines, *Front. Neuroendocrinol.*, 13, 95, 1992.
223. **Wetsel, W.C., Valencia, M.M., Merchenthaler, I., Liposits, Z., Lopez, F.J., Weiner, R.I., Mellon, P.L., and Negro-Vilar, A.**, Intrinsic pulsatile secretory activity of immortalized luteinizing hormone-releasing hormone-secreting neurons, *Proc. Natl. Acad. Sci. USA*, 89, 4149, 1992.
224. **Wilson, J.D., Griffin, J.E., George, F.W., and Leshin, M.**, The endocrine control of male phenotypic development, *Aust. J. Biol. Sci.*, 36, 101, 1983.
225. **Wislocki, G.B., and King, L.S.**, The permeability of the hypophysis and hypothalamus to vital dyes with a study of the hypophysial vascular supply, *Am. J. Anat.*, 58, 421, 1936.
226. **Witkin, J.W., and Demasio, K.A.**, Ultrastructural differences between smooth and thorny GnRH neurons, *Neuroscience*, 34, 777, 1990.
227. **Wray, S., and Hoffman, G.E.**, Postnatal morphological changes in rat LHRH neurons correlated with sexual maturation, *Neuroendocrinology*, 43, 93, 1986.
228. **Wray, S., Grant, P., and Gainer, H.**, Evidence that cells expressing luteinizing hormone-releasing hormone mRNA in the mouse are derived from progenitor cells in the olfactory placode, *Proc. Natl. Acad. Sci. USA*, 86, 8132, 1989.
229. **Wray, S.G., and Hoffman, G.E.**, Catecholamine innervation of LH-RH neurons: A developmental study, *Brain Res.*, 399, 327, 1986.
230. **Xia, L., Van Vugt, D., Alston, E.J., Luckhaus, J., and Ferin, M.**, A surge of gonadotropin-releasing hormone accompanies the estradiol-induced gonadotropin surge in the rhesus monkey, *Endocrinology*, 131, 2812, 1993.
231. **Yen, S.S.C., Tsai, C.C., Naftolin, F., Vandenberg, G., and Ajabor, L.**, Pulsatile patterns of gonadotropin-release in subjects with and without ovarian function, *J. Clin. Endocrinol. Metab.*, 34, 671, 1972.
232. **Thind, K.K., Boggan, J.E., and Goldsmith, P.C.**, Interactions between vasopressin- and gonadotropin-releasing hormone-containing neuroendocrine neurons in the monkey supraoptic nucleus, *Neuroendocrinology*, 53, 287, 1991 (reference added in revision).

Chapter 2

NEUROENDOCRINE CONTROL OF FEMALE PUBERTY: THE ROLE OF BRAIN GROWTH FACTORS

S.R. Ojeda and Y.J. Ma

TABLE OF CONTENTS

0-8493-2451-3/95/$0.00+.50

ABBREVIATIONS

CNS–central nervous system
DER–*Drosophila* epidermal growth factor receptor
E_2–estradiol
EGF–epidermal growth factor
EGFR–epidermal growth factor receptor
ERBB3–erb-B proto-oncogene type B3
flb–faint little ball
FSH–follicle-stimulating hormone
GABA–gamma aminobutyric acid
HB-EGF–heparin-binding epidermal growth factor-like growth factor
HER-2 to 4–human epidermal growth factor 2 to 4
kDa–kilodaltons
LH–luteinizing hormone
LHRH–luteinizing hormone-releasing hormone
ME–median eminence
MT–metallothionein
NDF–*neu* differentiation factor
NMDA–N-methyl-D-aspartate
NPY–neuropeptide Y
PGE_2–prostaglandin E_2
rho–rhomboid
RNase–ribonuclease
SDGF–schwannoma-derived growth factor
spi–spitz
TGFα–transforming growth factor alpha

INTRODUCTION

Much remains to be learned of the process by which the neuroendocrine reproductive system ages. While the detrimental consequences of aging on female reproduction are fairly well characterized, little is known about the mechanisms whose initial deterioration precipitates the loss of reproductive function. A similar paucity of information exists in the field of neuroendocrinology of sexual development, which has been extensively scrutinized, but with meager success in identifying the key components responsible for the initiation of puberty. Beyond this rather parenthetical similarity, other more substantive resemblances exist between the two processes, as the developmental activation of several events leading to the acquisition of reproductive competence during puberty is mirrored by the gradual conclusion of such events during aging. The ability of ovarian steroids to elicit a preovulatory surge of gonadotropins and of gonadotropins to maintain ovarian secretory output are prominent examples of this mirroring correspondence.

Because of these and other considerations, it may be contended that increasing our knowledge of the neuroendocrine events governing the initiation of puberty may lead to a better understanding of the phenomena underlying the process of reproductive aging. It is the purpose of this chapter to discuss the concept, derived from a series of recent studies conducted in our laboratory, that growth factors of glial origin are physiological components of the neuroendocrine process that controls the initiation of female puberty and contributes to the neuropathological events underlying sexual precocity of cerebral origin. We hope that the information presented here will stimulate new research efforts that may lead to a better understanding of the neuroendocrine changes responsible for reproductive aging. Since the chapter is not intended to be comprehensive, the interested reader is referred to other recent reviews for additional information.[61,63]

THE TRANSSYNAPTIC CONTROL OF LUTEINIZING HORMONE-RELEASING HORMONE (LHRH) SECRETION

The initiation of mammalian puberty ultimately requires the activation of neurons that produce LHRH, the neurohormone controlling sexual development. Extensive studies by several groups have demonstrated that the secretory activity of LHRH neurons is under the regulatory control of both excitatory and inhibitory neurotransmitter inputs and have identified the neuronal systems involved. The inhibitory control of LHRH secretion is affected by two main neuronal systems, one that utilizes opioid peptides as transmitters[7] and another that employs the inhibitory amino acid gamma aminobutyric acid (GABA).[50] Both neuronal networks are synaptically connected to LHRH neurons,[41,83] suggesting that at least some of their actions are exerted directly on LHRH neurons. While a loss in opioid tone may not play a major role in initiating the pubertal activation of LHRH secretion,[63,65] recent findings in nonhuman primates have demonstrated that release of GABA from the median eminence (ME) of the hypothalamus is higher in juvenile than in peripubertal monkeys and that pharmacological removal of GABAergic inputs to LHRH neurons increases LHRH output in juvenile animals.[23] These observations strongly suggest that GABAergic inputs to the LHRH neuronal network contribute to maintaining a low level of LHRH secretory activity in prepubertal animals and that a reduction in the strength of this influence may contribute to the rise in LHRH secretion that initiates puberty. It would be difficult, however, to envisage a neurosecretory system exclusively controlled by inhibitory inputs; in the absence of excitatory inputs, removal of any inhibitory tone would be of little consequence because the system would lack the influence of a driving force.

Indeed, evidence exists suggesting that LHRH neuronal function is under the stimulatory influence of at least three excitatory neurotransmitter systems. These include neuronal networks that utilize norepinephrine,[69] neuropeptide Y (NPY),[31] and excitatory amino acids[9,60,65] as neurotransmitters. Norepinephrine stimulates LHRH release via α_1 adrenoreceptors,[62] NPY via Y_1 receptors[31] and excitatory amino acids via both N-methyl-D-aspartate (NMDA) and non-NMDA transmembrane receptors.[9,19,60] As in the case of inhibitory neurotransmitters, the effect of at least two of these three excitatory neuronal systems on LHRH secretion may be exerted directly on LHRH neurons. While noradrenergic and NPY-ergic neurons have been shown to be synaptically connected to the LHRH network,[42,84] recent studies have failed to show the presence of NMDA receptors in LHRH neurons,[1] suggesting that these cells may not have the transmembrane signaling machinery to respond to excitatory amino acid stimulation. Importantly, at least two of these excitatory systems (catecholaminergic and NPY-ergic neurons) are endowed with estrogen receptors,[72,73] indicating that the facilitatory effects of the steroid on LHRH release may be exerted indirectly via activation of the neuronal circuitry associated with LHRH neurons.

Activation of each one of these excitatory systems occurs at the time of puberty[4,10,22,81] or results in advancement of sexual maturation when induced experimentally.[54,66,86] Conversely, inhibition of their actions by either immunoneutralization of the ligand,[53] pharmacological blockade of receptors[87] or inhibition of neurotransmitter synthesis[74] results in delayed puberty and/or blockade of the preovulatory surge of gonadotropins in immature rats, suggesting that all three systems are physiological components of the neuroendocrine process controlling the advent of sexual maturity.

Whether these systems form part of a biologically redundant organization or represent independent, but complementary neuronal circuits, is not known. Nevertheless, the coexistence of these inhibitory and excitatory neuronal networks has led to the conclusion that the transsynaptic control of puberty involves a simultaneous decrease in inhibitory tone and an increase in excitatory inputs to LHRH neurons.[63] As will be discussed below, evidence now exists that, in addition to this transsynaptic control, LHRH neurons are under the regulatory influence of transforming growth factor alpha (TGFα), a polypeptide of glial origin that appears to affect LHRH neuronal function via cell-cell signaling mechanisms different from those involved in synaptic communication. The fact that TGFα belongs to a large family of growth factors affecting cell growth and differentiation via interaction with membrane-spanning tyrosine kinase receptors, suggests that TGFα may be just one of several trophic molecules involved in the glial-neuronal regulatory control of LHRH neurons.

THE EPIDERMAL GROWTH FACTOR (EGF) FAMILY AND ITS RECEPTORS

TGFα is a member of the EGF family of growth factors.[51] This family includes EGF,[15] TGFα,[18,51] the poxy virus growth factors,[8,11,80] amphiregulin,[76] heparin-binding EGF-like growth factor (HB-EGF),[24] schwannoma-derived growth factor (SDGF),[34] the *Drosophila spitz* gene product (*spi*)[71] and the recently identified neuregulins.[25,49,89] While most members of the EGF family are the product of separate genes, the neuregulins — which include heregulins[25] and *neu* differentiation factor (NDF)[89] — are alternatively spliced products of a single gene[49] (for references and additional details see Marchionni *et al.*[49] and Wen *et al.*[89]). An important feature of the EGF family is that these proteins are synthesized as transmembrane precursors; as such, they can interact with neighboring cells bearing the appropriate receptors without the need of releasing the mature peptide.[51] This allows for a form of interaction termed "juxtacrine,"[51] which appears to be characteristic of membrane-anchored growth factors. Needless to say, the structural arrangement of the central nervous system (CNS) makes juxtacrine interactions a particularly effective mode of cell-to-cell communication. In addition, however, most cells producing EGF-like peptides can enzymatically cleave the extracellular domain of the precursor to yield the mature growth factor, which can then reach its target cells by either diffusion or the blood stream.

EGF and its congeners exert a variety of biological actions as disparate as mitogenesis, promotion of growth arrest, activation of differentiated functions, cell homing and adhesion,[15,51] most of which are not only relevant to the control of CNS development in general, but may also be of importance for the developmental control of the neuroendocrine brain in particular.

All members of the EGF family initiate their biological actions by interacting with a family of membrane-spanning receptor molecules endowed with intrinsic protein tyrosine kinase activity.[14,37,67,68,75] The first of these receptors to be isolated and cloned,[85,91] the EGF receptor (EGFR), is a 170-kDa glycosylated protein that recognizes EGF, TGFα, amphiregulin and HB-EGF as ligands. The EGFR has an extracellular domain that contains the amino acid motifs essential for ligand recognition; it is connected by a short, single transmembrane domain to an intracellular domain that contains the catalytic region of the receptor, (i.e., the tyrosine kinase domain). The other three members of the EGFR family, which include the human EGF receptor-2

(HER-2) or *neu* receptor,[75] HER-3 (also called ERBB3)[37,68] and HER-4,[67] share these same basic features and display a degree of sequence similarity to the EGFR that ranges from 40–45% in the extracellular domain to 60–75% in the kinase domain.[37,67,68,75] These EGFR relatives recognize neither EGF nor TGFα as ligands. They also fail to bind amphiregulin,[37,67,68] suggesting that they mediate the biological actions of an entirely different family of EGF-like growth factors.

Very recently, one of these hypothetical ligands was identified as a membrane-anchored protein with similarity to EGF in its extracellular domain.[89] The peptide, which as other members of the EGF family can be enzymatically cleaved to yield a secreted form, was shown to bind and activate the *neu* receptor without interacting with EGFR. Because of its ability to induce growth arrest of mammary tumor cells and induce their functional differentiation, it was termed *neu* differentiation factor (NDF). The fact that NDF forms part of a subfamily of *neu* ligands is indicated by the identification of several alternatively spliced products of the same gene that bind to and activate the *neu* receptor and are expressed in glial cells of the developing CNS.[49] The isolation of these growth factors from the CNS and the identification of glial cells as their site of synthesis suggest that glial cells may express a large family of genes encoding EGF- and EGFR-related polypeptides involved in brain development and the response of the nervous system to injury. Undoubtedly, the identification of those factors recognized by the HER-3 and HER-4 receptors will represent a significant step toward a more complete understanding of the paracrine/autocrine mechanisms controlling the development and adult function of the CNS.

EGFR AND ITS CONGENERS, EGF-LIKE PEPTIDES AND THE NEUROENDOCRINE BRAIN

All four members of the EGFR family are expressed in the brain.[17,67,68,78,88] A detailed examination of the presence of EGFR in the developing female hypothalamus demonstrated that the receptor is predominantly localized to glial cells, including tanycytes of the third ventricle, glial cells of the ME and astroglia of the optic chiasm.[46] This localization was observed in both rats[46] and nonhuman primates (Ma *et al.*, submitted). It would appear that the amount of receptor expressed per cell is relatively low because only a few cells bearing EGFR mRNA can be detected by hybridization histochemistry, whereas the mRNA is readily detectable by ribonuclease (RNase) protection assay or Northern blot analysis.[46] Cultures of hypothalamic astrocytes from newborn rats showed the presence of EGFR mRNA and biologically active receptor in these cells[47] indicating that, as reported in other brain regions,[78,88] hypothalamic astrocytes are targets for EGF and/or EGF-like peptides recognized by EGFR. At present no evidence exists for the presence of the other members of the EGFR family in the hypothalamus.

With regard to ligands recognized by EGFR or EGFR-like receptors, it is now clear that TGFα is expressed in the brain at levels one to three orders of magnitude higher than those of EGF.[32,39] While neither amphiregulin nor HB-EGF have been described in the CNS, NDF mRNA (which encodes the NDF ligand for the HER-2 receptor) has been shown to be present throughout the CNS, with particularly high levels in the spinal cord.[89] In the hypothalamus, as in the rest of the brain, the TGFα gene is much more highly expressed than that encoding EGF.[39,64] TGFα mRNA has been detected in the hypothalamus by a variety of methods, including RNA blot hybridization, hybridization histochemistry, and RNase protection assays;[38,39,45,64] the TGFα protein has been identified by immunohistochemistry.[38,45] Hybridization histochemistry and immunohistochemistry revealed that, within the hypothalamus, the TGFα gene is preferentially expressed in glial cells of the astroglial lineage, especially in those associated with hypothalamic nuclei, some of which are involved in the control of LHRH neuronal function.[45] TGFα mRNA and protein were abundant in tanycytes and glial cells of the ME, suggesting the involvement of the growth factor in neuroendocrine regulatory processes. As will be discussed below, one of the processes in which TGFα appears to participate is the control of LHRH secretion.

TGFα AND LHRH NEURONAL FUNCTION

The first evidence that EGF and/or related peptides may be involved in the control of LHRH secretion was provided by the demonstration that EGF stimulated luteinizing hormone (LH) release from hypothalamic-anterior pituitary co-incubates, but not from pituitary glands incubated in the absence of hypothalamic fragments.[55] A subsequent study demonstrated that both EGF and TGFα stimulated the release of LHRH from ME fragments *in vitro*, in a dose-related manner.[64] Since the ME of the hypothalamus does not contain the cell bodies of LHRH neurons, these experiments made it clear that TGFα/EGF could elicit LHRH secretion without affecting LHRH gene expression. While both growth factors increased prostaglandin E_2 (PGE_2) release, blockade of prostaglandin synthesis prevented the increase in LHRH release elicited by the polypeptides, indicating that the stimulatory effect of TGFα/EGF on LHRH secretion is mediated by prostaglandins. That the effect also requires the intermediacy of EGFR (or a related receptor) was shown by the ability of tyrphostin RG-50864, a blocker of EGFR tyrosine kinase activity,[92] to prevent the increase in LHRH release induced by TGFα.[64] The stimulatory effect of TGFα/EGF on LHRH release was most consistently seen after 2 hours of exposure to the growth factors, suggesting that they were not acting directly on the LHRH nerve terminals but rather through an extracellular mechanism involving cell-cell interactive events. The most likely candidates for this role are glial cells, because the postnatal brain EGFRs are predominantly found on astroglial cells[78,88] and astrocytes are a major source of prostaglandin production in the CNS.[33,56]

A detailed examination of EGFR in the hypothalamus of immature female rats demonstrated that LHRH neurons do not contain EGFR, but that their cell bodies, and in particular their nerve terminals in the ME, are in close association with glial cells and tanycytes expressing immunoreactive EGFR.[46] Other recent studies demonstrated that purified hypothalamic astrocytes respond to EGF and TGFα with PGE_2 release,[6] lending credence to the view that TGFα stimulates LHRH release by facilitating glial release of prostaglandins.[64]

The exceedingly low levels of EGF in the developing hypothalamus in comparison to those of TGFα suggest that the physiological ligand for EGFR in the neuroendocrine brain is TGFα. That TGFα may play a role in the neuroendocrine control of normal sexual development was suggested by the pattern of TGFα mRNA expression observed in the hypothalamus of immature female rats. TGFα mRNA levels were found to increase in a region-specific manner at about 12 days of life and during the day of the first preovulatory surge of gonadotropins,[45] i.e., at the two phases of development when gonadotropin secretion is increased.[63] As indicated above, TGFα mRNA was mainly localized to astroglial cells, with levels of expression that appeared to be higher in astrocytes associated with several neuronal nuclei, including those involved in the control of LHRH secretion, such as the suprachiasmatic and arcuate nucleus. Peak levels of TGFα mRNA were detected by RNase protection assay in the ME-arcuate nucleus and suprachiasmatic regions at the time of the first preovulatory surge of gonadotropins, suggesting that at least part of the increase was due to gonadal steroids. Experiments carried out to test this possibility demonstrated that estradiol (E_2), and in particular the sequential administration of E_2 and progesterone at ovulatory doses, increased TGFα mRNA levels in both regions.

The importance of this peripubertal increase in TGFα gene expression for the initiation of puberty was demonstrated by an experiment in which a blocker of EGFR tyrosine kinase was implanted directly into the ME to locally prevent the activation of EGFR that may result from a hypothetical increase in ligand availability. The treatment significantly delayed the onset of puberty,[45] suggesting that under normal circumstances the initiation of puberty requires ligand-dependent activation of EGFR and/or EGFR-like receptors in the ME of the hypothalamus. The hypothalamus of nonhuman primates also shows a pattern of TGFα mRNA expression that parallels the changes in gonadotropin output seen during pre- and peripubertal development, i.e., moderately high levels during neonatal days, a decline during the juvenile period (6 months to 2 years of age) and a marked increase at the expected time of puberty (around 3–4 years of age).[44] These findings strongly suggest that TGFα may not only be involved in the hypothalamic control of rodent puberty but may also have a role in the pubertal process of higher primates, including humans.

That a change in EGFR function does occur at the time of female puberty was shown by the finding of profound alterations in EGFR mRNA levels in the ME-arcuate nucleus region on the day of the first proestrus.[46] EGFR mRNA levels were lowest in the morning of this day as compared with prepubertal ages and returned to higher levels in the afternoon, at the time of the preovulatory surge of gonadotropins. Whether these changes are accompanied by similar alterations in gene expression of other receptors of the EGFR family and/or an increased production of specific TGFα congeners is not known, but it requires careful consideration because the TGFα/EGFR system may merely represent a portion of a much more complex array of EGF-related proteins involved in the neuroendocrine control of reproduction.

The presence of both TGFα and EGFR in astroglial cells and tanycytes of the third ventricle implies that TGFα affects LHRH neuronal function via autocrine/paracrine mechanisms involving first the activation of a glial function and then, as the result of this change, an increase in LHRH release. Culture of hypothalamic astrocytes isolated from neonatal rats demonstrated that indeed astrocytes express the TGFα gene and that administration of either TGFα or EGF results in up-regulation of TGFα gene expression.[47] This effect was inhibited by both a monoclonal antibody that prevents binding of the growth factor to its recognition site in the extracellular domain of EGFR[57] and a blocker of EGFR tyrosine kinase, suggesting that, as expected, the actions of TGFα on hypothalamic astrocytes are mediated by EGFR or a closely related receptor. Surprisingly, TGFα was unable to affect its own gene expression in cerebellar astrocytes.

In vitro exposure of hypothalamic astrocytes to E_2 also resulted in up-regulation of TGFα gene expression, indicating that the effect previously observed *in vivo* may have been directly exerted on glial cells. This inference was supported by the detection of estrogen receptor mRNA in hypothalamic astrocytes. Because cerebellar astrocytes did not show a response to E_2 nor detectable levels of estrogen mRNA, the results offered evidence for the concept that hypothalamic astrocytes are molecularly and functionally different from those of brain regions not involved in neuroendocrine regulation. This unexpected regional specialization appears to include expression of the estrogen receptor gene, the ability of TGFα to up-regulate expression of its own gene and the ability of E_2 to increase TGFα mRNA levels directly, without neuronal intermediacy.

Further evidence supporting a role for TGFα in the neuroendocrine control of female sexual development was provided by studies using transgenic mice bearing a human TGFα cDNA under the control of the mouse metallothionein (MT) 1 promoter (Ma *et al.*, submitted). Because the MT promoter is activated by heavy metals, the animals were treated with zinc chloride, either daily or acutely as a single injection, to enhance the transgene expression. Chronic administration of the zinc salt resulted in a complex pattern of reproductive changes. The timing of the first estrus was delayed, but the interval between the first estrus (an index of estrogen secretion) and the initiation of estrous cyclicity (an index of reproductive competence) was considerably shortened. Histological examination of the ovaries revealed a marked accumulation of small antral follicles in transgenic animals, suggesting that the delay in first estrus may have been related to either a deficiency in gonadotropin secretion or a reduced responsiveness of the ovary to normal circulating gonadotropin levels.

Measurement of serum LH and follicle-stimulating hormone (FSH) demonstrated that animals in which TGFα gene expression was chronically activated had reduced levels of LH and normal FSH values, suggesting that TGFα overexpression may result in inappropriate gonadotropin support to the ovary. Examination of the brain and ovaries by hybridization histochemistry after acute activation of the MT promoter revealed that (5 hours after s.c. injection of zinc chloride) the TGFα transgene was most highly expressed in the hypothalamus and granulosa cells of antral follicles, respectively. Basal release of LHRH from the ME was significantly enhanced when analyzed *in vitro* shortly after collection of the tissues. This increase was prevented by inhibition of EGFR tyrosine kinase activity, suggesting the participation of EGFR and/or related receptors in the process by which TGFα overexpression leads to enhanced LHRH release from ME nerve terminals.

In contrast to the ME, the ovaries of transgenic animals acutely exposed to zinc chloride had a significantly attenuated E_2 and androgen response to gonadotropins when challenged *in vitro* 5 hours after the injection of zinc chloride. To better define the physiological importance of this inhibition, ovaries from transgenic animals were transplanted to normal controls and the ovaries from these animals were transplanted to transgenic mice; all animals were then given zinc chloride in the drinking water. The results showed that normal ovaries, placed under the control of a transgenic hypothalamus, rapidly ovulated and initiated estrous cyclicity. On the other hand, acquisition of reproductive capacity was markedly delayed in normal animals bearing transgenic ovarian grafts. These results support the view that TGFα is a physiological component of the developmental process by which the hypothalamus controls the initiation of sexual maturity; they also lend credence to the concept[3] that, within the ovary, TGFα utilizes a regulatory paracrine loop to modulate the stimulatory effects of gonadotropins on follicular development and steroidogenesis.

TGFα AND EGFR IN SEXUAL PRECOCITY INDUCED BY HYPOTHALAMIC LESIONS

Most cases of human sexual precocity occur in females and are of cerebral origin. Despite its prevalence, the etiology of the syndrome is usually identified in only a minority of cases, so that about 70% of the patients remain with the diagnosis of idiopathic precocious puberty. Experimentally, female sexual precocity can be induced in rodents and nonhuman primates by lesions of the hypothalamus,[20,82] suggesting that at least a fraction of the human cases may be explained as due to microlesions affecting the neuroendocrine brain. Such a possibility is supported by findings made with the help of computer axial tomography that a number of patients with idiopathic sexual precocity have, in fact, discrete hypothalamic lesions.[13]

Because the response of the brain to injury includes the production of neurotrophic/mitogenic activities in the area surrounding the site of lesion,[59] we considered the possibility that the TGFα/EGFR system is one of the regulatory components activated after lesions of the hypothalamus able to induce precocious puberty, and that this activation contributes to the neuroendocrine mechanism by which the lesions induce sexual precocity. As expected, lesions of the anterior hypothalamus of early juvenile animals led to vaginal opening and ovulation within 7 days after the injury.[29]

Within 20 hours, many of the cells adjacent to the lesion site, including astrocytes, reacted with an increase in c-fos expression, implying genomic activation.[30] Notably, this activation was not shared by LHRH neurons, which instead suffered a loss of spininess, a morphological change indicative of reversal of the cells to a more immature functional condition.[26] Following these initial changes, there was the expected astrocytic reaction to injury, which was accompanied by activation of both TGFα and EGFR gene expression in reactive astrocytes surrounding the lesion site.[28,29] Activation of EGFR gene expression selectively increased the levels of a predominant mRNA transcript encoding the full-length, membrane-spanning receptor, but failed to affect those of a less abundant, alternatively spliced mRNA form encoding a truncated and presumably secreted form of EGFR.[28]

Coinciding with these changes, the spininess of LHRH neurons returned to normal,[30] and LHRH secretion began to increase, as judged by the changes in circulating LH levels and increase in uterine weight, an index of estrogen secretion. At no time were changes in LHRH mRNA levels observed, suggesting that lesion-dependent activation of LHRH secretion does not involve up-regulation of LHRH gene expression. Five to seven days after the lesion, ovarian estrogen output appears to increase to levels sufficiently high to trigger a precocious surge of gonadotropins, and ovulation takes place. It is only at the time of the precocious first proestrus that genomic activation first occurs in LHRH neurons, as determined by the selective increase in c-fos expression.[30]

That ligand-dependent activation of EGFR and perhaps also of other members of the EGFR family underlies the ability of hypothalamic lesions to advance puberty was suggested by the ability of an EGFR tyrosine kinase receptor blocker, infused directly into the site of the lesion, to prevent the advancing effect of the lesion on puberty.[29]

These observations further support the concept, discussed earlier in this chapter, that TGFα is a physiological component of the neuroendocrine mechanism that controls the onset of female puberty. They also suggest that the paracrine actions of TGFα on astroglial cells are exacerbated following injury to the hypothalamus because of the up-regulation of its own gene and that encoding its receptor, and that this change may contribute to the stimulation of LHRH secretion and to the precocious initiation of puberty. As discussed above, the stimulatory effect of TGFα on LHRH release may result from cell-to-cell interactive events involving TGFα-dependent glial production of prostaglandins and the direct action of these prostaglandins on neighboring LHRH neurons. If the validity of this concept is accepted, it may be envisioned that a focal derangement in glial activity resulting in increased production of EGF-like polypeptides and/or increased activation of EGFR-like receptors, may suffice to stimulate LHRH secretion and lead to sexual precocity, provided that the derangement takes place in the immediate vicinity of neuroendocrine LHRH neurons.

CONCLUSIONS AND PERSPECTIVES

Mammalian puberty is initiated by an increase in the secretory activity of LHRH neurons. This change appears to be initiated by a series of coordinated events that include an increase in transsynaptic excitatory inputs (excitatory amino acids, norepinephrine, NPY), a decrease in transsynaptic inhibitory control (GABA, opioids), and an increase in a facilitatory influence from glial cells mediated by a growth factor-dependent cell-to-cell interaction mechanism. One of the glial molecules involved in this process appears to be TGFα, which may exert at least part of its biological effects on LHRH neurons indirectly, via an EGFR-mediated activation of glial production of PGE_2.

While it is likely that the transsynaptic and paracrine components of this regulatory mechanism are complementary, the possibility also exists that they might be interactive. Thus, EGFRs have been recently described in synaptic fractions of the brain,[21] suggesting that the receptors (located on neuronal membranes themselves and/or on the membranes of glial cells ensheathing the synapses) contribute to the regulation of synaptic transmission. Direct evidence for this view comes from the observation that EGF selectively enhances the ability of excitatory amino acids to increase intracellular calcium via activation of NMDA receptors in hippocampal neurons, without affecting calcium mobilization mediated by non-NMDA receptors.[2]

The studies described in this article do not address the issue of how essential a role TGFα plays in the neuroendocrine process controlling sexual development. Disruption of the TGFα gene by homologous recombination did not result in any measurable alteration in reproductive development and/or mature reproductive function,[43,48] suggesting that either TGFα is irrelevant to reproductive development or, more likely, that molecules with similar biological activities are recruited as a compensatory mechanism in response to the TGFα deficiency.

Prominent examples of gene disruption resulting in either almost normal phenotypes,[12] or abnormalities much less severe than those predicted by the purported importance of the disrupted molecule, have been recently described.[27,40] The unexpectedly mild alterations resulting from the disruption of mammalian TGFα is strongly reminiscent of the situation of its *Drosophila* counterpart, *spi*, the product of the *spitz* gene,[71] which as TGFα, is widely expressed throughout neural and non-neural tissues. Mutation of the *spi* gene to the null state results in a much less severe phenotype than null mutations of one of its putative receptors, the *Drosophila* EGFR (DER). Null DER mutations of the *faint little ball* (*flb*) type are lethal, whereas weak *flb* mutations result in phenotypes similar to those caused by *spi* null mutations (for references see Rutledge *et al.*[71] and Zak *et al.*[93]). Interestingly, a common consequence of *flb* mutations of DER, *spi* mutations and mutations of rhomboid (rho; a transmembrane protein thought to be another *spi* receptor) is the failure of specific glial cells to migrate.[36,71,93]

The existence of related molecules able to functionally substitute mutated members within a family of genes raises the issue as to the role of EGFR in the neurobiology of sexual development. Do only EGFRs participate in the process, or are other related receptors also involved? We believe that this possibility deserves careful consideration, especially in light of the recently described interaction of EGFR with the HER-2 receptor through heterodimerization to either promote or inhibit cellular responses.[79] The ability of EGFR to transregulate a congener may be a feature shared by other members of the EGFR family, and thus of importance for the understanding of their role in the regulation of reproductive neuroendocrine function.

An important feature of membrane-anchored EGF-related polypeptides, including TGFα, is their ability to facilitate cell adhesion and promote cell homing.[51,70,89] They exert these functions via direct protein-to-protein interactions involving EGF-like repeats in their extracellular domains. Although the relevance of such cell surface events to the neuroendocrine control of sexual development is not known, it would not be unreasonable to expect their involvement in various aspects of the process. For instance, TGFα/EGFR and/or TGFα/TGFα interactions, as well as interactions between other members of the EGF ligand/EGFR families, may contribute to the changes in glial-LHRH neuron association observed during development and following steroid treatment.[35,77,90] Very importantly, these interactions may represent an important intercellular mechanism by which the structural rearrangements of the hypothalamus observed during prepubertal development and following steroid treatment[5,16,52,58] facilitate glial production of neuroactive molecules, such as prostaglandins, at the time of puberty. It would also be expected that key aspects of such an array of cell-to-cell communication events become functionally disrupted during aging, contributing to reproductive failure.

ACKNOWLEDGMENTS

This work was supported by NIH Grants HD25123, HD18185, RR00163, and the Medical Research Foundation of Oregon-9242. The authors wish to thank Diane Hill and Janie Gliessman for editorial assistance. This is publication no. 1916 of the Oregon Regional Primate Research Center, Beaverton, Oregon.

REFERENCES

1. **Abbud, R., and Smith, M.S.,** Do GnRH neurons express the gene for the NMDA receptor? *Prog. 75th Ann. Meet. Endocr. Soc.*, 1993, 346.
2. **Abe, K., and Saito, H.,** Epidermal growth factor selectively enhances NMDA receptor-mediated increase of intracellular Ca^{2+} concentration in rat hippocampal neurons, *Brain Res.*, 587, 102, 1992.
3. **Adashi, E.Y., and Resnick, C.E.,** Antagonistic interactions of transforming growth factors in the regulation of granulosa cell differentiation, *Endocrinology*, 119, 1879, 1986.
4. **Advis, J.P., Simpkins, J.W., Chen, H.T., and Meites, J.,** Relation of biogenic amines to onset of puberty in the female rat, *Endocrinology*, 103, 11, 1978.
5. **Arai, Y., and Matsumoto, A.,** Synapse formation of the hypothalamic arcuate nucleus during post-natal development in the female rat and its modification by neonatal estrogen treatment, *Psychoneuroendocrinology*, 3, 31, 1978.
6. **Berg-von der Emde, K., Ma, Y.J., Costa, M.E., and Ojeda, S.R.,** Hypothalamic astrocytes respond to transforming growth factor alpha (TGFα) with prostaglandin E_2 (PGE_2) release, *Soc. Neurosci. Abstr.*, 1993, 19.
7. **Bhanot, R., and Wilkinson, M.,** Opiatergic control of gonadotropin secretion during puberty in the rat, *Endocrinology*, 113, 596, 1983.
8. **Blomquist, M.C., Hunt, L.T., and Baker, W.C.,** Vaccinia virus 19-kilodalton protein: Relationship to several mammalian proteins, including two growth factors, *Proc. Natl. Acad. Sci. USA*, 81, 7363, 1984.
9. **Bourguignon, J.-P., Gerard, A., Fawe, L., Alvarez-Gonzalez, M.-L., and Franchimont, P.,** Neuroendocrine control of the onset of puberty: Secretion of gonadotrophin-releasing hormone from rat hypothalamic explants, *Acta Paediatr. Scand. Suppl.*, 372, 19, 1991.
10. **Bourguignon, J.-P., Gerard, A., Mathieu, J., Mathieu, A., and Franchimont, P.,** Maturation of the hypothalamic control of pulsatile gonadotropin-releasing hormone secretion at onset of puberty. I. Increased activation of N-methyl-D-aspartate receptors, *Endocrinology*, 127, 873, 1990.
11. **Brown, J.P., Twardzik, D.R., Marquardt, H., and Todaro, G.J.,** Vaccinia virus encodes a polypeptide homologous to epidermal growth factor and transforming growth factor, *Nature*, 313, 491, 1985.
12. **Büeler, H., Fischer, M., Lang, Y., Bluethmann, H., Lipp, H.-P., DeArmond, S.J., Prusiner, S.B., Aguet, M., and Weissmann, C.,** Normal development and behaviour of mice lacking the neuronal cell-surface PrP protein, *Nature*, 356, 577, 1992.
13. **Cacciari, E., Frejaville, E., Cicognani, A., Pirazzoli, P., Frank, G., Balsamo, A., Tassinari, D., Zappulla, F., Bergamaschi, R., and Cristi, G.F.,** How many cases of true precocious puberty in girls are idiopathic? *J. Pediatr.*, 102, 357, 1983.
14. **Carpenter, G.,** Receptors for epidermal growth factor and other polypeptide mitogens, *Ann. Rev. Biochem.*, 56, 881, 1987.
15. **Carpenter, G., and Cohen, S.,** Epidermal growth factor, *J. Biol. Chem.*, 265, 7709, 1990.
16. **Clough, R.W., and Rodriguez-Sierra, J.F.,** Synaptic changes in the hypothalamus of the prepubertal female rat administered estrogen, *Am. J. Anat.*, 167, 205, 1983.
17. **Coussens, L., Yang-Feng, T.L., Liao, Y.-C., Chen, E., Gray, A., McGrath, J., Seeburg, P.H., Libermann, T.A., Schlessinger, J., Francke, U., Levinson, A., and Ullrich, A.,** Tyrosine kinase receptor with extensive homology to EGF receptor shares chromosomal location with *neu* oncogene, *Science*, 230, 1132, 1985.
18. **Derynck, R.,** Transforming growth factor, *Cell*, 54, 593, 1988.
19. **Donoso, A.O., López, F.J., and Negro-Vilar, A.,** Glutamate receptors of the non-*N*-methyl-D-aspartic acid type mediate the increase in luteinizing hormone releasing hormone release by excitatory amino acid *in vitro*, *Endocrinology*, 126, 414, 1990.
20. **Donovan, B.T., and van der Werff ten Bosch, J.J.,** Precocious puberty in rats with hypothalamic lesions, *Nature*, 178, 745, 1956.

21. **Faúndez, V., Krauss, R., Holuigue, L., Garrido, J., and Gonzalez, A.,** Epidermal growth factor receptor in synaptic fractions of the rat central nervous system, *J. Biol. Chem.*, 267, 20363, 1992.
22. **Gore, A.C., and Terasawa, E.,** A role for norepinephrine in the control of puberty in the female rhesus monkey, *Macaca mulatta*, *Endocrinology*, 129, 3009, 1991.
23. **Mitsushima, D., Hei, D.L., and Teresawa, E.,** Gamma-aminobutyric acid is an inhibitory neurotransmitter restricting the release of luteinizing hormone-releasing hormone before the onset of puberty, *Proc. Natl. Acad. Sci. USA*, 91, 395, 1994.
24. **Higashiyama, S., Abraham, J.A., Miller, J., Fiddes, J.C., and Klagsbrun, M.,** A heparin-binding growth factor secreted by macrophage-like cells that is related to EGF, *Science*, 251, 936, 1991.
25. **Holmes, W.E., Sliwkowski, M.X., Akita, R.W., Henzel, W.J., Lee, J., Park, J.W., Yansura, D., Abadi, N., Raab, H., Lewis, G.D., Shepard, H.M., Kuang, W.-J., Wood, W.I., Goeddel, D.V., and Vandlen, R.L.,** Identification of heregulin, a specific activator of $p185^{erbB2}$, *Science*, 256, 1205, 1992.
26. **Jennes, L., Stumpf, W.E., and Sheedy, M.E.,** Ultrastructural characterization of gonadotropin-releasing hormone (GnRH)-producing neurons, *J. Comp. Neurol.*, 232, 534, 1985.
27. **Joyner, A.L., Herrup, K., Auerbach, B.A., Davis, C.A., and Rossant, J.,** Subtle cerebellar phenotype in mice homozygous for a targeted deletion of the En-2 homeobox, *Science*, 251, 1239, 1991.
28. **Junier, M.-P., Hill, D.F., Costa, M.E., Felder, S., and Ojeda, S.R.,** Hypothalamic lesions that induce female precocious puberty activate glial expression of the epidermal growth factor receptor gene: Differential regulation of alternatively spliced transcripts, *J. Neurosci.*, 13, 703, 1993.
29. **Junier, M.-P., Ma, Y.J., Costa, M.E., Hoffman, G., Hill, D.F., and Ojeda, S.R.,** Transforming growth factor alpha contributes to the mechanism by which hypothalamic injury induces precocious puberty, *Proc. Natl. Acad. Sci. USA*, 88, 9743, 1991.
30. **Junier, M.-P., Wolff, A., Hoffman, G.E., Ma, Y.J., and Ojeda, S.R.,** Effect of hypothalamic lesions that induce precocious puberty on the morphological and functional maturation of the luteinizing hormone-releasing hormone neuronal system, *Endocrinology*, 131, 787, 1992.
31. **Kalra, S.P., and Crowley, W.R.,** Neuropeptide Y: A novel neuroendocrine peptide in the control of pituitary hormone secretion, and its relation to luteinizing hormone, in *Frontiers in Neuroendocrinology, Vol. 13*, Ganong, W.F., and Martini, L., Eds., Raven Press, New York, 1992, 1.
32. **Kaser, M.R., Lakshmanan, J., and Fisher, D.A.,** Comparison between epidermal growth factor, transforming growth factor-α and EGF receptor levels in regions of adult rat brain, *Mol. Brain Res.*, 16, 316, 1992.
33. **Katsuura, G., Gottschall, P.E., Dahl, R.R., and Arimura, A.,** Interleukin-1 *beta* increases prostaglandin E_2 in rat astrocyte cultures: Modulatory effect of neuropeptides, *Endocrinology*, 124, 3125, 1989.
34. **Kimura, H., Fischer, W.H., and Schubert, D.,** Structure, expression and function of a schwannoma-derived growth factor, *Nature*, 348, 257, 1990.
35. **King, J.C., and Letourneau, R.J.,** Plasticity of LHRH terminals in male and female rats in response to gonadectomy, *Prog. 75th Ann. Meet. Endocr. Soc.*, 1993, 60.
36. **Klämbt, C., Jacobs, J.R., and Goodman, C.S.,** The midline of the *Drosophila* central nervous system: A model for the genetic analysis of cell fate, cell migration, and growth cone guidance, *Cell*, 64, 801, 1991.
37. **Kraus, M.H., Issing, W., Miki, T., Popescu, N.C., and Aaronson, S.A.,** Isolation and characterization of ERBB3, a third member of the ERBB/epidermal growth factor receptor family: Evidence for overexpression in a subset of human mammary tumors, *Proc. Natl. Acad. Sci. USA*, 86, 9193, 1989.

38. **Kudlow, J.E., Leung, A.W.C., Kobrin, M.S., Paterson, A.J., and Asa, S.L.,** Transforming growth factor-α in the mammalian brain, *J. Biol. Chem.*, 264, 3880, 1989.
39. **Lazar, L.M., and Blum, M.,** Regional distribution and developmental expression of epidermal growth factor and transforming growth factor α mRNA in mouse brain by a quantitative nuclease protection assay, *J. Neurosci.*, 12, 1688, 1992.
40. **Lee, K.-F., Li, E., Huber, L.J., Landis, S.C., Sharpe, A.H., Chao, M.V., and Jaenisch, R.,** Targeted mutation of the gene encoding the low affinity NGF receptor p75 leads to deficits in the peripheral sensory nervous system, *Cell*, 69, 737, 1992.
41. **Leranth, C., MacLusky, N.J., Sakamoto, H., Shanabrough, M., and Naftolin, F.,** Glutamic acid decarboxylase-containing axons synapse on LHRH neurons in the rat medial preoptic area, *Neuroendocrinology*, 40, 536, 1985.
42. **Leranth, C., MacLusky, N.J., Shanabrough, M., and Naftolin, F.,** Catecholaminergic innervation of luteinizing hormone-releasing hormone and glutamic acid decarboxylase immunopositive neurons in the rat medial preoptic area, *Neuroendocrinology*, 48, 591, 1988.
43. **Luetteke, N.C., Qiu, T.H., Peiffer, R.L., Oliver, P., Smithies, O., and Lee, D.C.,** TGFα deficiency results in hair follicle and eye abnormalities in targeted and waved-1 mice, *Cell*, 73, 263, 1993.
44. **Ma, Y.J., Costa, M.E., and Ojeda, S.R.,** Developmental expression of transforming growth factor alpha (TGFα) mRNA in the hypothalamus of female rhesus monkeys, *Prog. 75th Ann. Meet. Endocr. Soc.*, 1993, 537.
45. **Ma, Y.J., Junier, M.-P., Costa, M.E., and Ojeda, S.R.,** Transforming growth factor alpha (TGFα) gene expression in the hypothalamus is developmentally regulated and linked to sexual maturation, *Neuron*, 9, 657, 1992.
46. **Ma, Y.J., Hill, D.F., Junier, M., Costa, M.E., Felder, S.E., and Ojeda, S.R.,** Expression of epidermal growth factor rcceptor changes in the hypothalamus during the onset of female puberty, *Mol. Cell. Neurosci.*, 1994, in press.
47. **Ma, Y.J., Berg-von der Emde, K., Moholt-Siebert, M., Hill, D.F., and Ojeda, S.R.,** Region-specific regulation of transforming growth factor alpha (TGFa) gene expression in astrocytes of the neuroendocrine brain, *J. Neurosci.*, 1994, in press.
48. **Mann, G.B., Fowler, K.J., Gabriel, A., Nice, E.C., Williams, R.L., and Dunn, A.R.,** Mice with a null mutation of the TGFα gene have abnormal skin architecture, wavy hair, and curly whiskers and often develop corneal inflammation, *Cell*, 73, 249, 1993.
49. **Marchionni, M.A., Goodearl, A.D.J., Chen, M.S., Bermingham-McDonogh, O., Kirk, C., Hendricks, M., Danehy, F., Misumi, D., Sudhalter, J., Kobayashi, K., Wroblewski, D., Lynch, C., Baldassare, M., Hiles, I., Davis, J.B., Hsuan, J.J., Totty, N.F., Otsu, M., McBurney, R.N., Waterfield, M.D., Stroobant, P., and Gwynne, D.,** Glial growth factors are alternatively spliced erbB2 ligands expressed in the nervous system, *Nature*, 362, 312, 1993.
50. **Masotto, C., and Negro-Vilar, A.,** Activation of gamma-amino butyric acid B-receptors abolishes naloxone-stimulated luteinizing hormone release, *Endocrinology*, 121, 2251, 1987.
51. **Massague, J.,** Transforming growth factor α. A model for membrane-anchored growth factors, *J. Biol. Chem.*, 265, 21393, 1990.
52. **Matsumoto, A., and Arai, Y.,** Precocious puberty and synaptogenesis in the hypothalamic arcuate nucleus in pregnant mare serum gonadotropin (PMSG) treated immature female rats, *Brain Res.*, 129, 375, 1977.
53. **Minami, S., Frautschy, S.A., Plotsky, P.M., Sutton, S.W., and Sarkar, D.K.,** Facilitatory role of neuropeptide Y on the onset of puberty: Effect of immunoneutralization of neuropeptide Y on the release of luteinizing hormone and luteinizing hormone-releasing hormone, *Neuroendocrinology*, 52, 112, 1990.
54. **Minami, S., and Sarkar, D.K.,** Central administration of neuropeptide Y induces precocious puberty in female rats, *Neuroendocrinology*, 56, 930, 1992.

55. **Miyake, A., Tasaka, K., Otsuka, S., Kohmura, H., Wakimoto, H., and Aono, T.,** Epidermal growth factor stimulates secretion of rat pituitary luteinizing hormone *in vitro, Acta Endocrinol.*, 108, 175, 1985.
56. **Murphy, S., Pearce, B., Jeremy, J., and Dandona, P.,** Astrocytes as eicosanoid-producing cells, *Glia*, 1, 241, 1988.
57. **Murthy, U., Rieman, D.J., and Rodeck, U.,** Inhibition of TGFα-induced second messengers by anti-EGF receptor antibody-425, *Biochem. Biophys. Res. Commun.*, 172, 471, 1990.
58. **Naftolin, F., Garcia-Segura, L.M., Keefe, D., Leranth, C., MacLusky, N.J., and Brawer, J.R.,** Estrogen effects on the synaptology and neural membranes of the rat hypothalamic arcuate nucleus, *Biol. Reprod.*, 42, 21, 1990.
59. **Nieto-Sampedro, M., and Cotman, C.W.,** Growth factor induction and temporal order in central nervous system repair, in *Synaptic Plasticity*, Cotman, C.W., Ed., Gilford Press, New York, 1985, 407.
60. **Ojeda, S.R.,** The mystery of mammalian puberty: How much more do we know? *Perspect. Biol. Med.*, 34, 365, 1991.
61. **Ojeda, S.R., Dissen, G.A., and Junier, M.-P.,** Neurotrophic factors and female sexual development, in *Frontiers in Neuroendocrinology, Vol. 13*, Ganong, W.F., and Martini, L., Eds., Raven Press, New York, 1992, 120.
62. **Ojeda, S.R., and Urbanski, H.F.,** Intracellular regulatory mechanisms of LHRH secretion and the onset of female puberty, in *Neuroendocrine Control of the Hypothalamic-Pituitary System*, Imura, H., Ed., Japan Scientific Press, Tokyo, 1988, 49.
63. **Ojeda, S.R., and Urbanski, H.F.,** Puberty in the rat, in *The Physiology of Reproduction, 2nd edition, Vol. 2*, Knobil, E., and Neill, J.D., Eds., Raven Press, New York, 1994, 363.
64. **Ojeda, S.R., Urbanski, H.F., Costa, M.E., Hill, D.F., and Moholt-Siebert, M.,** Involvement of transforming growth factor α in the release of luteinizing hormone-releasing hormone from the developing female hypothalamus, *Proc. Natl. Acad. Sci. USA*, 87, 9698, 1990.
65. **Plant, T.M.,** Puberty in primates, in *The Physiology of Reproduction, 2nd edition, Vol. 2*, Knobil, E., and Neill, J., Eds., Raven Press, New York, 1994, 453.
66. **Plant, T.M., Gay, V.L., Marshall, G.R., and Arslan, M.,** Puberty in monkeys is triggered by chemical stimulation of the hypothalamus, *Proc. Natl. Acad. Sci. USA*, 86, 2506, 1989.
67. **Plowman, G.D., Culouscou, J.-M., Whitney, G.S., Green, J.M., Carlton, G.W., Foy, L., Neubauer, M.G., and Shoyab, M.,** Ligand-specific activation of HER4/p180^{erbB4}, a fourth member of the epidermal growth factor receptor family, *Proc. Natl. Acad. Sci. USA*, 90, 1746, 1993.
68. **Plowman, G.D., Whitney, G.S., Neubauer, M.G., Green, J.M., McDonald, V.L., Todaro, G.J., and Shoyab, M.,** Molecular cloning and expression of an additional epidermal growth factor receptor-related gene, *Proc. Natl. Acad. Sci. USA*, 87, 4905, 1990.
69. **Ramirez, V.D., Feder, H.H., and Sawyer, C.H.,** The role of brain catecholamines in the regulation of LH secretion: A critical inquiry, in *Frontiers in Neuroendocrinology, Vol. 8*, Martini, L., and Ganong, W.F., Eds., Raven Press, New York, 1984, 27.
70. **Rebay, I., Fleming, R.J., Fehon, R.G., Cherbas, L., Cherbas, P., and Artavanis-Tsakonas, S.,** Specific EGF repeats of notch mediate interactions with delta and serrate: Implications for notch as a multifunctional receptor, *Cell*, 67, 687, 1991.
71. **Rutledge, B.J., Zhang, K., Bier, E., Jan, Y.N., and Perrimon, N.,** The *Drosophila spitz* gene encodes a putative EGF-like growth factor involved in dorsal-ventral axis formation and neurogenesis, *Genes Dev.*, 6, 1503, 1992.
72. **Sar, M., Sahu, A., Crowley, W.R., and Kalra, S.P.,** Localization of neuropeptide-Y immunoreactivity in estradiol-concentrating cells in the hypothalamus, *Endocrinology*, 127, 2752, 1990.
73. **Sar, M., and Stumpf, W.E.,** Central noradrenergic neurons concentrate ^{3}H-oestradiol, *Nature*, 289, 500, 1981.
74. **Sarkar, D.K., Smith, G.C., and Fink, G.,** Effect of manipulating central catecholamines on puberty and the surge of luteinizing hormone and gonadotropin releasing hormone induced by pregnant mare serum gonadotropin in female rats, *Brain Res.*, 213, 335, 1981.

75. **Schechter, A.L., Hung, M.-C., Vaidyanathan, L., Weinberg, R.A., Yang-Feng, R.L., Francke, U., Ullrich, A., and Coussens, L.,** The *neu* gene: An *erb*-B-homologous gene distinct from and unlinked to the gene encoding the EGF receptor, *Science*, 229, 976, 1985.
76. **Shoyab, M., Plowman, G.D., McDonald, V.L., Bradley, J.G., and Todaro, G.J.,** Structure and function of human amphiregulin: A member of the epidermal growth factor family, *Science*, 243, 1074, 1989.
77. **Silverman, R.C., Gibson, M.J., and Silverman, A.-J.,** Relationship of glia to GnRH axonal outgrowth from third ventricular grafts in hpg hosts, *Exp. Neurol.*, 114, 259, 1991.
78. **Simpson, D.L., Morrison, R., deVellis, J., and Herschman, H.R.,** Epidermal growth factor binding and mitogenic activity on purified populations of cells from the central nervous system, *J. Neurosci. Res.*, 8, 453, 1982.
79. **Spivak-Kroizman, T., Rotin, D., Pinchasi, D., Ullrich, A., Schlessinger, J., and Lax, I.,** Heterodimerization of c-erbB2 with different epidermal growth factor receptor mutants elicits stimulatory or inhibitory responses, *J. Biol. Chem.*, 267, 8056, 1992.
80. **Stroobant, P., Rice, A.P., Gullick, W.J., Cheng, D.J., Kerr, I.M., and Waterfield, M.D.,** Purification and characterization of vaccinia virus growth factor, *Cell*, 42, 383, 1985.
81. **Sutton, S.W., Mitsugi, N., Plotsky, P.M., and Sarkar, D.K.,** Neuropeptide Y (NPY): A possible role in the initiation of puberty, *Endocrinology*, 123, 2152, 1988.
82. **Terasawa, E., Noonan, J.J., Nass, T.E., and Loose, M.D.,** Posterior hypothalamic lesions advance the onset of puberty in the female rhesus monkey, *Endocrinology*, 115, 2241, 1984.
83. **Thind, K.K., and Goldsmith, P.C.,** Infundibular gonadotropin-releasing hormone neurons are inhibited by direct opioid and autoregulatory synapses in juvenile monkeys, *Neuroendocrinology*, 47, 203, 1988.
84. **Tsuruo, Y., Kawano, H., Kagotani, Y., Hisano, S., Daikoku, S., Chichara, K., Zhang, T., and Yanaihara, N.,** Morphological evidence for neuronal regulation of luteinizing hormone-releasing hormone-containing neurons by neuropeptide Y in the rat septo-preoptic area, *Neurosci. Lett.*, 110, 261, 1990.
85. **Ullrich, A., Coussens, L., Hayflick, J.S., Dull, T.J., Gray, A., Tam, A.W., Lee, J., Yarden, Y., Libermann, T.A., Schlessinger, J., Downward, J., Mayes, E.L.V., Whittle, N., Waterfield, M.D., and Seeburg, P.H.,** Human epidermal growth factor receptor cDNA sequence and aberrant expression of the amplified gene in A431 epidermoid carcinoma cells, *Nature*, 309, 418, 1984.
86. **Urbanski, H.F., and Ojeda, S.R.,** Activation of luteinizing hormone-releasing hormone release advances the onset of female puberty, *Neuroendocrinology*, 46, 273, 1987.
87. **Urbanski, H.F., and Ojeda, S.R.,** A role for N-methyl-D-aspartate (NMDA) receptors in the control of LH secretion and initiation of female puberty, *Endocrinology*, 126, 1774, 1990.
88. **Wang, S.-L., Shiverick, K.T., Ogilvie, S., Dunn, W.A., and Raizada, M.K.,** Characterization of epidermal growth factor receptors in astrocytic glial and neuronal cells in primary culture, *Endocrinology*, 124, 240, 1989.
89. **Wen, D., Peles, E., Cupples, R., Suggs, S.V., Bacus, S.S., Luo, Y., Trail, G., Hu, S., Silbiger, S.M., Ben Levy, R., Koski, R.A., Lu, H.S., and Yarden, Y.,** *Neu* differentiation factor: A transmembrane glycoprotein containing an EGF domain and an immunoglobulin homology unit, *Cell*, 69, 559, 1992.
90. **Witkin, W.J., Ferin, M., Popilskis, S.J., and Silverman, A.-J.,** Effects of gonadal steroids on the ultrastructure of GnRH neurons in the rhesus monkey: Synaptic input and glial apposition, *Endocrinology*, 129, 1083, 1991.
91. **Xu, Y.-H., Ishii, S., Clark, A.J.L., Sullivan, M., Wilson, R.K., Ma, D.P., Roe, M.A., Merlino, G.T., and Pastan, I.,** Human epidermal growth factor receptor cDNA is homologous to a variety of RNAs overproduced in A431 carcinoma cells, *Nature*, 309, 806, 1984.
92. **Yaish, P., Gazit, A., Gilon, C., and Levitzki, A.,** Blocking of EGF-dependent cell proliferation by EGF receptor kinase inhibitors, *Science*, 242, 933, 1988.
93. **Zak, N.B., Wides, R.J., Schejter, E.D., Raz, E., and Shilo, B.-Z.,** Localization of the DER/*flb* protein in embryos: Implications on the *faint little ball* lethal phenotype, *Development*, 109, 865, 1990.

Chapter 3

NEUROENDOCRINE REGULATION OF OVARIAN FUNCTION

G.F. Erickson

TABLE OF CONTENTS

0-8493-2451-3/95/$0.00+.50

ABBREVIATIONS

αGs–$\alpha G_{stimulatory}$
BP–binding protein
cAMP–cyclic adenosine monophosphate
FSH–follicle-stimulating hormone
G proteins–guanine nucleotide-binding proteins
GH–growth hormone
GHRH–growth hormone-releasing hormone
GnRH–gonadotropin-releasing hormone
IGF–insulin-like growth factor
IGFBP–insulin-like growth factor binding protein
LH–luteinizing hormone
NGF–nerve growth factor
NPY–neuropeptide Y
PKA–protein-A kinase
TIC–theca interstitial cells
VIP–vasoactive intestinal peptide

INTRODUCTION

The purpose of this chapter is to discuss some new advances in our understanding of how ligands interact with receptors in the ovarian granulosa cells to cause folliculogenesis. Knowledge of this subject is important because it holds the key for understanding how a few follicles are selected to ovulate their eggs into the oviduct to be fertilized. This fundamental principle is the basis of fertility in the female mammal. In the ovary, the follicle cells have the ability to respond to an enormous amount of information. Almost all of this information is provided to the target cells via changes in the concentration of ligands, e.g., hormones, growth factors and neuromodulators. For example, the follicle-stimulating hormone (FSH) and luteinizing hormone (LH) initiate signal transduction pathways in ovarian follicle cells to promote growth and differentiation (steroidogenesis) by endocrine mechanisms. Locally produced polypeptide growth factors, such as inhibin, activin, and TGFβ, are associated with stimulation of folliculogenesis by FSH and LH. The emerging concept is that the growth factors act as autocrine/paracrine factors to modulate FSH and LH action.

It is clear that drugs and the aging process can affect the ovarian responses to these chemical signals. To our detriment, their effects on ovary homeostasis are almost always negative. Before we can fully understand the effects of drugs and aging on reproductive responses, we must understand 1) how the hypothalamic-pituitary system operates to control normal ovarian function and 2) how intrinsic polypeptides with growth factor activity modulate the activity of this system by autocrine/paracrine mechanisms. Here we will focus on how the pituitary gonadotropins (the endocrine system) and the insulin-like growth factor or IGF system (the autocrine/paracrine system) act and interact to control the normal processes of follicle growth and atresia.

STATEMENT OF THE PROBLEM

Physiologically, the female mammal secretes a fixed number of eggs during each cycle which is species specific. The central question in this issue is how a few follicles are selected to develop to the ovulatory stage while the vast majority of follicles and eggs (>99.9%) are destroyed by atresia.[9,19] In women, the process of folliculogenesis is very long (Figure 1). It begins when a cohort of primordial follicles is recruited to initiate growth.[9,19] Successive recruitments give rise to cohorts of developing follicles present in the ovaries. The dominant follicle in each cycle originates from a primordial follicle that was recruited to grow about one year earlier (Figure 1). It follows, therefore, that ligands, including neuromodulators and drugs, can enter the microenvironment of the follicle and effect changes in the somatic cells and egg that may not manifest themselves until one year later.

Selection of the dominant follicle is the last step in this long sequence of events. The actual decision to select the dominant follicle is made near the end of the luteal phase of the cycle.[19] At this time, each ovary contains a cohort

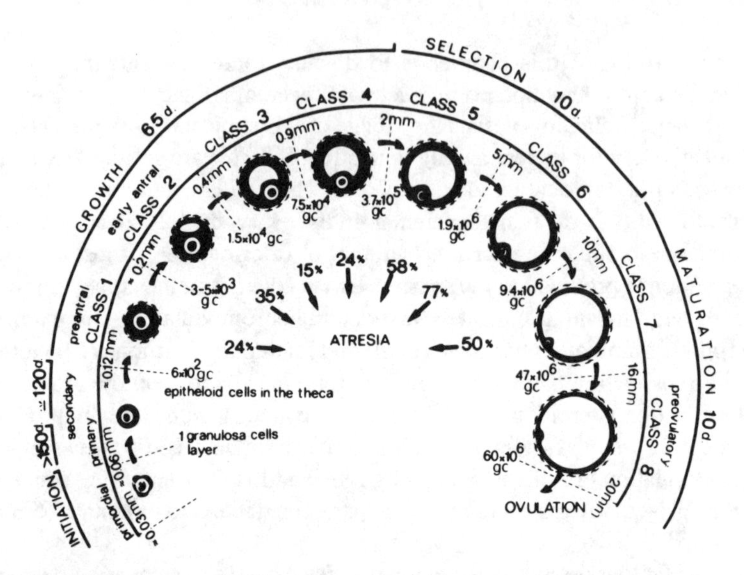

FIGURE 1. The time course of folliculogenesis in the human ovary. The time (days, d) required for a primordial follicle to grow and develop to the ovulatory stage is indicated. The follicle classes (Classes 1 to 8), their size (mm), their rate of atresia (%), and the number of granulosa cells (gc) are shown. (Used with permission from Gougeon, A., *Hum. Reprod.*, 1, 81, 1986.)

of small Graafian follicles (2–5 mm), all of which are rapidly growing; it is from this cohort that the dominant follicle is selected.[19] After selection, the mitotic activities of the granulosa and theca cells increase markedly.[32] The sustained capacity for rapid growth and cell division is the sine qua non of a dominant follicle. Those follicles that lose this property die by atresia. Atresia involves a programmed cell death mechanism called apoptosis.[22] To understand what determines selection and atresia, we must understand the basic signal processing mechanisms that are coupled to these cellular responses.

The Structure of the Follicle Compartment

A Graafian follicle is composed of several different cell types, all target cells for ligands. The follicle cells include the oocyte, granulosa, theca interna, and theca externa (Figure 2). The granulosa develop as a pseudostratified epithelium that is bound peripherally by a basal lamina. The basal lamina is an extracellular matrix of collagen, laminin, and fibronectin fibrils to which the membrana granulosa cells are attached. By virtue of the basal lamina, the granulosa and egg develop within an avascular microenvironment termed the antrum. The antrum contains follicular fluid, which is an exudate of plasma, plus various growth factors and neurotropins. The growth factors, which are produced by the follicle itself,[9] are potent ligands that regulate follicle growth and development by autocrine/paracrine mechanisms. Under normal conditions, the entrance and exit of ligands in follicular fluid are tightly regulated.[32] For example, FSH is sequestered by the dominant follicle, while LH is

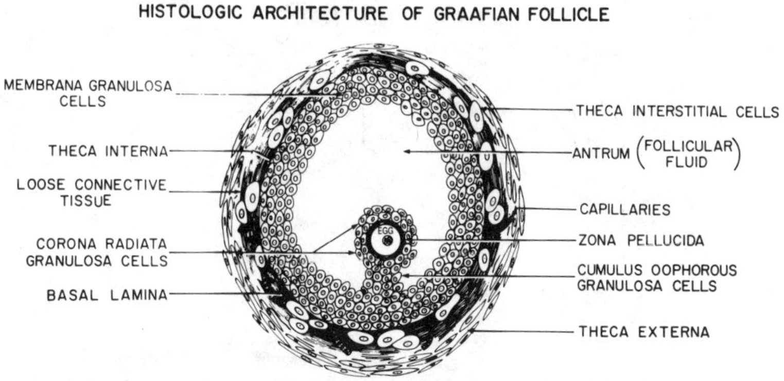

FIGURE 2. The anatomy of the Graafian follicle. (Reprinted from Erickson, G.F., The ovary: Basic principles and concepts, in *Endocrinology and Metabolism, 2nd edition,* Felig, P., Baxter, J.D., Broadus, A.E., and Frohman, L.A., Eds., McGraw-Hill, New York, 1987, with permission from McGraw-Hill, Inc.)

excluded until the late preovulatory stage.[32] It is believed that the tight regulation of the concentration of ligands in the follicular fluid is of critical importance for the generation of a healthy follicle. In this regard, if a drug enters the follicular fluid at some time during folliculogenesis, it has the potential to interact with receptors in the follicle cells, and the binding event could be transduced into an intracellular signal that negatively affects the fate of the egg and granulosa cells.

FSH is the single most important ligand in follicular fluid. All granulosa cells contain FSH receptors and the FSH-bound receptors generate intracellular signals that induce a temporal pattern of granulosa cytodifferentiation.[9] This FSH-induced program includes the stimulation of P450 aromatase ($P450_{arom}$), P450 side chain cleavage ($P450_{scc}$), and LH receptor gene transcription and translation.[38] The specific FSH responses must occur within the granulosa cells of the dominant follicle if the process of growth and development to the ovulatory stage is to be accomplished. The theca interna (Figure 2) is a loose connective tissue containing several layers of highly differentiated steroidogenic cells called the theca interstitial cells (TIC). The TIC are responsible for almost all the androgens produced by the ovary.[8] LH is the primary regulator of ovarian androgen biosynthesis.[8] LH receptors are present in all TIC and LH controls their differentiation into androgen-producing cells.[8]

The theca externa contains smooth muscle cells that are innervated by the nervous system.[9] There is a large body of evidence indicating that the ovarian nerves represent a major control system of folliculogenesis. For example, the rat ovaries contain a wide variety of nerve fibers, including the noradrenergic, cholinergic, vasoactive intestinal peptide (VIP), substance P, neuropeptide Y (NPY), gonadotropin-releasing hormone (GnRH), and nerve growth factor (NGF).[37] As such, specific neurotropin ligands modulate responses in the theca externa. Significantly, some neurotropins enter the follicular fluid where

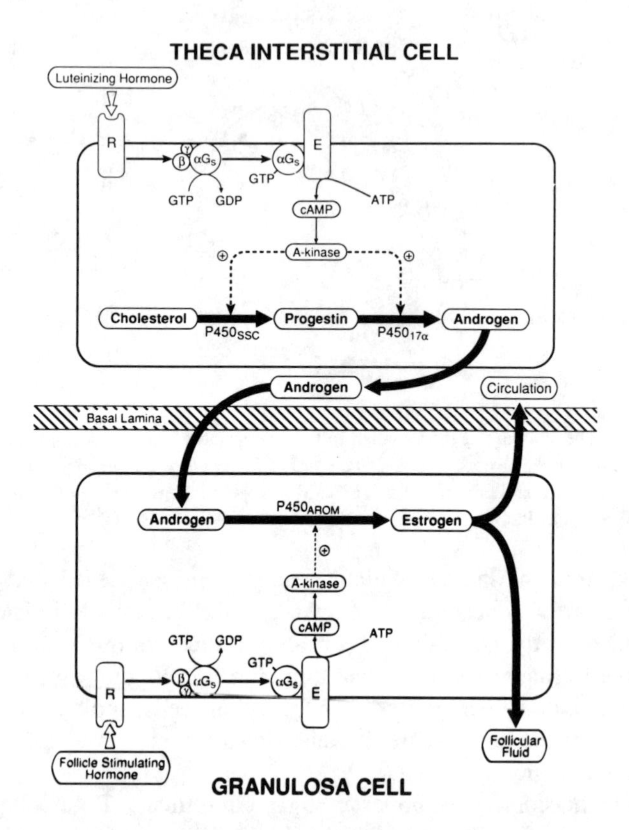

FIGURE 3. The two-gonadotropin/two-cell mechanism of follicle estrogen biosynthesis. (Reprinted from Erickson, G.F., and Yen, S.S.C., The polycystic ovary syndrome, in *The Ovary*, Adashi, E.Y., and Leung, D.C.K., Eds., Raven Press, New York, 1993, with permission from Raven Press.)

they affect the egg and granulosa cells.[9] The conclusion emerging from a large number of investigations is that the developing follicle is a target for a wide variety of neuromodulators. Because some drugs interact with specific neurotropic receptors, they have the potential to alter the way in which a given follicle grows and develops in response to stimulation by the neuromodulators.

The Two-Gonadotropin/Two-Cell Concept

Classical experiments have demonstrated that interaction of the follicle with FSH and LH leads to the synthesis of the hormone 17β-estradiol. Estradiol produced by the dominant follicles is responsible for the growth of the endometrium during the proliferative phase of the endometrial cycle. Consequently, follicle-derived estradiol is crucial for implantation and pregnancy. The mechanism by which estradiol is formed by the follicle is called the two-gonadotropin/two-cell concept (Figure 3). In this process, FSH and LH interact with transmembrane receptors in the granulosa and TIC, respectively, and the binding events are transduced into intracellular signals via the heterodimeric guanine nucleotide-binding proteins (G proteins). It can be seen (Figure 3) that the LH-bound receptor is coupled to the activation of the $\alpha G_{stimulatory}$ (αGs)/ cyclic adenosine monophosphate (cAMP)/protein-A kinase (PKA) pathway. This pathway leads to the stimulation of $P450_{scc}$ and $P450_{17\alpha}$ gene transcriptions and translations in TIC.[8] As such, the TIC synthesizes androgen hormones, especially androstenedione. The FSH-bound receptor activates the αGs/cAMP/PKA pathway in granulosa cells, and this leads to the stimulation of $P450_{arom}$ gene transcription and translation.[38,44] The $P450_{arom}$ causes the aromatization of androstenedione to 17β-estradiol (Figure 3). By virtue of its unique ability to induce $P450_{arom}$, FSH is obligatory for dominant follicle growth, and only those follicles that concentrate FSH in the follicular fluid are able to mature to the preovulatory stage and escape atresia.[9,32] It is obvious, therefore, that any ligand that antagonizes the FSH-induced signaling mechanism in the granulosa cells will result in anovulation and infertility in the female.

The Growth Factor Concept

Work carried out in animals and humans has provided important new information about how intrinsic proteins with growth factor activity couple the FSH- and LH-induced cellular responses in the granulosa and TIC, respectively. The evidence to date supports the novel theory that the actions of FSH and LH might be mediated by growth factors produced by the follicle itself. That is, the elaboration of intrinsic growth factors is physiologically coupled to the mechanisms by which FSH and LH evoke follicle selection and/or atresia. This theory is illustrated in Figure 4.

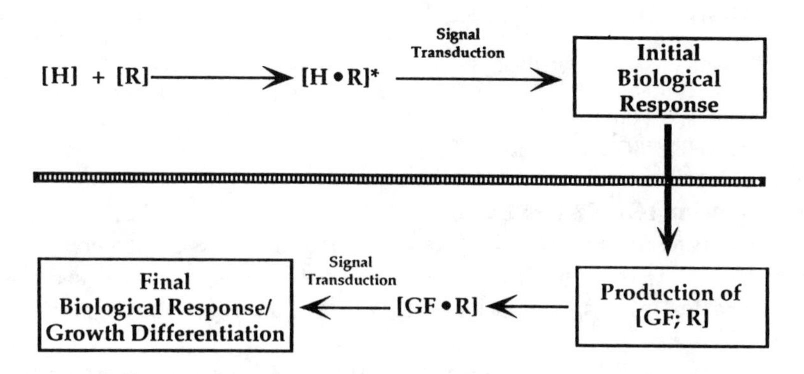

FIGURE 4. Current theory of the functional role of growth factors as autocrine/paracrine mediators of the endocrine system. ***Abbreviations:*** H, hormone; R, receptor; [H·R]*, active H·R complex after allosteric shape change; GF, growth factor.

Growth factors are proteins that regulate a wide variety of proliferative and developmental functions. All are ligands that interact with specific receptors in target cells, and the binding events initiate signal transduction pathways that modulate (either amplify or attenuate) cellular responses to endocrine hormones (Figure 4). With respect to the ovary, the search for intrinsic growth factors has been very productive, in both the rat and human. It is clear that all five families of growth factors are expressed in rat and human follicles (Table 1). The concept that emerges from these investigations is that the follicle produces all classes of growth factors, and that these regulatory proteins have the potential to influence the growth and development of follicle cells by autocrine/paracrine mechanisms.

For the purpose of limiting the scope of this section, we will focus our attention on the recent advances in the basic biology of the ovarian IGF system as an example of how a growth factor family regulates granulosa cytodifferentiation in rat and human follicles.

TABLE 1
The Growth Factor Families that are Expressed in the Rat and Human Follicle, with Some of the Major Ligands Listed

The Autocrine/Paracrine Control Systems	References	
	Rat	Human
1. IGF family (GHRH, GH, IGF-I, -II; IGFBPs)	2, 14	17
2. TGFβ family (TGFβ, PDGF, MIS, activin, inhibin; Follistatin)	16, 25, 45	6, 21, 24, 36, 43
3. EGF/TGFα family (EGF, TGFα)	31	7, 30, 39, 41
4. FGF family (basic FGF, acid FGF)	18	18
5. Cytokine family (IL, TNF, CSF, Interferon)	1	15

The Ovarian IGF System Example

There is a large body of evidence supporting the concept that a complete IGF system exists within developing follicles of laboratory animals and humans (Table 2). The system includes all aspects of the hypothalamic, pituitary and IGF axis.

In the rat ovary, the granulosa cells express the mRNA and protein for the hypothalamic peptide, growth hormone-releasing hormone (GHRH), and contain specific high affinity receptors for the GHRH ligand.[3,4] *In vitro*, administration of exogenous GHRH amplifies FSH-stimulated cAMP and estradiol production[4] and up-regulates GHRH receptors in the granulosa cells.[33] These results are consistent with the hypothesis that the GHRH-bound receptors

TABLE 2
The Intrinsic IGF System of the Rat and Human Follicle

The IGF System		References	
		Rat	Human
Growth hormone-releasing hormone:	(GHRH; GHRH receptor)	3, 4, 33	34, 35
Growth hormone:	(GH; GH receptor)	23, 28	29
Insulin-like growth factors:	(IGF-I; IGF-II)	2, 14	2, 17
IGF receptors:	(Type I; Type II)	26, 46	20, 47
IGF binding proteins:	(IGFBP-1, -2, -3, -4, -5, -6)	14, 42	17, 47

might mediate, in part, the FSH activation of granulosa cytodifferentiation. In human ovaries, immunoreactive GHRH has been detected.[34] Human follicles are apparently able to interact with GHRH because administration of exogenous GHRH to infertile patients has been reported to reduce the amount of FSH required for ovulation induction.[35] Thus the principal is emerging that intrinsic GHRH in mammalian ovaries might synergize with FSH to promote growth and differentiation of a dominant preovulatory follicle.

In the pituitary, GHRH stimulates growth hormone (GH) production and the GH that is secreted into the blood mediates the GHRH-dependent biological responses. As yet there is no evidence that the ovary produces GH; however, it is clear that receptors for GH are expressed. In the rat, GH receptors are widely distributed in the ovary, being present in oocytes, granulosa, theca interstitial, and the corpora lutea.[28] *In vitro*, interaction of GH with its receptor amplifies FSH-stimulated estradiol and progesterone production by rat granulosa cells.[5] The ability of exogenous GH to stimulate estradiol production by cultured human granulosa cells has also been reported.[29] Thus, GH enhances FSH signal transduction in granulosa cells similar to that observed with GHRH. Whether the GH responses in the ovary are coupled to intrinsic GHRH production/action is an interesting question that remains to be answered.

Physiologically, IGF-I mediates the biological responses of GH in the basic tissues of the body. Rat granulosa cells in healthy, but not atretic, follicles express relatively high levels of the IGF-I ligand. Furthermore, administration of exogenous GH stimulates IGF-I production by granulosa cells *in vivo*.[2,14] IGF-I receptors are present in granulosa cells,[26,46] and IGF-I bound receptor acts in synergy with FSH to stimulate maximal levels of granulosa estradiol, LH receptor, and progesterone production.[2] In humans, the IGF-I receptors are present in granulosa and theca cells,[47] and IGF-I has been found to be a potent stimulator of the differentiation of human follicle cells, especially the granulosa.[11] Thus, these findings support the conclusion that specific interactions

TABLE 3
Tissue-Specific Expression of mRNAs for the Family IGFBPs in Adult Rat Ovary Tissue

Tissue	IGFBP-1	IGFBP-2	IGFBP-3	IGFBP-4	IGFBP-5	IGFBP-6
Granulosa cells						
Healthy:						
Primordial	–	–	–	–	–	–
Primary	–	–	–	–	–	–
Secondary	–	–	–	–	–	–
Tertiary	–	–	–	–	–	–
Dominant	–	–	–	–	–	–
Preovulatory	–	–	–	–	–	–
Atretic:						
Preantral	–	–	–	+	±	–
Antral	–	–	–	+	±	–
Theca interstitial cells						
Secondary	–	+	–	–	–	–
Tertiary	–	+	–	–	–	–
Dominant	–	+	–	–	–	–
Preovulatory	–	+	–	±	–	–
Atretic	–	+	–	–	–	–
Secondary						
Interstitial cells	–	+	–	–	±	–
Corpora lutea	–	–	±	–	±	–
Oocyte	–	–	–	–	–	–
Theca externa	–	–	–	–	–	+
Stroma	–	–	–	±	–	+
Smooth muscle	–	–	–	–	–	+
Surface epithelium	–	+	–	–	+	–

Note: –, mRNA was undetectable; +, mRNA was detectable and the hybridization signal was strong; ±, there was heterogeneity in mRNA expression (i.e., some of the histological units contained the mRNA, others did not).

between GH and the granulosa cells generate the production of intrinsic IGF-I that then interacts synergistically with FSH to promote normal granulosa cytodifferentiation. Taken together, the evidence just discussed has led to the conclusion that there is an intrinsic ovarian IGF system that appears to play an important role in the physiological activities of follicle selection.

In the blood and interstitial fluids, the IGFs are bound to a set of binding proteins (BPs) called the insulin-like growth factor binding proteins (IGFBPs).[42] The significance of the IGFBPs is emphasized by the fact that they can modulate, either amplify or attenuate, the biological activity of the IGF-I and -II.[42] Consequently, the regulated production of the IGFBPs could have profound implications for determining the final bioactivity of the intrinsic IGF system in any tissue. In the rat and human, six insulin-like growth factor binding proteins (IGFBP-1, -2, -3, -4, -5 and -6) have been identified by molecular cloning.[42] Significantly, the rat[14] and human ovaries[47] express the IGFBPs in a tissue-specific manner. In the rat, IGFBP-2 is found predominantly in the theca and secondary interstitial cells, IGFBP-3 is found in corpora lutea, IGFBP-4 and -5 are found in atretic granulosa cells, and IGFBP-6 is found in the theca externa. IGFBP-1 is not expressed (Table 3). These findings have established the novel concept that there is inducible and tissue-specific expression of IGFBP-2 to -6 in the basic endocrine tissues of the rat ovary.

An important question concerns the role of thc IGFBPs in folliculogenesis. In this regard, our finding that IGFBP-4 and -5 mRNA are strongly expressed in granulosa cells during atresia (Figures 5 and 6), whereas no IGFBP is detected in the granulosa cells of the healthy follicle, is relevant (Table 3). The ability of atretic granulosa cells to express the IGFBP-4 and -5 genes is highly relevant because it suggests a possible role of inducible and tissue-specific expression of IGFBP-4 and -5 in the mechanisms of atresia. What might this role be? As just discussed, the dominant follicles strongly express IGF-I in granulosa cells throughout the course of their development.[14] Further, IGF-I is a potent stimulator of mitosis and differentiation of the granulosa cells. Thus, it seems likely that in the rat, IGF-I produced by the granulosa cells acts in an autocrine/paracrine fashion to help generate dominant follicles. Following this reasoning, the induction of IGFBP-4 and -5 expression in the granulosa cells would have far-reaching consequences for the ability of IGF-I to act on the follicle.

In this connection, we are beginning to get evidence for a most exciting and important function of the IGFBPs in the ovaries that fits with this idea. In rodents, IGFBP-2, -3, -4 and -5 have been shown to function as FSH antagonists *in vitro*.[14] That is, addition of exogenous IGFBP-2, -3, -4 or -5 to serum-free medium blocks the ability of a maximally effective dose of FSH to stimulate estradiol and progesterone production by rat granulosa cells.[14,27] Significantly, the mechanism by which these IGFBPs inhibit FSH action is by

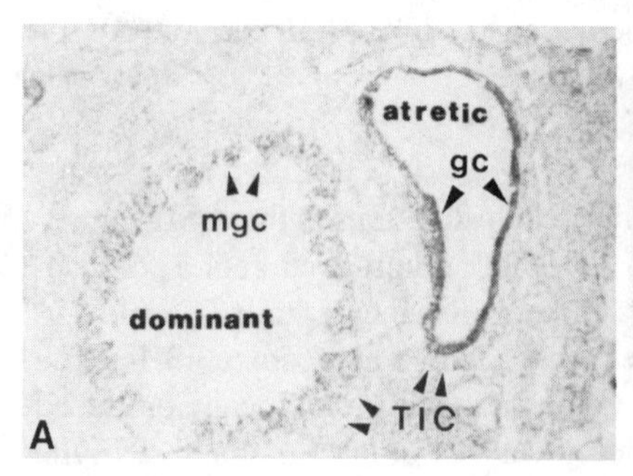

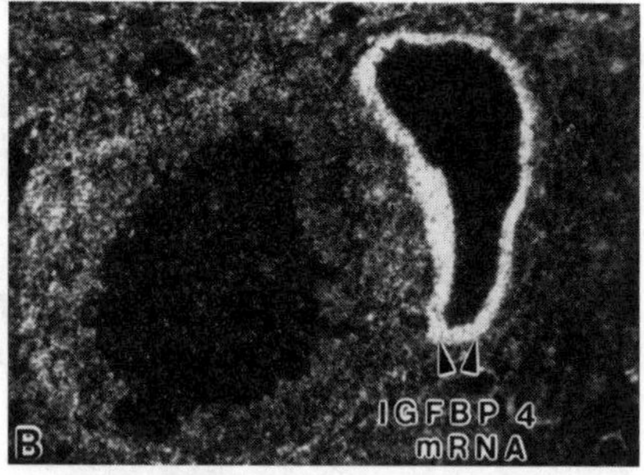

FIGURE 5. Photomicrograph showing the selective expression of IGFBP-4 mRNA in rat granulosa cells of atretic Graafian follicles. (*A*) Bright-field photomicrograph; (*B*) photomicrograph after hybridization with the antisense with RNA probe. ***Abbreviations:*** gc, granulosa cells; mgc, membrana granulosa cells; TIC, theca interstitial cells. (Revised from Erickson *et al.*[12])

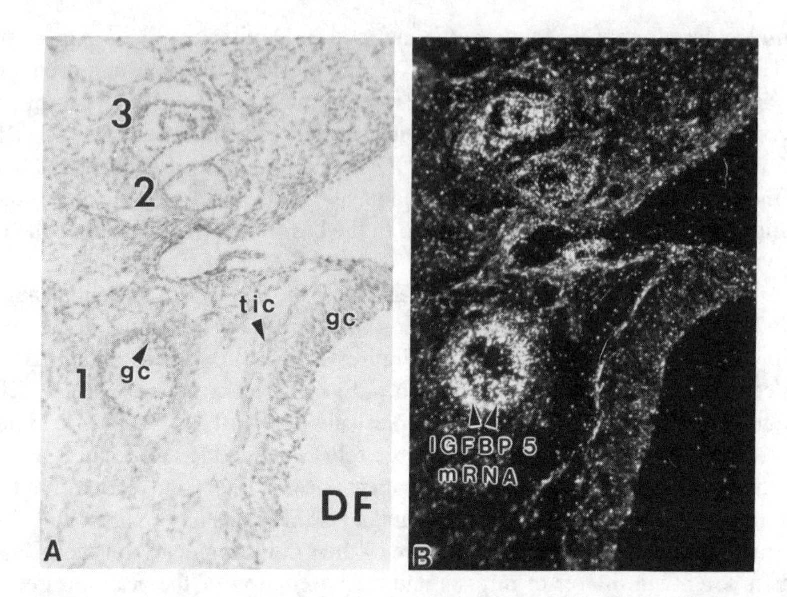

FIGURE 6. Photomicrograph showing the selective expression of IGFBP-5 mRNA in rat granulosa cells of atretic follicles. (*A*) Bright-field photomicrograph; (*B*) photomicrograph of section after hybridization with no antisense cRNA probe. ***Abbreviations:*** DF, dominant follicle 1, 2, 3, small atretic follicles; gc, granulosa cells; tic, theca interstitial cells. (Revised from Erickson *et al.*[13])

the intrinsic IGF-I produced by the granulosa cells, thereby preventing IGF-I receptor signal transduction.[14] Collectively these results have led to the exciting and important theory that locally produced IGF-I is an obligatory mediator of FSH action in rat granulosa cells (Figure 7), and that the expression of intrinsic IGFBP-4 and -5 acts to inhibit IGF-I action on the granulosa cells (Figure 7). As such, IGFBP-4 and -5 are classic FSH antagonists. Therefore, we have proposed that the activation of IGFBP-4 and -5 expression *in situ* initiates, facilitates or completes the process of atresia in rat ovaries by virtue of their ability to antagonize the actions of IGF-I induced by FSH.

IMPLICATIONS

The novel concept emerging from these investigations is that FSH-dependent folliculogenesis is mediated, in part, by the local production of a polypeptide growth factor, IGF-I, which acts in an autocrine/paracrine fashion to stimulate steroidogenesis. The discovery that IGFBPs antagonize FSH action by sequestering intrinsic IGF-I has led to the proposition that locally produced IGFBPs might lead to atresia. A physiological role of IGFBP-4 and -5 in the atretic process comes from our discovery that these two IGFBPs are selectively expressed in granulosa cells during normal atresia. Implicit in these findings is the concept that regulatory ligands that induce IGFBP-4 and -5 transcription and translation should promote follicle atresia. At this time, we

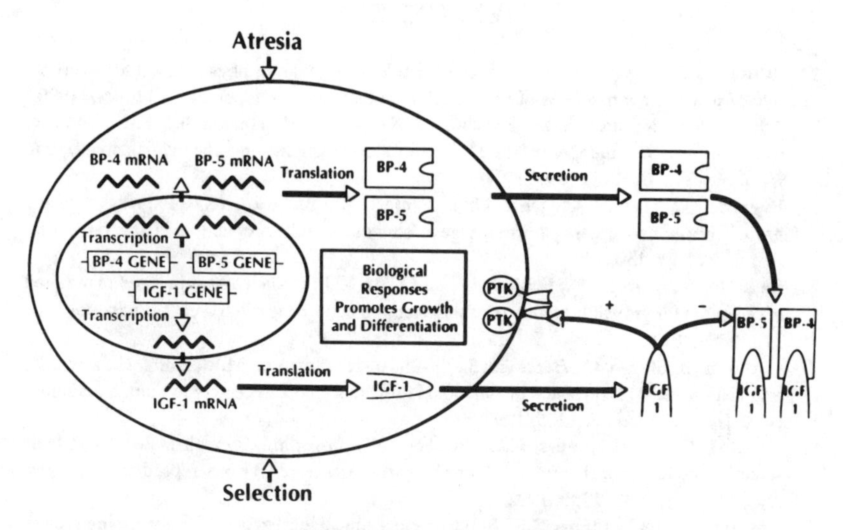

FIGURE 7. Interactions of IGF-I and the IGFBPs in causing selection and atresia.

know precious little about the regulation of IGFBP-4 and -5 production; however, because aging, drugs and neuromodulators all influence folliculogenesis, it is not unreasonable to propose that a potential mechanism by which they evoke their responses might be at the level of the intrinsic IGF system, especially the IGFBPs. We know that the other growth factor families are expressed in developing ovarian follicles. Hence, there could be specific interactions between drugs/aging and the inducible and tissue-specific expression of these other growth factor families in the follicle as well.

Further work in this area could lead to some surprising results. The current challenges are to understand how specific growth factor systems regulate folliculogenesis and how these interactions are integrated into the overall pattern of fertility and infertility. Such knowledge could provide the basis for formulating new hypotheses about how to treat human problems that are causal to aging and the abuse of chemical substances such as drugs and alcohol.

ACKNOWLEDGMENTS

I thank my colleagues Drs. Ling and Shimasaki for their valuable contributions to this work and Marta Murray for typing the manuscript. This work was supported by NICHD Research Grant R01 HD29008.

REFERENCES

1. **Adashi, E.Y.,** The potential relevance of cytokines to ovarian physiology: The emerging role of resident ovarian cells of the white blood cell series, *Endocr. Rev.*, 11, 454, 1990.
2. **Adashi, E.Y., Resnick, C.E., Hurwitz, A., Ricciarelli, E., Hernandez, E.R., Roberts, C.T., LeRoith, D., and Rosenfeld, R.,** Insulin-like growth factors: The ovarian connection, *Hum. Reprod.*, 6, 1213, 1991.
3. **Bagnato, A., Moretti, C., Ohnishi, J., Frajese, G., and Catt, K.J.,** Expression of the growth hormone-releasing hormone gene and its peptide production in the rat ovary, *Endocrinology*, 130, 1992.
4. **Bagnato, A., Moretti, C., Frajese, G., and Catt, K.J.,** Gonadotropin-induced expression of receptors for growth hormone-releasing factor in cultured granulosa cells, *Endocrinology*, 128, 2889, 1991.
5. **Carlsson, B., Bergh, C., Bentham, J., Olson , J.-H., Norman, M.R., Billig, H., Roos, P., and Hillensjo, T.,** Expression of functional growth hormone receptors in human granulosa cells, *Hum. Reprod.*, 7, 1205, 1992.
6. **Chegini, N., and Flanders, K.C.,** Presence of transforming growth factor-β and their selective cellular localization in human ovarian tissue of various reproductive stages, *Endocrinology*, 130, 1705, 1992.
7. **Chegini, N., and Williams, R.S.,** Immunocytochemical localization of transforming growth factors (TGFs) TGF-α and TGF-β in human ovarian tissues, *J. Clin. Endocrinol. Metab.*, 74, 973, 1992.
8. **Erickson, G.F.,** Normal regulation of ovarian androgen production, in *Semin. Reprod. Endocrinol.*, 4, 307, 1993.
9. **Erickson, G.F.,** The ovary: Basic principles and concepts, in *Endocrinology and Metabolism, 2nd edition*, Felig, P., Baxter, J.D., Broadus, A.E., and Frohman, L.A., Eds., McGraw-Hill, New York, 1987.
10. **Erickson, G.F., and Yen, S.S.C.,** The polycystic ovary syndrome in *The Ovary*, Adashi, E.Y., and Leung, D.C.K., Eds., Raven Press, New York, 1993.
11. **Erickson, G.F., Garzo, V.G., and Magoffin, D.A.,** Insulin-like growth factor 1 (IGF-I) regulates aromatase activity in human granulosa and granulosa luteal cells, *J. Clin. Endocrinol. Metab.*, 69, 716, 1989.
12. **Erickson, G.F., Nakatani, A., Ling, N., and Shimasaki, S.,** Cyclic changes in insulin-like growth factor binding protein-4 (IGFBP-4) in the rat ovary, *Endocrinology*, 130, 675, 1992.
13. **Erickson, G.F., Nakatani, A., Ling, N., and Shimasaki, S.,** Localization of insulin-like growth factor binding protein-5 (IGFBP-5) in adult rat ovaries, *Endocrinology*, 130, 1867, 1992.
14. **Erickson, G.F., Nakatani, A., Liu, X.-J., Shimasaki, S., and Ling, N.,** The role of IGF-I and the IGFBPs in folliculogenesis, in *Cellular and Molecular Mechanisms in Female Reproduction*, Findlay, J.K., Ed., Academic Press, 1994.
15. **Fukuoka, M., Yasuda, K., Emis, M., Fujiwara, H., Iwai, M., Takakura, K., Kanzaki, H., and Mori, T.,** Cytokine modulation of progesterone and estradiol secretion in cultures of luteinized human granulosa cells, *J. Clin. Endocrinol. Metab.*, 72, 254, 1992.
16. **Findlay, J.F.,** An update on the roles of inhibin, activin, and follistatin as local regulators of folliculogenesis, *Biol. Reprod.*, 48, 15, 1993.
17. **Giudice, L.C.,** Insulin-like growth factors and ovarian follicular development, *Endocr. Rev.*, 13, 641, 1992.
18. **Gospodarowicz, D., Ferrara, N., Schweigere, L., and Neufeld, G.,** Structural characterization and biological functions of fibroblast growth factor, *Endocr. Rev.*, 8, 95, 1987.
19. **Gougeon, A.,** Dynamics of follicular growth in the human: A model from preliminary results, *Hum. Reprod.*, 1, 81, 1986.
20. **Hernandez, E.R., Hurwitz, A., Vera, A., Pellicer, A., Adashi, E.Y., LeRoith, D., and Roberts, C.T.,** Expression of the genes encoding the insulin-like growth factors and their receptors in the human ovary, *J. Clin. Endocrinol. Metab.*, 74, 419, 1992.

21. **Hillier, S.G.,** Regulatory functions for inhibin and activin in human ovaries, *J. Endocrinol.*, 131, 171, 1991.
22. **Hurwitz, A., and Adashi, E.Y.,** Ovarian follicular atresia is an apoptotic process: A paradigm for programmed cell death in endocrine tissues, *Mol. Cell Endocrinol.*, 84, C19, 1992.
23. **Hutchinson, L.A., Findlay, J.K., and Herington, A.C.,** Growth hormone and insulin-like growth factor-I accelerate PMSG-induced differentiation of granulosa cells, *Mol. Cell Endocrinol.*, 55, 61, 1988.
24. **Kim, J.H., Seibel, M.M., MacLaughlin, D.T., Donahoe, P.K., Ransil, B.J., Hermetz, P.A., and Richards, C.J.,** The inhibitory effects of Müllerian-inhibiting substance in epidermal growth factor induced proliferation and progesterone production in human granulosa-luteal cells, *J. Clin. Endocrinol. Metab.*, 75, 911, 1992.
25. **Knecht, M., Feng, P., and Catt, K.T.,** Transforming growth factor-β. Autocrine, paracrine, and endocrine effects in ovarian cells, *Semin. Reprod. Endocrinol.*, 7, 12, 1989.
26. **Levy, M.J., Hernandez, E.R., Adashi, E.Y., Stillman, R.J., Roberts, C.T., and LeRoith, D.,** Expression of the insulin-like growth factor (IGF)-I and -II and the IGF-I and -II receptor genes during postnatal development of the rat ovary, *Endocrinology*, 131, 1202, 1992.
27. **Liu, X.-J., Malkowski, M., Guo, Y.-L., Erickson, G.F., Shimasaki, S., and Ling, N.,** Development of specific antibodies to rat insulin-like growth factor binding proteins (IGFBP-2 to -6): Analysis of IGFBP production by rat granulosa cells, *Endocrinology*, 132, 1176, 1993.
28. **Lobie, P.E., Breipohl, W., Garcia, J., Aragon, G., and Waters, M.J.,** Cellular localization of the growth hormone receptor/binding protein in the male and female reproductive systems, *Endocrinology*, 126, 2214, 1990.
29. **Mason, H.D., Martikainen, H., Beard, R.W., Anyaoku, V., and Franks, S.,** Direct gonadotropic effect of growth hormone on oestradiol production by human granulosa cells *in vitro*, *J. Endocrinol.*, 126, R1, 1990.
30. **Maruo, T., Ladiness-Llave, C.A., Samoto, T., Matsuo, H., Manalo, A.S., Ito, H., and Mochizuki, M.,** Expression of epidermal growth factor and its receptor in the human ovary during follicular growth and regression, *Endocrinology*, 132, 924, 1993.
31. **May, J.V., and Schomberg, D.W.,** The potential relevance of epidermal growth factor (EGF) and transforming growth factor alpha (TGFα) to ovarian physiology, *Semin. Reprod. Endocrinol.*, 7, 1, 1989.
32. **McNatty, K.P., Smith, D.M., Osathanondh, R., and Ryan, K.J.,** The human antral follicle: Functional correlates of growth and atresia, *Ann. Biol. Anim. Physiol.*, 19, 1547, 1979.
33. **Moretti, C., Bagnato, A., Solan, N., Frajese, G., and Catt, K.J.,** Receptor-mediated actions of growth hormone-releasing factor in granulosa cell differentiation, *Endocrinology*, 127, 2117, 1990.
34. **Moretti, C., Fabbri, A., Gnessi, L., Bonifacio, V., Bolotti, M., Arizzi, M., Nazzicone, Q., and Spera, G.,** Immunohistochemical localization of growth hormone-releasing hormone in human gonads, *J. Endocrinol. Invest.*, 13, 301, 1990.
35. **Moretti, C., Fabbri, A., Gnessi, L., Forni, L., Fraioli, F., and Frajese, G.,** GHRH stimulates follicular growth and amplifies FSH-induced ovarian folliculogenesis in women with anovulatory infertility, in *Reproductive Medicine: Medical Therapy, Vol. 875*, Frajese, G., Steinberg, E., and Rodriguez-Rigau, L.J., Eds., Elsevier Science Publishers, New York, 1989, 103.
36. **Mulheron, G.W., Bossert, N.L., Lapp, J.A., Walmer, D.K., and Schomberg, D.W.,** Human granulosa-luteal and cumulus cells express transforming growth factors-beta type 1 and type 2 mRNA, *J. Clin. Endocrinol. Metab.*, 74, 458, 1992.
37. **Ojeda, S.R., Dissen, G.A., and Junier, M.-P.,** Neurotrophic factors and female sexual development, in *Frontiers of Neuroendocrinology, Vol. 13*, Ganong, W.F., and Martini, L., Eds., Raven Press, New York, 1992, 120.

38. **Richards, J.S.,** Gonadotropin-regulated gene expression in the ovary, in *The Ovary*, Adashi, E.Y., and Leung, P.C.K., Eds., Raven Press, New York, 1993, 93.
39. **Richardson, M.C., Gadd, S.C., and Masson, G.M.,** Augmentation by epidermal growth factor of basal and stimulated progesterone production by human luteinized granulosa cells, *J. Endocrinol.*, 121, 397, 1989.
40. **Roby, K.F., Weed, J., Lyles, R., and Terranova, P.F.,** Immunological evidence for a human ovarian tumor necrosis factor-α, *J. Clin. Endocrinol. Metab.*, 71, 1096,1990.
41. **Steinkampf, M.P., Mendelson, C.R., and Simpson, E.R.,** Effects of epidermal growth factor and insulin-like growth factor 1 on the levels of mRNA encoding aromatase P450 of human ovarian granulosa cells, *Mol. Cell Endocrinol.*, 59, 93, 1988.
42. **Shimasaki, S., and Ling, N.,** Identification and molecular characterization of insulin-like growth factor binding proteins (IGFBP-1,-2,-3,-4,-5,-6), *Prog. Growth Factor Res.*, 3, 243, 1991.
43. **Voutilainen, R., and Miller, W.L.,** Human Müllerian inhibitory factor messenger ribonucleic acid is hormonally regulated in the fetal testis and adult granulosa cells, *Mol. Endocrinol.*, 1, 604, 1987.
44. **Whitelaw, P.F., Smyth, C.D., Howles, C.M., and Hillier, S.G.,** Cell-specific expression of aromatase and LH receptor mRNAs in rat ovary, *J. Mol. Endocrinol.*, 9, 309, 1992.
45. **Ying, S.-Y.,** Inhibins, activins and follistatins: Gonadal proteins modulating the secretion of follicle-stimulating hormone, *Endocr. Rev.*, 9, 267, 1988.
46. **Zhou, J., Chin, E., and Bondy, C.,** Cellular pattern of insulin-like growth factor-I (IGF-1) and IGF-I receptor gene expression in the developing and mature ovarian follicle, *Endocrinology*, 179, 3281, 1991.
47. **Zhou, J., and Bondy, C.,** Anatomy of the human ovarian insulin-like growth factor system, *Biol. Reprod.*, 48, 467, 1993.

Chapter 4

NEUROENDOCRINE REGULATION OF TESTICULAR FUNCTION

M.P. McGuinness and M.D. Griswold

TABLE OF CONTENTS

0-8493-2451-3/95/$0.00+.50

ABBREVIATIONS

CRE–cyclic AMP response element
FSH–follicle-stimulating hormone
GnRH–gonadotropin-releasing hormone
LH–luteinizing hormone
LHRH–luteinizing hormone-releasing hormone

INTRODUCTION

The production of mature spermatozoa requires a complex network of interactions between various cells and numerous endocrine and paracrine factors. The principal cell types involved in the regulation of spermatogenesis are the gonadotrophs, located in the pituitary, and the Sertoli cells and Leydig cells of the testis. Communication between the pituitary and the testis comprises a classical feedback system referred to as the pituitary-testicular axis (Figure 1). In brief, the pituitary gonadotrophs secrete the gonadotropins, follicle-stimulating hormone (FSH) and luteinizing hormone (LH), which are transported through the circulatory system to the testis. In the testis, FSH binds to receptors on Sertoli cells and is thought to play a major role in initiating the

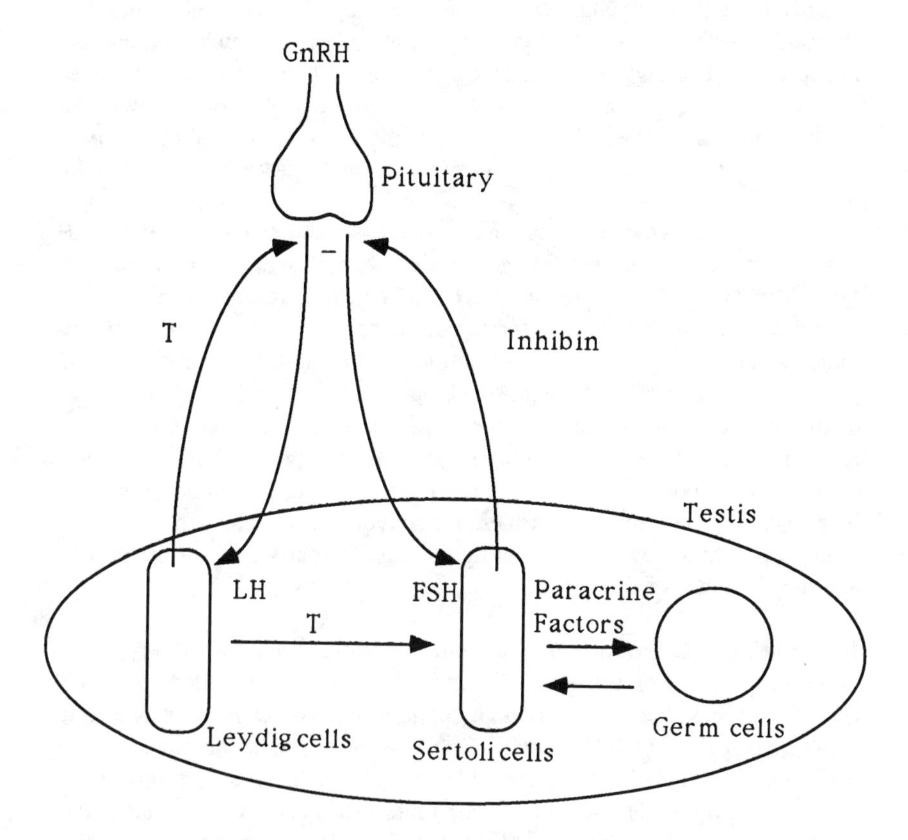

FIGURE 1. Components of the normal pituitary-testicular axis. GnRH acts on the pituitary to stimulate the release of the gonadotropins. In this figure the testis is represented by the large oval. LH acts on the Leydig cells to stimulate testosterone (T) synthesis and secretion while FSH acts on Sertoli cells. T stimulates spermatogenesis through actions on peritubular and Sertoli cells and inhibits LH release through a classical feedback pathway. Inhibin can negatively regulate FSH release from the pituitary whereas the development of germ cells to sperm is a result of the biological activities of the Sertoli cells. Sertoli cells and germ cells communicate through a number of paracrine factors.

first wave of spermatogenesis at puberty.[39] LH binds to Leydig cells and stimulates the production of testosterone,[42] which also plays an important role in initiation and maintenance of spermatogenesis. In response, inhibin and activin, produced by Sertoli cells, and testosterone, from the Leydig cells, are secreted and influence the production of the gonadotropins. Interruptions of this carefully balanced pituitary-testicular axis as a result of aging and drug abuse are detrimental to spermatogenesis. The basic elements of the normal pituitary-testicular axis with the important feedback regulators intact are discussed in this chapter.

THE GONADOTROPINS

FSH and LH are glycoproteins (MW 33,000 and 28,000, respectively) that share structural similarity to thyroid-stimulating hormone and chorionic gonadotropin.[15] These glycoproteins are comprised of two noncovalently linked polypeptides. The α-subunit is common to each of these glycoproteins whereas the β-subunit is unique.[37] The formation of the heterodimer within the endoplasmic reticulum of the gonadotrophs is essential for activity of the gonadotropins.

The cDNAs and the genes for the common α-subunit and both FSH and LH β-subunits have been determined for a number of species, including bovine,[32,51,107] rat,[14,36,48] and human.[91,101,108] The gonadotropins are encoded by separate genes localized on different chromosomes.[37] The α-subunit in humans consists of approximately 89 amino acids, and the processed protein product contains two carbohydrate moieties.[15] The primary amino acid sequence of human FSH-β and LH-β are approximately 117 and 119 residues, respectively, and these glycoproteins also contain two carbohydrate moieties.[15] The carbohydrate moieties present on the glycoprotein hormones are heterogeneous, and removal of these carbohydrates can result in the loss of biological activity, as well as altering the intracellular sorting of the protein and the clearance of this glycoprotein.[37]

Hypothalamic Regulation of FSH and LH Release from Pituitary

The regulation of the synthesis and release of the gonadotropins is a complex process because individual gonadotrophs are capable of secreting either FSH, LH, or both FSH and LH.[5,70] In addition, gonadotropin-releasing hormone (GnRH), which is secreted from the hypothalamus, stimulates the secretion of both LH and FSH.[37] For years it has been known that LH is released in a pulsatile fashion.[35,92] FSH displays less frequent and irregular pulses with small amplitudes.[56] Although GnRH influences the pulsatile release of LH, its effect on pulsatile release of FSH is less clear.[56] The frequency of GnRH release is apparently involved in the differential regulation of gonadotropin secretion. One pulse per hour of GnRH is capable of maintaining normal gonadotropin secretion in adult rhesus monkeys bearing a hypothalamic lesion.[77] Decreasing the frequency of the GnRH pulse lowers the secretion of LH and increases the secretion of FSH.[20,77,112] In addition, pulsatile FSH release cannot be generated by the exogenous or endogenous levels of GnRH sufficient to evoke pulsatile release of LH.[64]

The basic principles described above do not explain the mechanism by which differential secretion of LH and FSH occurs. Several factors, including feedback mechanisms directed at the unique gonadotrophs, could influence differential secretion. Another possible explanation is that the organization or location of secretory granules within a gonadotroph could facilitate the differential release of FSH and LH. Constant fusion of GnRH causes a biphasic pattern of LH release from the pituitary of males.[11] The initial release, peaking at 30 minutes, may represent the release of LH stored in granules situated near the plasma membrane. A second rise begins after 90 minutes of GnRH stimulation and continues for 4 hours. This could result from secretory granules located more distal relative to the plasma membrane. In contrast, FSH shows only a progressive rise in response to GnRH. Thus, the storage of LH and FSH within gonadotrophs and the frequency of GnRH release from the hypothalamus could influence the release of the gonadotropins.

The release of GnRH from the hypothalamus is most likely regulated to some extent by both the testis and the pituitary. In rats, castration leads to increased levels of LH, FSH and GnRH. Moreover, treatment of castrated rats with testosterone implants lowers LH levels but does not affect FSH or GnRH levels.[37] Details of the hypothalamic-pituitary axis are covered at greater length in chapter 1 of this volume.

THE TESTIS

Spermatogenesis

The production of sperm occurs within the convoluted seminiferous tubules of the testis. In humans, these tubules are 150–250 μm in diameter and 30–70 cm in length. The seminiferous epithelium is a complex stratified epithelium composed of Sertoli cells and spermatogenic cells. The process of spermatogenesis can be divided into different stages based on the association of germ cells at a particular phase of maturation.[54] The cycle is a continuum of developmental stages but the arbitrary number of recognizable and definable stages varies among species. For example, 14 stages have been described for rats,[22] whereas in mice there are 12,[71] and in humans there are only 6.[21] The cycle of the seminiferous epithelium refers to the sequence of changes occurring in a given area of the seminiferous tubule between two successive appearances of a given stage. The synthesis and secretion of several Sertoli cell proteins vary with the stage of the cycle of the seminiferous epithelium.[67,68,93,113] Because Sertoli cell function varies relative to the stage of the cycle of the seminiferous epithelium, and all stages are present within the normal testis at any given moment, the Sertoli cells most likely vary their ability to respond to endocrine or paracrine factors. This is supported by the observation that the level of mRNA for the FSH receptor varies during the cycle of seminiferous epithelium.[45]

Sertoli Cells

In the adult, the Sertoli cells have a complex irregular shape with cytoplasmic processes surrounding the developing spermatogenic cells. The ultrastructure of these cells is typical of a polarized epithelial cell, but somewhat atypical of a protein-secreting cell (for review, see Russell[88]). The rough endoplasmic reticulum is sparse and located in the basal portion of the cell. The abundance of smooth endoplasmic reticulum suggests that these cells could be involved in steroid synthesis and/or metabolism. Sertoli cells also possess receptors for FSH and testosterone and thus mediate the action of these hormones on spermatogenesis.

Sertoli cells also regulate the extracellular environment of the maturing germ cells. Tight junctions, which form between Sertoli cells as the testis matures, physically divide the seminiferous tubules into a basal and an adluminal compartment. Spermatogonia and preleptotene spermatocytes are located in the basal compartment of the seminiferous tubule and the more mature germ cells are located in the adluminal compartment. The ionic composition of the extracellular milieu in each compartment is different, and this difference is thought to be essential for the regulation of early and late steps in spermatogenesis.[1,103] Sertoli cells can regulate the composition of the extracellular milieu by secretion of proteins into either the basal or the adluminal compartment. In addition, Sertoli cells transport ions from the basal compartment, across the lymph-testis barrier to germ cells in the adluminal compartment. For example, there is substantial experimental support for a proposed model for the role of transferrin in iron transport from the lymph, across Sertoli cells to the developing germinal epithelium.[69] Thus, Sertoli cells could differentially mediate the effect of endocrine and paracrine factors on germ cells in the basal and adluminal compartments of the seminiferous epithelium.

Actions of FSH on Sertoli Cells

The receptor for FSH is a member of the G-protein receptor family, which includes the receptors for thyroid-stimulating hormone and LH (for review, see Griswold[39]). In the rat, the extracellular domain or N-terminal portion is 348 amino acids. Seven helices composed of 264 amino acids comprise the putative transmembrane domain. There is a short intracellular or C-terminal domain of 63 amino acids. There are three potential N-glycosylation sites and the molecular weight of the complete structure is estimated to be 70 to 80 kDa.[39] The structure of the gene for the FSH receptor is complex, consisting of 10 exons and a minimum size of 85 kb.[44]

The binding of FSH to its receptor results in an increase in intracellular cyclic AMP levels. Cyclic AMP activates various protein kinases which phosphorylate numerous proteins. In particular, protein kinase-A is capable of activating the cyclic AMP response element (CRE) binding protein, a transcriptional activator.[30] In addition, there is evidence that FSH can alter intracellular calcium levels by a pathway that does not involve adenylate cyclase or protein kinase C.[38] Thus, FSH acts on Sertoli cell function at the post-translational level, but the phosphorylation of unique proteins probably results in changes that are transcriptional, translational and modify the activities of existing proteins.

The response of Sertoli cells to FSH varies relative to the age of the animal. In fetal and early postnatal rats, proliferation of Sertoli cells essentially ceases on postnatal day 14 or 15.[24,41,73,99] During postnatal development, FSH can stimulate the proliferation of Sertoli cells.[40,74] In contrast, the suppression of FSH levels in the rat results in a reduced number of Sertoli cells.[47,105] Inhibition or stimulation of Sertoli cell replication in postnatal rats causes a decrease or increase, respectively, in sperm output in the adult.[26,76] Because Sertoli cells can support only a limited number of germ cells, the neonatal proliferation of Sertoli cells is vital to the production of adequate numbers of mature sperm to ensure fertility in the adult.[8,89,90]

As Sertoli cell replication ceases, these cells undergo functional and morphological changes. In immature rats, FSH is required for the formation of tight junctions between Sertoli cells and for the initiation of spermatogenesis.[34,78] In addition, there is an increase in synthesis of androgen-binding protein, plasminogen activator, transferrin and sulfated glycoproteins 1 and 2.[39]

The role of FSH in the adult is less clear because it is generally accepted that testosterone alone can maintain the production of mature sperm (i.e., qualitatively normal spermatogenesis).[10,95] The controversy arises as to whether or not testosterone alone produces quantitatively normal spermatogenesis. In primates including man, the effect of exogenous FSH on sperm production is easily demonstrated, whereas the role of FSH in adult rodents is difficult to demonstrate.[39] Numerous studies in which hypophysectomized rats were given testosterone implants resulted in decreased sperm production.[13,19,25,84] However, when testosterone is administered to rats in which Leydig cells are destroyed by treatment with ethylene dimethanesulfonate, normal testicular weight is maintained, suggesting that germ cell numbers are maintained at normal levels.[96,97] In contrast, when FSH alone is administered to hypophysectomized rats, spermatogenesis can only be maintained for brief periods.[23] When rats are immunized against GnRH, both LH and FSH levels decrease to undetectable levels.[3] Testosterone implants alone are capable of restoring nearly normal spermatogenesis in these animals.[3] Moreover, FSH has been shown to augment the effects of LH or testosterone on spermatogenesis in rat,[100] ram,[27] cynomolgus monkey [109] and man.[60] However, The exact role of FSH in maintaining spermatogenesis in the adult has not been elucidated, and it is likely that the function of FSH varies relative to the stages of spermatogenesis, which would make determining its role in the adult more difficult.

It is generally accepted that the FSH receptor is present only on Sertoli cells. However, there is evidence that indicates that this receptor may also be present on spermatogonia.[75] In addition, the administration of FSH to cynomolgus or rhesus monkeys causes an increase in the number of type B spermatogonia.[104,110] Although the effect of FSH on proliferation of spermatogonia may be mediated by Sertoli cells, the direct effect of FSH on spermatogonia cannot be ruled out at this time. Moreover, Sertoli cells may mediate the effect of FSH on Leydig cells by increasing Leydig cell numbers and size, the number of LH receptors and the ability to produce testosterone.[17,18,72]

Actions of LH on Leydig Cells

The Leydig cells are located within the interstitium of the testis and are the target of LH action. These cells display ultrastructural features characteristic of steroid-secreting cells. In general, Leydig cells contain an abundance of smooth endoplasmic reticulum and lipid droplets. The mitochondria usually have tubular cristae and contain the enzymes essential for side chain cleavage of cholesterol and the production of testosterone. LH is capable of stimulating the synthesis of testosterone as well as androstenedione and dehydroepiandrosterone by Leydig cells.[12,42]

Similar to the receptor for FSH, the LH receptor is a member of the G-protein receptor family. Purification of this protein from rats indicates that the receptor is a single protein with a molecular mass of 93 kDa.[87] The receptor itself is composed of 674 amino acids (75 kDa). There is a cytoplasmic C-terminal domain of 68 amino acids, 7 putative transmembrane domains, and a 341-amino-acid extracellular N-terminal domain.[65] The coding region of the rat gene is approximately 60 kb and consists of 11 exons and 10 introns.[53] LH exerts its action on Leydig cells via cyclic AMP and the elevation of calcium via the stimulation of phospholipase C.[2,49]

Testicular Regulation of the Gonadotropins

The concept that the testis produces a factor that inhibits the release of gonadotropins from the pituitary was proposed based on the observation that bilateral orchidectomy caused a marked increase in mRNAs encoding the α- and β-subunits of FSH and LH and increased the circulating levels of these hormones.[4] In general, testosterone replacement lowers the levels of circulating LH but not that of FSH. Thus, the regulation of LH secretion mediated by the testes has been attributed to testosterone secreted from Leydig cells. This observation also led to the search for the factor from the seminiferous epithelium termed "inhibin," which specifically regulates the levels of circulating FSH. Many years of work and hundreds of publications on the inhibin activity ultimately led to the characterization of the protein responsible.[6,86] During purification of inhibin by high-pressure liquid chromatography, side fractions showing FSH-stimulating activities were also observed.[58,83] It is now known that Sertoli cells produce inhibin and activin, which can specifically regulate the levels of FSH while having little effect on LH levels. In general, inhibin is capable of suppressing FSH secretion whereas activin has been shown to stimulate the secretion of FSH.[29,80,82,114] However, the effect of inhibin on FSH may be age dependent. A significant increase in FSH occurs when immature male rats, but not adult male rats, are passively immunized against the α-subunit of inhibin.[28,81] However, FSH levels increase when testosterone levels are abolished in adult males treated with ethylene dimethanesulfonate.[80] Therefore, testosterone and inhibin may function together in the adult male to regulate levels of FSH.

Inhibin and Activin

The term inhibin is now applied to two structurally related heterodimers that share a common α-subunit but a distinct β-subunit (β_A or β_B). In contrast, the term activin refers to two β-dimers ($\beta_A\beta_B$ and $\beta_B\beta_B$). These proteins are composed of two disulfide-linked chains with a molecular weight of 32,000. The amino acid sequence also has a similarity to that of transforming growth factor-β.[63] Inhibin cDNAs have been cloned and sequenced from humans, and the genes for the α and β_B subunits are found on different ends of the same chromosome, whereas the β_A gene is encoded on a different chromosome.[114] In addition to its production in the testis, inhibin is present in the hypothalamus, pituitary, placenta, adrenal glands, bone marrow and kidney, and probably displays different functions in these tissues.[37]

In the testis, inhibin is produced by Sertoli cells and Leydig cells and inhibits both the synthesis and release of FSH by gonadotrophs.[31,98] In addition, intratesticular levels of inhibin increase in response to the presence of FSH. Because inhibin affects the release of FSH but not LH, inhibin could also block the release of an FSH stimulatory factor from the hypothalamus.[59]

Although the major function of inhibin produced by the testis is to decrease FSH production by the pituitary, inhibin, through paracrine-type actions, may also influence other testicular events, including Leydig cell steroidogenesis.[46,57] Although a direct effect of LH on inhibin production in Leydig cells has not been demonstrated, LH has been shown to stimulate the release of inhibin from granulosa cells in the ovary after these cells are treated with FSH for 2 days to stimulate production of LH receptors.[114] Leydig cells may also influence Sertoli cell production of inhibin.[31,98] However, since Leydig cells have also been shown to produce inhibin and activin,[55,79,85,94] the relevance of the influence of Sertoli cells and Leydig cells on inhibin production remains unclear.

In addition to its influence on the pituitary, inhibin may also have local effects on germ cells within the testis.[106] In particular, inhibin blocks DNA synthesis in spermatogonia and reduces the number of spermatogonia in the testis.[33,106] In contrast, activin stimulates proliferation of spermatogonia *in vitro*.[62] The binding of inhibin to germ cells and the synthesis of inhibin are both stage-dependent events.[9] Recently, male and female mice homozygous for a deleted α-subunit gene were created.[63] These animals are susceptible to the development of gonadal tumors. Thus, inhibin may be a negative regulator of gonadal cell proliferation.

Testosterone

The putative role of testosterone in spermatogenesis has been elucidated through studies involving hypophysectomy, subcutaneous testosterone implants, and immunization against GnRH or the gonadotropins. The receptor to this androgen is present on Sertoli cells and peritubular myoid cells. Similar to FSH, the role of testosterone changes with age. Testosterone levels are low until puberty.[66] At the onset of puberty testosterone is required for the initiation of spermatogenesis, and in the adult its presence is essential for the maintenance of spermatogenesis. It is important to note that the molecular details by which testosterone influences the actions of Sertoli cells and/or myoid cells in the adult have not been elucidated.

As mentioned above, testosterone is also involved in regulating the release of pituitary gonadotropins. Exogenous testosterone treatment increases pituitary concentrations of FSH while lowering those of LH.[52] In addition, pituitary cell cultures isolated from male rats increase their secretion of FSH when these cells are treated with testosterone.[50] Moreover, the administration of testosterone to castrated rats decreases the levels of mRNA for α- and β-subunits of LH to precastration levels.[111] This effect is apparently mediated by the hypothalamus.[37] In contrast, testosterone acts directly on the pituitary to increase mRNA for the β-subunit of FSH.[37] Thus, testosterone can affect gonadotropin secretion through the hypothalamus or directly on the gonadotrophs.

Seasonal Breeders

Seasonal breeders, such as hamsters, are sexually active during the spring and fall. During the winter months, when the duration of daylight is shorter, these animals typically hibernate. In the laboratory, these animals can be kept in their reproductive phase by exposing them to long photoperiods of 12.5 hours per day or longer. Transferring these animals to short photoperiods (less than 12 hours per day) results in the repression of the testis, gonadal atrophy and functional quiescence. In these short-day animals, testicular weight is reduced by at least 80%, spermatogenesis ceases, and plasma levels of gonadotropins and testosterone are reduced.[7]

The influence of daylight on seasonal breeders is apparently mediated by the pineal gland, an endocrine gland located above the diencephalon and posterior to the third ventricle in the brain. The pineal secretes an indolamine hormone, melatonin, which apparently alters hypothalamic control of the anterior pituitary. In seasonal breeders, melatonin concentrations increase in the pineal and plasma during short photoperiods.[7] Interestingly, the effects of a short day on reproduction can be prevented by removing the pineal gland, while injections of melatonin mimic the effects of a short photoperiod.[102] In particular, exposure to short days leads to the suppression of FSH and subsequently LH and prolactin with a profound reduction in the incidence of LH secretion. A reduced noradrenergic input to hypothalamic neurons secreting LH-releasing hormone (LHRH) could mediate the reduced secretion of LH during short photoperiods.[7] The response to LHRH of male hamsters maintained on a short photoperiod is similar to that of hamsters maintained on a long photoperiod. Thus, the response of the pituitary to LHRH is unaffected by day length. The release of LHRH is apparently inhibited without affecting LHRH synthesis.

Reduction of levels of prolactin mRNA is the earliest response of pituitary to light deprivation.[61] Treatment of regressed testis with prolactin can restore normal gonadal weight, plasma testosterone levels and fertility. This effect of prolactin is most likely due to its ability to increase binding of LH in the testis and plasma levels of FSH and LH.[7] Interestingly, treatment with prolactin prior to short-day exposure does not prevent regression unless GnRH treatment is also included.[16] Moreover, prolactin is necessary for the presence of LH receptors in Leydig cells.

SUMMARY

The regulation of spermatogenesis involves a classical interaction between the testis and pituitary that includes fundamental roles for the hypothalamus and pineal gland. The cellular components of the testis, which are the direct targets of pituitary action and are involved in feedback regulation of the pituitary, are the Leydig and Sertoli cells. Although the germinal cells display local paracrine effects on Sertoli cell function,[43] there is no evidence that germ cells directly affect gonadotropin levels. While the changes in the pituitary-testis interactions as a result of development have been well described, the effect of aging and drug abuse has been examined only in limited studies in a few species.

REFERENCES

1. **Amann, R., Killian, G., and Benton, A.,** Differences in the electrophoretic characteristics of bovine rete testis fluid and plasma from the cauda epididymis, *J. Reprod. Fertil.*, 35, 321, 1973.
2. **Ascoli, M., and Segaloff, D.,** On the structure of the luteinizing hormone/chorionic gonadotropin receptor, *Endocr. Rev.*, 10, 27, 1989.
3. **Awoniyi, C., Zirkin, B., Chandrashekar, V., and Schlaff, W.,** Exogenously administered testosterone maintains spermatogenesis quantitatively in adult rats actively immunized against GnRH, *Endocrinology*, 130, 3283, 1992.
4. **Badger, T., Wilcox, C., Meyer, E., Bell, R., and Cicero, T.,** Simultaneous changes in tissue and serum levels of luteinizing hormone, follicle-stimulating hormone, and luteinizing hormone/follicle-stimulating hormone releasing factor after castration in the male rat, *Endocrinology*, 102, 136, 1978.
5. **Baker, B., Pierce, J., and Cornell, J.,** The utility of antiserums to the subunits of TSH and LH for immunochemical staining of the rat hypophysis, *Am. J. Anat.*, 135, 251, 1972.

6. **Bardin, C., Morris, P., and Chan, C.,** Testicular inhibin: Structure and regulation by FSH, androgen and EGF, in *Inhibin Non-Steroidal Regulation of Follicle Stimulating Hormone Secretion*, Burger, H.G., *et al.*, Eds., Raven Press, New York, 1987, 179.
7. **Bartke, A., and Steger, R.,** Seasonal changes in the function of the hypothalamic-pituitary-testicular axis in the Syrian hamster, *Proc. Soc. Exp. Biol. Med.*, 199, 139, 1992.
8. **Berndtson, W., and Thompson, T.,** Changing relationships between testis size, Sertoli cell number and spermatogenesis in Sprague-Dawley rats, *J. Androl.*, 11, 429, 1990.
9. **Bhasin, S., Krummen, L., Swerdloff, R., Morelos, B., Kim, W., diZerega, G., Ling, N., Esch, F., Shimasaki, S., and Toppari, J.,** Stage dependent expression of inhibin α- and β-B subunits during the cycle of the rat seminiferous epithelium, *Endocrinology*, 124, 987, 1989.
10. **Boccabella, A.,** Reinitiation and restoration of spermatogenesis with testosterone propionate and other hormones after long-term and post-hypophysectomy regression period, *Endocrinology*, 72, 787, 1963.
11. **Bremmer, W., and Paulsen, C.,** Two pools of luteinizing hormone in human pituitary: Evidence from constant administration of luteinizing hormone-releasing hormone, *J. Clin. Endocrinol. Metab.*, 39, 811, 1974.
12. **Brinck-Johnsen, T., and Eik-Nes, K.,** Effect of human chorionic gonadotrophin on the secretion of testosterone and 4-androstene-3,17-dione by the canine testis, *Endocrinology*, 61, 676, 1957.
13. **Buhl, A., Cornette, J., Kirton, K., and Yuan, Y.-D.,** Hypophysectomized male rats treated with polydimethylsiloxane capsules containing testosterone: Effects on spermatogenesis, fertility and reproductive tract concentrations on androgens, *Biol. Reprod.*, 27, 183, 1982.
14. **Burnside, J., Buckland, P., and Chin, W.,** Isolation and characterization of the gene encoding the α-subunit gene of the rat pituitary glycoprotein hormones, *Gene*, 70, 67, 1988.
15. **Catt, K., and Rierce, J.,** Gonadotropic hormones of the adenohypophysis, in *Reproductive Endocrinology: Physiology, Pathophysiology, and Clinical Management, 2nd edition*, Yen, S.S.C., and Jaffe, R.B., Eds., W.B. Saunders, Philadelphia, 1986, 75.
16. **Chen, H., and Reiter, R.,** The combination of twice daily luteinizing hormone-releasing factor administration and renal pituitary homografts restores normal reproductive organ size in male hamsters with pineal-mediated gonadal atrophy, *Endocrinology*, 106, 1382, 1980.
17. **Chen, Y., Payne, A., and Kelch, R.,** FSH stimulation of Leydig cell function in the hypophysectomized immature rat, *Proc. Soc. Exp. Biol. Med.*, 153, 473, 1976.
18. **Chen, Y., Shaw, M., and Payne, A.,** Steroid and FSH action on LH receptors and LH-sensitive testicular responsiveness during sexual maturation of the rat, *Mol. Cell Endocrinol.*, 8, 291, 1977.
19. **Chowdhury, A., and Tcholakian, R.,** Effects of various doses of testosterone propionate on intratesticular and plasma testosterone levels and maintenance of spermatogenesis in adult hypophysectomized rats, *Steroids*, 34, 151, 1979.
20. **Clarke, I., Cummins, J., Findlay, J., Burman, K., and Doughton, D.,** Effects on plasma luteinizing hormone and follicle-stimulating hormone of varying the frequency and amplitude of gonadotropin-releasing hormone pulses in ovariectomized ewes with hypothalamo-pituitary disconnection, *Neuroendocrinology*, 39, 214, 1984.
21. **Clermont, Y.,** The cycle of the seminiferous epithelium in man, *Am. J. Anat.*, 112, 35, 1963.
22. **Clermont, Y., and Harvey, S.,** Duration of the cycle of the seminiferous epithelium of normal, hypophysectomized and hypophysectomized-hormone treated albino rats, *Endocrinology*, 76, 80, 1965.
23. **Clermont, Y., and Harvey, S.,** Effects of hormones on spermatogenesis in the rat, *CIBA Found. Colloq. Endocrinol.*, 16, 173, 1967.
24. **Clermont, Y., and Perey, B.,** Quantitative study of the cell population of the seminiferous tubule in immature rats, *Am. J. Anat.*, 100, 241, 1957.
25. **Collins, P., and Tsang, W.,** Quantitative assessment of the gametogenic and androgenic properties of testicular steroids in hypophysectomized rats, *J. Reprod. Fertil.*, 75, 285, 1985.

26. **Cooke, P., Hess, R., Porcelli, J., and Meisami, E.**, Increased sperm production in adult rats after transient neonatal hypothyroidism, *Endocrinology,* 129, 244, 1991.
27. **Courot, M., Hocherau de Reviers, M., and Monet-Kuntz, C.**, Endocrinology of spermatogenesis in the hypophysectomized ram, *J. Reprod. Fertil.*, (Suppl.) 26, 165, 1979.
28. **Culler, M., and Negro-Vilar, A.**, Passive immunoneutralization of endogenous inhibin: Sex-related differences in the role of inhibin during development, *Mol. Cell Endocrinol.*, 58, 263, 1988.
29. **de Jong, F., and Robertson, D.**, Inhibin: 1985 update on action and purification, *Mol. Cell Endocrinol.*, 68, 555, 1988.
30. **Deutsch, P., Hoeffler, J., Jameson, J., Lin, L., and Habener, J.**, Structural determinants for transcriptional activation by cAMP-responsive DNA elements, *J. Biol. Chem.*, 263, 18466, 1988.
31. **Drummond, A., Risbridger, G.P., and de Kretser, D.**, The involvement of Leydig cells in the regulation of inhibin secretion by the testis, *Endocrinology,* 125, 1941, 1989.
32. **Erwin, C., Croyle, M., Donelson, J., and Maurer, R.**, Nucleotide sequence of cloned complementary deoxyribonucleic acid for the subunit of bovine pituitary glycoprotein hormones, *Biochemistry,* 22, 4856, 1983.
33. **Franchimont, P., Croze, F., Demoulin, A., Bologne, R., and Hustin, J.**, Effect of inhibin on rat deoxyribonucleic acid (DNA) synthesis *in vivo* and *in vitro*, *Acta Endocrinol.*, 98, 312, 1981.
34. **Fritz, I.**, Sites of actions of androgens and follicle stimulating hormone on cells of the seminiferous tubule, in *Biochemistry Actions of Hormones,* Litwack, G., Ed., Academic Press, New York, 1978, 249.
35. **Gay, V., and Sheth, N.**, Evidence for periodic release of LH in castrated male and female rats, *Endocrinology,* 90, 158, 1972.
36. **Gharib, S., Roy, A., Wierman, M., and Chin, W.**, Isolation and characterization of the gene encoding the β-subunit of rat follicle-stimulating hormone, *DNA,* 8, 339, 1989.
37. **Gharib, S., Wierman, M., Supnik, M., and Chin, W.**, Molecular biology of the pituitary gonadotropins, *Endocrinol. Rev.*, 11, 177, 1990.
38. **Grasso, P., and Reichert, L.**, Follicle-stimulating hormone receptor-mediated uptake of $^{45}Ca^{2+}$ by cultured rat Sertoli cells does not require activation of cholera toxin- or pertussis toxin-sensitive guanine nucleotide binding proteins or adenylate cyclase, *Endocrinology,* 127, 1248, 1990.
39. **Griswold, M.**, Actions of FSH on mammalian Sertoli cells, in *The Sertoli Cell,* Russell, L.D., and Griswold, M.D., Eds., Cache River Press, Clearwater, 1993, 493.
40. **Griswold, M., Mably, E., and Fritz, I.**, Stimulation by follicle stimulating hormone and dibutyryl cyclic AMP of incorporation of ^{3}H-thymidine into nuclear DNA of cultured Sertoli cell-enriched preparations from immature rats, *Curr. Top. Mol. Endocrinol.*, 2, 413, 1975.
41. **Griswold, M., Solari, A., Tung, P., and Fritz, I.**, Stimulation by FSH of DNA synthesis and of mitosis in culture Sertoli cells prepared from testes of immature rats, *Mol. Cell. Endocrinol.*, 7, 151, 1977.
42. **Hall, P., and Eik-Nis, K.**, The influence of gonadotrophins *in vivo* upon the biosynthesis of androgens by homogenate of rat testis, *Biochim. Biophys. Acta,* 71, 438, 1963.
43. **Han, I., Sylvester, S., Kim, K., Schelling, M., Vekateswaran, S., Banckaert, V., McGuinness, M., and Griswold, M.**, Basic fibroblast growth factor (bFGF) is a testicular germ cell product which may regulate Sertoli cell function, *Mol. Endocrinol.*, 7, 889, 1993.
44. **Heckert, L., Daley, I., and Griswold, M.**, Structural organization of the follicle-stimulating hormone receptor gene, *Mol. Endocrinol.*, 6, 70, 1992.
45. **Heckert, L., and Griswold, M.**, Expression of follicle-stimulating hormone receptor mRNA in rat testes and Sertoli cells, *Mol. Endocrinol.*, 5, 670, 1991.
46. **Hsueh, A., Dahl, K., Vaughan, J., Tucker, E., Pivier, J., Bardin, C., and Vale, W.**, Heterodimers and homodimers of inhibin subunits have different paracrine action in the modulation of luteinizing hormone-stimulated androgen biosynthesis, *Proc. Natl. Acad. Sci. USA,* 84, 5082, 1987.

47. **Huhtaniemi, I., Nevo, N., Amsterdam, A., and Naor, Z.,** Effect of postnatal treatment with a gonadotropin-releasing hormone antagonist on sexual maturation of male rats, *Biol. Reprod.*, 35, 501, 1986.
48. **Jameson, J., Chin, W., Hollenberg, A., Chang, A., and Habener, J.,** The gene encoding the β-subunit of rat luteinizing hormone, *J. Biol. Chem.*, 259, 15474, 1984.
49. **Johnson, G., and Dhanasekaran, N.,** The G-protein family and their interaction with receptors, *Endocrinol. Rev.*, 10, 317, 1989.
50. **Kennedy, J., and Cappel, S.,** Direct pituitary effects of testosterone and luteinizing hormone-releasing hormone upon follicle-stimulating hormone: Analysis by radioimmuno- and radioreceptor assay, *Endocrinology,* 116, 741, 1985.
51. **Kim, K., Gordon, D., and Maurer, R.,** Nucleotide sequence of the bovine gene for follicle-stimulating hormone β-subunit, *DNA,* 7, 227, 1988.
52. **Kingsley, T., and Bogdanove, E.M.,** Direct feedback of androgens: Localize effects of intrapituitary implants of androgens on gonadotrophic cells and hormone stores, *Endocrinology,* 93, 1398, 1973.
53. **Koo, Y., Ji, I., Slaughter, R., and Ji, T.,** Structure of the luteinizing hormone receptor gene and multiple exons of the coding sequence, *Endocrinology,* 128, 2297, 1991.
54. **Leblond, C.P., and Clermont, Y.,** Definition of the stages of the cycle of the seminiferous epithelium in the rat, *Ann. N. Y. Acad. Sci.*, 55, 548, 1952.
55. **Lee, W., Mason, A., Schwall, R., Szonyi, E., and Mather, J.,** Secretion of activin by interstitial cells in the testis, *Science,* 243, 396, 1989.
56. **Levine, J., Conaghan, L., Luderer, U., and Strobl, F.,** Hypothalamic regulation of FSH secretion, in *Follicle Stimulating Hormone: Regulation of Secretion and Molecular Mechanisms of Action*, Hunzicker-Dunn, M., and Schwartz, N.B., Eds., Springer-Verlag, New York, 1993, 66.
57. **Lin, T., Calkins, J., Morris, P., Vale, W., and Bardin, C.,** Regulation of Leydig cell function in primary culture by inhibin and activin, *Endocrinology,* 125, 2134, 1989.
58. **Ling, N., Ying, S.-Y., Ueno, N., Esch, F., Enoroy, L., and Guillemin, R.,** Isolation and partial characterization of a Mr 32,000 protein with inhibin activity from porcine follicular fluid, *Proc. Natl. Acad. Sci. USA,* 82, 7217, 1985.
59. **Lumpkin, M., Negro-Vilar, A., Frachimont, P., and McCann, S.,** Evidence for a hypothalamic site of action of inhibin to suppress FSH release, *Endocrinology,* 108, 1101, 1981.
60. **Mancini, R.,** Effect of different types of gonadotropins on the induction and restoration of spermatogenesis in the human testis, *Acta Eur. Fertil.*, 1, 401, 1969.
61. **Massa, J., and Blask, D.,** An early pineal-induced suppression of pituitary prolactin mRNA levels in light-deprived male hamsters, *Neuroendocrinology,* 50, 506, 1989.
62. **Mather, J., Attie, K., Woodruff, T., Rice, G., and Phillips, D.,** Activin stimulates spermatogonial proliferation in germ-Sertoli cell cocultures from immature rat testis, *Endocrinology,* 127, 3206, 1990.
63. **Matzuk, M., Finegold, M., Su, J.-G., Hsueh, J., and Bradley, A.,** α-Inhibin is a tumour-suppressor gene with gonadal specificity in mice, *Nature,* 360, 313, 1992.
64. **McCann, S.,** Regulation of secretion of follicle-stimulating hormone and luteinizing hormone, in *Handbook of Physiology IV,* Geiger, S.R., Ed., American Physiological Society, Washington, D.C., 1974, 489.
65. **McFarland, K., Sprengel, R., Phillips, H., Kohler, M., Rosemblit, N., Nikolics, K., Segaloff, D., and Seeburg, P.,** Lutropin-choriogonadotropin receptor: An unusual member of the G-protein-coupled receptor family, *Science,* 245, 494, 1989.
66. **Miyachi, Y., Mieschlag, E., and Lipsett, M.,** The secretion of gonadotropins and testosterone by the neonatal male rat, *Endocrinology,* 92, 1, 1972.
67. **Morales, C., Alcivar, A., Hecht, N., and Griswold, M.,** Specific mRNAs in Sertoli and germinal cells of testes from stage synchronized rats, *Mol. Endocrinol.*, 3, 725, 1989.
68. **Morales, C., Hugly, S., and Griswold, M.,** Stage-dependent levels of specific mRNA transcripts in Sertoli cells, *Biol. Reprod.*, 36, 1035, 1987.

69. **Morales, C., Sylvester, S., and Griswold, M.,** Transport of iron and transferrin synthesis by the seminiferous epithelium of the rat *in vivo*, *Biol. Reprod.*, 37, 995, 1987.
70. **Moriarty, G.,** Immunocytochemistry of the pituitary glycoprotein hormones, *J. Histochem. Cytochem.*, 24, 846, 1976.
71. **Oakberg, E.,** Duration of spermatogenesis in the mouse and timing of stages of the cycle of the seminiferous epithelium, *Am. J. Anat.*, 99, 507, 1956.
72. **Odell, W., Swerdloff, R., Jacobs, H., and Hescox, M.,** FSH induction of sensitivity to LH: One cause of sexual maturation in the male rat, *Endocrinology*, 92, 160, 1973.
73. **Orth, J.,** Proliferation of Sertoli cells in fetal and postnatal rats: A quantitative autoradiographic study, *Anat. Rec.*, 203, 485, 1982.
74. **Orth, J.,** The role of follicle-stimulating hormone in controlling Sertoli cell proliferation in testes of fetal rats, *Endocrinology*, 115, 1248, 1984.
75. **Orth, J., and Christensen, A.,** Localization of ^{125}I-labeled FSH in the testes of hypophysectomized rats by autoradiography at the light and electron microscope levels, *Endocrinology*, 101, 262, 1977.
76. **Orth, J., Gunsalus, G., and Lamperti, A.,** Evidence from Sertoli cell-depleted rats indicates that spermatid numbers in adults depends on numbers of Sertoli cells produced during perinatal development, *Endocrinology*, 122, 787, 1988.
77. **Pohl, C., Richardson, D., Hitchison, J., Germak, J., and Knobil, E.,** Hypophysiotropic signal frequency and the functioning of the pituitary-ovarian system in the rhesus monkey, *Endocrinology*, 112, 2076, 1983.
78. **Posalaky, Z., Meyer, R., and McGinley, D.,** The effects of follicle-stimulating hormone (FSH) on Sertoli cell junctions *in vitro*: A freeze-fracture study, *J. Ultrastruc. Res.*, 74, 241, 1981.
79. **Risberger, G., Clements, J., Robertson, D., Drummond, A., Muir, J., Berger, H., and de Kretser, D.M.,** Immuno- and bioactive inhibin and inhibin alpha-subunit expression in rat Leydig cell cultures, *Mol. Cell Endocrinol.*, 66, 119, 1989.
80. **Risbridger, G., Robertson, D., and deKretser, D.,** Current perspectives of inhibin biology, *Acta Endocrinol.*, 122, 673, 1990.
81. **Rivier, C., Cajander, S., Vaughan, J., Hsueh, A., and Vale, W.,** Age dependent changes in physiological action content and immunostaining of inhibin in male rats, *Endocrinology*, 123, 120, 1988.
82. **Rivier, C., Vale, W., and Rivier, J.,** Studies of the inhibin family: A review, *Horm. Res.*, 28, 104, 1987.
83. **Rivier, J., Spiess, J., McClintock, R., Vaughan, J., and Vale, W.,** Purification and partial characterization of inhibin from porcine follicular fluid, *Biochem. Biophys. Res. Commun.*, 133, 120, 1985.
84. **Robaire, B., and Zirkin, B.,** Hypophysectomy and simultaneous testosterone replacement: Effects on male rat reproductive tract and epididymal 4-5 a reductase and 3α-hydroxysteroid dehydrogenase, *Endocrinology*, 109, 1225, 1981.
85. **Roberts, V., Meunier, H., Sawchenko, P., and Vale, W.,** Differential production and regulation of inhibin subunits in rat testicular cell types, *Endocrinology*, 125, 2350, 1989.
86. **Robertson, D., Foulds, L., Leversha, L., Morgan, F., Hearn, M., Burger, H., Wetterhall, R., and de Kretser, D.,** Isolation of inhibin from bovine follicular fluid, *Biochem. Biophys. Res. Commun.*, 126, 220, 1985.
87. **Rosemblit, N., Ascoli, M., and Segaloff, D.,** Characterization of an antiserum to the rat luteal luteinizing hormone/chorionic gonadotropin receptor, *Endocrinology*, 123, 2284, 1988.
88. **Russell, L.,** Sertoli cell structure, in *The Sertoli Cell*, Russell, L.D., and Griswold, M.D., Eds., Cache River Press, Clearwater, 1993, 1.
89. **Russell, L., and Peterson, R.,** Determination of the elongate spermatid-Sertoli cell ratio in various mammals, *J. Reprod. Fertil.*, 70, 635, 1984.
90. **Russell, L., Ren, H., Sinha Hikim, S., Schulze, W., and Sinha Hikim, A.,** A comparison study in twelve mammalian species with respect to key morphometric parameters related to volume densities and volumes of selected testis components, *Am. J. Anat.*, 188, 21, 1990.

91. **Sairam, M.,** Primary structure of the ovine pituitary follitropin α-subunit, *Biochem. J.*, 197, 535, 1981.
92. **Santen, R., and Bardin, C.,** Episodic luteinizing hormone secretion in man: Pulse analysis, clinical interpretation, physiologic mechanisms, *J. Clin. Invest.*, 52, 2617, 1973.
93. **Shabanowitz, R., DePhilip, R., Crowell, J., Tres, L., and Kierzenbaum, A.,** Temporal appearance and cyclic behavior of Sertoli cell-specific secretory proteins during the development of the rat seminiferous tubule, *Biol. Reprod.*, 35, 745, 1986.
94. **Shaha, C., Morris, P., Chen, C., Vale, W., and Bardin, C.,** Immunostainable inhibin subunits are in multiple types of testicular cells, *Endocrinology*, 125, 1941, 1989.
95. **Sharpe, R.,** Testosterone and spermatogenesis, *J. Endocrinol.*, 113, 1, 1987.
96. **Sharpe, R., Donachie, K., and Cooper, I.,** Re-evaluation of the intratesticular level of testosterone required for quantitative maintenance of spermatogenesis in the rat, *J. Endocrinol.*, 117, 19, 1988.
97. **Sharpe, R., Fraser, H., and Ratnasooriya, W.,** Assessment of the role of Leydig cell products other than testosterone in spermatogenesis and fertility in adult rats, *Int. J. Androl.*, 11, 507, 1988.
98. **Sharpe, R., Kerr, J., and Maddocks, S.,** Evidence for a role of the Leydig cells in control of the intratesticular secretion of inhibin, *Mol. Cell. Endocrinol.*, 60, 243, 1988.
99. **Steinberger, A., and Steinberger, E.,** Replication pattern of Sertoli cells in maturing rat testis *in vivo* and in organ culture, *Biol. Reprod.*, 4, 84, 1971.
100. **Steinberger, E.,** Hormonal control of mammalian spermatogenesis, *Physiol. Rev.*, 51, 1, 1971.
101. **Talmadge, K., Vamvakopoulos, N., and Fiddes, J.,** Evolution of the genes for the β-subunits of human chorionic gonadotropin and luteinizing hormone, *Nature*, 307, 37, 1984.
102. **Tamarkin, L., Westrom, W., Hamil, A., and Goldman, B.,** Effect of melatonin on the reproductive system of male and female Syrian hamsters: A diurnal rhythm in sensitivity to melatonin, *Endocrinology*, 99, 1534, 1976.
103. **Tuck, R., Setchell, B., Waites, G., and Young, G.,** The composition of fluid collected by micropuncture and catheterization from the seminiferous tubules and rete testis of rats, *Eur. J. Physiol.*, 318, 225, 1970.
104. **van Alphen, M., van de Kant, H., and de Rooij, D.,** Follicle-stimulating hormone stimulates spermatogenesis in the adult monkey, *Endocrinology*, 123, 1449, 1988.
105. **van Den Dungen, H., van Dieten, J., van Rees, G., and Shoemaker, J.,** Testicular weight, tubular diameter and number of Sertoli cells in rats are decreased after early prepubertal administration of an LHRH antagonist: The quality of spermatozoa is not impaired, *Life Sci.*, 46, 1081, 1990.
106. **van Dissel-Emiliani, F., Grootenhuis, A., de Jong, F., and de Rooij, D.,** Inhibin reduces spermatogonial numbers in testes of adult mice and Chinese hamsters, *Endocrinology*, 125, 1899, 1989.
107. **Virgin, J., Silver, B., Thomason, A., and Nilson, J.,** The gene for the subunit of bovine luteinizing hormone encodes a gonadotropin mRNA with an unusually short 5'-untranslated region, *J. Biol. Chem.*, 260, 7072, 1985.
108. **Watkins, P., Eddy, R., Beck, A., Wellucci, V., Leveron, B., Tanzi, R., Gusella, J., and Shows, T.,** DNA sequence and regional assignment of the human follicle-stimulating hormone β-subunit gene to the short arm of human chromosome 11, *DNA*, 6, 205, 1987.
109. **Weinbauer, G., Behre, H., Fingscheidt, U., and Nieschlag, E.,** Human FSH exerts a stimulatory effect on spermatogenesis, testicular size and serum inhibin levels in the GnRH antagonist-treated nonhuman primate (*Macaca fasicularis*), *Endocrinology*, 129, 1831, 1991.
110. **Weinbauer, G., and Nieschlag, E.,** Peptide and steroid regulation of spermatogenesis in primates, *Ann. N.Y. Acad. Sci.*, 637, 107, 1991.
111. **Wierman, M., Gharib, S., LaRovere, J., Badger, T., and Chin, W.,** Selective failure of androgens to regulate follicle-stimulating hormone β mRNA levels in the male rat, *Mol. Endocrinol.*, 2, 492, 1988.

112. **Wildt, L., Hausler, A., Marshall, G., Hutchison, J., Plant, T., Belchetz, P., and Knobil, E.,** Frequency and amplitude of gonadotropin-releasing hormone stimulation and gonadotropin secretion in the rhesus monkey, *Endocrinology,* 109, 376, 1981.
113. **Wright, W., Parvinen, M., Musto, N., Gunsalus, G., Phillips, D., Mather, J., and Bardin, C.,** Identification of stage-specific proteins synthesized by rat seminiferous tubules, *Biol. Reprod.,* 29, 257, 1983.
114. **Ying, S.-Y.,** Inhibins, activins, and follistatins: Gonadal proteins modulating secretion of follicle-stimulating hormone, *Endocr. Rev.,* 9, 267, 1988.

Chapter 5

GENOTYPIC INFLUENCES ON REPRODUCTION AND REPRODUCTIVE AGING IN MICE: INFLUENCES FROM ALLELES IN THE *H-2* COMPLEX AND OTHER LOCI

C.E. Finch and J.F. Nelson

TABLE OF CONTENTS

0-8493-2451-3/95/$0.00+.50

ABBREVIATIONS

hsp–heat shock protein
LH–luteinizing hormone
Mhc–major histocompatibility complex
RFLP–restriction fragment length polymorphism
rRNA–ribosomal ribonucleic acid

Note: BRCA1, C3H, CBA/HeJ, DBA, etc. are formal designations of mice strains.

INTRODUCTION: WHY STUDY THE GENETICS OF REPRODUCTION AND REPRODUCTIVE AGING?

It is difficult to find many reports on the Mendelian genetics of reproduction in adult laboratory rodents. On one hand, there is considerable interest in mutants such as the hypogonadal mouse, which lacks GnRH-reproducing neurons in the hypothalamus,[15] and the *t*-alleles, which cause abnormal development and embryo death.[20] However, our main reproductive journals publish relatively few reports on allelic influences that have quantitative, non-lethal effects on reproduction. Of course, these journals feature molecular approaches to the physiology of reproduction. In contrast, there is a large amount of literature on the quantitative genetics of reproduction in cattle, fowl, and other domestic animals.

Laboratory rodents show numerous differences between inbred lines in reproductive parameters such as mating preferences through urinary pheromones, onset of puberty, and litter size.[15,20,23] These quantitative traits were largely recognized *post facto* in strains of rodents that were bred for other traits, such as the incidence of tumors. However, as described below, comparison of inbred strains does not assure an allelic basis for differences.

Our interests in the genetics of reproduction arose in connection with research into mechanisms of reproductive aging. The phenomena of reproductive aging in female mammals present some of the most broadly generalizable changes, because ovarian oocyte pools are finite and decline irreversibly in virtually all mammals that have been carefully studied.[11,14,28,40] The canonical mid-life depletion of ovarian oocytes contrasts with other changes of aging that vary widely between species. For example, benign prostatic hypertrophy, which is extremely common after age 60, has only been described as common during aging in one other species, the domestic dog.[11] Thus, there are genetic differences in reproductive aging between species.

Age-related disorders in reproduction are a major public health concern with an enormous impact on the economy that will increase further with more generalized availability of health insurance. During mid-life there is an accelerating increase of life-threatening and debilitating disorders, such as reproductive tract cancers, bone fractures due to acceleration of osteoporosis by menopause, and fetal aneuploidy. We argue here that there is a neglected subject: the quantitative genetics of allelic influences on reproductive functions in laboratory rodent models that could lead to markers for identifying common risk factors for dysfunctions of reproductive aging in human populations. There has been recent success in locating genes for susceptibility to reproductive tract cancers, e.g., the breast cancer susceptibility gene *BRCA1*, which maps to chromosome 17.[19] However, little attention has been given to the possibility of common genetic risk factors for other outcomes of aging in reproduction-related functions. For example, few clinical texts on osteoporosis do more than mention possible racial differences in its risks.

Evolutionary biology also pays great attention to quantitative details of reproduction. Numerous reproductive parameters are the ongoing target of natural selection, including the age of first reproduction, the timing of ovulation, and the incidence of fetal abnormalities. Each of these parameters of the reproductive schedule may be selected by local conditions, as indicated by differences between populations. It is considered likely, according to evolutionary theory, that species differences in life span are, for the most part, a direct result of selection for the reproductive schedule.[4,32,41]

Three examples are briefly described. Under arid conditions, field populations of the Hawaiian fly, *Drosophila mercatorum,* show an increase in genotype that has earlier onset of reproduction, but a shorter life expectancy; the basis for this are genes that alter the amplification of rRNA cistrons and indirectly affect the levels of juvenile hormone.[7,11,37] In the laboratory *D. melanogaster*, artificial selection for early reproduction yielded populations with shorter life spans, whereas selection for reproduction at later ages increased the life span; these shifts depended on the use of outbred fly stocks with extensive genetic variations.[32] In fish, natural populations of the guppy, *Poecilia reticulata (=Lebistes r.),* have different reproductive schedules (age of first reproduction and numbers of offspring) according the predation by another fish that prefer adult prey; the different reproductive schedules were maintained in laboratory populations and demonstrated a genetic basis during breeding studies.[30,31]

In conclusion, the genetics of reproduction and the consequences of reproductive aging are of keen interest to numerous scientific communities: evolutionary biology; basic and clinical medicine; gerontology; and health care economics.

LIMITS OF GENETICS IN PHENOTYPIC VARIABILITY

At the outset, it is key to recognize that there are extensive, nongenetic sources of individual variations (even in highly inbred rodent strains) that must be taken into account in experimental design. Figure 1 shows longitudinal records of estrous cycle frequency during aging in C57BL/6J mice, which display major differences in estrous cycle length and in the onset during aging of infertility and acyclicity. These variations are widely observed in mouse strains, despite brother-sister mating for 50 years that eliminated all variations in histocompatibility genes.[20]

Many sources of nongenetic variations can be considered. During development in vertebrates, it is well recognized that cell lineage assignments are statistical, rather than absolutely determinate as is the case for the highly determinate lineage of nematodes that give rise to individuals with identical cell numbers in gonads and other organs.[6] In contrast, individual mice in several inbred strains examined vary about twofold in the numbers of ovarian oocytes at birth (Figure 2). Individual differences in the numbers of oocytes are a plausible factor in the variable age when the ovary is exhausted of

ERRATA

The Reproductive Neuroendocrinology of Aging and Drug Abuse

ISBN 0-8493-2451-3

During the course of production of the above-mentioned book, the editors' names were erroneously transposed. The cover and title page should read:

Edited by

Dipak K. Sarkar

Charles D. Barnes

ERRATA

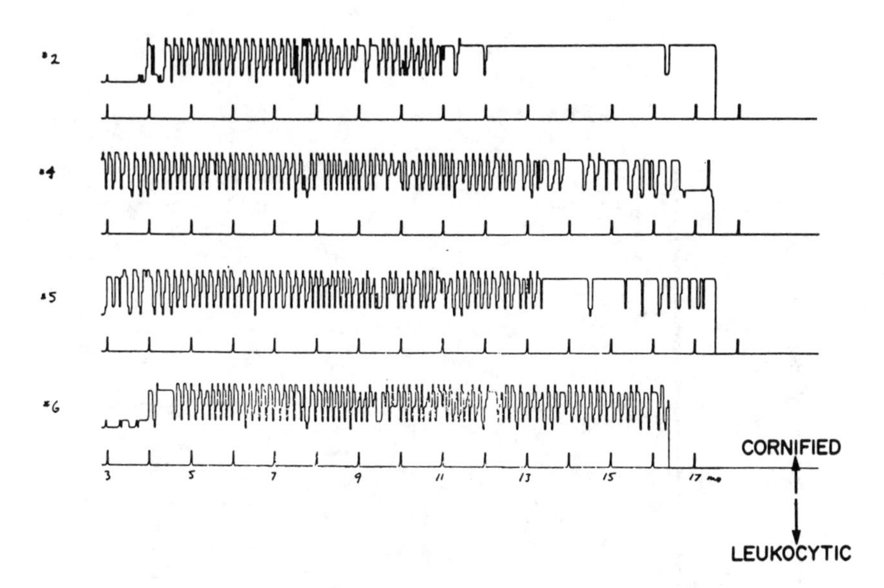

FIGURE 1. Estrous cycle histories of individual C57BL/6J mice as determined from their vaginal smear cytology. The position of the horizontal line indicates the type of vaginal smear as scaled according to a gradient from leukocytes (*bottom*) to cornified epithelial cells (*top*). (Reprinted from Finch *et al.*, Studies on ovarian-hypothalamic-pituitary interactions during reproductive aging in C57BL/6J mice, in *Brain-Endocrine Interaction, IV: Neuropeptides in Development and Aging (Peptides)* Vol. 1, Suppl. 1, Scott, D., and Sladek, J.L., Jr., Eds., with kind permission from Pergamon Press Ltd., Headington Hill Hall, Oxford OX3 0BW, UK.)

oocytes. Variations in the numbers of cells have been analyzed in depth for a few types of neurons, which show a Poisson distribution that is consistent with a fixed probability for survival of daughter cells during clonal descent.[42]

Another source of variation in reproductive parameters is the effect of the gender on the fetal neighbor, which in rodents has important influences in estrous cycle frequency in the young adult,[38] in the size of the last litter and the age when fertility ceases.[39,40] The basis for these intrauterine influences from neighboring fetuses is still obscure. Yet other epigenetic influences on reproductive development may arise through pheromones, maternally transmitted viruses, and nutrients. Finally, it is our experience that cohorts of mice from Jackson Laboratories can differ in estrous cycle frequency.[27] These variations could reflect sampling biases from developmental influences in cell number or fetal neighbors, or they could represent subtle effects from variations due to season, trace nutrients in commercial chow, or animal handling.

The experimentalist is thus confronted with identical genotypes that give rise during development to extensive phenotypic variations. Experiments to identify chromosomal loci contributing to the genetic components of these variations therefore utilize the distributions of values found in individuals as well as in the population sample. As an example, Figure 3 shows the modal

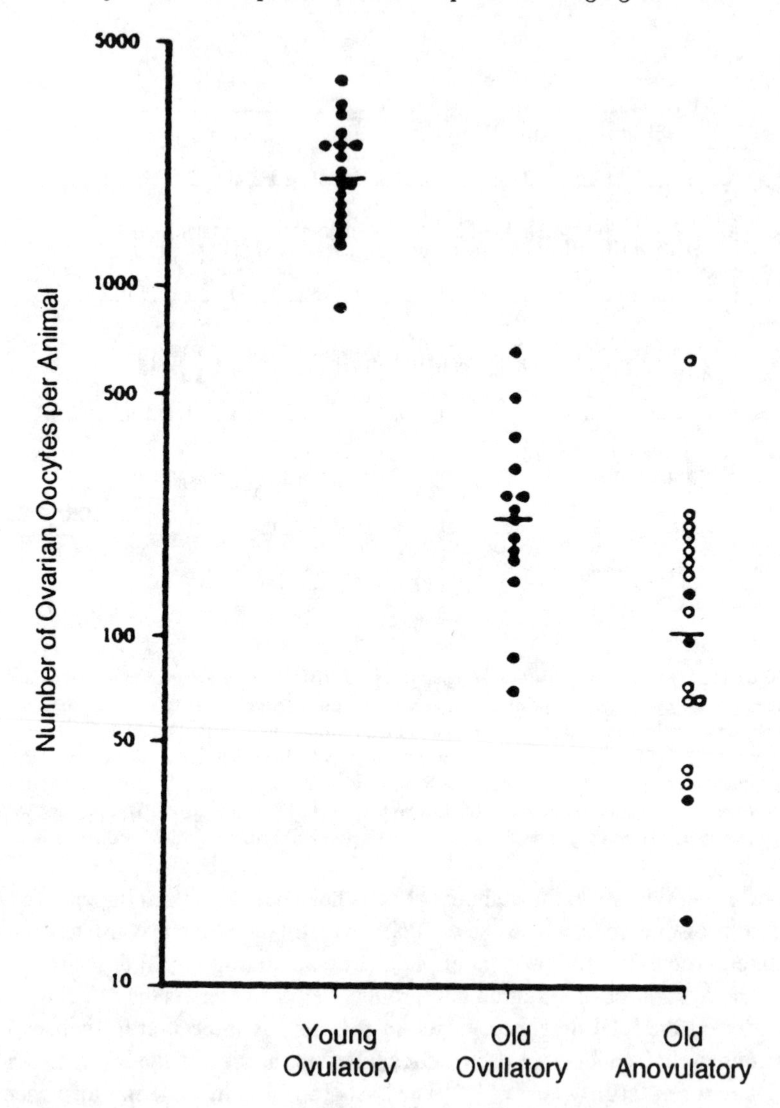

FIGURE 2. Oocyte numbers in C57BL/6J at the ages of 4–5 months (young) and 13–14 months (old). Closed circles represent mice that show cyclic changes in vaginal cytology, including irregular cycles. Open circles represent mice that are consistently acyclic. (Used with kind permission of *Biology of Reproduction;* from Gosden *et al.*[13])

frequency of estrous cycle lengths in young mice of common mouse strains. The figure shows that C3H mice have predominantly 4-day estrous cycles, whereas C57BL/6J mice have equal proportions of 4- and 5-day cycles, and the C57BL/10J strain has predominantly 5-day cycles.[24,29]

The main approach to resolving the genetic basis for quantitative variations in reproductive traits is the genetic crossing to make F1 hybrids, followed by back-crossing to follow the cosegregation of genetic markers with reproduc-

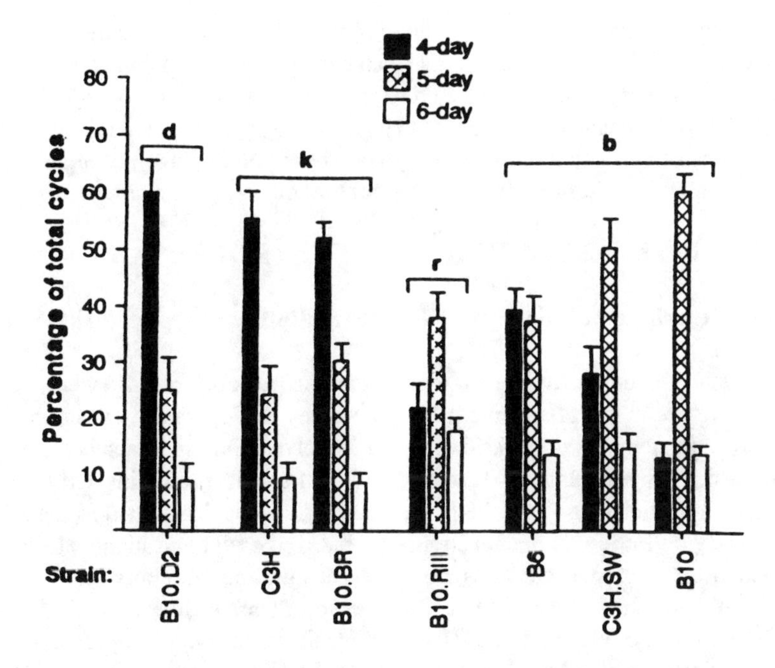

FIGURE 3. Mouse strains that show different estrous cycle length distributions as young adults; the small letters refer to the *H-2* haplotypes. (From Lerner *et al.*[24] with kind permission of Blackwell Scientific Publications Limited.)

tive traits. This is necessary because inbred mice differ at many loci. Thus, the CBA/HeJ and C57BL/6J strains have different alleles in 41% of catalogued genes.[a,11] Moreover, the mapped genes are still a small proportion of the 50,000–100,000 genes estimated to be present in any mammal.

We also note the difference between rare mutants and common genetic polymorphisms. In population genetics, genetic variations are less frequent than about one percent, are considered to be in a sporadic category of rare mutants, and, by their relative rarity, are less important in responses to selection than more common genes. While familial blood lipid disorders that promote early onset vascular disease are relatively rare (e.g., familial hypercholesterolemia is 0.1%), familial Alzheimer's disease may be tenfold higher.[9]

One hypothesis about the higher prevalence of some later-onset hereditary conditions is that they are co-selected because they enhance some feature of fitness. There is some evidence, for example, that Huntington's disease, an

---

[a] The designation of a strain, such as C57BL/6J (where J signifies that it was obtained directly from the Jackson Laboratory), does not inform about allelic differences within the numerous sublines available from other sources. For example, C57BL/6Nnia was derived 40 years ago from C57BL/6J mice and has been maintained at the NIH since 1951 (C57BL/6N). A further subline was developed under contract by the NIA at the Charles River Laboratories (C57BL/6Nnia). This issue is also pertinent in comparing studies on strains raised in different countries.

autosomal dominant with mid-life onset of neurodegeneration, confers greater fecundity.[2] Huntington's, thus, may be an example of antagonistic pleiotropy, in which a genotype may be selected on the basis of fitness in early adulthood, with delayed adverse consequences that are not strongly selected against. The frequency of genes that influence osteoporosis and other outcomes of aging is unknown and could fall into either the mutant or polymorphism category.

THE *H-2* LOCUS AND THE ESTROUS CYCLE

Brief Description of the Major Histocompatibility Complex (Mhc) and the *H-2*

The first example we give of genetic approaches concerns influences from alleles in the major histocompatibility complex (Mhc), which considerably narrows the numbers of genes that may be involved. One approach is to study strains differing in relatively few loci, such as the congenic strains derived by immunogeneticists to study the Mhc, which in mice is designated as the *H-2* locus. By definition, congenic strains are back-crossed to eliminate all but a small portion of one chromosome. Several hundred congenic strains are available, in which the *H-2* locus (the mouse's Mhc) was back-crossed into other strains. For example, the B10.Br congenic strain contains the *H-2* locus from C57Br/SgSn, which was inserted into the chromosome of the C57BL/10Sn by back-crossing; thus, B10.Br differ from B10 in having the *H-2* from the Br donors but with virtually all other chromosomal loci being identical.

The *H-2* locus contains about 100 genes within 2–3 centimorgans of map length (about 3,000,000 nt) that code for the classic class I and II transplantation antigens.[3,20] These sets of alleles from different donors are identified by a superscript in a lower case font (e.g., the B10 is $H\text{-}2^b$, the Br is $H\text{-}2^k$, and the B10.Br is $H\text{-}2^k$). These allele sets are collectively called the Mhc haplotype, a designation which does not specify which genes and alleles differ. The Mhc genes in most mammals examined contain an unusual number of genetic polymorphisms (allelic differences).[20]

Besides the influences of the Mhc alleles on graft acceptance through class I and II glycoproteins, the *H-2* locus in mice contains many loci that influence reproduction among other hormonal functions.[23] For example, *H-2* alleles quantitatively influence the production of testosterone by Leydig cells in young mice through interaction of the class I Mhc glycoproteins with the luteinizing hormone (LH) receptor.[36] Class I alleles also influence cAMP generated by insulin and glucagon receptors.[21] An Mhc I peptide also influences glucose transport and interacts with IGF-I receptors.[16]

Moreover, the progesterone receptor may be subject to Mhc influences through the gene for the 70-kDa heat shock protein (hsp70), which is located in the Mhc. The unliganded progesterone receptor interacts physically with hsp70 molecule, as well as with hsp90, which is not encoded within the Mhc.[33] It remains to be shown whether naturally occurring allelic variations in hsp70 alter its interactions with the progesterone receptor or with hsp90.

Other Mhc-associated genes with varying allelic differences code for enzymes that influence hormone metabolism (21-hydroxylase involved in steroid synthesis and the mixed function oxidase system; the latter maps outside of the Mhc proper). Thus, the highly polymorphic Mhc shows quantitative traits that influence reproduction, metabolism, and the recognition of antigens. The cluster of Mhc genes thus contains genetic variants with influences on many aspects of function that are subject to selection because of direct influences on fitness. Elsewhere, it is proposed that the Mhc serves as a *life history gene complex* (C. Finch and M. Rose, manuscript in preparation).

H-2 Haplotypes and Estrous Cycle Length

As shown in Figure 3, the frequency distribution of estrous cycles varies among congenic strains according to the *H-2* haplotype.[24] The $H\text{-}2^d$ and $H\text{-}2^k$ haplotypes appear to favor shorter cycles. These associations are further strengthened by the observation that the k haplotype has predominantly 4-day cycles on both the C57BL/10 and C3H backgrounds, which implies that the numerous allelic differences outside of the *H-2* have less influence on cycle frequency than those within the *H-2*. Many further studies can be done with *H-2* recombinants that differ in subregions of the *H-2* complex. As noted below, there are also influences of genes located outside of the *H-2*, since the C57BL/6J and C57BL/10J strains have different cycle frequencies, yet both are $H\text{-}2^b$ haplotypes.

An F1 Back-Cross Study

Back-crossing was used to study estrous cycle lengths, using convenient serological markers to identify the *H-2* haplotype. The *longer* cycle length of the parental strain was dominant in F1 hybrids (B10 × B10.Br).[24] Using classical genetics protocols, we then demonstrated that the *shorter* cycle length cosegregated with the $H\text{-}2^k$ haplotype in back-crosses of the F1 hybrids with parental strains; the $H\text{-}2^b$ homozygotes and $H\text{-}2^b/H\text{-}2^k$ heterozygotes both had longer cycles (Figure 4).

INFLUENCES ON THE ESTROUS CYCLE FROM NON-*H-2* LOCI

Alleles outside of the *H-2* complex may also influence cycle length. For example, B6 and B10 are both $H\text{-}2^b$ yet differ in cycle length and frequency, and showed dominance of longer cycles in F1 hybrids.[24] The involvement of numerous genes in quantitatively varying reproductive traits is consistent with findings of the dominance of *shorter* cycles in the F1 hybrid of DBA/2J × C57BL/6J (Nelson *et al.* [29] in contrast to Lerner *et al.*;[24] the parental strains were $H\text{-}2^d$ and $H\text{-}2^b$, respectively).

Other crosses of these strains with C3H/HeJ ($H\text{-}2^k$) distinguish allelic influences on cycle regularity and frequency, as well as on cycle length. For example, whereas the frequency of regular cycles of C57BL/6J mice was three

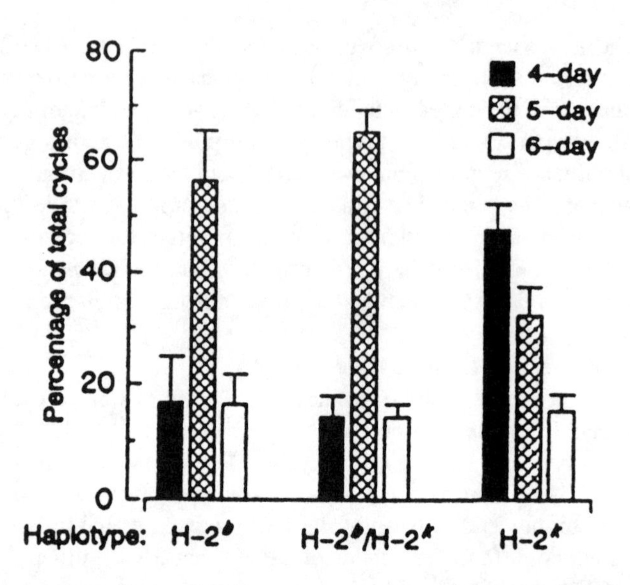

FIGURE 4. Genetics of estrous cycle lengths, as revealed directly by back-crossing of *H-2* genotypes that had different modal cycle lengths as young mice. The F1 hybrids of *H-2*^{b} and *H-2*^{k} showed dominance of 5-day cycle lengths. These mice were then back-crossed to the parental strains and the *H-2* haplotype was determined from blood proteins with antisera that identify the *H-2* haplotype. Only *H-2*^{k} haplotypes had a modal frequency of 4-day cycle lengths. The b/k heterozygotes were 5-day modal, as observed in the F1 hybrids. Thus, the influences of alleles in the *H-2* locus are stronger in determining the estrous cycle lengths than those on other chromosomes, which are randomly assorted in the F2 generation. (Reprinted from Lerner *et al.*[24] with the kind permission of Blackwell Scientific Publications Limited.)

times higher than that of DBA/2J and C3H/HeJ mice, average cycle length of C57BL/6J mice was longer, not shorter, than that of the two other strains.[29] In the absence of differences in irregular or acyclic intervals, cycle frequency should be inversely related to *cycle length*. The lack of this inverse relationship indicates that the more frequent irregular and acyclic intervals must account for the reduced frequency of regular cycles in the DBA and C3H strains, a result that was observed. The results also indicate that at least some of the genes specifying cycle length differ from those specifying cycle frequency.

Further evidence for differential gene regulation of cycle length and cycle frequency was indicated by studies of F1 hybrids.[29] In the DBA/C57 crosses, genetic variance for cycle frequency was additive; however, cycle length was overdominant in these crosses. Also, in the C57/C3H crosses, genetic variance for cycle frequency showed dominance of the C57 complement, but dominance of the C3H complement was shown for cycle length. Further analysis of non-*H-2* genes could use the newly available strategy of DNA microsatellites that are used as RFLP's for small genomic regions.

REPRODUCTIVE AGING

Associations with *H-2* Alleles

Fertility Patterns During Aging

Influences of Mhc alleles on aging were suggested by *H-2* congenic differences in life span[34] and immunological aging.[25] In an exploratory study, *H-2* congenics on the B10 background showed modest differences in the age at last litter, but larger differences in the numbers of litters, with an ordering that paralleled their life spans: B10.F<B10<B10.Br<B10.RIII[22] (Figure 5). B10.F mice failed reproductively at the earliest age and were the shortest lived. This

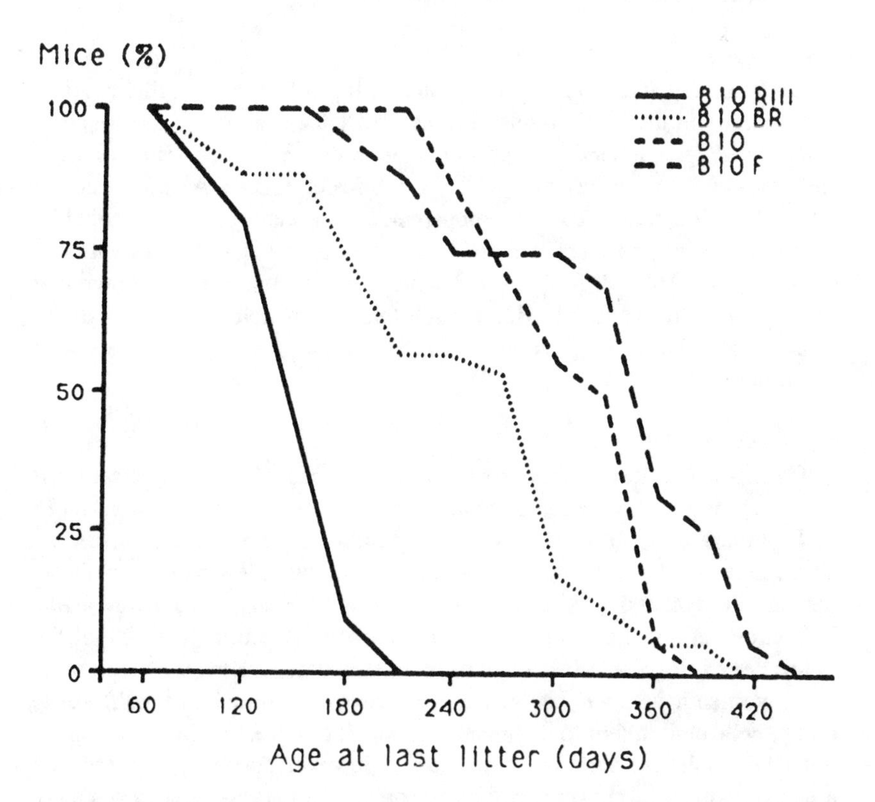

FIGURE 5. Genetics of reproductive aging, as shown by *H-2* congenic strains that differed in the *H-2* allele set as introduced by back-crossing into the B10 strain. The relative order of age at last litter among these strains corresponds to their life spans. The mice were mated continuously. (Used with kind permission of *Biology of Reproduction;* from Lerner *et al.*[22])

strain is atypical in its early onset of leukemia and coat graying that are due to a xenotropic murine leukemia virus.[11,43]

The RIII strain shows a greater fecundity at early ages. In contrast to results from selection of *Drosophila* (Section II), the RIII strain showed greater fecundity at earlier ages *and* at later ages.[22] RIII mice were selected for their low incidence of cancer as well as passively for their fecundity (David Harrison, personal communication). These selection criteria cannot be compared to other protocols that were designed to select for fertility at specific ages. In summary, many genotypic strains can be assembled under conditions that would not have occurred in nature. It must still be shown that these differences segregate genetically with *H-2* haplotypes during back-crossing. Such tests have not yet been applied to issues of aging, but are made practical by easy identification of *H-2* haplotypes.

Oocyte Loss

Ovarian follicular pools and the rates of loss of primary follicles differ substantially in different strains of mice.[8,18] Of the few strains examined in detail, CBA mice have the fastest oocyte loss, which is consistent with their earlier loss of fertility (Figure 6). The F1 hybrids of (CBA × A) have a slower rate of loss, like that of the A strain parent, a dominance effect that could be interpreted as hybrid vigor. Several of the mouse strains with different rates of oocyte loss[18] also differed in *H-2* haplotypes. It remains to be shown that *H-2* are specifically involved, although this is plausible in view of allelic influences on the class I glycoprotein interactions with gonadotropin receptors mentioned above.

Osteopenia

Osteopenia, or loss of bone from any cause, appears to be a universal trend during aging in both men and women that is accelerated by menopause. In both genders, diet, physical activity, and smoking have important influences. There may also be important genetic factors, although few hereditary disorders are known, and they are rare in mice[15] and humans.[26] There are a few indications of genetic influences on normal bone structure. The size of the radius bone is more similar in monozygous than dizygous twins.[35] Moreover, black women have fewer age-related fractures than Caucasians,[5] a difference that may be consequent to the greater thickness of bones in black women, as judged from the vertebral size at puberty.[12] However, specific genetic factors are unknown. A strong genetic influence on osteopenia was recently reported. In Caucasian women, common alleles of the vitamin D receptor influence the amount of bone present in young adults and hence the reserve available before the critical threshold for spontaneous fractures is reached during age-related osteopenia.[26a]

In exploratory studies, Finch's laboratory examined some of the inbred strains that showed differences in reproductive aging for possible genetic

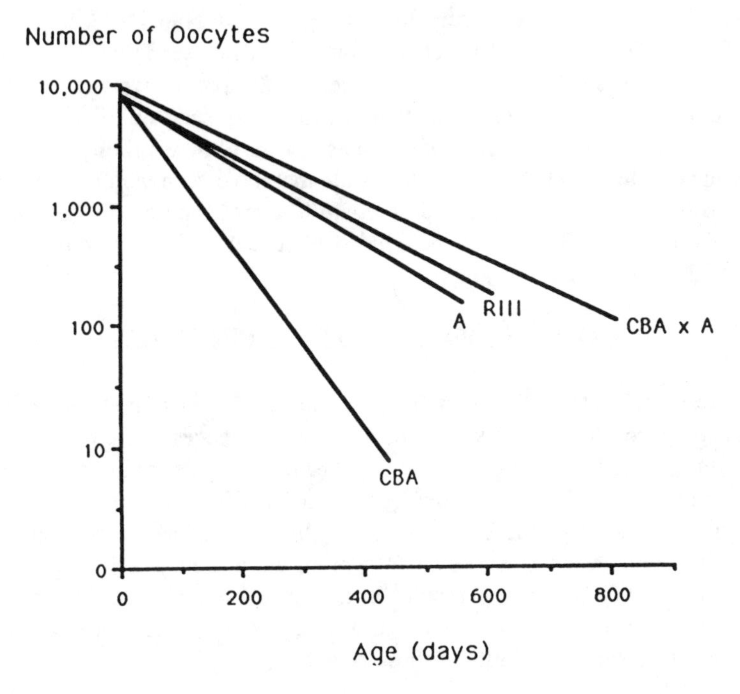

FIGURE 6. Mouse strain differences in the rates of ovarian oocyte loss with aging. These strains have different *H-2* haplotypes: A, *H-2*[a]; CBA, *H-2*[a]; RIII, *H-2*[r]. However, there are also numerous allelic differences at other locations throughout the chromosomes that could cause these strain differences in rates of oocyte loss. Study of congenic strains (as in Figures 4 and 5) could be useful. (Redrawn from Jones and Krohn[18] [Figure 6.11 in Finch[11]] with publisher's permission, all rights reserved, University of Chicago Press.)

influences on bone density during aging and responses to ovariectomy. In brief, effects of *H-2* haplotype were observed. For example, the cross-sectional area of the femur occupied by solid bone was about 10% less in B10 mice than in B10.Br or B10.RIII. In all strains, ovariectomy caused about 10% decrease in the relative amount of solid bone that was maintained during aging (unpublished observations).

These observations indicate potential effects from the *H-2* locus that serve as precedents for endocrine-mediated effects on osteopenia in humans. Even a small difference in bone density of young individuals can have large consequences at later ages by altering the reserve strength above a threshold for spontaneous fractures. The Mhc in humans is already implicated in a maturity onset disease, ankylosing spondelytis, which is associated with particular Mhc haplotypes.[1] Further pursuit of these studies is hoped for, despite doubts by many bone researchers about the value of rodent models for ovariprival features of osteopenia and osteoporosis.[17]

Possible Associations of Reproductive Aging with Non-*H-2* Loci

The emphasis in this brief chapter on *H-2* influences is not intended to slight the other loci that may influence details of reproductive aging and that we anticipate will eventually be mapped in rodents and other species. Rat strains have marked differences in the age of onset of acyclicity and in the vaginal cytologies of the postcycling state that have been tabulated.[40] These strains have not been considered in regard to their Mhc (the RT1 complex). Relatively little effort might be needed to reveal major new clues to the genetics of reproductive aging.

CONCLUSIONS AND PERSPECTIVES

These studies from the laboratories of the authors show the application of genetic approaches to analyzing quantitative differences in the reproductive schedule, including reproductive aging. The role of genetics in breast cancer is already established and is implied by racial differences in osteoporotic fractures. In view of the great emphasis placed on genetic risk factors for Alzheimer's disease, it would seem appropriate to now consider a focused effort to identify genetic risk factors for age-related disorders of the reproductive tract in men as well as in women. We hope that the examples presented above give a sense of reality to these possibilities.

ACKNOWLEDGMENTS

The experimental work summarized herein was supported by grants to C.E. Finch from the NIA (AG04419; AG7909) and to J.F. Nelson from the University of Texas Health Science Center at San Antonio, Texas.

REFERENCES

1. **Ahearn, J.M., Calomiris, J.J., Wigley, F.M., Jabs, D.A., Bias, W.B., and Hochberg, M.C.,** Characterization of the class I HLA 9:2-kb PVU II restriction fragment length polymorphism: Linkage to HLA-A and lack of disease association, *Arthritis Rheum.*, 33, 870, 1989.
2. **Albin, R.,** Antagonistic pleiotropy, mutation accumulation, and human genetic disease, *Genetica*, 91, 287, 1993.
3. **Campbell, R.D., and Troswdale, J.,** Map of the human MHC, *Immunol. Today*, 14, 349, 1993.
4. **Charlesworth, B.,** *Evolution in Age-Structured Populations*, Cambridge University Press, Cambridge, 1980.
5. **Commings, S.R., Kelsey, J.L., Nevitt, N.C., and O'Dowd, K.J.,** Epidemiology of osteoporosis and osteoporotic fractures, *Epidemiol. Rev.*, 7, 178, 1985.
6. **Davidson, E.H.,** Spatial mechanisms of gene regulation in metazoan embryos, *Development*, 113, 1, 1991.
7. **DeSalle, R., Slightom, J., and Zimmer, E.,** The molecular through ecological genetics of abnormal abdomen. 2. Ribosomal DNA polymorphism is associated with the abnormal abdomen syndrome in *Drosophila mercatorum*, *Genetics*, 112, 861, 1986.

8. **Faddy, J.M., Jones, E.C., and Edwards, R.G.,** An analytical model for ovarian follicle dynamics, *Exp. Zool.*, 197, 173, 1983.
9. **Farrer, L.A., Myers, R.H., Cupples, L.A., St. George-Hyslop, P., Bird, T.D., Rossor, M.N., Mullan, M.J., Polinsky, R., Nee, L., Heston, L., Van Broeckhoven, C., Martin, J.-J., Crapper-McLachlan, D., and Growdon, J.H.,** Transmission and age at onset patterns in familial Alzheimer's disease: Evidence for heterogeneity, *Neurology*, 40, 395, 1990.
10. **Finch, C.E., Felicio, L.S., Flurkey, K., Gee, D.M., Mobbs, C., Nelson, J.F., and Osterburg, H.H.,** Studies on ovarian-hypothalamic-pituitary interactions during reproductive aging in C57BL/6J mice, in *Brain-Endocrine Interaction, IV: Neuropeptides in Development and Aging (Peptides) Vol. 1, Suppl. 1,* Scott, D., and Sladek, J.L., Jr., Eds., ANKHO Int., Fayetteville, NY, 1980, 163.
11. **Finch, C.E.,** *Longevity, Senescence, and the Genome*, University of Chicago Press, Chicago, 1990.
12. **Gilsanz, V., Roe, T.F., Mora, S., Costin, G., and Goodman, W.G.,** Changes in vertebral bone density in black girls and white girls during childhood and puberty, *N. Engl. J. Med.*, 325, 1597, 1991.
13. **Gosden, R.G., Laing, S.C., Felicio, L.S., Nelson, J.F., and Finch, C.E.,** Imminent oocyte exhaustion and reduced follicular recruitment mark the transition to acyclicity in aging C57BL/6J mice, *Biol. Reprod.*, 28, 255, 1983.
14. **Gosden, R.G.,** *The Biology of Menopause: The Cause and Consequences of Ovarian Aging*, Academic Press, New York, 1985.
15. **Green, M.C.,** *Genetic Variants and Strains of the Laboratory Mouse,* Gustaf Fischer Verlag, New York, 1981.
16. **Hsu, D., and Olefsky, J.M.,** Effect of an MHC Class I peptide on ILGF-I receptor internalization and biological signaling, *Endocrinology*, 133, 1247, 1993.
17. **Jee, W.S.S.,** The aged rat model for bone biology studies, in *Cells and Materials Supplement 1*, Scanning Microscopy International, Chicago, AMF O'Hare, 1991, 1.
18. **Jones, E.C., and Krohn, P.L.,** The relationships between age, numbers of oocytes, and fertility in virgin and multiparous mice, *J. Endocrinol.*, 21, 469, 1961.
19. **King, M.-C.,** Linkage of early-onset familial breast cancer to chromosome 17q21, *Science*, 250, 1684, 1990.
20. **Klein, J.,** *Natural History of the Major Histocompatibility Complex*, John Wiley and Sons, New York, 1986.
21. **Lafuse, W., and Edidin, M.,** Influence of the mouse major histocompatibility complex, *H-2*, on liver adenylate cyclase activity and on glucagon binding to liver cell membranes, *Biochemistry*, 19, 49, 1980.
22. **Lerner, S.P., Anderson, C.P., Walford, R.L., and Finch, C.E.,** Genotypic influences on reproductive aging of inbred female mice: Effects of *H-2* and non-*H-2* alleles, *Biol. Reprod.*, 38, 1035, 1988.
23. **Lerner, S.P., and Finch, C.E.,** The major histocompatibility complex and reproductive functions, *Endocr. Rev.*, 112, 78, 1991.
24. **Lerner, S.P., Anderson, C.P., Harrison, D., Walford, R.L., and Finch, C.E.,** Polygenic influences on the length of oestrous cycles in mice involve alleles, *Int. J. Immunogenet.*, 19, 361, 1992.
25. **Meredith, P.J., and Walford, R.L.,** Effect of age on response to T-and B-cell mitogens in mice congenic at the *H-2* locus, *Immunogenetics*, 5, 109, 1977.
26. **McKusick, V.A.,** *Mendelian Inheritance in Man, 7th edition*, Johns Hopkins University Press, Baltimore, 1986.
26a. **Morrison, N.A., Qi, J.C., Tokita, A., Kelly, P.J., Crofts, L., Nguyen, T.V., Sambrook, P.N., and Eisman, J.A.,** Prediction of bone density from vitamin D receptor alleles, *Nature*, 367, 284, 1994.
27. **Nelson, J.F., Felicio, L.S., Randall, P.K., Sims, C., and Finch, C.E.,** A longitudinal study of estrous cyclicity in aging C57BL/6J mice: 1. Cycle frequency, length and vaginal cytology, *Biol. Reprod.*, 27, 327, 1982.

28. **Nelson, J.F., and Felicio, L.S.,** Reproductive aging in the female: An etiological perspective, *Rev. Biol. Res. Aging,* 2, 251, 1985.
29. **Nelson, J.F., Karelus, K., Felicio, L.S., and Johnson, T.E.,** Genetic influences on oestrous cyclicity in mice: Evidence that cycle length and frequency are differentially regulated, *J. Reprod. Fertil.,* 94, 261, 1992.
30. **Reznick, D.N., Bryga, H., and Endler, J.A.,** Experimentally induced life-history evolution in a natural population, *Nature,* 346, 257, 1990.
31. **Reznick, D.N.,** Life history evolution in guppies (*Poecilia reticulata*): Guppies as a model for studying the evolutionary biology of aging, *Exp. Gerontol.,* in press.
32. **Rose, M.R.,** *Evolutionary Biology of Aging,* Oxford University Press, New York, 1991.
33. **Smith, D.F., Sternsgard, B.A., Welch, W.J., and Toft, D.O.,** Assembly of progesterone receptor with heat shock proteins and receptor activation are ATP mediated events, *J. Biol. Chem.,* 267, 1350, 1992.
34. **Smith, G.S., and Walford, R.L.,** Influence of the main histocompatibility complex on aging in mice, *Nature,* 270, 727, 1977.
35. **Smith, G.S., Walford, R.L., and Mickey, M.R.,** Lifespan and incidence of cancer and other diseases in selected long-lived inbred mice and the F_1 hybrids, *J. Natl. Cancer Inst.,* 50, 1195, 1973.
36. **Solano, A.R., Sanchez, M.L., Sardonons, M.L., Dada, L., and Podesta, E.J.,** Luteinizing hormone triggers a molecular association between its receptor and the major histocompatibility complex class I antigen to produce cell activation, *Endocrinology,* 122, 2080, 1988.
37. **Templeton, A.R., Hollocher, H., Lawler, S., and Johnston, J.S.,** The ecological genetics of abnormal abdomen in *Drosophila melanogaster,* in *Ecological and Evolutionary Genetics of Drosophila,* Barker, J.S.F., and Starmer, W.T., Eds., Academic Press, San Diego, 1990.
38. **vom Saal, F.S., and Bronson, F.H.,** Sexual characteristics of adult female mice are correlated with their blood testosterone levels during prenatal development, *Science,* 208, 597, 1980.
39. **vom Saal, F.S., and Moyer, C.L.,** Prenatal effects on reproductive capacity during aging in female mice, *Biol. Reprod.,* 32, 1111, 1985.
40. **vom Saal, F.S., Nelson, J.F., and Finch, C.E.,** The natural history of reproductive humans, laboratory rodents, and selected other vertebrates, in *The Physiology of Reproduction,* Knobil, E., and Neill, J.D., Eds., Raven Press, New York, 1994, 1213.
41. **Williams, G.C.,** Pleiotropy, natural selection, and the evolution of senescence, *Evolution,* 11, 398, 1957.
42. **Winklbauer, R., and Hausen, P.,** Development of the lateral line system in *Xenopus laevis.* 2. Cell multiplication and organ formation in the supraorbital system, *J. Embryol. Exp. Morphol.,* 76, 283, 1983.
43. **Yetter, R.A., Hartley, J.W., and Morse, H.C.,** *H-2*-linked regulation of xenotropic murine leukemia virus expression, *Proc. Natl. Acad. Sci. USA,* 80, 505, 1983.

Chapter 6

NEUROENDOCRINE CONTROL OF REPRODUCTION IN AGING RATS AND HUMANS

J. Meites

TABLE OF CONTENTS

0-8493-2451-3/95/$0.00+.50

ABBREVIATIONS

CA–catecholamine
DA–dopamine
FSH–follicle-stimulating hormone
GnRH–gonadotropin-releasing hormone
LH–luteinizing hormone
MAO–monoamine oxidase
NE–norepinephrine
PRL–prolactin

INTRODUCTION

Aging can be defined as a progressive decline in body functions with time, associated with a decreased ability to maintain homeostasis. The decline in body functions begins well before the first half of the maximum life span in animals and man. Shock[16] reported that in humans, a progressive decrease occurs from 30 to 80 years of age in basal metabolism, cardiovascular function, vital capacity, nerve conduction velocity, renal function, and muscle work capacity. Immune and other body functions also decline with age. Similar decreases have been reported in animals. This chapter will compare the decline in reproductive functions with age in rats and man, and it will describe the neuroendocrine mechanisms largely responsible for these changes.

REPRODUCTIVE DECLINE

Rats

Female rats begin to exhibit estrous cycles of 4 or 5 days' duration at 35–40 days of age. Spermatogenesis is observed in male rats at 60–70 days of age. At about 6–7 months of age, female rats begin to show a decline in fertility as evidenced by a decrease in the number of pups born during successive pregnancies. At about 7–10 months of age, the cycles become irregular and lengthened due to an increase in the estrous phase of the cycle.[6,11] The proestrous surge of luteinizing hormone (LH) in these irregularly cycling rats does not reach the amplitude seen in younger rats.[8] At 10–15 months of age, most rats cease to undergo estrous cycles and enter into a persistent estrous state characterized by the presence of well-developed ovarian follicles but no ovulations. These rats exhibit constant sexual receptivity to males. By 15–20 months of age, some rats ovulate and develop corpora lutea that secrete progesterone for lengthy periods of time. In the final stage of life, between 2–3 years of age, most rats become anestrus, and their ovaries are shrunken and contain only small undeveloped follicles with little evidence of estrogen secretion as indicated by the atrophic reproductive tract. Many of these anestrous rats have prolactin (PRL)-secreting pituitary tumors (prolacinomas) and many mammary tumors.[6,11] Male rats exhibit a decrease in testosterone secretion and in spermatogenesis with age.

What are the causes for the decrease in fertility and loss of cycling in aging rats? All components of the neuroendocrine system show a decline in the aging rat. The cyclic surge in release of gonadotropin-releasing hormone (GnRH) from the hypothalamus every 4 or 5 days ceases; the cyclic elevation in secretion of gonadotropins every 4 or 5 days by the pituitary is absent; and the ovaries receive less stimulation from pituitary gonadotropins. The pituitary also becomes less responsive to GnRH and the gonads to gonadotropic hormone stimulation. This may be due to faults that develop in receptors or postreceptor mechanisms. However, the major cause for the reproductive decline in rats is due to dysfunctions that develop in the hypothalamus. It has been demonstrated that when the ovaries from old, non-cycling rats are transplanted to young ovariectomized rats, most of the young rats resumed cycling. However, when the ovaries from young rats were transplanted to old ovariectomized rats, cycling did not resume.[1] Also, when the pituitary of old rats was transplanted underneath the median eminence of young hypophysectomized rats, many of the young rats resumed cycling.[14] Thus neither the ovaries nor pituitary of old rats is primarily responsible for loss of cycling.

Electrical simulation of the preoptic area, an area essential for GnRH release, induced ovulation in old, constant-estrous rats. Injection of a large dose of epinephrine-in-oil also was effective.[2] The observation that preoptic stimulation produced ovulation suggested that sufficient GnRH was present in the hypothalamus of old rats to induce ovulation but that the stimulus for its release was lacking. The demonstration that epinephrine was also effective suggested that the missing stimulus might be a catecholamine (CA). Since norepinephrine (NE) was known to promote LH release in rats, drugs that elevated brain NE levels were administered to old, constant-estrous rats, and these induced LH release and resumption of estrous cycles.[10,15,18] Effective drugs included L-dopa, the precursor of the CAs, and iproniazid, an inhibitor of monoamine oxidase (MAO) which is the major enzyme that catabolizes CAs.

When the two principal brain CAs were assayed in the hypothalamus of old and young rats of both sexes, both NE and dopamine (DA) were found to be significantly lower in old than in young rats.[10,12] Several causes are believed to account for the decrease in hypothalamic CAs in aging rats. During metabolism of CAs by MAO, "free radicals" (hydrogen peroxide, superoxide anions, hydroxyl radicals, highly reactive quinones) are formed that damage neurons in the hypothalamus and other brain areas. This is believed to account in part for the significant decrease of neurons observed in the arcuate nucleus, medial preoptic area, and ventromedial and lateral hypothalamus of old rats. There is also evidence that in old rats, MAO is increased in different brain areas, including the hypothalamus of old rats, whereas tyrosine hydroxylase, the rate-limiting enzyme for synthesis of CAs, is decreased. In addition, the action of estrogen over long periods of time has been shown to damage hypothalamic neurons. This is believed to account for the extension of cycling in rats after ovariectomy early in life, followed by grafting of fresh ovaries many months later.[1] Ovariectomy prevents damage to hypothalamic neurons by estrogen. This probably also accounts for the significantly longer period of estrous cycles in multiparous than in virgin rats. The action of progesterone during each pregnancy is believed to protect hypothalamic neurons from damage by estrogen.[7,8] Administration of progesterone to virgin rats was shown to extend the period of cycling to the same life period as in multiparous rats. It is also possible that unrestrained estrogen action over a long period of time may damage ovarian follicles.

Other neurotransmitters in the hypothalamus in addition to the CAs may be involved in the reproductive decline. Acetylcholine, brain opiates, and other neurotransmitters can influence release of the GnRH, and secretion of these may be altered with age. However, little is known of their possible influence at present.

In aging male rats, the reduction of GnRH and gonadotropin secretion leads to a decrease in testosterone secretion and spermatogenesis. There is also decreased responsiveness of the pituitary to GnRH and of the testes to gonadotropin stimulation.[4,10,12] Spermatogenic arrest, tubular degeneration, and folding of the basement membrane were observed in some areas of the testes in aging rats. A decline in sexual interest also occurs in old male rats when placed together with young females, but this is only partially related to the decline in testosterone secretion.[3] Other factors, including health, nutrition, genetic expression, and early sexual experiences, can influence sexual behavior in old animals (and humans). Healthy male rats may remain fertile well into old age.

Humans

Women

The menopause in women in the United States occurs at about 50 years of age and is characterized by loss of cyclic ovarian function. The few follicles that remain in the ovaries at menopause are usually not responsive to gonadotropic hormone stimulation, and they disappear entirely in the postmenopausal period. About 800,000 follicles may be present in the ovaries at birth, and these are progressively reduced to about 4,000 at menopause as a result of atresia, gonadotropic hormone stimulation, and perhaps as a result of chronic estrogen action.[8,11] Estrogen secretion begins to decrease several years prior to menopause mainly due to loss of follicles. This results in shortened, irregular menstrual cycles and a resultant elevation in gonadotropic hormone secretion, particularly follicle-stimulating hormone (FSH). The rise in FSH is probably augmented by a decrease in inhibin secretion by the ovaries, since inhibin normally inhibits FSH secretion.

In the postmenopausal period, secretion of FSH and LH is elevated further as a result of the continuing decrease in estrogen secretion. Whereas most of the estrogen prior to the menopause is in the form of estradiol, the small amount of estrogen in the postmenopausal state is in the biologically less active form of estrone, which is derived mainly from conversion of adrenal androstenedione. When a woman reaches approximately 80 years of age, there is a reduction in the amounts of FSH and LH secreted by the pituitary, probably due to decreased release of GnRH from the hypothalamus.

It is clear that in contrast to the rat, the decline and ultimate cessation of menstrual cycles in women is due primarily to ovarian failure, to loss of ovarian follicles and to the decrease in estrogen secretion rather than to defects in hypothalamic or pituitary function. This does not mean, however, that there are no changes in hypothalamic and pituitary function with age in women. It has been reported that the doses of estrogen required to reduce gonadotropin secretion in postmenopausal women is much greater than in young castrated women, probably reflecting a change in hypothalamic and/or pituitary function. There is also evidence that the pituitary gonadotropic response to GnRH administration is lower in postmenopausal than in young cycling women.[4] Hypothalamic CAs have been reported to decline with age both in men and women,[5] but it is unknown whether this influences hypothalamic and pituitary hormone secretion in humans. The decline in hypothalamic NE activity appears to be due mainly to loss (about 25%) of neurons in the locus coeruleus, which is the major source of NE in the hypothalamus. However, the role of NE in control of GnRH release in the primate hypothalamus remains to be established.

Men

In aging men, as in aging women, the primary cause of decline in reproductive function is due to defects that develop in the gonads rather than in the hypothalamus or pituitary. There is a decrease in testosterone secretion by the testes beginning at 40–50 years of age, and this results in a rise in FSH and LH secretion, which is not nearly as great as in postmenopausal women. The reduction in testicular inhibin secretion also contributes to the rise in FSH. The decrease in testosterone secretion is associated with loss of Leydig cells (which secrete testosterone). This has been attributed to atherosclerotic changes in blood vessels in the testes. With the loss of Leydig cells, fewer cells are available to respond to gonadotropic hormone stimulation.[11]

A reduction in spermatogenesis also has been reported in aging men.[13] Sperm were found to decrease from 68.5% in the sixth decade to 48% in the eighth decade of life. In another study, areas of normal spermatogenesis were observed in more than 70% of men over 70 years of age. A decline in sexual interest and performance also occurs with age that is only partly related to the decrease in testosterone secretion.[3] Sexual interest and performance are also influenced by health, medications, hormonal deficiencies, stress, mental disorders, alcohol intake, etc. Despite the downward trend in testicular function with age, healthy men may continue to remain fertile and reproduce even into old age. Records exist of men in their 80s and 90s who have produced offspring.[4]

Although little is yet known of hypothalamic changes with age in humans, there are several indications of decreased capacity to secrete hormones related to reproduction. Thus there is loss of a circadian rhythm in testosterone secretion and blunting of the morning peak of testosterone secretion, suggesting the appearance of defects in the hypothalamic mechanisms regulating rhythms.[13] Many rhythmic activities are lost or blunted with age. There is also reduction in hypothalamic CAs,[5] but again the relationship of this decrease to the decline in reproductive functions is not clear. It has been reported that the pituitary response to GnRH progressively decreases with age,[4] suggesting that defects develop in receptors or postreceptor mechanisms in gonadotropic cells. Some cases of hypogonadism in elderly men may have their origin in the hypothalamus, pituitary, or both.[13]

PROLACTIN

Rats

In female rats, PRL secretion shows a rise beginning at about the time of the constant estrous period and continues to rise thereafter.[1,12] This is due mainly to the reduction in DA activity in the hypothalamus, since DA is well established as the principal inhibitor of PRL secretion. Unrestrained estrogen action during the constant estrous state also promotes PRL secretion, both via the hypothalamus and by a direct action on the pituitary. Primarily as a result of the decline in DA activity but also because of the stimulatory influence of estrogen on PRL secretion, many PRL-secreting pituitary and mammary tumors develop in old rats. PRL is believed to be the most important hormone involved in development of mammary tumors in rats,[17] although estrogen also has a role. In old Sprague-Dawley female rats, the mammary tumor incidence may reach 80%, whereas about 40% of Long-Evans female rats develop mammary tumors. Most of these are benign fibroadenomas. Most pituitary and mammary tumors in female rats are seen between 2 and 3 years of age when estrogen secretion is low (the anestrous stage), indicating that the decline in hypothalamic DA activity is mainly responsible for development of these tumors. Administration of dopaminergic drugs such as bromocryptin, other ergot drugs, or drugs that elevate hypothalamic DA activity (l-dopa, iproniazid, deprenyl) to old female rats has induced regression of both pituitary and mammary tumors.[10,12] Chronic administration of bromocryptin to C_3H female mice, in whom the incidence of spontaneous adenocarcinomas normally reaches 80–90%, prevented the occurrence of mammary cancers in these mice.[17]

Humans

PRL secretion does not show any significant change in women approaching menopause, but in the postmenopausal period there is a decline in PRL secretion due to the marked reduction in estrogen secretion. PRL has not been demonstrated to have a prominent role in development of breast cancer in women. PRL-secreting microadenomas of the pituitary are common in both men and women of all ages, but these have not been shown to be related to hypothalamic DA activity. Ergot drugs are widely used to induce regression of PRL-secreting pituitary tumors in man.[9]

SUMMARY

There are profound differences in the causes of the reproductive decline in the rat as compared to humans. In the rat, defects that develop in the hypothalamus are primarily responsible for loss of cycling in females and decrease in testosterone secretion in males. The principal hypothalamic defect appears to be the decrease in NE activity, which results in reduced release of GnRH, leading to lower pituitary gonadotropin secretion and depressed gonadal function. Pituitary responsiveness to GnRH stimulation and gonadal responsiveness to gonadotropic action are also reduced. The progressive decline in hypothalamic DA activity results in elevation of PRL secretion, leading to development of numerous mammary and PRL-secreting pituitary tumors. Administration of drugs that raise CA levels can reinitiate estrous cycles and induce regression of mammary and pituitary tumors in old rats.

In humans, defects that develop in the gonads are primarily responsible for the decline in reproductive functions, although defects may also occur in the hypothalamus and pituitary. In women approaching the menopause and in the postmenopausal state, estrogen secretion decreases due to loss of ovarian follicles and reduced responsiveness to gonadotropic hormones. The decrease in estrogen leads to elevated FSH and LH secretion. PRL secretion is reduced in the postmenopausal period due to reduced estrogen secretion. In aging males, testosterone secretion declines primarily due to loss of Leydig cells, and this results in elevated gonadotropin secretion. The greater increase in FSH than in LH probably is due to reduced inhibin secretion. There is some reduction in spermatogenesis, although healthy males may remain fertile well into old age. No significant change occurs in PRL secretion with age.

REFERENCES

1. **Aschheim, P.,** Aging in the hypothalamic-hypophyseal ovarian axis in the rat, in *Hypothalamus, Pituitary, and Aging,* Everitt, A.V., and Burgess, J.A., Eds., Charles C Thomas, Springfield, IL, 1976, 376.
2. **Clemens, J.A., Amenomori, Y., Jenkins, T., and Meites, J.,** Effects of hypothalamic stimulation, hormones and drugs on ovarian function in old female rats, *Proc. Soc. Exp. Biol. Med.,* 132, 561, 1969.
3. **Davidson, J.M., Gray, G.D., and Smith, E.R.,** The sexual psychoneuroendocrinology of aging, in *Neuroendocrinology of Aging,* Meites, J., Ed., Plenum Press, New York, 1983, 221.
4. **Harman, S.M., and Talbert, G.B.,** Reproductive aging, in *Handbook of the Biology of Aging (2),* Finch, C.E., and Schneider, E.L., Eds., Van Nostrand Reinhold, New York, 1985, 457.
5. **Hornykiewicz, O.,** Neurotransmitter changes in human brain during aging, in *Modification of Cell to Cell Signals During Normal and Pathological Aging,* Stefano, S., and Battaini, F., Eds., Springer Verlag, Berlin-Heidelberg, 1987, 169.
6. **Huang, H.H., and Meites, J.,** Reproductive capacity of aging female rats, *Neuroendocrinology,* 17, 289, 1975
7. **Lu, J.K.H., LaPolt, P.S., Nass, T.E., Matt, D.W., and Judd, H.L.,** Relation of circulating estradiol and progesterone to gonadotropin secretion and estrous cyclicity in aging female rats, *Endocrinology,* 116, 1953, 1985.
8. **Lu, J.K.H., Matt, D.W., and LaPolt, P.S.,** Modulatory effects of estrogens and progestins on female reproductive aging, in *Ovarian Secretion and Cardiovascular and Neurological Function,* Naftolin, F., and DeCherney, A.H., Raven Press, New York, 1990, 297.
9. **Martin, J.B., and Reichlin, S.,** *Clinical Neuroendocrinology (2),* F.A. Davis Co., Philadelphia, 1987, 7.
10. **Meites, J.,** Aging: Hypothalamic catecholamines, neuroendocrine-immune interactions, and dietary restriction, *Proc. Soc. Exp. Biol. Med.,* 195, 304, 1990.
11. **Meites, J., and Lu, J.K.H.,** Relation of neuroendocrine function to reproductive aging, in *Oxford Reviews of Reproductive Biology, Vol. 16,* Oxford University Press, Oxford, Great Britain, 1994, 216.

12. **Meites, J., Goya, R., and Takahashi, S.,** Why the neuroendocrine system is important in aging processes, *Exper. Gerontology*, 22, 1, 1987.
13. **Morley, J.E., and Kaiser, F.E.,** Testicular function in the aging male, in *Endocrine Function and Aging*, Armbrecht, H.J., Coe, R.M., and Wongsurawat, N., Eds., Springer Verlag, New York, 1990, 99.
14. **Peng, M.T.,** Changes in hormone uptake and receptors in the hypothalamus during aging, in *Neuroendocrinology of Aging*, Meites, J., Ed., Plenum Press, New York, 1983, 61.
15. **Quadri, S.K., Kledzik, G.S., and Meites, J.,** Reinitiation of estrous cycles in old constant-estrous rats by central acting drugs, *Neuroendocrinology*, 11, 148, 1973.
16. **Shock, N.,** System integration, in *Handbook of the Biology of Aging*, Finch, C.D., and Hayflick, L., Eds., Van Nostrand Reinhold, New York, 1977, 639.
17. **Welsch, C.W., and Nagasawa, H.,** Prolactin and mammary tumorigenesis: A review, *Cancer Res.*, 37, 951, 1977.
18. **Wise, P.M.,** Alterations in proestrous LH, FSH, and prolactin surges in middle-aged rats, *Proc. Soc. Exp. Biol. Med.*, 169, 348, 1982.

Chapter 7

THE ROLES OF REGULATORY AND MODULATORY PROCESSES IN THE DECLINE OF REPRODUCTIVE BEHAVIOR OF MALES

K.C. Chambers

TABLE OF CONTENTS

0-8493-2451-3/95/$0.00+.50

ABBREVIATIONS

GnRH–gonadotropin-releasing hormone
ORPRC–Oregon Regional Primate Research Center
PCPA–p-chlorophenylalanine
LH–luteinizing hormone

INTRODUCTION

Perhaps one of the most reliable and pervasive changes found in the reproductive system of aging male mammals is the decline in sexual behavior. Although age-related changes in sexual behavior have not been found in all species, e.g., golden hamsters,[140] decreases in sexual interest or performance have been reported in most species studied. These include $CB6F_1$ and CBF_1 mice,[32,68] rats,[54,83,84] guinea pigs,[57] bulls,[5] Japanese macaques,[41] rhesus macaques,[111,125] and humans.[34,77,90,91] The causes of this decline have remained elusive. It is clear that gaining a knowledge of the processes regulating reproductive behavior is a necessary first step in determining the etiology of the age-related decline. But a complete determination must include an understanding of the physiological, environmental, and social factors that play a modulatory role in arousal and performance. In this chapter the discussion will be divided into regulatory processes, which may play a causal role in the decline, and modulatory factors, which can alter the rate of decline.

Sexual behavior is not an isolated set of behaviors. It is the result of the continual integration of sensory stimuli and motor responses. The primary sensory stimuli, which include auditory, olfactory, tactile, and visual stimulation from the potential partner, initiate and maintain arousal and lead to the appropriate motor responses. The motor responses themselves provide sensory stimulation that contributes to the maintenance of arousal. Conceivably, any deficit in sexual behavior could be the result of failure of an essential sensory system or systems, a problem with the integration of sensory information and motor responses, or a degeneration of the motor pathways. A modulatory factor that slows the rate of decline probably provides some compensatory action, and a factor that accelerates the rate could act on a degenerating system or a system that normally is spared from the effects of aging.

The discussion in this chapter will focus on the research done by the author in collaboration with Charles Phoenix. The specific behavioral changes that occur in aging male Fischer 344 rats and rhesus macaques and the relationship of these changes with the endocrine and neural changes that also have been found will be discussed. In addition, analyses of the work of other investigators who have studied these species as well as mice, guinea pigs, and humans will be included. Although all of these species share many of the fundamental elements of the regulatory processes, there are significant differences. The possibility that the cause of the decline involves different aspects of the regulatory processes for different species must be kept in mind.

SPECIFIC CHANGES IN REPRODUCTIVE BEHAVIORS

Aging Male Rats

The following behaviors were measured in Fischer 344 male rats: mount (when the male places his front paws on the lower back of the female and his

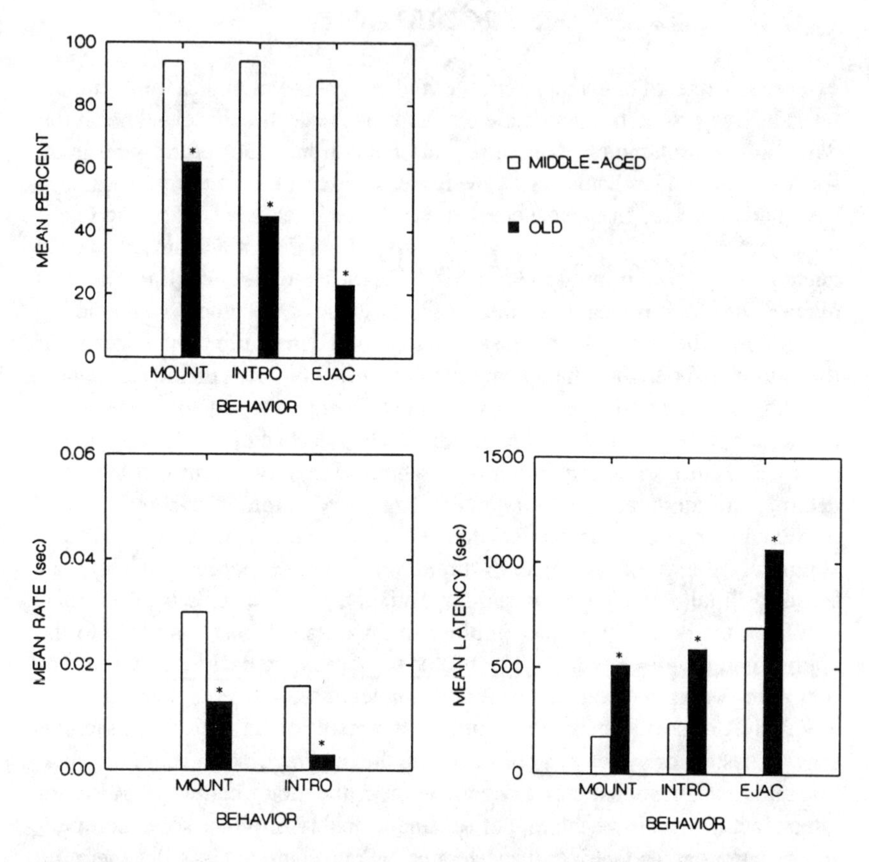

FIGURE 1. Mean of several measures of sexual behavior of middle-aged and old male rats. *Differed significantly from middle-aged males, $p < .05$. ***Abbreviations***: EJAC, ejaculation; INTRO, intromission. (Adapted from Chambers and Phoenix.[17])

back paws on the floor and exhibits pelvic thrusting), intromission (insertion of the erect penis into the vagina), and ejaculation (a deep pelvic thrust accompanied by a prolonged clasp of the female and spasm-like contractions of the thigh). For both rats and rhesus macaques, the following computations were made: rate, frequency of a given behavior per minute; percent tests, percentage of tests during which a given behavior occurred; and latency, time from the beginning of the test to the occurrence of a given behavior.

In studies using rats as the animal model, the males designated as old were at least 21 months old, those designated as middle-aged were between 9 and 19 months old, and those designated as young were between 2.5 and 5.5 months old. All males had been given several tests of sexual behavior in the same test arena and with the same testing procedures. The older males were either retired breeders or had been tested extensively when younger and were sexually proficient at that time.

We consistently have found fewer ejaculations exhibited in a given test and a lower percentage of tests with ejaculation in old male rats than in young males.[17,18,23,128] In one study, we examined a number of behavioral measures of old and middle-aged males that were still sexually active.[17] We found that the old males had significantly lower percentages of tests with mounts, intromissions, and ejaculations, lower rates of mounting and intromitting, and longer latencies to mount, intromit, and ejaculate than middle-aged males (see Figure 1).

Aging Rhesus Males

We have measured 4 behaviors in our tests of sexual behavior of rhesus macaque males. Contact is when the male places his hands on the lower back or hips of a female and she is not in a present posture (standing with four limbs extended and tail deviated), mount is when the male places his hands on the hips of a female and clasps her ankles or calves with his feet, intromission is the insertion of the erect penis into the vagina, and ejaculation is a pause after rapid pelvic thrusting that is accompanied by spasm-like contractions of the thigh and jerking vertical movement of the tail.

Most of the males used in these experiments were born in the wild and purchased as adults. Estimates of their ages upon arrival at the Oregon Regional Primate Research Center (ORPRC) were based on comparisons of their general appearance, sexual maturity, dental state, and body weights with those of rhesus males of known age born at the ORPRC. All males designated as old were at least 19 years old, those designated as middle-aged were between 13 and 18 years old, and those designated as young were between 8 and 12 years old. All males had been given numerous tests of sexual behavior in the same facilities with the same procedures. The older males had been tested when younger and had been sexually active at that time. In each experiment reported in this chapter, a span of several years separated older males from younger males.

In our most recent study, 6 old males had significantly lower percentages of tests with intromissions and ejaculations, lower rates of contacts, mounts, and intromissions, and longer latencies to mount, intromit, and ejaculate than 6 young males (see Figure 2).[21] We also have found decreases in the percentages of tests with contacts and mounts in some of our other studies.[16,22,115] The particular behavioral measures that have differed between young and older males have varied across studies, but we have consistently found that older males achieve fewer ejaculations in a given test or a lower percentage of tests with ejaculations than do young males.[15,16,21,22,113-115,120]

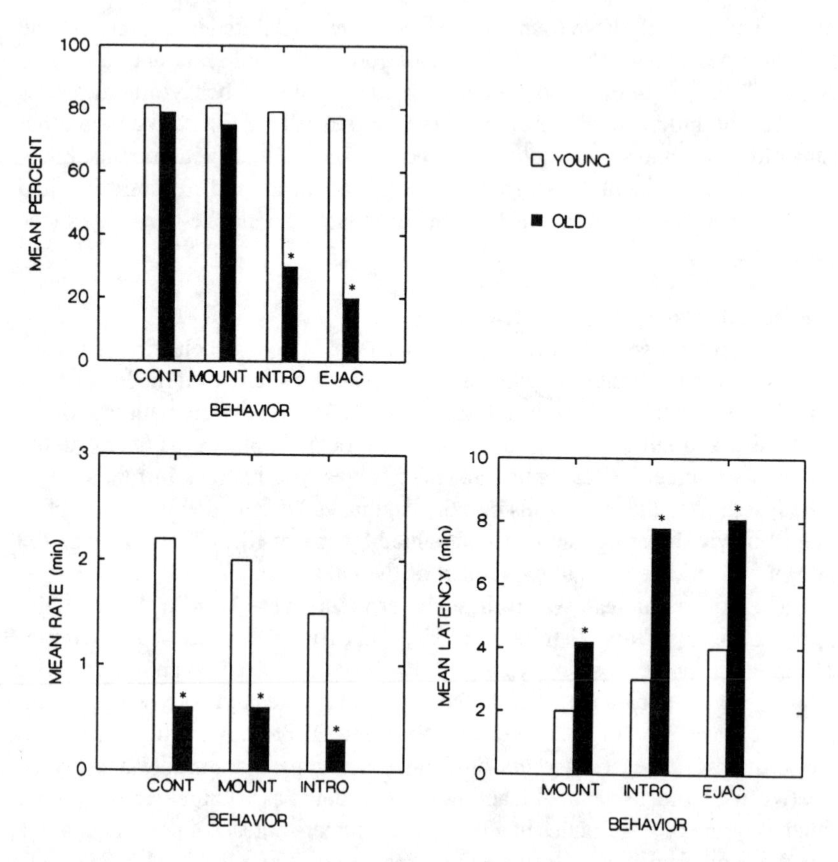

FIGURE 2. Mean of several measures of sexual behavior of young and old rhesus males. *Differed significantly from young males, $p < .05$. ***Abbreviations***: EJAC, ejaculation; INTRO, intromission. (Adapted from Chambers and Phoenix.[21])

REGULATORY PROCESSES

Relationship Between Serum Testosterone and Behavior

Testosterone is essential for optimal performance in most mammalian species studied. After gonadectomy, most males experience a decline in sexual interest and activity, although the rate of decline and the baseline levels of performance eventually reached vary widely.[62] It is not surprising that a decrease in levels of circulating testosterone was one of the first hypotheses suggested to account for the decrements in sexual performance of aging males.[10,98]

Decreased Testosterone Levels

There are dozens of reports of decreases in circulating testosterone levels in rats (see Figure 3).[11,17,24,44,46,52,76,121] Young males but not old males show a diurnal pattern of testosterone concentration, and the lower testosterone

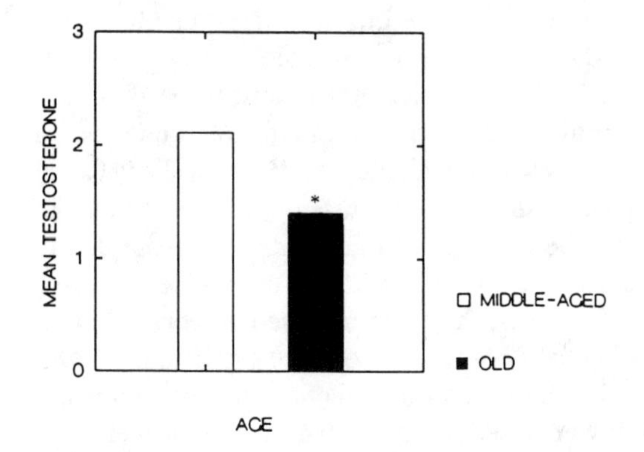

FIGURE 3. Mean serum testosterone (ng/ml) levels of middle-aged and old male rats. *Differed significantly from middle-aged males. (Adapted from Chambers and Phoenix.[17])

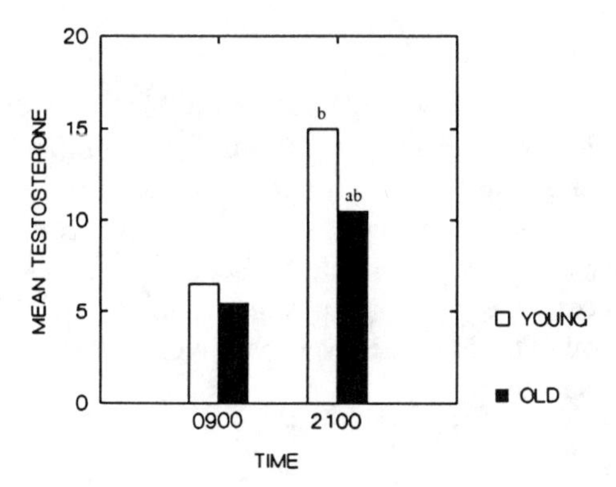

FIGURE 4. Mean serum testosterone (ng/ml) levels of young and old rhesus males at 0900 and 2100 hours. [a]Differed significantly from young males, $p < .05$. [b]Differed significantly from the corresponding 0900 hours value, $p < .05$. (Adapted from Chambers and Phoenix.[21])

levels of old males do not appear to be tied to any part of the day/night cycle.[97] Old males also fail to show the broad transient oscillations found in young males.[138]

Reductions in testosterone pulse amplitude have been reported in CBF_1 mice[31] and decreased testosterone levels have been found in guinea pigs.[123] However, no changes in circulating testosterone levels have been reported in $CB6F_1$ mice[26] nor in golden hamsters.[140]

In rhesus males, testosterone levels are higher in the evening than they are in the morning.[13,22,49,96,107] Afternoon levels of testosterone are higher than the morning but lower than the evening levels.[13] In seven separate studies, we have failed to find significant age differences in the morning testosterone levels.[12,13,15,21,22,114,120] These results were based on the means of two to eight blood samples from animals that were so habituated to the bleeding procedure that many of them voluntarily extended their legs for the venipuncture. At night, both old and young males showed a significant elevation in testosterone levels. The testosterone levels of old males tended to be lower than those of young males at night in all of our studies,[13,21,22,114] but we found significant differences in only one of our studies (see Figure 4).[21]

The blunting of the circadian rhythm in old rhesus males also has been reported by Kaler *et al.*[74] at the ORPRC and is similar to what has been reported in healthy elderly men.[9,36,48,89,142] Kaler and his colleagues examined the daily pulsatile pattern of testosterone secretion in old and young rhesus males by measuring plasma levels every 20 minutes for 24 hours. They found no age-related differences in the mean 24-hour testosterone concentrations, in the mean concentration during the 12 hours of light, in the pulse amplitude during the 12 hours of light and the 12 hours of dark, or in the number of pulses during the 12 hours of light. However, the mean testosterone concentration and the number of pulses during the 12 hours of dark were significantly lower in old males than in young males. In men, mean 24-hour testosterone concentrations were significantly lower in elderly men than in young men, and although both age groups exhibited rhythms, the maximum excursion was less in elderly men than young men.[142]

Association Between Testosterone Levels and Performance Levels

Although testosterone levels clearly decrease in aging males of a number of different species, this change alone does not account for the decline in sexual behavior. This conclusion is supported by two lines of evidence. First, across species, investigators consistently have failed to find correlations between changes in circulating testosterone levels and changes in the level of sexual performance. Second, elevating testosterone levels by giving testosterone exogenously fails to restore behavior.

Endogenous Testosterone Levels

No significant correlations have been found between testosterone levels and different parameters of sexual behavior in male rats followed longitudinally from 7 months to 27 months of age.[34] The age-related decreases in sexual behavior of CBF_1 male mice are not associated with the decreases in testosterone levels, and, although sexual performance declines in aging $CB6F_1$ male mice, testosterone levels do not.[26,31]

In one study we found significant correlations between the percentage of tests with mounts and erections and serum testosterone levels in old rhesus males when both behavior and hormone were measured at 0900 hours.[22] However, in other studies we failed to find significant correlations between sexual performance at 0900 hours and testosterone levels at 0900, 1300, and 2100 hours or the magnitude of circadian excursion of testosterone levels.[13,21,114] We have found significant correlations between testosterone levels and the rates of contacts, mounts, and intromissions and the percentages of tests with intromissions and ejaculations when both hormone and behavior were measured at 2100 hours.[22] Since, as mentioned above, we have found significant correlations that have not been replicated in subsequent studies, it is possible that these significant correlations also are spurious. Until this study is replicated, the significance of these correlations remains unclear.

In humans, no significant correlations have been found between circulating testosterone levels and performance measures or sexual interest.[34] When old men are separated according to their level of sexual activity, the more sexually active men have higher testosterone levels than those less active.[143] These results are similar to those reported in rats; testosterone levels of old male rats that exhibited at least mounts and intromissions were higher than those that did not.[43] The higher level of testosterone is most likely the consequence rather than the cause of the sexual activity. Testosterone levels have been reported to increase as a result of sexual activity in younger males.[75]

Exogenous Testosterone Treatment

Exogenous testosterone treatment has been shown to restore sexual performance of young adult gonadectomized males to pregonadectomy levels in a number of different species.[4,119,124] However, exogenous testosterone treatment consistently has failed to restore sexual behavior in aging males.

In our experiments with aging male rats, injections of testosterone propionate in gonadectomized old males that resulted in supraphysiological serum levels of testosterone increased several measures of sexual behavior, e.g., percentage of tests with mounts, intromissions, and ejaculations and mount rate (see Figure 5). However, only the percentage of tests with mounts were restored to the level found in young males (either untreated intact or testosterone-treated gonadectomized). In other studies, testosterone treatment failed to eliminate age differences between middle-aged and young male rats in probability of mating and latency measures.[54] Decreased erectile capacity in middle-aged male rats, however, was restored by testosterone treatment, which suggests that decreased testosterone levels may be a major causal factor in this age-related change.[53]

Testosterone treatment has no effect on sexual performance in guinea pigs[69] or rhesus macaques.[116] Injections of testosterone propionate that resulted in supraphysiological levels of serum testosterone in intact old rhesus males failed to significantly increase any measure of sexual behavior (see Figure 6).

Decreased Sensitivity to Testosterone

Another testosterone-related hypothesis that has been proposed to account for the decline in sexual behavior in aging males is that the target tissues, central or peripheral, controlling behavior are less responsive to testosterone.[17,69,70,83] Reduced sensitivity is an ambiguous term that could be explained by a number of different models. For example, it could be the result of decreases in the availability of testosterone or the active metabolites of testosterone at the receptor site, decreases in the total available cellular receptors, qualitative changes in the properties of the receptor, such as a decrease in affinity of receptor for the hormone or a change in hormone specificity, or changes in the hormone-receptor-nuclear chromatin interaction.

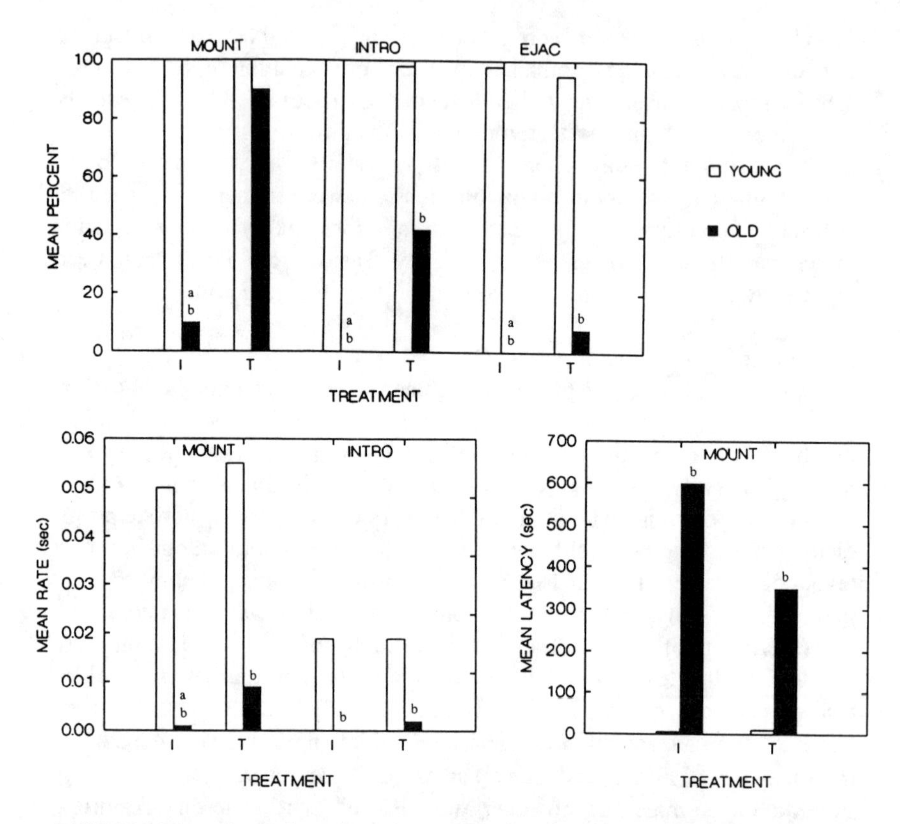

FIGURE 5. Mean of several measures of sexual behavior of young and old male rats when intact (I) or gonadectomized and treated with testosterone propionate (T). [a]Differed significantly from old T males. [b]Differed significantly from young I and T males. ***Abbreviations***: INTRO, intromission; EJAC, ejaculation. (Adapted from Chambers and Phoenix.[18])

There is substantial evidence that suggests testosterone is a prohormone that activates sexual behavior in rodents only after metabolic conversion to another hormone.[145] Testosterone can be metabolized to dihydrotestosterone and estradiol in central neural tissues.[85,102,134] Dihydrotestosterone generally does not activate behavior when administered systemically; estradiol can stimulate behavior although it usually is less effective than a combination of estradiol and dihydrotestosterone.[3,33,92] The accepted hypothesis is that testosterone is aromatized to estradiol in brain cells and then estradiol binds to the estrogen receptor to activate sexual behavior in male rats. Dihydrotestosterone, reduced from testosterone, clearly plays a role in sexual behavior by stimulating peripheral tissues, but it also may bind to brain androgen receptors to affect sexual performance.[3,92] There is evidence to support the involvement of androgen receptors in the androgen-induced regulation of aromatase activity.[127] Sexual behavior probably is mediated primarily by estrogen receptors in mice as well, but androgen most likely mediates behavior in guinea pigs and rhesus macaques. Dihydrotestosterone is fully effective in stimulating sexual

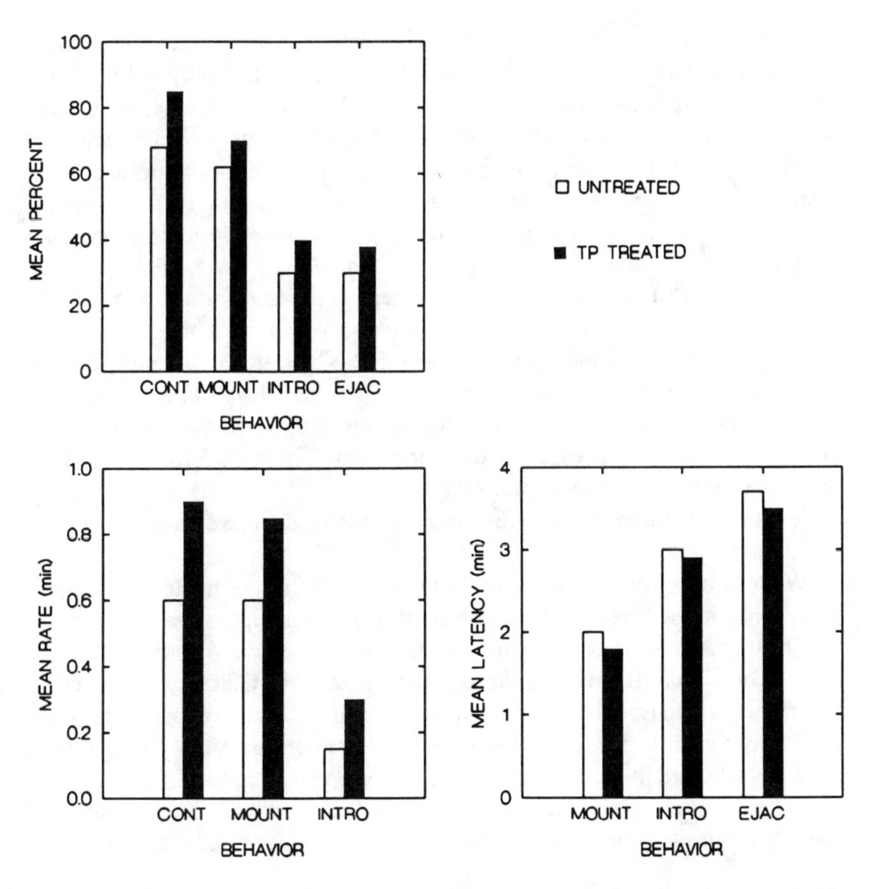

FIGURE 6. Mean of several measures of sexual behavior of old rhesus males when treated with testosterone propionate (TP) and when untreated. There were no significant differences. ***Abbreviations***: CONT, contact; INTRO, intromission; EJAC, ejaculation. (Adapted from Phoenix and Chambers.[116])

behavior, whereas estradiol only stimulates mounting in rhesus monkeys and is ineffective in activating any sexual behaviors in guinea pigs.[2,109,110,112,144]

Sexually dimorphic forebrain neural circuitry, which includes the preoptic area and the amygdala, is thought to regulate male sexual behavior and to coordinate sensory information and motor responses.[63,129] The preoptic area integrates sensory information and activates motor pathways; the amygdala, specifically the corticomedial division, receives afferent olfactory input and sends efferent projections through the stria terminalis to the medial preoptic area.[80] Lesions of the medial preoptic area abolish sexual behavior in experienced male rats[64] and lesions of the corticomedial amygdala produce deficits in precopulatory investigation of the female partner and copulatory behavior.[47,61,93] These two areas have high concentrations of androgen receptor, aromatase activity and estrogen receptors in male rats,[60,85,126,131,133,136,139] and neural implantation of testosterone and estradiol into the preoptic area activates sexual behavior in gonadectomized males.[25,72]

In collaboration with Charles Roselli and Jan Thornton, the author embarked on a series of studies in Fischer 344 male rats to determine whether there are age-related changes in the nuclear binding of testosterone, aromatase activity, and the testosterone metabolite, estradiol, in the preoptic area and amygdala, and, if present, whether these changes could be eliminated by maintaining circulating testosterone at levels characteristic of young adult males.[23,128] Unoccupied receptors, referred to as cytosolic receptors, can be recovered in cytosols after subcellular fractionation of tissue homogenates. After binding to a steroid hormone, the receptors are transformed and exhibit a high affinity for nuclear chromatin where they act to change genomic transcription and subsequent protein synthesis.[51] These steroid hormone-bound receptors, referred to as nuclear receptors, can be salt extracted from cell nuclei. Although there have been studies reporting decreases in cytosolic androgen and estrogen receptor concentration or binding affinity, no studies have examined nuclear receptor concentration.[55,58] Sexually active young males and old males whose sexual activity had decreased were used as subjects.

We found significant decreases in the concentration of nuclear androgen receptors and the level of aromatase activity in the preoptic area (see Figure 7). The testosterone levels of young and old testosterone-treated gonadectomized males tended to be higher than those of the young intact males. Consequently, the aromatase activity levels and nuclear androgen concentrations were generally higher in the young testosterone-treated males than the young intact males (see Figure 8). Testosterone treatment elevated aromatase activity to the same levels in young and old gonadectomized males. It did not elevate nuclear androgen receptor concentrations of old gonadectomized males to the levels

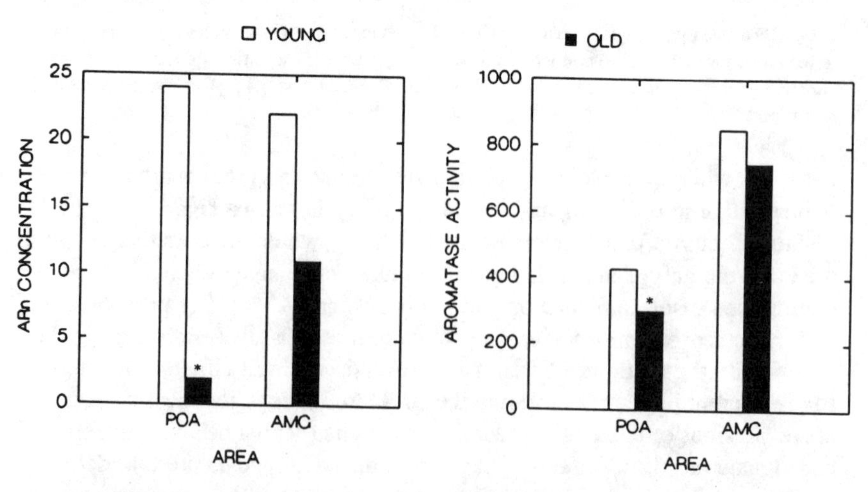

FIGURE 7. Mean androgen binding (^{3}H-R1881 bound, fmol/mg DNA) to nuclear androgen receptors (ARn) and mean aromatase activity (fmol ^3H_2O/homg protein) in the preoptic area (POA) and amygdala (AMG) of young and old male rats. *Differed significantly from young males, $p < .01$. (Adapted from Chambers *et al.*[23])

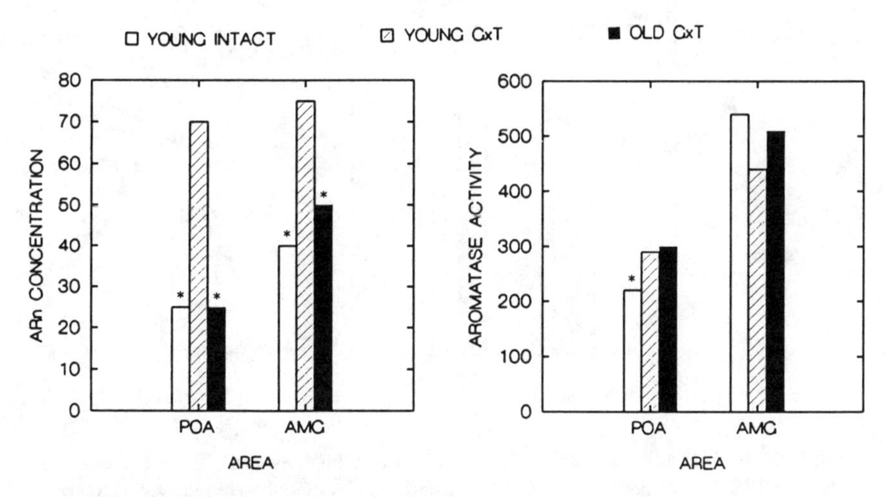

FIGURE 8. Mean androgen binding (^{3}H-R1881 bound, fmol/mg DNA) to nuclear androgen receptors (ARn) and mean aromatase activity (fmol ^3H_2O/homg protein) in the preoptic area (POA) and amygdala (AMG) of young intact males rats and young and old gonadectomized male rats treated with testosterone (GxT). *Differed significantly from young GxT males, $p < .05$. (Adapted from Chambers, *et al.*[23])

found in young gonadectomized males. Testosterone treatment, however, did increase nuclear androgen receptor concentrations of the old gonadectomized males to the levels found in young intact males. The sexual performance of the old testosterone-treated gonadectomized males was not increased to the level displayed by either of the young groups. Thus, the reduced capacity of testosterone treatment to elevate nuclear androgen receptor concentrations in old males may have some significance for other systems, but it is not important for behavior.

These data suggest that the decreases in the brain androgen binding in the preoptic area and amygdala do not contribute to the behavioral decline, and that the capacity of these neural areas to aromatize androgen is not a limiting factor for androgen responsiveness in aging male rats. That reduced aromatization of testosterone may not play a role in other neural areas mediating sexual behavior is supported by our finding that injections of estradiol and dihydrotestosterone are no more effective than injections of testosterone in restoring sexual behavior in aged male rats.[18]

As was true for androgen binding and aromatase activity, the concentration of nuclear estrogen receptors in the amygdala was lower in old than young male rats (see Figure 9).[128] However, testosterone treatment did not restore nuclear estrogen receptor concentrations in old males. Both sexual behavior and nuclear estrogen receptor concentrations in the amygdala were lower in testosterone-treated gonadectomized old males than either gonadectomized testosterone-treated young males or intact young males. It is unlikely that the altered nuclear estrogen receptor levels reflect a loss of total estrogen receptors since we have found equivalent levels of cytosolic estrogen receptor

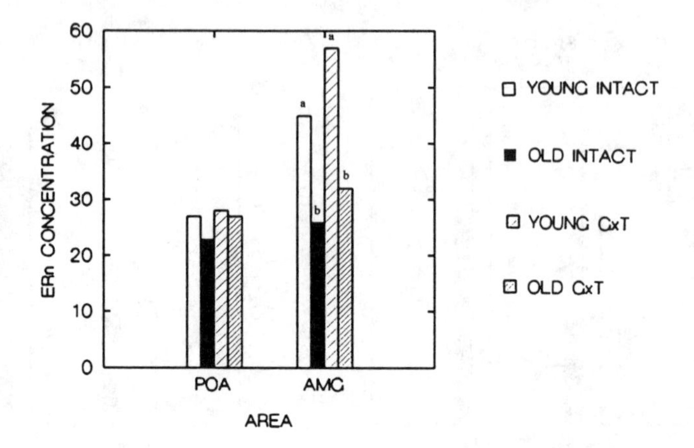

FIGURE 9. Mean estrogen binding (^{3}H-estradiol bound, fmol/mg DNA) to nuclear estrogen receptors (ERn) in the preoptic area (POA) and amygdala (AMG) of young and old intact male rats and young and old gonadectomized male rats treated with testosterone (GxT). Bars with dissimilar letters are significantly different, $p < .05$. (Adapted from Roselli *et al.*[128])

concentrations in young and old males after gonadectomy when cytosolic estrogen receptor levels account for the major portion of all estrogen receptor levels. The decrease in nuclear estrogen receptor levels may be due to defects in the ability of the estrogen receptor to tightly bind nuclear chromatin.

We have suggested that the decreases in nuclear estrogen receptor concentrations in the amygdala may contribute to the decreases in sexual behavior of aging male rats by reducing the transmission of olfactory information from the female partner to the preoptic area.[128] Studies assessing the possible decline in responsiveness to olfactory cues from the female partner, the time course of decline in behavior, olfactory sensitivity, and estrogen receptor levels and the degree of association between levels of behavior, olfactory sensitivity, and estrogen receptor levels will need to be performed to test the viability of this hypothesis. Also, the importance of this finding for other species, such as the rhesus macaque, that are not as dependent on olfactory cues for successful sexual performance and whose sexual behavior is mediated by androgen receptors rather than estrogen receptors, will need to be determined.

Age-Related Changes in Other Hormones

Estradiol and Dihydrotestosterone

Blood levels of estradiol and dihydrotestosterone are low in many male mammals, and it is unclear whether the circulating levels of these hormones contribute to behavioral regulation. In any event, alterations in the levels of either of these hormones cannot account for the decreases in sexual behavior in aging male rats or rhesus macaques.

Plasma estradiol levels have been reported to increase in male rats between 6 and 24 months of age.[86] We also found increased estradiol levels in old male rats in one study, although subsequent studies have revealed no differences between young and old males.[18,23,128] Since increased estradiol levels are associated with increased sexual behavior in young adult males, however, this change does not contribute to the age-related behavioral decline. In rhesus males, we consistently have found no age differences in mean estradiol levels at 0900, 1300, or 2100 hours and no correlations between estradiol and sexual behavior in old males.[13,21,22,120]

No age differences have been found in dihydrotestosterone levels of old and young male rats[23] or rhesus males.[13,125] There also are no age-related differences in the diurnal pattern of dihydrotestosterone levels and no correlation between dihydrotestosterone levels and sexual behavior in old rhesus males.

GnRH

There is some evidence to suggest that GnRH may have a direct neurohumoral influence on the neural regulation of male sexual behavior. GnRH shortens ejaculation latencies in intact male rats and in gonadectomized males treated with testosterone and stimulates mounting behavior when injected intraventricularly.[39,100,101] In addition, administration of a GnRH analog that antagonizes functions of the native peptide increases the rate of decline in sexual behavior after gonadectomy.[40] There is decreased GnRH content in the median eminence of aged male rats as measured by radioimmunoassay and immunocytochemistry, decreased hypothalamic GnRH secretion, and decreased numbers of neurons in the medial preoptic area that express the GnRH gene.[56,67,71,122,135] It is possible that the decreases in neural GnRH contribute to the age-related decline in sexual behavior of rats.

GnRH does not appear to play a facilitatory role in rhesus macaques. We have found that GnRH injections lengthen ejaculation latency in young males and lower intromission rate in aging males.[117] Whether there are age-related changes in GnRH secretion in rhesus males remains unknown.

Prolactin

Although it is unclear what role prolactin plays in the regulation of male sexual behavior, high levels of prolactin have a suppressive effect on behavior. Hyperprolactinemia decreases sexual performance in young adult rats and mice.[37] Sexually impotent men with hyperprolactinemia induced by pituitary tumors report increases in sexual potency after tumor removal or treatment with the prolactin-secretion inhibitor, bromocriptin.[38] It is possible that elevated prolactin contributes to the behavioral decline in aging male rats and men because increases in prolactin secretion have been reported in these two species.[94] However, the levels of prolactin reported in old rats and men are lower than the levels that induce decreases in sexual performance in young adults of both of these species.[34] Also, some studies have not found substantial increases in prolactin levels in rats and men.[34] In addition, elevated prolactin levels do not contribute to decreases in sexual behavior of aging rhesus males since the prolactin levels and diurnal rhythms of old and young males do not differ and there is no correlation between performance and prolactin levels.[21]

Age-Related Changes in Neurotransmitters

A number of different neurotransmitters have been implicated in the control of male sexual behavior. Dopaminergic and noradrenergic activity have been found to facilitate certain aspects of sexual behavior and serotonergic activity is generally inhibitory.[1,28-30,45,87,88,141] Recent evidence indicates that sexual performance can be increased in aging male rats given drugs that facilitate noradrengergic and dopaminergic transmission and block serotonergic transmission.

Yohimbine, a drug that increases noradrenergic transmission by blocking inhibitory α-2 adrenoreceptors, when administered systemically, increased the percentage of middle-aged rats exhibiting intromissions and ejaculations and decreased ejaculation latencies, and when given intracerebroventricularly, it decreased mount latencies and frequencies.[130,137] Since it did not elevate behavior to the level of young males treated with yohimbine, it may be that there is age-related deterioration in the part of the noradrenergic system that mediates sexual behavior. Both PCPA, an inhibitor of serotonin synthesis, and methysergide, a serotonin receptor blocker, increased the frequencies of mounting and intromitting in old male rats.[106] However, because the performance of these males was not compared to that of young males, it is difficult to interpret these results with respect to the possibility that increases in serotonergic transmission contribute to the age-related deficit in sexual behavior.

The most compelling evidence that alterations in neurotransmitter systems contribute to the decline in sexual behavior comes from studies of the dopaminergic system. Apomorphine, a dopaminergic receptor agonist, has differential effects on young and middle-aged male rats.[27] Before apomorphine treatment, young males exhibited shorter ejaculation latencies than middle-aged males. Apomorphine treatment further decreased the latencies in young males but had no effect in aging males. This suggests that there may be age-related deficits in the part of the dopaminergic system that facilitates male sexual behavior. Knoll[78] has suggested that the decline in male sexual behavior is due largely to age-related deficits in the nigrostriatal dopaminergic neurons. The nigrostriatal system appears to be part of the neural circuitry controlling male sexual behavior since lesions in this system result in deficits in sexual arousal and behavior in young adult males.[6] In aging male rats, there are reductions in the concentration of dopamine in the striatum, an area that contains the terminals of nigrostriatal dopaminergic neurons.[35] There also is a loss of striatal dopamine receptors, in particular D_2 receptors, which is due in part to reductions in receptor-containing neurons and changes in D_2 receptor gene expression.[59,65,73,95,105] When (-)deprenyl, a dopaminergic agonist that selectively facilitates the activity of the nigrostriatal dopaminergic neurons, is given to old, sexually sluggish male rats, the percentage of males exhibiting ejaculations and the percentage of tests with ejaculations increases.[78,79]

We have studied the effects of yohimbine and (-)deprenyl on the sexual performance of young and old sexually sluggish rhesus males.[20] Yohimbine significantly decreased ejaculation latencies in the young males but had no effect on latencies in old males. These results are consistent with those of the rat studies and support the suggestion that there are age-related deficits in the noradrenergic system mediating sexual behavior. We found no significant effects of (-)deprenyl in young or old males. There was a tendency, however, for more young and old males to ejaculate when under (-)deprenyl treatment (80% and 67%, respectively) than saline treatment (60% and 33%, respectively). In a personal communication, Joseph Knoll suggested that (-)deprenyl may require longer treatment time, e.g., as long as a year, to be effective in primates. Additional research will need to be done to determine the possible involvement of the nigrostriatal dopaminergic system in the sexual decline of rhesus males and the efficacy of (-)deprenyl in reversing that decline.

MODULATORY FACTORS

Exposure to Testosterone

It has been suggested that some hormones themselves may contribute to the initial deterioration of their brain target cells during aging.[81] Although it is unclear whether hormones initiate the decline, there is evidence to support the hypothesis that the rate of decline is associated with the amount of exposure to the hormone. Age-related changes in the hippocampus of middle-aged male rats are correlated with plasma levels of adrenocorticoids, and administration of adrenocorticoids in moderate doses over a number of months accelerates the age-related elevation in hippocampal astrocyte reactivity.[82] Glial reactivity in the arcuate nucleus increases with age in female rats, and there are impairments in spontaneous and experimentally induced LH surges in female mice.[7,42,132] Brief exposures to high levels of estradiol accelerate these changes whereas ovariectomy delays them.

These data suggested that the amount of exposure to testosterone may be associated with the rate of decline in male sexual behavior. We set about to determine whether altering the amount of exposure to testosterone in adulthood and during perinatal development would alter the age-related decrease in sexual performance.

Adulthood

Three groups of middle-aged male rats were used: intact males, castrated males continuously implanted with testosterone, and castrated males implanted with testosterone only around the time of behavioral testing.[17] The males were castrated or sham castrated when 13 months old and were tested when 15, 19, and 23 months old. The testosterone levels of the treated males were maintained at significantly higher blood levels than found in the intact males, i.e., testosterone levels were twice as high during the first two tests and 20 times higher during the third test. The two testosterone-treated groups did

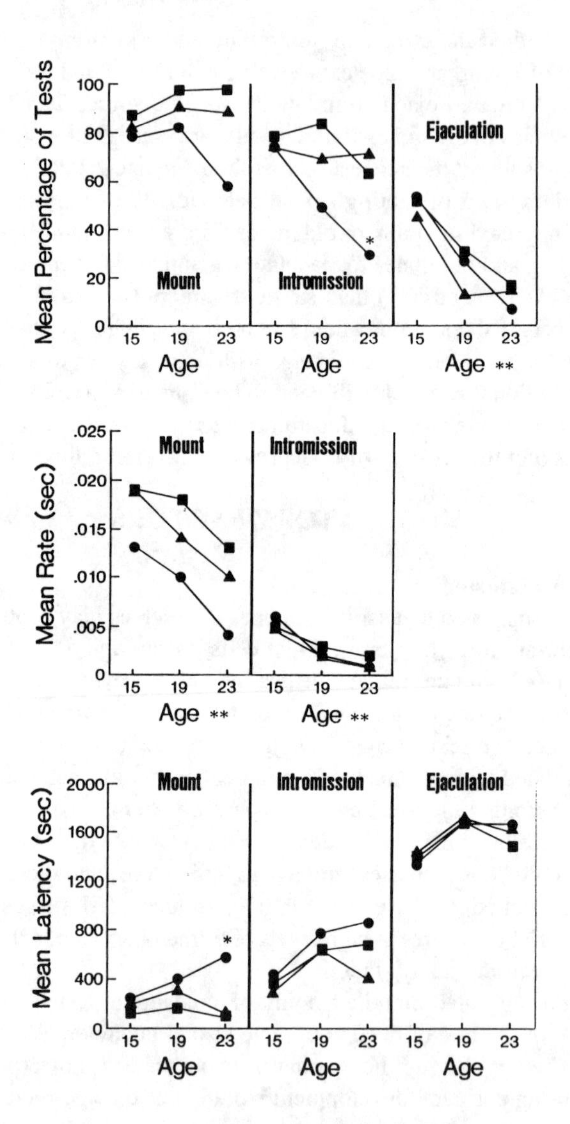

FIGURE 10. Mean of several measures of sexual behavior of intact male rats and castrated male rats given continuous or intermittent (only during testing) treatment of testosterone propionate from 13 to 23 months of age and tested when 15, 19, and 23 months old. *Differed significantly across time for the intact males only. *Note:* ● represents 12 intact male rats, ■ represents 10 castrated male rats with continuous testosterone, and ▲ represents 10 castrated male rats with intermittent testosterone. **Differed significantly across time for all three groups and the rate of change for the three groups did not differ. (Adapted from Chambers and Phoenix.[17])

not differ in the rate of decline in behavior (see Figure 10). For all three groups of males, the percentage of tests with ejaculations and the rates of mounting and intromitting decreased significantly across time. The percentage of tests with intromissions decreased and the mount latency increased in the intact males but not in either of the testosterone-treated males.

We had access to two groups of rhesus males that had been matched for age and sexual performance when they were 8 to 12 years old and were 18 to 22 years old at the time of this study.[15] One of the groups of males was gonadectomized and the other was left intact. During the 10-year interval, the gonadectomized males had been tested at widely spaced intervals under limited treatment with testosterone propionate, dihydrotestosterone, estradiol, 19-hydroxy-testosterone, and 19-hydroxy-dihydrotestosterone. The intact males also had been tested periodically during the 10 years but without exogenous hormone treatments. Thus the gonadectomized males had intermittent exposure to testosterone and its metabolites and the intact males had continuous exposure. In this study, we injected the old gonadectomized males with testosterone propionate and compared their performance levels to those of the old intact males and young intact males.

We found no significant differences in the sexual performance of the two groups of old males (see Figure 11). However, when the performance of the two groups of old males was compared to that of young males, differences between the intact and gonadectomized old males emerged. Whercas the old intact males had lower percentages of tests with intromissions and ejaculations and slower rates of intromitting than the young males, the old gonadectomized males differed from the young males only in percentage of tests with ejaculation. Thus the intromission measures fell in between those of the old and young intact males. This suggests that the rate of decline in intromissions is slower in the old gonadectomized males. But whether the rate of decline is slower because they only had intermittent exposure to androgens is unclear. The testosterone levels of the old gonadectomized males were 5 times higher than those of the young and old intact males. Although we have failed to find significant increases in any measures of sexual behavior in old intact males treated with equivalent doses of testosterone, there was a tendency for the means of several measures to be higher in the males when under testosterone supplements.[116] The performance of these males was not compared to that of young males. It is possible that the intromission measures of old intact males given testosterone treatment would also fall in between those of young intact and old untreated intact males.

The results of these two studies lend very little support to the hypothesis that increased exposure to testosterone during adulthood accelerates behavioral aging. This conclusion is similar to that of other investigators who found that gonadectomy had no effect on the age-related increase in microglial and astrocytic activity in the arcuate nucleus of male rats, and testosterone supplements induced only modest increases in microglial activity.[8,132]

Perinatal Period

Although the amount of testosterone during adulthood has little effect on the rate of decline of sexual behavior, the amount of testosterone during the perinatal period has substantial effects. In this study, two treatment conditions were used, androgenization and control. In the androgenization condition, 5

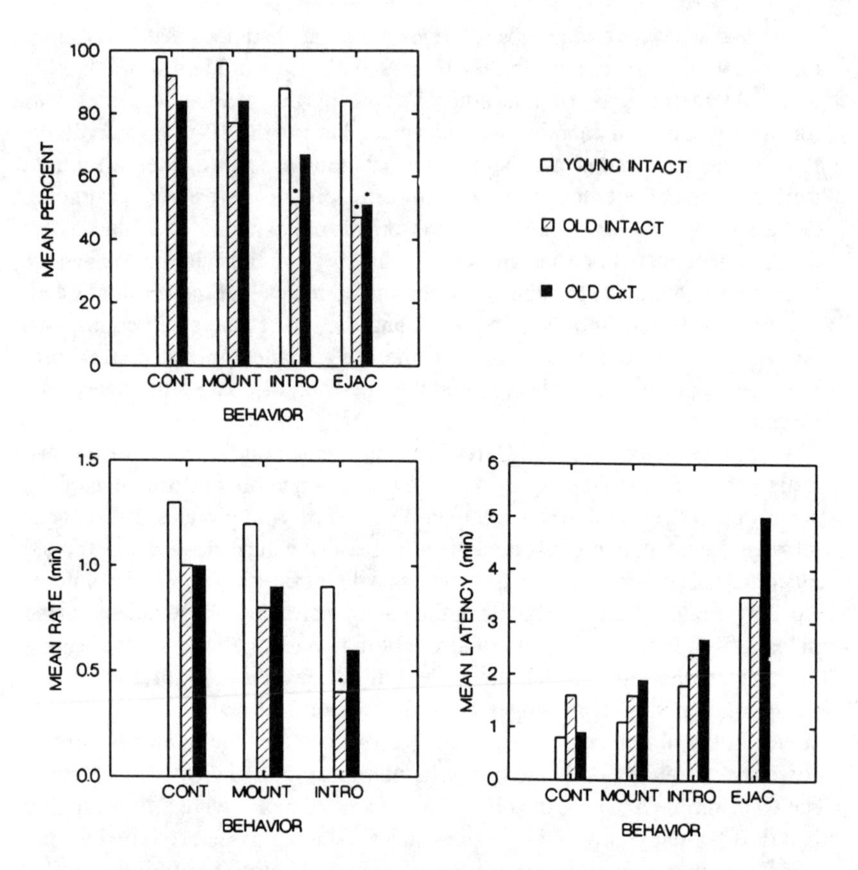

FIGURE 11. Mean of several measures of sexual behavior of young and old intact rhesus males and old gonadectomized rhesus males treated with testosterone propionate (GxT). *Differed significantly from young intact males. ***Abbreviations***: CONT, contact; INTRO, intromission; EJAC, ejaculation. (Adapted from Chambers and Phoenix.[15])

pregnant female rats were injected with 1 mg of testosterone (dissolved in sesame oil) propionate once a day on days 16 to 20 of pregnancy. The offspring of these females were then injected with 0.5 mg of testosterone propionate 24 and 72 hours after birth. In the control condition, 5 pregnant female rats and their offspring were injected with sesame oil only.

Eight androgenized and nine control males were selected from across each of the five respective litters. When the males were about 2.5 months old, they were gonadectomized and implanted with silastic capsules filled with testosterone. Beginning one week later, they were given a series of 6 weekly tests. They were given another series of tests while under testosterone treatment when they were 12 and 17 months old.

The androgenized and control males did not exhibit any differences in any of the measures of sexual behavior when young (3 months old). Only the androgenized males showed a significant decrease in the percentage of tests

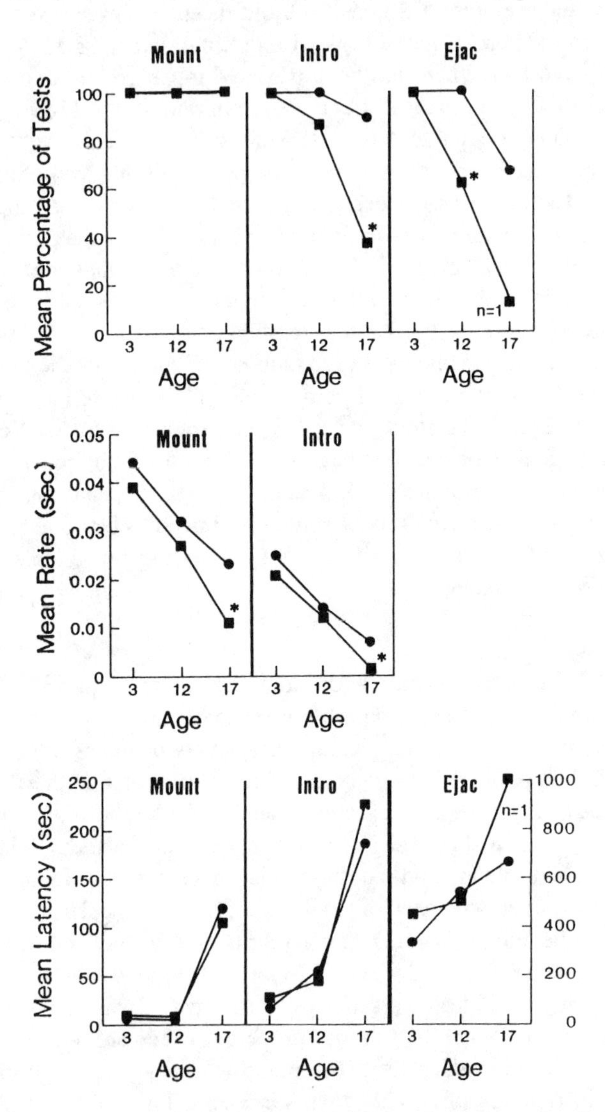

FIGURE 12. Mean of several measures of sexual behavior of male rats when 3, 12, and 17 months of age. They had been treated with oil (●, n = 9) or testosterone propionate (■, n = 8) during the perinatal period, and castrated and treated with testosterone when adults. *Differed significantly from perinatal oil males, $p < .05$.

with intromissions from 3 to 17 months of age (see Figure 12). Although both groups exhibited significant decreases in the percentage of tests with ejaculation and mean rates of mounting and intromitting, the rate of decline was greater for the androgenized males than for the control males. The two groups did not differ in the rate of increase in latencies to mount and intromit.

There are striking differences in the rate of decline of individual rats, rhesus macaques, and men.[15,21,78] Some of our old rhesus males showed only small decreases in the percentage of tests with ejaculation across a 10-year period (e.g., 70% to 60%), whereas others exhibited much greater decreases (e.g., 70% to 10%). The frequency of ejaculation in one of our 27-year-old males was greater than or equal to that of 50% of our young males. It is tempting to hypothesize that the amount of testosterone present during the perinatal period is associated with the rate of decline in sexual behavior during aging. However, the increased amount of testosterone present during the perinatal period may produce other changes not found in normal aging that contribute to the normal age-related decline and thus accelerate it.

It is interesting that a similar acceleration of reproductive aging has been observed in female rats and mice given submasculinizing doses of testosterone or estradiol during the perinatal period.[42,50,99,103] These females experience puberty and regular ovulatory cycles but their transition to acyclicity occurs at an earlier age than normal females. There also are striking parallels between the changes that occur in these females and the changes that occur in normally aging females. Clearly, additional studies will need to be done to determine the similarities in the reproductive changes of androgenized males and those of normally aging males.

The Female Partner

It is well known that male and female rhesus macaques exhibit strong sexual partner preferences.[66] These preferences can be so strong that under certain testing situations, the performance levels of males are lower when tested with less preferred ovariectomized females treated with estradiol than when tested with untreated preferred females.[19] Both male and female rhesus macaques living in the wild have multiple sexual partners. Thus, in our experiments with young and old rhesus macaques, we gave the males one or two tests each week with each of ten different females chosen at random from a group of potential partners. Over a 4-year period, we noticed that there was one female with whom all of our old males ejaculated, and if an old male ejaculated only once during a series of tests, it was always with this female. This suggested to us that the stimulus properties of fewer females are capable of arousing the old males. But it is possible that a decreased capacity to ejaculate in a series of weekly tests also contributes to the reduction in performance. To determine whether either of these possibilities contributed to the behavioral decline, we gave old males 9 biweekly tests with the highly preferred female 1339 and 9 biweekly tests with a different female during each test.[16] We compared the performance of the old males under each of these conditions with that of young males tested biweekly with the same 9 females. We found the performance levels of the old males tested with female 1339 was at least as high as that of the young males (see Figure 13). Indeed, the percentage of tests with mounts and ejaculations and the rates of mounting, intromitting, and ejaculating were higher in the old males tested with female

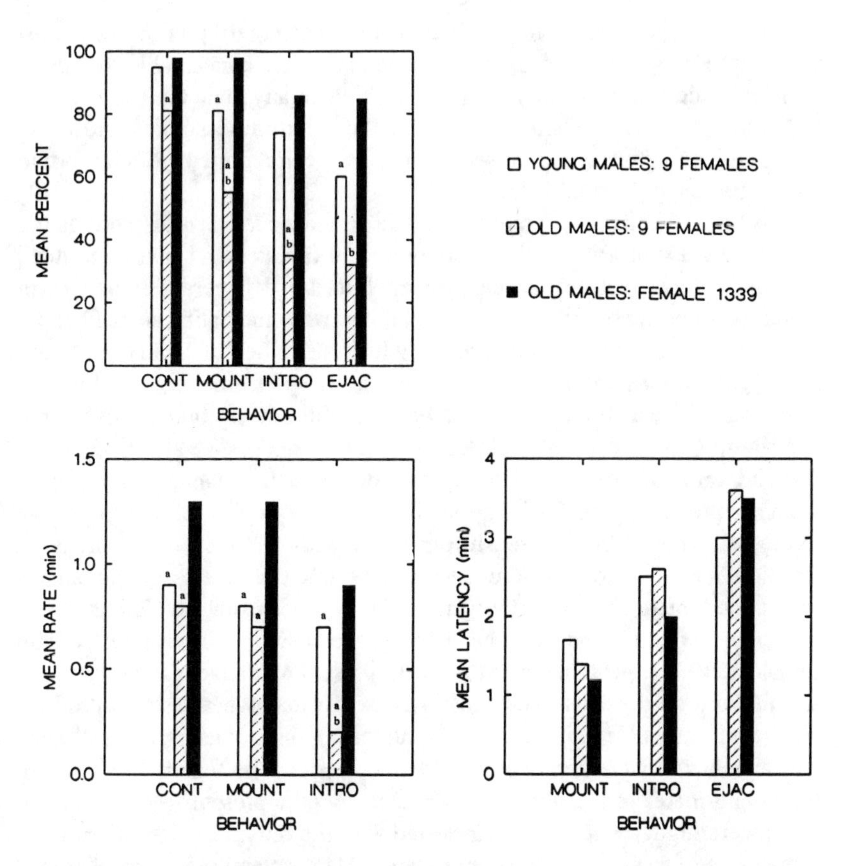

FIGURE 13. Mean of several measures of sexual behavior of young and old rhesus males given 1 test each with 9 different females and the same old males given 9 separate tests with preferred female 1339. [a]Differed significantly from the old males tested with female 1339. [b]Differed significantly from the young males. ***Abbreviations***: CONT, contact; INTRO, intromission; EJAC, ejaculation. (Adapted from Chambers and Phoenix.[16])

1339 than in the young males tested with 9 different females. All percent and rate measures were higher when the old males were tested with female 1339 than with 9 different females.

These studies demonstrate that the males retain their capacity to ejaculate in a series of weekly tests. They further suggest that the female partner has a powerful modulatory effect on the rate of decline in sexual performance. Although the behavior of old males is changed in that they engage in sexual encounters with fewer females than when young, the level of performance can be maintained if they have access to a preferred female. This access, however, does not prevent further reductions in behavior. A few years after this study was conducted, some of our old males ceased responding sexually to female 1339 as well. It is important to note that female 1339 is not unique. We subsequently have found other females that are highly preferred.[118] It also is worth pointing out that some females that are not highly preferred by most males are highly preferred by at least one male. It is conceivable that these females could also raise the performance levels of any male that has a strong preference for them. There are parallels between these studies and those examining aging men. Marital satisfaction has been reported as an important factor influencing sexual behavior.[104,108]

It is likely that a highly preferred female provides more sensory stimulation to activate sexual arousal than less preferred females and that this added stimulation compensates, at least for a while, for degenerative changes in either the same sensory pathways or other pathways mediating sexual behavior. We have attempted to determine which attributes of highly preferred females contribute to the maintenance of sexual performance in old males. Because 1339 was both highly familiar and old, novelty and youth are not contributing factors (see also Chambers and Phoenix[14]). We also have found that endocrine profiles and olfactory cues do not distinguish highly preferred from nonpreferred females.[16,118] One factor that does play an important role is the behavior of the female. In our study with 1339 and the 9 different females, we measured the frequencies of female presents to male contacts (when the female presents after a male contact) and sexual invitations (when the female exhibits hand slaps, head ducks, and head bobs). When tested with female 1339, the percentage of tests with present to contacts and invitations and the ratio of presents to contacts was greater than when tested with the 9 different females (see Figure 14). It is interesting that some measures of male sexual behavior were higher in the old males tested with female 1339 than in the young males tested with the 9 females, and the present-to-contact measures were higher when 1339 was tested with the old males than when the 9 females were tested with the young males. Also, several measures of sexual behavior were higher in the young males than the old males when both were tested with the 9 females, and the percentage of tests with invitations and the present-to-contact ratio were higher when the 9 females were tested with the young males than the old males.

In subsequent studies, we have observed a similar association between the levels of sexual responses in the female and the levels of performance in a group of young and a group of old males. In one study, there was a tendency for the ratios of presents to contacts and the rates of sidles (a new measure of female sexual interest that we have defined as moving obliquely in an unobtrusive manner toward the male partner[115]) to be higher when the females were tested with young males than with old males and, in another study, these ratios and rates were significantly higher when the females were tested with young males.[21,120] These data strongly suggest that the behavior of the female modulates performance levels in aging rhesus males.

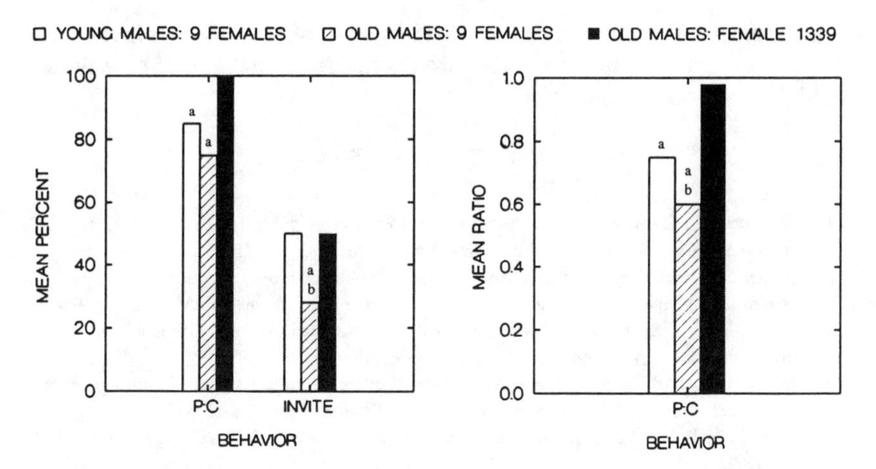

FIGURE 14. Mean of several measures of sexual behavior of 9 different females when tested with young and old males and preferred female 1339 when tested 9 separate times with the same old males. [a]Differed significantly from female 1339. [b]Differed significantly from the 9 females when tested with the young males. ***Abbreviations***: P:C, presents to contacts. (Adapted from Chambers and Phoenix.[16])

CONCLUSIONS

There are a number of sensory systems controlling and modulating male sexual behavior, and there are a number of different motor responses that comprise this sequence of behaviors that terminates with the ejaculatory response. Although the task of finding the cause of the age-related decline in male sexual behavior would be simplified if only one essential part of the complex network mediating sexual behavior was involved, the accumulation of evidence points to the involvement of a number of parts of the network. It is likely that some of the degenerative changes initiate the decline while others contribute to it at later times. Even decreases in testosterone levels, which clearly do not cause the decline, contribute to the progressive decrease. It also is probable that some of the same degenerative changes will be found across different species, but other changes may be unique to a particular species. For example, although degenerative changes in the olfactory system may prove to play a critical role in rodents, it is unlikely that it will be crucial in species such as rhesus macaques and men in which olfactory stimuli play a minor role in stimulating sexual responses.

There are a number of promising candidates that may contribute to the decline in sexual behavior. These include decreases in the concentration of estrogen receptors, decreased transmission in the noradrenergic system and deficits in the nigrostriatal dopaminergic system. Research is needed to determine the association between the time sequence of changes in these systems and the different measures of sexual behavior. In addition, there are two powerful modulators, the amount of testosterone present during perinatal development and the female partner, that induce substantial changes in the rate of decline of sexual behavior. Research is needed to determine how these modulators produce their effects.

REFERENCES

1. **Ahlenius, S., Larsson, K., and Svensson, L.,** Further evidence for an inhibitory role of central 5-HT in male rat sexual behavior, *Psychopharmacology (Berlin)*, 68, 217, 1980.
2. **Alsum, P., and Goy, R.W.,** Actions of esters of testosterone, dihydrotestosterone or estradiol on sexual behavior in castrated male guinea pigs, *Horm. Behav.*, 5, 207, 1974.
3. **Baum, M.J., and Vreeburg, J.T.M.,** Copulation in castrated male rats following combined treatment with estradiol and dihydrotestosterone, *Science*, 182, 283, 1973.
4. **Beach, F.A., and Holz-Tucker, A.M.,** Effects of different concentrations of androgen upon sexual behavior in castrated male rats, *J. Comp. Physiol. Psychol.*, 42, 433, 1949.
5. **Bishop, M.W.H.,** Aging and reproduction in the male, *J. Reprod. Fertil. (Supplement)*, 12, 65, 1970.
6. **Brackett, N.L., Iuvone, P.M., and Edwards, D.A.,** Midbrain lesions, dopamine and male sexual behavior, *Behav. Brain Res.*, 20, 231, 1986.
7. **Brawer, J.R., Schipper, H., and Naftolin, F.,** Ovary-dependent degeneration in the hypothalamic arcuate nucleus, *Endocrinology*, 107, 274, 1980.
8. **Brawer, J.R., Schipper, H., and Robaire, B.,** Effects of long term androgen and estradiol exposure on the hypothalamus, *Endocrinology*, 112, 194, 1983.
9. **Bremner, W.J., Vitiello, M.V., and Prinz, P.N.,** Loss of circadian rhythmicity in blood testosterone levels with aging in normal men, *J. Clin. Endocrinol. Metab.*, 56, 1278, 1983.
10. **Brown-Sequard, C.E.,** The effects produced on man by subcutaneous injections of a liquid obtained from the testicles of animals, *Lancet*, 2, 105, 1889.
11. **Bruni, J.F., Huang, H.-H., Marshall, S., and Meites, J.,** Effects of single and multiple injections of synthetic GnRH on serum LH, FSH and testosterone in young and old male rats, *Biol. Reprod.*, 17, 309, 1977.
12. **Chambers, K.C., Hess, D.L., and Phoenix, C.H.,** Relationship of free and bound testosterone to sexual behavior in old rhesus males, *Physiol. Behav.*, 27, 615, 1981.
13. **Chambers, K.C., and Phoenix, C.H.,** Diurnal patterns of testosterone, dihydrotestosterone, estradiol, and cortisol in serum of rhesus males: Relationship to sexual behavior in aging males, *Horm. Behav.*, 15, 416, 1981.
14. **Chambers, K.C., and Phoenix, C.H.,** Sexual behavior in old male rhesus monkeys: Influence of familiarity and age of female partners, *Arch. Sex. Behav.*, 11, 299, 1982.
15. **Chambers, K.C., and Phoenix, C.H.,** Sexual behavior in response to testosterone in old long-term castrated rhesus males, *Neurobiol. Aging*, 4, 223, 1983.
16. **Chambers, K.C., and Phoenix, C.H.,** Restoration of sexual performance in old rhesus macaques when paired with a preferred female partner, *Int. J. Primatol.*, 5, 287, 1984.
17. **Chambers, K.C., and Phoenix, C.H.,** Testosterone and the decline of sexual behavior in aging male rats, *Behav. Neural Biol.*, 40, 87, 1984.
18. **Chambers, K.C., and Phoenix, C.H.,** Testosterone is more effective than combined dihydrotestosterone and estradiol in activating sexual behavior in old male rats, *Neurobiol. Aging*, 7, 127, 1986.
19. **Chambers, K.C., and Phoenix, C.H.,** Differences among ovariectomized female rhesus macaques in the display of sexual behavior without and with estradiol treatment, *Behav. Neurosci.*, 101, 303, 1987.
20. **Chambers, K.C., and Phoenix, C.H.,** Apomorphine, deprenyl, and yohimbine fail to increase sexual behavior in rhesus male, *Behav. Neurosci.*, 103, 816, 1989.
21. **Chambers, K.C., and Phoenix, C.H.,** Sexual behavior and serum levels of prolactin, testosterone, and estradiol in young and old rhesus males, *Physiol. Behav.*, 52, 13, 1992.

22. **Chambers, K.C., Resko, J.A., and Phoenix, C.H.,** Correlation of diurnal changes in hormones with sexual behavior and age in male rhesus macaques, *Neurobiol. Aging*, 3, 37, 1982.
23. **Chambers, K.C., Thornton, J.E., and Roselli, C.E.,** Age-related deficits in brain androgen binding and metabolism, testosterone, and sexual behavior of male rats, *Neurobiol. Aging*, 12, 123, 1991.
24. **Chan, S.W.C., Leathem, J.H., and Esashi, T.,** Testicular metabolism and serum testosterone in aging male rats, *Endocrinology*, 101, 128, 1977.
25. **Christensen, L.W., and Clemens, L.G.,** Intrahypothalamic implants of testosterone or estradiol and resumption of masculine sexual behavior in long-term castrated male rats, *Endocrinology*, 95, 984, 1974.
26. **Chubb, C., and Desjardins, C.,** Testicular function and sexual activity in senescent mice, *Am. J. Physiol.*, 247, E569, 1984.
27. **Clark, J.T., and Smith, E.R.,** Effects of apomorphine on sexual behavior in young and middle-aged rats, *Neurobiol. Aging*, 8, 153, 1987.
28. **Clark, J.T., Smith, E.R., and Davidson, J.M.,** Enhancement of sexual motivation in male rats by yohimbine, *Science*, 225, 847, 1984.
29. **Clark, J.T., Smith, E.R., and Davidson, J.M.,** Evidence for the modulation of sexual behavior by alpha-adrenoceptors in male rats, *Neuroendocrinology*, 41, 36, 1985.
30. **Clark, J.T., Smith, E.R., and Davidson, J.M.,** Testosterone is not required for the enhancement of sexual motivation by yohimbine, *Physiol. Behav.*, 35, 517, 1985.
31. **Coquelin, A., and Desjardins, C.,** Luteinizing hormone and testosterone secretion in young and old male mice, *Am. J. Physiol.*, 243, E257, 1982.
32. **Craigen, W., and Bronson, F.H.,** Deterioration of the capacity for sexual arousal in aged male mice, *Biol. Reprod.*, 26, 869, 1982.
33. **Davidson, J.M.,** Effects of estrogen on the sexual behavior of male rats, *Endocrinology*, 84, 1365, 1969.
34. **Davidson, J.M., Gray, G.D., and Smith, M.S.,** The sexual psychoendocrinology of aging, in *Neuroendocrinology of Aging*, Meites, J., Ed., Plenum Press, New York, 1983, 221.
35. **Demarest, K.T., Riegle, G.D., and Moore, K.E.,** Characteristics of dopaminergic neurons in the aged male rat, *Neuroendocrinology*, 31, 222, 1980.
36. **Deslypere, J.P., and Vermeulen, A.,** Leydig cell function in normal men: Effect of age, life-style, residence, diet, and activity, *J. Clin. Endocrinol. Metab.*, 59, 955, 1984.
37. **Doherty, P.C., Bartke, A., and Smith, M.S.,** Differential effects of bromocriptine treatment on LH release and copulatory behavior in hyperprolactinemic male rats, *Horm. Behav.*, 15, 436, 1981.
38. **Donovan, B.T.,** *Hormones and Human Behavior*, Cambridge University Press, Cambridge, MA, 1985, 121.
39. **Dorsa, D.M., and Smith, E.R.,** Facilitation of mounting behavior in male rats by intracranial injection of luteinizing hormone-releasing hormone, *Regul. Pept.*, 1, 147, 1980.
40. **Dorsa, D.M., Smith, E.R., and Davidson, J.M.,** Endocrine and behavioral effects of continuous exposure of male rats to a potent LHRH agonist: Evidence for CNS actions of LHRH, *Endocrinology*, 109, 729, 1981.
41. **Eaton, G.G.,** Longitudinal studies of sexual behavior in the Oregon troop of Japanese macaques, in *Sex and Behavior*, McGill, T.E., Dewsbury, D.A., and Sachs, B.D., Eds., Plenum Press, New York, 1978, 35.
42. **Finch, C.E., Felicio, L.S., Flurkey, K., Gee, D.M., Mobbs, C., Nelson, J.F., and Osterburg, H.H.,** Studies on ovarian-hypothalamic-pituitary interactions during reproductive aging in C57BL/6J mice, *Peptides*, 1, 163, 1980.
43. **Frankel, A.I.,** Plasma testosterone levels are higher in old mating rats than in old non-mating rats, *Exp. Gerontol.*, 19, 345, 1984.
44. **Frankel, A.I., and Mock, E.J.,** Testis vein testosterone falls in the aging rat: Refutation of "Dilution" hypothesis, *J. Androl.*, 3, 113, 1982.

45. **Gessa, G.L., and Tagliamonte, A.**, Possible role of brain serotonin and dopamine in controlling male sexual behavior, *Adv. Biochem. Psychopharmacol.*, 11, 217, 1974.
46. **Ghanadian, R., Lewis, J.G., and Chisholm, G.D.**, Serum testosterone and dihydrotestosterone changes with age in rat, *Steroids,* 25, 753, 1975.
47. **Giantonio, G.W., Lund, N.L., and Gerall, A.A.**, Effect of diencephalic lesions and rhinencephalic lesions on the male rat's sexual behavior, *J. Comp. Physiol. Psychol.*, 73, 38, 1970.
48. **Goldman, J., Wajchenberg, B.L., Liberman, B., Nery, M., Achando, S., and Germek, O.A.**, Contrast analysis for the evaluation of the circadian rhythms of plasma cortisol, androstenedione, and testosterone in normal men and the possible influence of meals, *J. Clin. Endocrinol. Metab.*, 60, 164, 1985.
49. **Goodman, R.L., Hotchkiss, J., Karsch, F.J., and Knobil, E.**, Diurnal variations in serum testosterone concentrations in the adult male rhesus monkey, *Biol. Reprod.*, 11, 624, 1974.
50. **Gorski, R.A.**, Influence of age on the response to paranatal administration of a low dose of androgen, *Endocrinology*, 82, 1001, 1968.
51. **Gorski, J., and Hansen, J.C.**, The "one and only" step model of estrogen action, *Steroids*, 49, 461, 1987.
52. **Gray, G.D.**, Changes in the levels of luteinizing hormone and testosterone in the circulation of aging male rats, *J. Endocrinol.*, 76, 551, 1978.
53. **Gray, G.D.**, Age-related changes in penile erections and circulating testosterone in middle-aged male rats, in *Parkinson's Disease II: Aging and Neuroendocrine Relationships*, Finch, C.E., Potter, D.E., and Kenny, A.D., Eds., Plenum Press, New York, 1978, 149.
54. **Gray, G.D., Smith, E.R., Dorsa, D.M., and Davidson, J.M.**, Sexual behavior and testosterone in middle-aged male rats, *Endocrinology*, 109, 1597, 1981.
55. **Greenstein, B.D.**, Androgen receptors in the rat brain, anterior pituitary gland and ventral prostrate gland: Effects of orchidectomy and aging, *J. Endocrinol.*, 81, 75, 1979.
56. **Gruenewald, D.A., and Matsumoto, A.M.**, Age-related decreases in serum gonadotropin levels and gonadotropin-releasing hormone gene expression in the medial preoptic area of the male rat are dependent upon testicular feedback, *Endocrinology*, 129, 2442, 1991.
57. **Grunt, J.A., and Young, W.C.**, Consistency of sexual behavior patterns in individual male guinea pigs following castration and androgen therapy, *J. Comp. Physiol. Psychol.*, 46, 138, 1953.
58. **Haji, M., Kato, K.-I., Nawata, H., and Ibayashi, H.**, Age-related changes in the concentrations of cytosol receptors for sex steroid hormones in the hypothalamus and pituitary gland of the rat, *Brain Res.*, 204, 373, 1981.
59. **Han, Z., Kuyatt, B.L., Kochman, K.A., DeSouza, E.B., and Roth, G.S.**, Effect of aging on concentrations of D_2-receptor-containing neurons in the rat striatum, *Brain Res.*, 498, 299, 1989.
60. **Handa, R.J., Roselli, C.E., Horton, L., and Resko, J.A.**, The quantitative distribution of cytosolic androgen receptors in microdissected areas of the male rat brain: Effects of estrogen treatment, *Endocrinology*, 121, 233, 1987.
61. **Harris, V.S., and Sachs, B.D.**, Copulatory behavior in male rats following amygdaloid lesions, *Brain Res.*, 86, 514, 1975.
62. **Hart, B.L.**, Gonadal androgen and sociosexual behavior of male mammals: A comparative analysis, *Psychol. Bull.*, 81, 383, 1974.
63. **Hart, B.L., and Leedy, M.G.**, Neurobiological basis of male sexual behavior, in *Handbook of Behavioral Neurobiology: Volume 7, Reproduction*, Adler, N., Pfaff, D., and Goy, R.W., Eds., Plenum Press, New York, 1985, 3.
64. **Heimer, L., and Larsson, K.**, Impairment of mating behavior in male rats following lesions in the preoptic-anterior hypothalamic continuum, *Brain Res.*, 3, 248, 1967.
65. **Henry, J.M., Filburn, C.R., Joseph, J.A., and Roth, G.S.**, Effect of aging on striatal dopamine receptor subtypes in Wistar rats, *Neurobiol. Aging*, 7, 357, 1986.
66. **Herbert, J.**, Sexual preference in the rhesus monkey *Macaca mulatta* in the laboratory, *Anim. Behav.*, 16, 120, 1968.

67. **Hoffman, G.E., and Sladek, J.R., Jr.**, Age-related changes in dopamine, LHRH and somatostatin in the rat hypothalamus, *Neurobiol. Aging*, 1, 27, 1980.
68. **Huber, M.H.R., Bronson, F.H., and Desjardins, C.**, Sexual activity of aged male mice: Correlation with level of arousal, physical endurance, pathological status, and ejaculatory capacity, *Biol. Reprod.*, 23, 305, 1980.
69. **Jakubczak, L.F.**, Effects of testosterone propionate on age differences in mating behavior, *J. Gerontol.*, 19, 458, 1964.
70. **Jakubczak, L.F.**, Age, endocrines, and behavior, in *Endocrines and Behavior*, Gitman, L., Ed., Thomas, Springfield, 1967, 231.
71. **Jarjour, L.T., Handelsman, D.J., and Swerdloff, R.S.**, Effects of aging on the *in vitro* release of gonadotropin-releasing hormone, *Endocrinology*, 119, 1113, 1986.
72. **Johnston, P., and Davidson, J.M.**, Intracerebral androgens and sexual behavior in the male rat, *Horm. Behav.*, 3, 345, 1972.
73. **Joseph, J.A., Berger, R.E., Engel, B.T., and Roth, G.S.**, Age-related changes in the nigrostriatum: A behavioral and biochemical analysis, *Gerontology*, 33, 643, 1978.
74. **Kaler, L.W., Gliessman, P., Hess, D.L., and Hill, J.**, The androgen status of aging male rhesus macaques, *Endocrinology*, 119, 566, 1986.
75. **Kamel, F., and Frankel, A.I.**, Hormone release during mating in the male rat: Time course, relation to sexual behavior, and interaction with handling procedures, *Endocrinology*, 103, 2172, 1978.
76. **Kinoshita, Y., Higashi, Y., Winters, S.J., Oshima, H., and Troen, P.**, An analysis of the age-related decline in testicular steroidogenesis in the rat, *Biol. Reprod.*, 32, 309, 1985.
77. **Kinsey, A.C., Pomeroy, W.B., and Martin, C.E.**, *Sexual Behavior in the Human Male*, W.B. Saunders, Philadelphia, 1948, 235.
78. **Knoll, J.**, The striatal dopamine dependency of life span in male rats. Longevity study with (-)deprenyl, *Mech. Aging Devel.*, 46, 237, 1988.
79. **Knoll, J., Yen, T.T., and Dallo, J.**, Long-lasting, true aphrodisiac effect of (-)deprenyl in sexually sluggish old male rats, *Mod. Probl. Pharmacopsychiatry*, 19, 135, 1983.
80. **Krettek, J.E., and Price, J.L.**, Amygdaloid projections to subcortical structures within the basal forebrain and brainstem in the rat and cat, *J. Comp. Neurol.*, 178, 225, 1978.
81. **Landfield, P.W.**, An endocrine hypothesis of brain aging and studies on brain-endocrine correlations and monosynaptic neurophysiology during aging, in *Parkinson's Disease II: Aging and Neuroendocrine Relationships*, Finch, C.E., Potter, D., and Kenny, A., Eds., Plenum Press, New York, 1978, 179.
82. **Landfield, P.W., Waymire, J.C., and Lynch, G.**, Hippocampal aging and adrenocorticoids: Quantitative correlations, *Science*, 202, 1098, 1978.
83. **Larsson, K.**, Conditioning and sexual behavior in the male albino rat, *Acta Psych. Gothoburgensia*, 1, 1, 1956.
84. **Larsson, K.**, Sexual activity in senile male rats, *J. Gerontol.*, 13, 136, 1958.
85. **Lieberburg, I., and McEwen, B.S.**, Brain cell nuclear retention of testosterone metabolites 5α-dihydrotestosterone and estradiol-17β in adult rats, *Endocrinology*, 100, 588, 1977.
86. **Lupo-di Prisco, C., and Dessi-Fulgheri, F.**, Endocrine and behavioral modifications in aging male rats, *Horm. Res.*, 12, 149, 1980.
87. **Malmnas, C.O.**, Effects of LSD-25, clonidine and apomorphine on copulatory behavior in the male rat, *Acta Physiol. Scand., Supplement*, 395, 96, 1973.
88. **Malmnas, C.O.**, Dopaminergic reversal of the decline after castration of rat copulatory behavior, *J. Endocrinol.*, 73, 187, 1977.
89. **Marrama, P., Carani, C., Baraghini, G.F., Volpe, A., Zini, D., Celani, M.F., and Montanini, V.**, Circadian rhythm of testosterone and prolactin in the aging, *Mauritas*, 4, 131, 1982.
90. **Martin, C.E.**, Sexual activity in the aging male, in *Handbook of Sexology*, Money, J., and Musaph, H., Eds., Elsevier/North-Holland Biomedical Press, Amsterdam, 1977, 813.
91. **Masters, W.H., and Johnson, V.E.**, *Human Sexual Response*, Little and Brown, Boston, 1966, 262.

92. **McDonald, P., Beyer, C., Newton, F., Brien, B., Baker, R., Tan, H.S., Sampson, C., Kitching, P., Greenhill, R., and Pritchard, D.**, Failure of 5α-dihydrotestosterone to initiate sexual behavior in the castrated male rat, *Nature*, 227, 964, 1970.
93. **McGregor, A., and Herbert, J.**, Differential effects of excitotoxic basolateral and corticomedial lesions of the amygdala on the behavioral and endocrine responses to either sexual or aggression-promoting stimuli in the male rat, *Brain Res.*, 574, 9, 1992.
94. **Meites, J., Steger, R.W., and Huang, H.H.H.**, Relation of neuroendocrine system to the reproductive decline in aging rats and human subjects, *Fed. Proc.*, 39, 3168, 1980.
95. **Mesco, E.R., Joseph, J.A., Blake, M.J., and Roth, G.S.**, Loss of D_2 receptors during aging is partially due to decreased levels of mRNA, *Brain Res.*, 545, 355, 1991.
96. **Michael, R.P., Setchell, K.D.R., and Plant, T.M.**, Diurnal changes in plasma testosterone and studies on plasma corticosteroids in nonanesthetized male rhesus monkeys (*Macaca mulatta*), *J. Endocrinol.*, 63, 325, 1974.
97. **Miller, A.E., and Riegle, G.D.**, Temporal patterns of serum luteinizing hormone and testosterone and endocrine response to luteinizing hormone releasing hormone in aging male rats, *J. Gerontol.*, 37, 522, 1982.
98. **Minnick, R.S., Warden, C.J., and Arieti, S.**, The effect of sex hormones on the copulatory behavior of senile white rats, *Science*, 103, 749, 1946.
99. **Mobbs, C.V., Kannegieter, L.S., and Finch, C.E.**, Delayed anovulatory syndrome induced by estradiol in female C57BL/6J mice: Age-like neuroendocrine, but not ovarian, impairments, *Biol. Reprod.*, 32, 1010, 1985.
100. **Moss, R.L.**, Actions of hypothalamic-hypophysiotropic hormones on the brain, *Ann. Rev. Physiol.*, 41, 617, 1979.
101. **Myers, B.M., and Baum, M.J.**, Facilitation of copulatory performance in male rats by naloxone: Effects of hypohysectomy, 17 α-estradiol, and luteinizing hormone-releasing hormone, *Pharmacol. Biochem. Behav.*, 12, 365, 1980.
102. **Naftolin, F., and Ryan, K.J.**, The metabolism of androgens in central neuroendocrine tissue, *J. Steroid Biochem.*, 6, 993, 1975.
103. **Nelson, J.F., Felicio, L.S., Osterberg, H.H., and Finch, C.E.**, Altered profiles of estradiol and progesterone associated with prolonged estrous cycles and persistent vaginal cornification in aging C57BL/6J mice, *Biol. Reprod.*, 24, 784, 1981.
104. **Newman, G., and Nichols, C.R.**, Sexual activities and attitudes in older persons, *J. Am. Med. Assoc.*, 173, 33, 1960.
105. **O'Boyle, K.M., and Waddington, J.L.**, Loss of rat striatal dopamine receptors with aging is selective for D_2 but not D_1 sites: Association with increased non-specific binding of the D_1 ligand [^{3}H]piflutixol, *Euro. J. Pharmacol.*, 105, 171, 1984.
106. **Peng, M.T., Yang, F.J., Mu, S.C., Hsu, H.K., Wang, P.S., and Lue, S.Y.**, Rejuvenation of sexual behavior of aging rats with neurotransmitters, in *Interdisciplinary Topics in Gerontology, Vol. 24*, von Hahn, H.P., Ed., S. Karger, Basel, Switzerland, 1988, 134.
107. **Perachio, A.A., Alexander, M., Marr, L.D., and Collins, D.C.**, Diurnal variations of serum testosterone levels in intact and gonadectomized male and female rhesus monkeys, *Steroids*, 29, 21, 1977.
108. **Pfeiffer, E., Verwoerdt, A., and Wang, H.**, The natural history of sexual behavior in a biologically advantaged group of aged individuals, *J. Gerontol.*, 24, 193, 1969.
109. **Phoenix, C.H.**, Effects of dihydrotestosterone on sexual behavior of castrated male rhesus monkeys, *Physiol. Behav.*, 12, 1045, 1974.
110. **Phoenix, C.H.**, Sexual behavior of castrated male rhesus monkeys treated with 19-hydroxytestosterone, *Physiol. Behav.*, 16, 305, 1976.
111. **Phoenix, C.H.**, Factors influencing sexual performance in male rhesus monkeys, *J. Comp. Physiol. Psychol.*, 91, 697, 1977.
112. **Phoenix, C.H., and Chambers, K.C.**, Sexual behavior in adult gonadectomized female pseudohermaphrodite, female, and male rhesus macaques (*Macaca mulatta*) treated with estradiol benzoate and testosterone propionate, *J. Comp. Physiol. Psychol.*, 96, 823, 1982.
113. **Phoenix, C.H., and Chambers, K.C.**, Sexual behavior and serum hormone levels in aging rhesus males: Effects of environmental change, *Horm. Behav.*, 18, 206, 1984.

114. **Phoenix, C.H., and Chambers, K.C.**, Sexual deprivation and its influence on testosterone levels and sexual behavior of old and middle-aged rhesus males, *Biol. Reprod.*, 31, 480, 1984.
115. **Phoenix, C.H., and Chambers, K.C.**, Sexual solicitation by middle-aged and old rhesus females, *Neurobiol. Aging*, 7, 173, 1986.
116. **Phoenix, C.H., and Chambers, K.C.**, Testosterone therapy in young and old rhesus males that display low levels of sexual activity, *Physiol. Behav.*, 43, 479, 1988.
117. **Phoenix, C.H., and Chambers, K.C.**, Sexual performance of old and young male rhesus macaques following treatment with GnRH, *Physiol. Behav.*, 47, 513, 1990.
118. **Phoenix, C.H., Jensen, J.N., and Chambers, K.C.**, Stimulus qualities of a preferred female partner and sexual behavior of old rhesus males, *Physiol. Behav.*, 38, 673, 1986.
119. **Phoenix, C.H., Slob, A.K., and Goy, R. W.**, Effects of castration and replacement therapy on sexual behavior of adult male rhesus, *J. Comp. Physiol. Psychol.*, 84, 472, 1973.
120. **Phoenix, C.H., Walther, A.M., Jensen, J.N., and Chambers, K.C.**, The effect of human chorionic gonadotropin on serum levels of testosterone, estradiol, and sexual behavior in young and old rhesus males, *Physiol. Behav.*, 46, 647, 1989.
121. **Pirke, K.M., Vogt, H.-J., and Geiss, M.**, *In vitro* and *in vivo* studies on Leydig cell function in old rats, *Acta Endocrinol.*, 89, 393, 1978.
122. **Rice, G.E., Cho, G., and Barnea, A.**, Aging-related reduced release of LH-releasing hormone from hypothalamic granules, *Neurobiol. Aging*, 4, 217, 1983.
123. **Rigaudiere, N., Pelardy, G., Robert, A., and Delost, P.**, Changes in the concentrations of testosterone and androstenedione in the plasma and testis of the guinea-pig from birth to death, *J. Reprod. Fertil.*, 48, 291, 1976.
124. **Riss, W., and Young, W.C.**, The failure of large quantities of testosterone propionate to activate low drive male guinea pigs, *Endocrinology*, 54, 232, 1954.
125. **Robinson, J.A., Scheffler, G., Eisele, S.G., and Goy, R.W.**, Effects of age and season on sexual behavior and plasma testosterone and dihydrotestosterone concentrations of laboratory-housed male rhesus monkeys (*Macaca mulatta*), *Biol. Reprod.*, 13, 203, 1975.
126. **Roselli, C.E., Horton, L.E., and Resko, J.A.**, Distribution and regulation of aromatase activity in the rat hypothalamus and limbic system, *Endocrinology*, 117, 2471, 1985.
127. **Roselli, C.E., and Resko, J.A.**, Androgens regulate brain aromatase activity in adult male rats through a receptor mechanism, *Endocrinology*, 114, 2183, 1984.
128. **Roselli, C.E., Thornton, J.E., and Chambers, K.C.**, Age-related deficits in brain estrogen receptors and sexual behavior of male rats, *Behav. Neurosci.*, 107, 202, 1993.
129. **Sachs, B.D., and Meisel, R.L.**, The physiology of male sexual behavior, in *The Physiology of Reproduction*, Knobil, E., and Neill, J., Eds., Raven Press, New York, 1988, 1393.
130. **Saito, T.R., Hokao, R., Aoki, S., Chiba, N., Terada, M., Saito, M., Ohbutsu, A., Amao, H., Wakafuji, Y., and Sugiyama, M.**, Central effects of yohimbine on copulatory behavior in aged male rats, *Exp. Anim.*, 40, 337, 1991.
131. **Sar, M., and Stumpf, W.E.**, Autoradiographic localization of radioactivity in the rat brain after the injection of 1,2-^{3}H-testosterone, *Endocrinology*, 92, 251, 1973.
132. **Schipper, H., Brawer, J.R., Nelson, J.F., Felicio, L.S., and Finch, C.E.**, Role of the gonads in the histologic aging of the hypothalamic arcuate nucleus, *Biol. Reprod.*, 25, 413, 1981.
133. **Selmonoff, M.K., Brodkin, L.D., Weiner, R.I., and Siiteri, P.K.**, Aromatization and 5α-reduction of androgens in discrete hypothalamic and limbic regions of the male and female rat, *Endocrinology*, 101, 841, 1977.
134. **Sholiton, L.J., Marnell, R.T., and Werk, E.E.**, Metabolism of testosterone-4-^{14}C by rat brain homogenates and subcellular fractions, *Steroids*, 8, 265, 1966.
135. **Simpkins, J.W., Estes, K.S., Kalra, P.S., and Kalra, S.P.**, Alterations in hypothalamic neurotransmitters contribute to age-related decline in reproductive function in the male rat, in *Male Reproduction and Fertility*, Negro-Vilar, A., Ed., Raven Press, New York, 1983, 95.

136. **Simerly, R.B., Chang, C., Muramatsu, M., and Swanson, L.W.,** The distribution of androgen and estrogen receptor mRNA-containing cells in the rat brain: An *in situ* hybridization study, *J. Comp. Neurol.*, 294, 76, 1990.
137. **Smith, E.R., and Davidson, J.M.,** Yohimbine attenuates aging-induced sexual deficiencies in male rats, *Physiol. Behav.*, 47, 631, 1990.
138. **Steiner, R.A., Bremner, W.J., Clifton, D.K., and Dorsa, D.M.,** Reduced pulsatile luteinizing hormone and testosterone secretion with aging in the male rat, *Biol. Reprod.*, 31, 251, 1984.
139. **Stumpf, W.E., and Sar, M.,** Anatomical distribution of estrogen, androgen, progestin, corticosteroid and thyroid hormone target sites in the brain of mammals: Phylogeny and ontogeny, *Am. Zool.*, 18, 435, 1978.
140. **Swanson, L.J., Desjardins, C., and Turek, F.W.,** Aging of the reproductive system in the male hamster: Behavioral and endocrine patterns, *Biol. Reprod.*, 26, 791, 1982.
141. **Tagliamonte, A.P., Tagliamonte, G.L., and Gessa, G.L.,** Reversal of pargyline-induced inhibition of sexual behavior in male rats by p-chlorophenylalanine, *Nature*, 230, 244, 1971.
142. **Tenover, J.S., Matsumoto, A.M., Clifton, D.K., and Bremner, W.J.,** Age-related alterations in the circadian rhythms of pulsatile luteinizing hormone and testosterone secretion in healthy men, *J. Gerontol. Med. Sci.*, 43, 163, 1988.
143. **Tsitouras, P.D, Martin, C.E., and Harman, S.M.,** Relationship of serum testosterone to sexual activity in healthy elderly men, *J. Gerontol.*, 37, 288, 1982.
144. **Wallis, C.J., and Luttge, W.G.,** Maintenance of male sexual behavior by combined treatment with oestrogen and dihydrotestosterone in CD-1 mice, *J. Endocrinol.*, 66, 257, 1975.
145. **Whalen, R.E., Yahr, P., and Luttge, W.G.,** The role of metabolism in hormonal control of sexual behavior, in *Handbook of Behavioral Neurobiology: Vol. 7, Reproduction*, Adler, N., Pfaff, D., and Goy, R.W., Eds., Plenum Press, New York, 1985, 609.

Chapter 8

INTERACTIONS BETWEEN OVARIAN AND NEUROENDOCRINE FUNCTION DURING AGING

J.K.H. Lu and P.S. LaPolt

TABLE OF CONTENTS

0-8493-2451-3/95/$0.00+.50

ABBREVIATIONS

17α-OH-P–17α-hydroxy-progesterone
A–androstenedione
E2–estradiol
FSH–follicle-stimulating hormone
GnRH–gonadotropin-releasing hormone
HCG–human chorionic gonadotropin
LH–luteinizing hormone
P–progesterone
PE–persistent-estrous (state)
RIA–radioimmunoassay
SHBG–sex hormone-binding globulin
T–testosterone

INTRODUCTION

Infertility in women of advanced age is a growing clinical concern in the United States and other western countries, reflecting the increasing numbers of women who delay childbirth due to careers, later age of marriage, and other reasons. The effects of aging on oocyte quality and uterine function,[30] perhaps secondary to altered patterns of ovarian follicle development and steroidogenesis, render many women in their 40s less fertile, despite the occurrence of ovulatory menstrual cycles.[12] With further advanced age and the complete cessation of menstrual cyclicity, postmenopausal women experience additional physiological changes that result in increased incidences of hot flushes, osteoporosis, and cardiovascular disease, as well as other ailments. Biomedical research examining the endocrine, cellular, and molecular mechanisms underlying age-related infertility and the clinical consequences of menopause is thus critical for the development of clinical strategies to alleviate the effects of reproductive aging in the female. This chapter will review changes in neuroendocrine, ovarian, and pregnancy functions during the aging transition in the female rat model and in the human. It will also consider the possible neuroendocrine basis of menopausal hot flushes.

REPRODUCTIVE DECLINES IN AGING FEMALE RATS

Ovarian and Neuroendocrine Functions in Aging Female Rats

Although female rats exhibit relatively short (4–5 days) ovulatory cycles, the basic interactions and mechanisms governing neuroendocrine and ovarian functions in the rat are in general similar to that of humans. In addition, age-related changes in ovarian and neuroendocrine function appear to be similar between the two species. Thus, the female rat has provided a useful and relevant model to study mechanisms underlying the decline in fertility with age. During regular estrous cycles in the rat, basal gonadotropin levels stimulate growth, maturation, and steroidogenesis of antral follicles. Increasing estradiol (E2) production from developing follicles on diestrous day 2 and proestrus act upon the hypothalamus and pituitary to elicit preovulatory luteinizing hormone (LH) and follicle-stimulating hormone (FSH) surges on proestrus,[46] presumably involving hypothalamic release of gonadotropin-releasing hormone (GnRH) in response to E2 stimulation.[42] The gonadotropin surges result in increased circulating progesterone (P) levels on the afternoon of proestrus, followed by ovulation occurring early in the morning of estrus. Under the luteinizing effects of the gonadotropin surges, the ruptured follicles form corpora lutea, which produce substantial amounts of P during diestrous days 1 and 2 of the subsequent estrous cycle. In addition to the preovulatory surges of LH and FSH, there is a secondary GnRH-independent surge of FSH early in the morning of estrus, which is believed to play a role in selection of preovulatory follicles for the next cycle.

Reproductive aging in middle-aged (9- to 12-month-old) multiparous rats is characterized by a gradual transition from regular 4-day-long cycles to lengthened, irregular cyclicity,[26] as also occurs in perimenopausal women.[43] Lengthened cycles in the rat are due to both extended periods of diestrus, with low plasma E2 levels, and extended estrus, characterized by elevated E2 but the absence of preovulatory gonadotropin surges. These observations indicate that alterations of ovarian steroidogenesis and diminished positive feedback responsiveness to E2 contribute to lengthened, irregular cyclicity in aged rats.

Interestingly, the transition from regular to irregular cycles occurs considerably earlier in virgin than in multiparous rats,[22,29] suggesting a beneficial effect of repeated pregnancies on maintaining regular neuroendocrine and ovulatory function. While changes in ovarian and endocrine functions are similar in middle-aged cyclic rats and perimenopausal women, the anovulatory, persistent-estrous (PE) state that follows irregular cyclicity in the rat is distinguished by the accumulation of cystic follicles, moderately elevated E2 levels, and the absence of positive feedback responses,[26] and thus differs significantly from the hypoestrogenic menopausal condition. Following the PE state, aging rats spontaneously experience repeated ovulations and extended maintenance of the corpora lutea due to elevated prolactin levels, a phenomenon referred to as repetitive pseudopregnancies. Finally, the oldest female rats exhibit constant anestrus characterized by low ovarian steroid levels, with ovaries containing little or no follicular or luteal tissue. Study of these aged, anestrous rats is complicated by the common presence of pituitary pathology and/or tumors.

As in the human,[2,40] there is a marked decline in the numbers of ovarian follicles during aging in the rat.[27] Presumably, there is some minimal threshold of follicle number below which it is difficult to maintain normal patterns of follicle development, steroidogenesis, oocyte maturation, and/or feedback interaction with the neuroendocrine system. There is a constant, gonadotropin-independent recruitment of resting primordial follicles to begin their growth, resulting in a gradual, continuous depletion of the ovarian follicular pool. The majority of these recruited follicles become atretic, with only a few being selected (presumably by FSH early in the cycle)[13,14] for development to the preovulatory stage.

The ovarian and/or endocrine substances that trigger a resting primordial follicle to begin growth are unknown. Significantly, experimental manipulations that decrease the follicle pool hasten reproductive senescence,[3,32] whereas delaying the rate of follicle depletion by food restriction or early hormonal treatments lengthens the reproductive life span.[21,36] The decreased follicle pool in cyclic middle-aged rats results in fewer developing follicles per cycle[39] and is associated with a decline in circulating inhibin bioactivity,[9] increased plasma FSH during the period of follicle selection,[9,10] and accelerated follicle growth and follicular E2 production.[24] While fewer developing follicles are available for selection in regularly cyclic middle-aged than in young rats early in the cycle,[39] the prolonged secondary FSH surge on estrus in middle-aged rats[9,10] apparently decreases atresia, resulting in similar numbers of preovulatory follicles in middle-aged and young females.[39] The significance of an altered pattern of follicle development and steroidogenesis on the progression of reproductive aging is not clear but may include deleterious effects on oocyte quality[4] and neuroendocrine functions[25] resulting from excessive E2 exposure.

In addition to changes at the ovarian level, clear alterations in neuroendocrine function precede the transition from regular to irregular estrous cycles. The hallmark of neuroendocrine aging in the rat is a decline in preovulatory LH surge magnitude on proestrus.[6,34,51] Attenuated LH surges are believed to reflect decreased positive feedback responses to E2 stimulation and altered diurnal patterns of hypothalamic neurotransmitter activity regulating GnRH secretion.[52] However, there is considerable heterogeneity regarding the temporal onset of neuroendocrine aging in regularly cyclic rats. Whereas some middle-aged rats show a significant decline in LH surge magnitude, other females of the same age display LH surge magnitudes comparable to those in young animals. Nass *et al.*[34] demonstrated that middle-aged rats that display attenuated LH surges soon become irregularly cyclic or anovulatory, whereas those middle-aged females with normal preovulatory LH surge profiles maintain regular cycles for at least 2 months. Thus, the age-related decline in LH surge magnitude in middle-aged rats is both a marker of neuroendocrine aging and a reliable predictor of the imminent loss of regular ovulatory function.

Fertility and Fecundity

As in the human, increased maternal age in the rat is associated with decreased ability to achieve and maintain pregnancy. The decline in fertility in middle-aged rats occurs more rapidly than the decline in regular estrous cyclicity.[28] A study by Matt *et al.*[28] demonstrated that by 12 months of age, only 20% of those females maintaining regular estrous cycles were fertile, compared with 93% fertility in young (4-month-old) animals. Few irregularly cyclic or PE females achieved pregnancy after mating, indicating that regular estrous cyclicity is essential, but not sufficient, for fertility in middle-aged rats. The decline in fertility occurs significantly earlier in regularly cyclic virgin than multiparous rats, suggesting some beneficial effect of repeated pregnancies in delaying reproductive senescence.[22,29]

Advanced maternal age is also associated with a marked decrease in fecundity in middle-aged rats. Even among regularly cyclic, fertile females, litter size decreased from 12 pups per litter at 4 months of age to an average of 1 pup per litter at 12 months.[28] In addition, the litter sizes of those few irregularly cyclic females that were fertile were consistently smaller than those of regularly cyclic rats of the same age.[29] The decline in litter size with age was associated with reduced numbers of implantation sites, rather than postimplantation failures, suggesting that changes in embryonic development and/or maternal environment during the pre- and/or peri-implantation period contribute to decreased fecundity and fertility with advanced maternal age.[28]

Ovulation and Early Embryogenesis

Although the basis of age-related declines in fertility and fecundity are not completely understood, the decrease in numbers of implanting embryos appears to be related to a gradual decline in ovulation rate[7,29] and fewer numbers of blastocysts available for implantation on day 5 of pregnancy.[7,29] Our recent findings suggest that those middle-aged rats with attenuated LH surges have lower numbers of ovulating ova, compared with young rats and middle-aged females with normal LH surge profiles. Thus, a decline in LH surge magnitude may have immediate effects on the numbers of ovulated oocytes available for fertilization, embryonic development, and implantation.

Additional studies also demonstrate that the decline in numbers of blastocysts available for implantation in middle-aged pregnant rats is associated with altered patterns of preimplantation embryonic development. Fertilization rates are similar between young and middle-aged rats, as are the total numbers of embryos seen between days 2 and 5 of pregnancy.[7] However, embryos from 10- and 13-month-old females display a significantly delayed pattern of cleavage and an increased incidence of morphological abnormalities.[7] Thus, the decrease in numbers of blastocysts available for implantation on day 5 of pregnancy in aging rats appears to primarily reflect delayed and abnormal patterns of preimplantation embryonic development.

The mechanisms underlying altered patterns of preimplantation embryonic development are not clearly understood. Studies by Peluso *et al.*[38] demonstrate that in middle-aged females there is a high incidence of ultrastructural abnormalities in oocytes within preovulatory follicles, suggesting a decline in oocyte quality with age. Butcher and colleagues demonstrated that excessive exposure of preovulatory oocytes to intrafollicular estrogen is deleterious to oocyte quality.[4] These findings suggest that higher E2 content in developing follicles of middle-aged cyclic rats[24] may contribute to a decline in oocyte quality and to altered patterns of preimplantation embryonic development during aging. In addition, there is evidence that changes in the hormonal patterns during early gestation, specifically increased E2/P ratios, negatively influence the pattern of early embryonic development and implantation in middle-aged female rats.[20] Thus, age-related changes in follicle development and ovarian steroidogenesis, both prior to ovulation and during early pregnancy, may contribute to impaired oocyte quality, altered patterns of embryonic development, and decreased fertility and fecundity in middle-aged rats.

Influence of Estradiol (E2) and Progesterone (P) on Reproductive Declines During Aging

Several reports in the literature indicate that aging of the reproductive neuroendocrine system results partly from the cumulative effects of estrogen exposure during an animal's lifetime. Ovariectomy of young female rats results in the maintenance of positive feedback responses to E2 at middle-age.[1] Conversely, treatments of young females with estrogen implants abolish the positive feedback response.[25] As mentioned above, age-related declines in fertility and fecundity occur considerably earlier in virgin than multiparous female rats. Furthermore, continued caging of retired breeder females with fertile males delays the onset of reproductive senescence, compared with retired breeders that do not experience further pregnancies.[35] Interestingly, treatments of young virgin and middle-aged retired breeders with P implants, which decrease E2 production and mimic the endocrine environment of pregnancy in the rat, are also effective in delaying the loss of regular cyclicity and fertility,[6] whereas concomitant treatments with P and E2 implants abolish this effect.[23] During P implant treatments, ovarian production of estrogen is suppressed for a long period, compared with the cyclic increases in E2 seen every few days in control animals.[22] These findings are consistent with the hypothesis that exposure to endogenous estrogen levels during repetitive estrous cycles hastens the onset of neuroendocrine and reproductive dysfunction in aging female rats, although the mechanisms responsible for such effects are not known. Interestingly, the effects of P implants on delaying reproductive senescence may also involve influences on the ovarian follicle pool size. P implant treatments significantly delay the loss of ovarian follicles, such that females treated with successive P implants early in life have larger numbers of primordial follicles at 8 months of age than do control animals.[21] Thus, exposure to gonadal steroids may influence the onset of reproductive aging by effects at both neuroendocrine and ovarian levels.

REPRODUCTIVE AGING IN WOMEN

Neuroendocrine Regulation of Gonadotropin Secretion and the Menstrual Cycle

Ovarian Function During the Menstrual Cycle

In humans, menarche occurs around the age of 12–13 years. The early episodes of menstrual bleeding in teenage girls reflect mainly increased ovarian follicular activity and E2 secretion and are not associated with gonadotropin surges or ovulations. Within one year following menarche,[33] most teenagers begin to show monthly ovulatory menstrual cycles, resulting from maturation of the neuroendocrine mechanism responsible for the positive feedback effect of E2 on gonadotropin secretion. In young adult women, regular menstrual cycles and ovulations are repeated approximately once every month during the reproductive years, roughly between 14 and 50 years of age. Toward the menopausal transition, however, menstrual cycles in middle-aged women become irregular and are often anovulatory.[43] By 50 years of age, most women have lost menstrual and ovulatory function and become menopausal.[17]

During the normal menstrual cycle, the ovary undergoes cyclic, orderly changes in morphology and function that are closely regulated by pituitary secretion of gonadotropins. In turn, both steroid hormones (E2 and P) and peptides (inhibin and activin) produced by the ovary during the cycle modulate gonadotropin secretion through both positive and negative feedback mechanisms. Counting the first day of menstruation as day 1 of the cycle, E2 production by the developing follicles, and hence, the circulating level of E2, only increases slowly during the first week of the cycle. Beginning on day 7 or 8 of the cycle, ovarian secretion of E2 increases markedly and rapidly, mainly coming from the growing Graafian follicle destined for ovulation. During the follicular phase (about the first 2 weeks) of the cycle, pituitary secretion of gonadotropins not only stimulates follicle growth and E2 production, but also ensures oocyte maturation through the growing follicle and its paracrine secretions. Ovarian secretion of E2 reaches a peak one day before the midcycle LH surge. After the peak and before ovulation, a partial luteinization of the preovulatory follicle under LH action results in a transient fall in E2 secretion. Nonetheless, a steady and sustained rise in plasma E2 during the late follicular phase of the cycle acts through the hypothalamic-pituitary axis to elicit preovulatory surges of both LH and FSH, resulting in the final maturation and rupture of the follicle. Under LH action, the ruptured follicle is luteinized to form a corpus luteum, which produces large amounts of both E2 and P. Corpus luteum secretion of E2 and P reaches a maximum about 5–7 days after ovulation, and both E2 and P are essential for preparing the uterine endometrium properly for implanting the blastocyst.

Following implantation, human chorionic gonadotropin (HCG) produced by early placental tissue prevents the corpus luteum from regressing. However, if implantation does not occur, regression of the corpus luteum results in a return of plasma E2 and P to low basal values, and menstruation occurs. Compared with E2, the ovary produces less estrone than E2 throughout the menstrual cycle. Both the ovary and the adrenal gland secrete androstenedione (A), and most estrone and testosterone (T) in the circulation comes from enzymatic conversion of A. During the follicular phase of the cycle, secretion of P by the growing follicles is minimal. At the midcycle, the LH surge causes a partial luteinization of the preovulatory follicle, resulting in a small rise in circulating P but a large increase in 17α-hydroxy-progesterone (17α-OH-P). Following ovulation, the functioning corpus luteum produces some 17α-OH-P, in addition to large amounts of both E2 and P.

Neuroendocrine Regulation of Gonadotropin Secretion During the Menstrual Cycle

Pituitary secretion of gonadotropins is mainly controlled by the action of GnRH, which is synthesized and released by specific neurons in the hypothalamus. Hypothalamic GnRH is discharged in a pulsatile fashion into the pituitary portal circulation, resulting in pulsatile patterns of both LH and FSH release from the gonadotroph cells. Also, through the different phases of the menstrual cycle, both E2 and P in the circulation modulate pituitary secretion of gonadotropins through negative and positive feedback mechanisms, presumably by changing the amplitude and, to a lesser extent, frequency of hypothalamic GnRH release or by direct actions on pituitary gonadotrophs.

During the luteal phase of the cycle, high levels of both E2 and P in the circulation exert a potent inhibition on gonadotropin secretion, resulting in low basal levels of plasma LH and FSH. However, toward the end of the luteal phase, cessation of corpus luteum function decreases plasma E2 and P to low levels, thereby relieving a strong negative feedback on gonadotropins. These lead to a preferential rise in FSH secretion, which stimulates follicle development and growth. This slight increase in FSH secretion continues throughout the first week of the follicular phase of the next cycle, while LH secretion remains low. Beginning on day 7 or 8 of the cycle, a steady and marked rise in plasma E2 resulting from the growing Graafian follicle reduces FSH secretion considerably, while secretion of LH gains a small but steady increase under the action of E2. At midcycle, the sustained rise in circulating E2 during the late follicular phase acts through the neuroendocrine mechanism(s) within the hypothalamic-pituitary system to elicit both LH and FSH surges. In humans and other primates, whether the midcycle gonadotropin surges are mediated by an E2-elicited concomitant increase in hypothalamic release of GnRH remains a controversial issue.[19] During the midcycle LH surge, there is also a small rise in P secretion by the preovulatory follicle, which may enhance and prolong both the LH and FSH surges through a facilitatory action with E2.

Ovarian Function and Gonadotropin Secretion During the Menopausal Transition and After the Menopause

From Young Adulthood to Perimenopause

Young healthy adult women continue to exhibit regular monthly menstrual and ovulatory function associated with cyclic increases in ovarian follicular activity and in steroid and peptide secretion. During each cycle, a number of small follicles are recruited for development and growth, but usually only one of these is selected, matures fully, and becomes the preovulatory follicle at the midcycle, while all others undergo atresia. The total number of follicles in the ovaries of a healthy young woman is estimated to be around 400,000, but the follicular pool decreases with age.[2,40] It has been well established that while gonadotropins are essential for supporting follicle growth and maturation prior to ovulation, within the follicular pool there is a constant recruitment of small, resting follicles entering the growth pool even in the absence of gonadotropins.[16] It is interesting to note that, among the large stock of primordial follicles (about 400,000) in the ovaries of a young adult, only 0.1% will be actually involved in ovulatory function throughout her entire reproductive life (approximately 25 years),[12] while the great majority of the ovarian follicles are lost by atresia. In fact, a dramatic reduction in the follicular pool size occurs in the ovaries during postnatal and prepubertal years.

It is estimated that the total number of primordial follicles in the prepubertal ovaries decreases from approximately 2,000,000 to 400,000, representing an 80% drop within a period of 12–13 years even before sexual maturation and the initiation of menstrual and ovulatory cycles.[2,40] It has been further revealed recently that, beginning around 40–42 years of age in women, the rate of follicle depletion from the ovarian pool is accelerated.[41] Although the mechanism(s) underlying an accelerated follicular depletion in middle-aged women is not known, it is well recognized that women over 40 years of age often experience great difficulty in achieving fertility despite regular menstrual cyclicity.[12]

The perimenopause represents the transitional phase from regular menstrual cyclicity to the permanent cessation of menstrual cycle (i.e., menopause) and is characterized by menstrual irregularity. During the menopausal transition, changes in steroid hormone secretion resulting from altered function of the aging ovary occur. It has been reported that, in regularly menstruating women over the age of 45 years, the length of the follicular but not luteal phase is reduced as compared to that of the normal cycle in young adults.[43] In these older women, plasma levels of E2 during the follicular and luteal phases and at midcycle are lower than those in younger individuals, while no consistent differences in plasma P are observed. The consistent reduction in circulating E2 is associated with a striking elevation of plasma FSH throughout the cycle while LH values remain in the normal range.[43] These hormonal profiles suggest a significant reduction in the number of growing follicles and their associated E2 and inhibin production.

In perimenopausal women, hormonal changes associated with follicle maturation and corpus luteum function often occur even in the presence of high, menopausal levels of both LH and FSH with diminished secretion of E2 and P. In some cases, vaginal bleeding occurs during a fall in circulating E2 with no associated rise in plasma P. Menstrual cycles of variable length during the menopausal transition are likely due to irregular maturation of residual follicles left in the aging ovary with diminished responsiveness to gonadotropin stimulation (thus, a diminished circulating E2), or to anovulatory vaginal bleeding that follows E2 withdrawal without evidence of corpus luteum function.

Menopause and Postmenopause

The menstrual irregularity in perimenopausal women is soon followed by the menopause at an average age of 50 years. The menopause is characterized by a permanent cessation of menstrual and ovulatory function and associated with major changes in both ovarian and pituitary hormone secretion.[18,44,50] Following menopause, a few small follicles may still be seen in the ovaries of some but not most women,[2] but these residual follicles show limited or no response to gonadotropins. In postmenopausal women, a diminished E2 production by the aging ovary contributes primarily to the dramatic reduction in circulating E2. These low E2 levels are clearly less than those found in any phase of the normal menstrual cycle in young adults and are comparable to those seen in young women following bilateral oophorectomy.[18] After menopause, plasma levels of estrone are usually higher than those of E2, and most estrone in the circulation comes from peripheral conversation of androstenedione

(A) while most circulating E2 is derived from conversion of either estrone or testosterone (T). A is the principal androgen secreted by the developing follicles in young ovaries. With menopause, plasma levels of A are reduced to about half of that in young women. In contrast, the postmenopausal ovary continues to secrete T, probably originating from the hilar cells and the luteinized stromal cells, making the plasma levels of T only slightly lower in postmenopausal women than in young adults. In the absence of follicular activity and the corpus luteum, it is obvious that there is diminished secretion of P from the postmenopausal ovary.

During the normal cycle in young women, pituitary secretion of gonadotropins is under direct control of episodic hypothalamic GnRH release and through feedback actions by ovarian steroid hormones and peptides. Following menopause, the minimal amounts of both E2 and P present in the circulation are insufficient to exert a normal negative feedback mechanism inhibiting the synthesis and release of gonadotropins. Under these conditions, the pituitary secretion of both LH and FSH are markedly enhanced, with LH and FSH pulses of large amplitude detected once every 60–90 minutes in the circulation. This enhanced pituitary release of both LH and FSH presumably reflects augmented GnRH output from specific hypothalamic neurons in the face of diminished circulating E2 and P. Since the secretion of FSH is under dual inhibition by both ovarian E2 and inhibin, in addition to hypothalamic GnRH stimulation, a greater increase in FSH than LH secretion after menopause is presumably due to diminished production of both E2 and inhibin by the aging ovary.

Neuroendocrine Aspects of Menopausal Hot Flushes

Changes in Neuroendocrine and Other Body Functions After Menopause

The average woman goes through menopause at about 50 years of age.[17] Assuming a life expectancy of 78 years for women in the United States,[45] most women can anticipate experiencing 28 years of postmenopausal and postreproductive life. This prolonged period of over one-third of the total life span is not only associated with diminished ovarian E2, P and inhibin secretion but also with major changes in body functions, including neuroendocrine mechanisms within the hypothalamic-pituitary axis. A few examples of these are enhanced pituitary gonadotropin secretion, increased turnover rates for dopamine and norepinephrine in the hypothalamus, instability of core temperature maintenance through hypothalamic control, osteoporosis, cardiovascular disease, vaginal epithelium dryness, atrophy of most reproductive tissues (breasts, vagina, sex skin, etc.), sleep disturbances, and hot flushes. It is interesting to note that most of these abnormal body functions are directly related to a sudden decrease in circulating steroid hormones, particularly E2, in postmenopausal women, following a life-long exposure of tissues/cells to circulating E2 in sufficient amounts (an "estrogen withdrawal" response). In this regard, hormone replacement therapy with estrogen remains the most common means of alleviating postmenopausal symptoms.

Menopausal Hot Flushes as Physiological Events

With menopause, there is a diminished E2 secretion, and 65–76% of women over 50 years of age will experience an episodic disturbance consisting of sudden flushing and perspiration, referred to as a hot flush or flash.[49] It has been estimated that, of those older women having hot flushes, 82% will continue to experience such episodic disturbances for more than 1 year,[49] and 35–50% will experience the symptoms for longer than 5 years.[15]

Usually, a hot flush begins with a feeling of pressure in the head, similar to a headache, which increases in intensity until sudden perspiration occurs. The hot flush is associated with a subjective feeling of heat or burning of the skin in the areas affected, followed immediately by an outbreak of perspiration. The sweating affects the entire body but is most obvious over the head, neck, upper chest, and back. Hot flushes may occur once every hour, and each episode lasts for 4–10 minutes. Each episode of the hot flush involves cutaneous vasodilation, sweating, a decrease in core temperature, and an increase in pulse rate. The vasodilation begins about 1 minute after the onset of the subjective feeling of flush and continues for about 8 minutes, resulting in a measurable change in skin temperature particularly on the fingers and toes. On the finger, the average increase in skin temperature is about 4°C with the maximum being 9°C.[47] A decrease in skin resistance, resulting from sweating, begins on average 45 seconds following the onset of the subjective flush, and reaches its maximum within 4 minutes. As body heat is lost by cutaneous vasodilation and perspiration, body core temperature declines by about 0.2°C, beginning about 4 minutes after the initiation of the subjective flush and lasting for 30 minutes. From these sequences of events, it is clear that, while the subjective symptoms of the flush occur briefly, the physiologic changes associated with the flush continue many minutes thereafter. Most patients having hot flushes complain of night sweats and insomnia, and the occurrence of hot flushes and waking episodes during the night are closely temporally related.

Menopausal Hot Flushes as Neuroendocrine Events

Physical and psychological symptoms associated with hot flushes affect millions of both postmenopausal and perimenopausal women daily, and symptoms similar to these also occur in many premenopausal young women with premature ovarian failure or after oophorectomy or GnRH analog treatment.[8] In order to develop desirable medical treatments for older women (and young women with a diminished E2 secretion) and to improve the quality of their lives, the pathophysiology and neuroendocrine mechanisms of hot flushes should be an active area for biomedical investigations.

The mechanisms underlying the repetitive occurrence of hot flushes in postmenopausal women are not known. Since hot flushes only occur in women after spontaneous cessation of ovarian E2 secretion or following oophorectomy, it is logical to postulate that hot flushes result from neuroendocrine changes associated with the loss of ovarian function, specifically the decrease in ovarian E2 secretion and/or the enhanced pituitary gonadotropin secretion.

In an extensive study,[11] it was observed that the women with frequent, severe hot flushes had significantly lower mean body weight (percent ideal weight) and plasma levels of E2 and total estrogen than women who had never experienced the symptoms, suggesting that body size and its effects on endogenous estrogen metabolism may be a factor involved in the occurrence of hot flushes in some patients but not in others. In the human circulation, E2 can either bind to sex hormone-binding globulin (SHBG) or serum albumin or remain in a free form, and the portion of E2 not bound to the SHBG is the fraction that is transported into the brain, including the hypothalamus.[37] Results from the above-mentioned study also revealed that the mean value of non-SHBG-bound plasma E2 in symptomatic women was only 50% of that in older patients who had never experienced a hot flush,[11] suggesting that the non-SHBG-bound circulating E2 is a potential determinant for the occurrence of hot flushes.

It is well recognized that, with menopause, there is a dramatic increase in gonadotropin secretion with LH and FSH pulses in large amplitude, while many of these older women experience repetitive episodes of hot flushes. A landmark study[48] conducted in 1978 at the UCLA Medical Center revealed important insights into the potential neuroendocrine mechanisms underlying the occurrence of hot flushes in women with diminished circulating E2. During a 48-hour study in which frequent, consecutive blood samples were collected from 6 postmenopausal women, these investigators recorded 34 hot flushes (confirmed by both subjective and objective criteria) and identified 31 pulses of LH release by radioimmunoassay (RIA). Among these, 26 LH pulses (84%) had a close temporal relationship with the occurrence of hot flushes, and there was a good correlation between plasma LH levels and the increase in finger skin temperature. Although both LH and FSH secretion increased markedly in these older women, a correlation between FSH pulses and hot flushes was not seen. These findings suggest that the neuroendocrine mechanisms responsible for initiating pulsatile LH release may also be involved in triggering thermoregulatory events resulting in hot flushes. In patients with pituitary insufficiency and hypoestrogenism, hot flushes were also observed despite low plasma LH and no pulsatile release.[31] In postmenopausal women, administration of a potent agonist of GnRH abolished the pulsatile release of LH but not the occurrence of hot flushes.[8] These observations indicate that it is the hypothalamic mechanisms related to LH release, rather than LH secretion per se, that are responsible for the onset of hot flushes.

Evidence indicates that, in rhesus monkeys, GnRH of hypothalamic origin fluctuates in the pituitary portal vein blood,[5] and these fluctuations are thought to be involved in the pulsatile release of LH from the pituitary. The arcuate nucleus of the mediobasal hypothalamus is believed to be the center governing the episodic release of GnRH, and the hypothalamic discharge of GnRH is probably mediated by specific neurotransmitter inputs to the GnRH neurons.

Several neurohumors such as norepinephrine, dopamine, opioids, and prostaglandins can influence LH release, presumably through their direct or indirect effects on hypothalamic GnRH release. Based on these discussions, it is logical to propose that hypothalamic GnRH release or the neurohumors that influence GnRH release may somehow alter the set point of the hypothalamic thermoregulatory centers to trigger hot flushes, under persistently low circulating E2 conditions. Although the precise mechanisms for this thermoregulatory dysfunction are not known, the present working hypothesis is consistent with the observations that both norepinephrine turnover rate and LH secretion are markedly enhanced under hypoestrogenic conditions.

CONCLUSIONS

It is apparent that age-related declines in fertility and regular ovulatory cyclicity in perimenopausal women and the middle-aged rat model are associated with significant alterations in both ovarian and neuroendocrine functions. Whether such changes initially occur in the ovary or neuroendocrine system has been the subject of much debate, especially in the rat model, due to the close interactive relationships between these tissues. Indeed, the challenge currently faced is to further elucidate the mechanisms by which the aging neuroendocrine system and ovary interact with one another to influence each other's functions, ultimately resulting in the loss of reproductive capacity.

As early as 1975, Sherman and Korenman[43] suggested that elevations in follicular phase FSH levels in perimenopausal women were secondary to decreased inhibin production by a dwindling number of developing follicles. In the middle-aged regularly cyclic rat, increased FSH levels early in the cycle[9,10] appear to be related to decreased circulating inhibin bioactivity.[9] However, the detailed cellular and molecular mechanisms by which inhibin gene expression may be altered in the aging ovary, and the consequences of increased FSH stimulation on follicular growth, ovarian steroidogenesis, oocyte quality, and the endocrine interactions mediating reproductive functions are not clear. Similarly, although a decline in LH surge magnitude remains the hallmark of neuroendocrine aging in the rat, several intriguing questions regarding this phenomenon remain. Attenuated LH surges are believed to be secondary to diminished GnRH secretion from the hypothalamus resulting from diminished positive feedback responsiveness to E2[52] and altered patterns of hypothalamic catecholamine turnover rates.[52] However, further details concerning the bases of these changes remain elusive. In addition, more detailed studies of preovulatory gonadotropin surge profiles in perimenopausal women are required to clarify whether similar declines in LH surge magnitude occur in women, as is suggested by some of the available data,[43,44] and the immediate impact, if any, of attenuated LH surges on the processes of ovulation, oocyte maturation, luteinization, and gestation.

Studies of aging animals have highlighted our lack of knowledge regarding such processes as the initiation of follicle recruitment (and hence, depletion), the regulation of normal embryonic development, and the neuroendocrine mechanisms underlying postmenopausal hot flushes. Continuing advances in these fields will contribute to new medical therapies for dealing with the clinical manifestations of reproductive aging and further our knowledge of these basic biological processes.

REFERENCES

1. **Blake, C.A., Elias, K.A., and Huffman, L.J.,** Ovariectomy of young adult rats has a sparing effect on the ability of aged rats to release luteinizing hormone, *Biol. Reprod.*, 28, 575, 1983.
2. **Block, E.,** Quantitative morphological investigations of the follicular system in women: Variations in different phases of the sexual cycle, *Acta Endocrinol.*, 8, 33, 1951.
3. **Butcher, R.L.,** Effect of reduced ovarian tissue on cyclicity, basal hormone levels and follicular development in old rats, *Biol. Reprod.*, 32, 315, 1985.
4. **Butcher, R.L., and Pope, R.S.,** Role of estrogen during prolonged estrous cycles of the rat on subsequent embryonic death and development, *Biol. Reprod.*, 21, 491, 1979.
5. **Carmel, P.W., Araki, S., and Ferin, M.,** Pituitary stalk portal blood collection in rhesus monkeys: Evidence for pulsatile release of gonadotropin-releasing hormone (GnRH), *Endocrinology*, 99, 243, 1976.
6. **Cooper, R.L., Conn, P.M., and Walker, R.F.,** Characterization of the LH surge in middle-aged female rats, *Biol. Reprod.*, 23, 611, 1980.
7. **Day, J.R., LaPolt, P.S., Morales, T.H., and Lu, J.K.H.,** An abnormal pattern of embryonic development during early pregnancy in aging rats, *Biol. Reprod.*, 41, 933, 1989.
8. **DeFazio, J., Meldrum, D., Laufer, L., Vale, W., Rivier, J., Lu, J., and Judd, H.,** Introduction of hot flashes in premenopausal women treated with a long-acting GnRH agonist, *J. Clin. Endocrinol. Metab.*, 56, 445, 1983.
9. **DePaolo, L.V.,** Age-associated increases in serum follicle-stimulating hormone levels on estrus are accompanied by a reduction in the ovarian secretion of inhibin, *Exp. Aging Res.*, 13, 3, 1987.
10. **DePaolo, L.V., and Chappel, S.C.,** Alterations in the secretion and production of follicle-stimulating hormone precede age-related lengthening of estrous cycles in rats, *Endocrinology*, 118, 1127, 1986.
11. **Erlik, Y., Meldrum, D.R., and Judd, H.L.,** Estrogen levels in postmenopausal women with hot flashes, *Obstet. Gynecol.*, 59, 403, 1982.
12. **Gosden, R.G.,** *The Biology of Menopause: The Causes and Consequences of Ovarian Aging*, Academic Press, New York, 1985.
13. **Hirshfield, A.N., and Midgley Jr., A.R.,** The role of the FSH surge in the selection of preovulatory follicles, *Biol. Reprod.*, 19, 597, 1978.
14. **Hoak, D.C., and Schwartz, N.B.,** Blockade of recruitment of ovarian follicles by suppression of the secondary surge of follicle-stimulating hormone with porcine follicular fluid, *Proc. Natl. Acad. Sci. (USA)*, 77, 4953, 1980.
15. **Jaszmann, L., Van Lith, N.D., and Zaat, J.C.A.,** The perimenopausal symptoms, *Med. Gynecol. Soc.*, 4, 268, 1969.
16. **Jones, E.C., and Krohn, P.L.,** The effect of hypophysectomy on age changes in the ovaries of mice, *J. Endocrinol.*, 21, 497, 1961.
17. **Judd, H.L.,** Pathophysiology of menopausal hot flushes, in *Neuroendocrinology of Aging*, Meites, J., Ed., Plénum Press, New York, 1983, 173.
18. **Judd, H.L., Judd, G.E., Lucas, W.E., and Yen, S.S.C.,** Endocrine function of the postmenopausal ovary: Concentration of androgens and estrogens in ovarian and peripheral vein blood, *J. Clin. Endocrinol. Metab.*, 39, 1020, 1974.

19. **Knobil, E., and Plant, T.M.,** Neuroendocrine control of gonadotropin secretion in the female rhesus monkey, in *Frontiers in Neuroendocrinology,* Ganong, W.F., and Martini, L., Eds., Raven Press, New York, 1978, 249.
20. **LaPolt, P.S., Day, J.R., and Lu, J.K.H.,** Effects of estradiol and progesterone on early embryonic development in aging rats, *Biol. Reprod.,* 43, 843, 1990.
21. **LaPolt, P.S., and Lu, J.K.H.,** Effects of increased circulating progesterone on ovarian follicular loss and reproductive aging, *20th Annual Meeting of the Society for the Study of Reproduction,* Abstr., 1987, 380.
22. **LaPolt, P.S., Matt, D.W., Judd, H.L., and Lu, J.K.H.,** The relation of ovarian steroid levels in young female rats to subsequent estrous cyclicity and reproductive function during aging, *Biol. Reprod.,* 35, 1131, 1986.
23. **LaPolt, P.S., Yu, S.M., and Lu, J.K.H.,** Early treatment of young female rats with progesterone decelerates the aging-associated reproductive decline: A counteraction by estradiol, *Biol. Reprod.,* 38, 987, 1988.
24. **Lerner, S.P., Meredith, S., Thayne, W.V., and Butcher, R.L.,** Age-related alterations in follicular development and hormonal profiles in rats with 4-day estrous cycles, *Biol. Reprod.,* 42, 633, 1990.
25. **Lu, J.K.H., Gilman, D.P., Meldrum, D.R., Judd, H.L., and Sawyer, C.H.,** Relationship between circulating estrogens and the central mechanisms by which ovarian steroids stimulate luteinizing hormone secretion in aged and young rats, *Endocrinology,* 108, 836, 1981.
26. **Lu, J.K.K, Hopper, B.R., Vargo, T.M., and Yen, S.S.C.,** Chronological changes in sex steroid, gonadotropin and prolactin secretion in aging female rats displaying different reproductive states, *Biol. Reprod.,* 21, 193, 1979.
27. **Mandl, A.M., and Shelton, M.,** A quantitative study of oocytes in young and old nulliparous laboratory rats, *J. Endocrinol.,* 13, 444, 1959.
28. **Matt, D.W., Lee, J., Sarver, P.L., Judd, H.L., and Lu, J.K.H.,** Chronological changes in fertility, fecundity and steroid hormone secretion during consecutive pregnancies in aging rats, *Biol. Reprod.,* 34, 478, 1986.
29. **Matt, D.W., Sarver, P.L., and Lu, J.K.H.,** Relation of parity and estrous cyclicity to the biology of pregnancy in aging female rats, *Biol. Reprod.,* 37, 421, 1987.
30. **Meldrum, D.R.,** Female reproductive aging-ovarian and uterine factors, *Fertil. Steril.,* 59, 1, 1993.
31. **Meldrum, D.R., Erlik, Y., Lu, J.K.H., and Judd, H.L.,** Objectively recorded hot flashes in patients with pituitary insufficiency, *J. Clin. Endocrinol. Metab.,* 52, 684, 1981.
32. **Meredith, S., and Butcher, R.L.,** Role of decreased numbers of follicles on reproductive performance in young and aged rats, *Biol. Reprod.,* 32, 788, 1985.
33. **Metcalf, M.G., Skidmore, D.S., Lowry, G.F., and Mackenzies, J.A.,** Incidence of ovulation in the years after the menarche, *J. Endocrinol.,* 97, 213, 1983.
34. **Nass, T.E., LaPolt, P.S., Judd, H.L., and Lu, J.K.H.,** Alterations in ovarian steroid and gonadotrophin secretion preceding the cessation of regular oestrous cycles in aging female rats, *J. Endocrinol.,* 100, 43, 1984.
35. **Nass, T.E., LaPolt, P.S., and Lu, J.K.H.,** Effects of prolonged caging with fertile males on reproductive functions in aging female rats, *Biol. Reprod.,* 27, 609, 1982.
36. **Nelson, J.F., Gosden, R.G., and Felicio, L.S.,** Effect of dietary restriction on estrous cyclicity and follicular reserves in aging C57BL/6J mice, *Biol. Reprod.,* 32, 515, 1985.
37. **Pardridge, W.M., and Mietus, L.J.,** Transport of steroid hormones through the rat blood-brain barrier: Primary role of albumin-bound hormone, *J. Clin. Invest.,* 64, 145, 1979.
38. **Peluso, J.J., England-Charlesworth, C., and Hutz, R.,** Effect of age and of follicular aging on the preovulatory oocyte, *Biol. Reprod.,* 22, 999, 1980.
39. **Peluso, J.J., Steger, R.W., Huang, H., and Meites, J.,** Pattern of follicular growth and steroidogenesis in the ovary of aging cycling rats, *Exp. Aging Res.,* 5, 319, 1979.
40. **Richardson, S.J., and Nelson, J.F.,** Follicular depletion during the menopausal transition, *Ann. N.Y. Acad. Sci.,* 592, 13, 1990.

41. **Richardson, S.J., Senikas, V., and Nelson, J.F.**, Follicular depletion during the menopausal transition: Evidence for accelerated loss and ultimate exhaustion, *J. Clin. Endocrinol. Metab.*, 65, 1231, 1987.
42. **Sarkar, D.K., Chiappa, S.A., and Fink, G.**, Gonadotrophin-releasing hormone surge in proestrous rats, *Nature*, 264, 462, 1976.
43. **Sherman, B.M., and Korenman, S.G.**, Hormonal characteristics of the human menstrual cycle throughout reproductive life, *J. Clin. Invest.*, 5, 699, 1975.
44. **Sherman, B.M., West, J.H., and Korenman, S.C.**, The menopausal transition: Analysis of LH, FSH, estradiol and progesterone concentrations during menstrual cycles of older women, *J. Clin. Endocrinol. Metab.*, 42, 629, 1976.
45. **Smith, D.W.E.**, Is greater female longevity a general finding among animals? *Biol. Rev.*, 64, 1, 1989.
46. **Smith, M.S., Freeman, M.E., and Neill, J.D.**, The control of progesterone secretion during the estrous cycle and early pseudopregnancy in the rat: Prolactin, gonadotropin and estradiol levels associated with rescue of the corpus luteum of pseudopregnancy, *Endocrinology*, 96, 219, 1975.
47. **Tataryn, I.V., Lomax, P., Bajoreck, J.G., Chesarek, W., Meldrum, D.R., and Judd, H.L.**, Postmenopausal hot flushes: A disorder of thermoregulation, *Maturitas*, 2, 101, 1980.
48. **Tataryn, I.V., Meldrum, D.R., Lu, K.H., Frumar, A.M., and Judd, H.L.**, LH, FSH, and skin temperature during the menopausal hot flash, *J. Clin. Endocrinol. Metab.*, 49, 152, 1979.
49. **Thompson, B., Hart S.A., and Druno, D.**, Menopausal age and symptomology in general practice, *J. Biol. Sci.*, 5, 71, 1973.
50. **Vermeulen, A.**, The hormonal activity of the postmenopausal ovary, *J. Clin. Endocrinol. Metab.*, 34, 730, 1976.
51. **Wise, P.M.**, Alterations in proestrous LH, FSH, and prolactin surges in middle-aged rats, *Proc. Soc. Exp. Biol. Med.*, 169, 348, 1982.
52. **Wise, P.M.**, Estradiol-induced daily luteinizing hormone and prolactin surges in young and middle-aged rats: Correlations with age-related changes in pituitary responsiveness and catecholamine turnover rates in microdissected brain areas, *Endocrinology*, 115, 801, 1984.

Chapter 9

INTERACTIONS BETWEEN STRESS AND IMMUNE SIGNALS ON THE HYPOTHALAMIC-PITUITARY-ADRENAL AXIS OF THE RAT: INFLUENCE OF DRUGS

C. Rivier

TABLE OF CONTENTS

0-8493-2451-3/95/$0.00+.50

ABBREVIATIONS

ACTH–adrenocorticotropic hormone
BBB–blood-brain barrier
CNS–central nervous system
CRF–corticotropin-releasing factor
HPA–hypothalamic-pituitary-adrenal
IL–interleukin
IL-1β–interleukin-1β
i.p.–intraperitoneal
i.v.–intravenous
LH–luteinizing hormone
LHRH–luteinizing hormone-releasing hormone
LPS–lipopolysaccharides
MPOA–median preoptic area
OT–oxytocin
OVLT–organum vasculosum of the stria terminalis
POMC–pro-opiomelanocortin
PVN–paraventricular nucleus
VP–vasopressin

INTRODUCTION

The maintenance of homeostasis is essential for the survival of mammalian organisms. Thus, following exposure to a cognitive stimuli (such as a physical threat), noncognitive signals (such as pathogens), or drugs, necessary changes in metabolic, behavioral and endocrine functions must be made in order to restore the consistency of the *milieu interieur*. Two of the systems that play a primary role in this regard are the immune and the neuroendocrine systems—among the latter, the hypothalamic-pituitary-adrenal (HPA) axis is one of the most important. This suggests that the immune and the HPA axes probably communicate and interact with each other so as to provide an integrated response to stressful stimuli.

This chapter describes the response of the HPA axis to three types of threats to homeostasis: physical stress (represented by exposure to mild electroshocks), alcohol (an example of a drug), and interleukin-1β (IL-1β; an example of an immune stimulus). The chapter also discusses the role played by corticotropin-releasing factor (CRF) in modulating this response, and the interactions between two stimuli on the activity of the HPA axis.

OVERVIEW OF THE REGULATION OF THE HYPOTHALAMIC-PITUITARY-ADRENAL AXIS

Adrenocorticotropic hormone (ACTH) is released by cells of the anterior pituitary, called corticotrophs. A variety of factors can act on these cells, including peptides, bioamines, steroids and other neurosecretagogues.[39,56] In 1981, our laboratory reported the isolation and characterization of CRF, a 41-amino-acid peptide that stimulates the release of pro-opiomelanocortin (POMC)-like peptides such as ACTH.[57] While CRF is presently regarded as the primordial factor responsible for increased secretion of ACTH in response to a wide variety of stimuli, this peptide interacts with other hormones and can play an important permissive role in their stimulatory effects on the HPA axis. These factors include peptides such as vasopressin (VP) and oxytocin (OT), bioamines such as catecholamines, and adrenal steroids such as corticosterone.[10,28,30] The role played by each of these compounds appears to be determined, at least in part, by the nature and/or intensity of the signals to which the organism is exposed (see below).

EFFECT OF PHYSICAL STRESS

The ability of stressful physical stimuli to stimulate the activity of the HPA axis and to release ACTH and corticosteroids is well recognized, and (in the case of mild electroshocks), is illustrated in Figure 1.

The importance of CRF in this process was demonstrated by the ability of immunoneutralization of endogenous CRF to significantly blunt the stimulatory action of a variety of physical stresses, such as exposure to ether vapors,

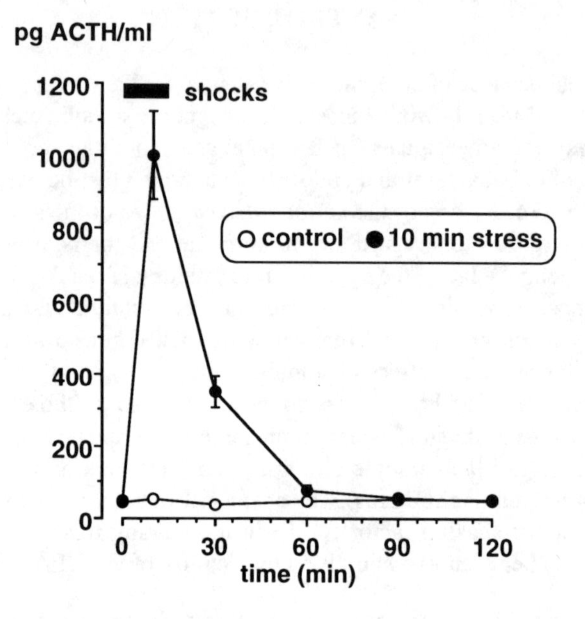

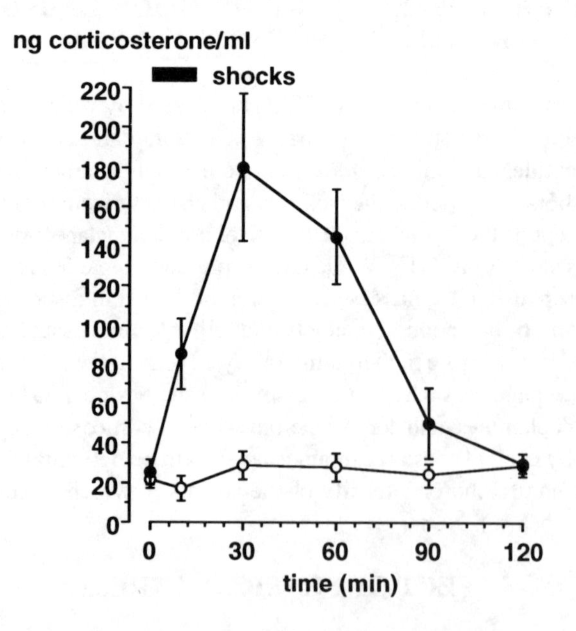

FIGURE 1. Effect of a 10-minute exposure to mild electroshocks (horizontal bar) on ACTH and corticosterone secretion in intact male rats. Each point represents the mean ± SEM of 6 animals. (Taken from Rivier, C., in *Neurobiology and Neuroendocrinology of Stress*, Brown, M., Rivier, C., and Koob, G., Eds., with permission of Marcel Dekker, Inc., New York, 1991.)

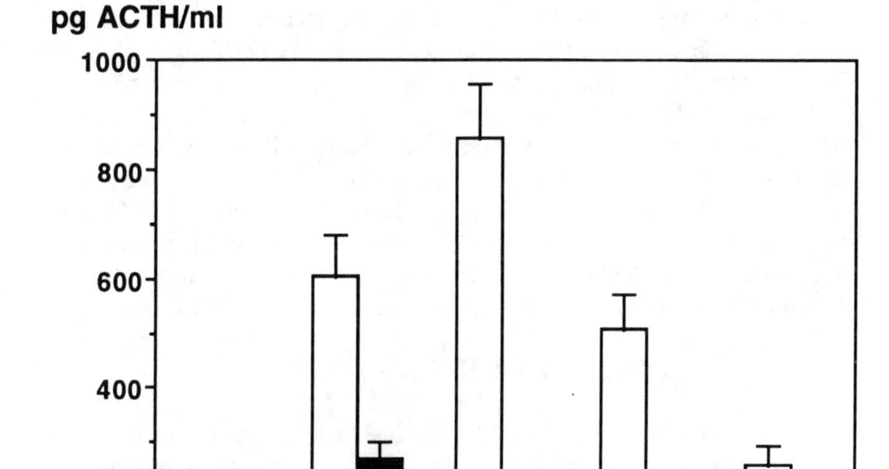

FIGURE 2. Effect of prior (1–2 minutes) i.v. injection of normal sheep serum (NSS, open bars) or anti-CRF serum (closed bars) on ACTH released in response to various stimuli. Ether = ether vapors; shocks = mild electroshocks delivered to the paws; immo = immobilization in a small cylinder (3 × 7"); transfer = transfer to a novel cage. Each bar represents the mean ± SEM of 6 intact male rats. (Modified from Rivier[34] with the permission of Alan R. Liss, Inc., New York.)

shocks, immobilization or transfer to a foreign environment.[49] As illustrated in Figure 2, however, plasma ACTH levels often do not return to basal levels in rats injected with the antiserum prior to exposure to the stress. This suggests that other factors, such as VP, OT and/or catecholamines, also played a role.[43] For example, in rats exposed to ether vapors, ACTH release is believed to be modulated by CRF, as well as by VP secreted in response to anoxia. Accordingly in this model, removal of endogenous CRF decreases ACTH levels, but only the additional blockade of VP receptors produces a near complete return to basal levels.[34,44] In contrast, exposure to immobilization appears to release ACTH primarily through a CRF-dependent mechanism, as removal of endogenous CRF totally blocks the activation of the corticotrophs. Other approaches used to investigate the role played by CRF have included experiments showing that various stresses increase CRF release into the portal circulation,[29] and CRF mRNA levels in the paraventricular nucleus (PVN) of the hypothalamus.[18,19] We have also reported that electrolytic lesions of the PVN, a process that removes most CRF perikarya, significantly blunt ACTH release in response to shocks.[31] Thus, CRF originating from the hypothalamus appears essential as a modulator of the increased release of ACTH in response to physical stresses.

Prior exposure of the HPA axis to stimuli often results in an altered response to subsequent stimuli, a phenomenon caused by changes in pituitary CRF receptors, in the activity of CRF cell body, and/or unaltered negative feedback exerted by corticosteroids.[10,41] Thus, it seemed reasonable to expect that exposure of rats to stimuli that result in increased CRF release would alter the response of the HPA axis to subsequent stimuli that also rely on endogenous CRF. We have therefore examined, first, the importance of CRF in mediating ACTH secretion in response to drugs such as alcohol and immune stimuli such as interleukins (ILs), and second, possible interactions between the stimulatory influence of some of these stimuli on the HPA axis.

EFFECT OF INTERLEUKIN-1

Clinical and laboratory studies had indicated that infectious processes are often accompanied by activation of the HPA axis. This suggested that the occurrence of immune activation was conveyed to the brain, which then coordinated the endocrine events necessary to restore homeostasis.[5] We know that upon presentation of an antigen, activated macrophages manufacture proteins called cytokines or ILs.[12] While these proteins exert many stimulatory effects on adjacent immune cells, they can enter the circulation and reach distant organs, including endocrine structures.[6] Cytokines are therefore considered part of the communication network that links the immune and the neuroendocrine systems.

Infectious processes are often difficult to control in a laboratory setting; furthermore, infected rats can experience signs of distress that might in themselves activate the HPA axis. We have therefore found it necessary to develop paradigms that allow the investigation of the events normally occurring in an organism exposed to pathogens, but that do not cause infection per se. One tool often used experimentally to mimic the events that characterize the early phase of immune activation (called the "acute-phase response")[11] is endotoxin. Endotoxins are lipopolysaccharides (LPS) contained in the wall of gram-negative bacteria and cause the release of cytokines/ILs. When we started to investigate possible functional relationships between the immune and the neuroendocrine axes, we first determined the ability of LPS to release ACTH. As illustrated in Figure 3, the intravenous (i.v.) injection of LPS to rats bearing indwelling i.v. cannulas, produced dose-related increases in plasma ACTH levels. We then established that this stimulatory effect was at least partly dependent on endogenous cytokines. This hypothesis was tested by injecting mice first with antibodies directed against IL-1 receptors, then with LPS. Indeed, neutralization of the receptors blunts LPS-induced ACTH re-

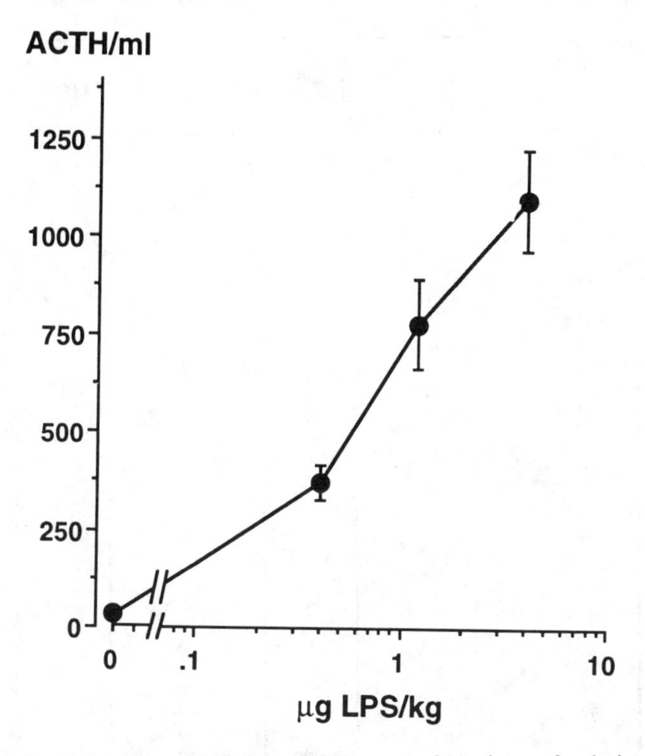

FIGURE 3. Dose-related increases in plasma ACTH measured 30 minutes after the i.v. injection of endotoxin (LPS). Each point represents the mean ± SEM of 5 intact male rats.

lease, providing evidence that endogenous IL-1 modulates the ability of endotoxin to activate the HPA axis.[47] The next step was to show that IL-1 itself could release ACTH, and we reported that, indeed, exogenous IL-1 causes dose-related increases in plasma ACTH levels.[50]

Subsequent experiments were devoted to the mechanisms that mediate this stimulatory action. Despite early reports to the contrary,[4] it is now well established that pituitary cells cultured in the absence of endotoxin do not acutely respond to ILs by releasing ACTH.[27,54] Prolonged exposure to cytokines will, however, allow this response to develop,[21] whereas preparations that maintain the cyto-architectural structure of pituitary cells (such as perifusion systems) are also reported to allow some degree of acute stimulatory action of cytokines on ACTH release.[8] Nevertheless, there is strong agreement that during the first 2–4 hours following peripheral IL-1 administration to intact rodents, structures other than the pituitary represent the primary sites of action of the cytokine. Indeed, the present consensus is that CRF represents the most important mediator of the acute response of the HPA axis to cytokines.[37] This is supported by experiments showing that the administration of IL-1 releases CRF into the portal circulation,[51] that CRF immunoneutralization significantly interferes with IL-1-induced ACTH secretion[51,55] (Figure 4), that destruction of hypothalamic CRF cell bodies by lesions of the PVN blunts the effect of IL-1,[31] and that administration of IL-1 causes significant increases in

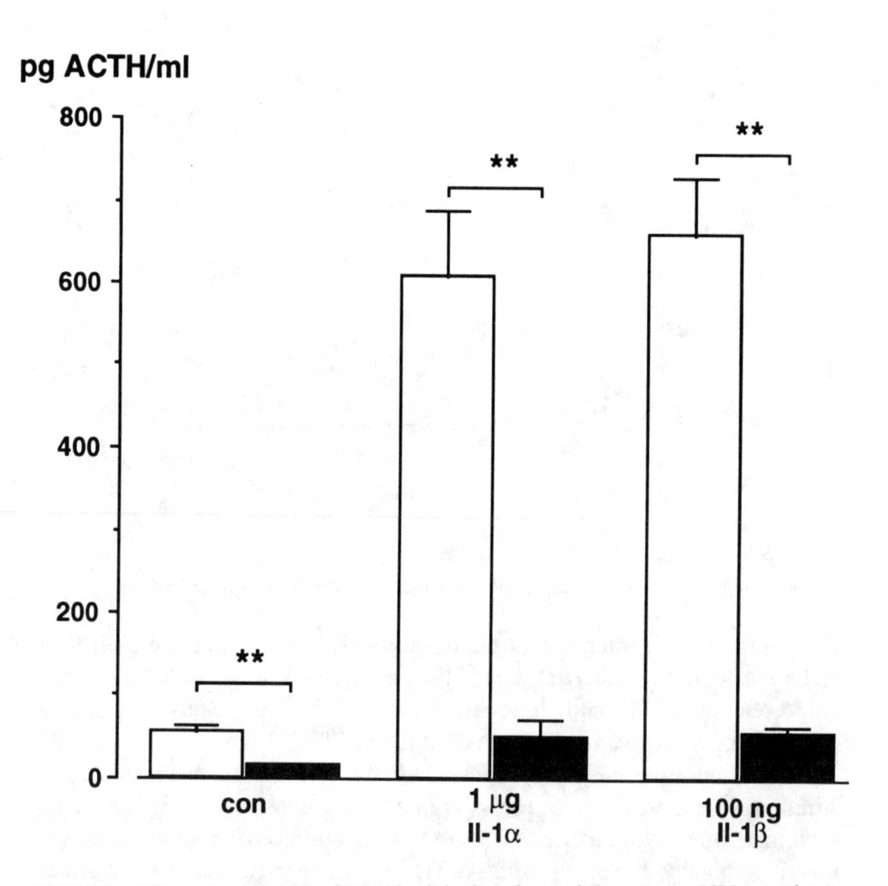

FIGURE 4. Effect of the prior (–2 minutes) i.v. injection of normal sheep serum (NSS, open bars) or anti-CRF serum (closed bars) on ACTH secretion induced by the i.v. injection of IL-1α or IL-1β. Each bar represents the mean ± SEM of 5 animals. $^{**}p < 0.01$ from rats injected with NSS. Blood samples were obtained 15 minutes after administration of the cytokines or their vehicle. (Modified from Rivier;[36] used with permission from S. Karger AG, Basel.)

CRF mRNA levels in this brain area.[16,53] Whether ILs can act directly at the level of the hypothalamus or depend on the activation of afferent circuitries remains an open question. Indeed, there is evidence for the involvement of neurosecretagogues in mediating the ability of IL-1 to act on CRF perikarya. We and others have shown, for example, that prostaglandins,[15,45] central[13,14] (but not peripheral)[50] catecholamines, and adrenal steroids,[58] modulate the release of ACTH following peripheral injection of IL-1. In contrast, opiates do not appear to be involved.[37]

One of the questions still debated is whether peripheral cytokines can cross the blood-brain barrier (BBB) and reach the hypothalamus. Although there is published evidence to support this contention,[2,3] one might argue that such a passage actually may not be necessary.[7] Indeed, alternative mechanisms involve the transfer of cytokines through fenestrated portions of the BBB (such as the organum vasculosum of the stria terminalis, OVLT);[20,52] stimulation of "relay mechanisms" through cells (such as astrocytes) that reside in the close vicinity of the BBB; effects at the level of the endothelium of the third ventricle;[17] activation of catecholamine-dependent pathways that impinge on the hypothalamus;[9,59] and stimulation (through mechanisms that, although not entirely elucidated, may involve peptides[26]) of the synthesis of cytokines within the CNS itself.[1,22] Although there is some evidence for a role played by each of these mechanisms, we have repeatedly failed to observe any effect of circulating cytokines on a hypothalamic structure adjacent to the OVLT, namely the median preoptic area (MPOA). Indeed, exogenous injection of even large doses of IL-1β does not cause any changes in luteinizing hormone (LH) release, whereas central administration of a small amount of this cytokine dramatically inhibits LH secretion.[35,40] This latter effect appears to be primarily mediated by the ability of IL-1β to decrease the activity of luteinizing hormone-releasing hormone (LHRH) cell bodies within the MPOA.[32] We would therefore argue that if peripheral cytokines could cross the BBB, we should have observed decreases in plasma LH levels following injection of peripheral IL-1β. Further support for the (at least partial) inability of peripheral IL-1 to directly activate CNS structures protected by the BBB stems from our observation that only the central but not the i.v. injection of IL-1β augments c-*fos* immunoreactivity within CRF-positive cells of the hypothalamus.[33] This observation led us to conclude that the acute stimulatory effect of peripherally injected IL-1β on ACTH release is secondary to stimulation of CRF nerve terminals in the median eminence, and not to the direct and immediate activation of CRF perikarya in the PVN.

EFFECT OF ALCOHOL AND INTERACTIONS BETWEEN ALCOHOL, PHYSICAL STRESS AND CYTOKINES

The acute peripheral administration of alcohol induces dose-related increases in plasma ACTH and corticosterone levels[46] (Figure 5). The observation that prior administration of anti-CRF serum significantly blunts these effects (Figure 6) and that long-term alcohol treatment increases CRF mRNA levels in the PVN,[48] suggested that the stimulatory action of alcohol on the HPA axis was dependent on endogenous CRF. As could be anticipated from earlier studies showing that exposure to CRF interferes with the subsequent response of the pituitary to this peptide (see above), we observed that rats injected with alcohol (either acutely or for several days) show a diminished release of ACTH in response to either exogenous CRF or exposure to physical stress (Figure 7). Indeed, we have previously reported significant interactions

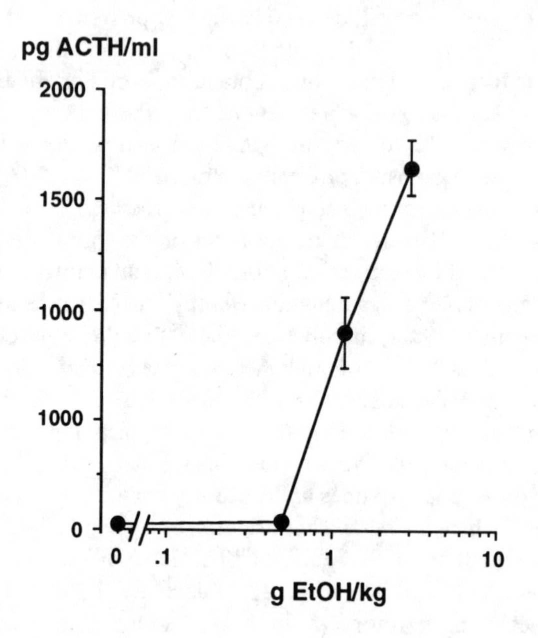

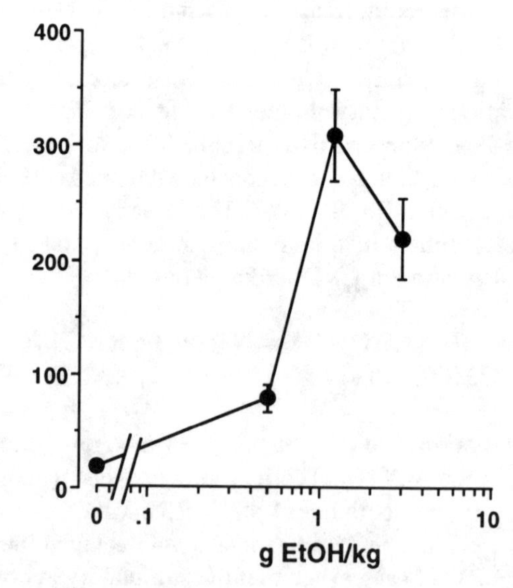

FIGURE 5. Dose-related increases in plasma ACTH and corticosterone levels in intact male rats injected with alcohol (EtOH, i.p.). Blood samples were obtained 20 minutes after treatment. Each point represents the mean ± SEM of 5 rats.

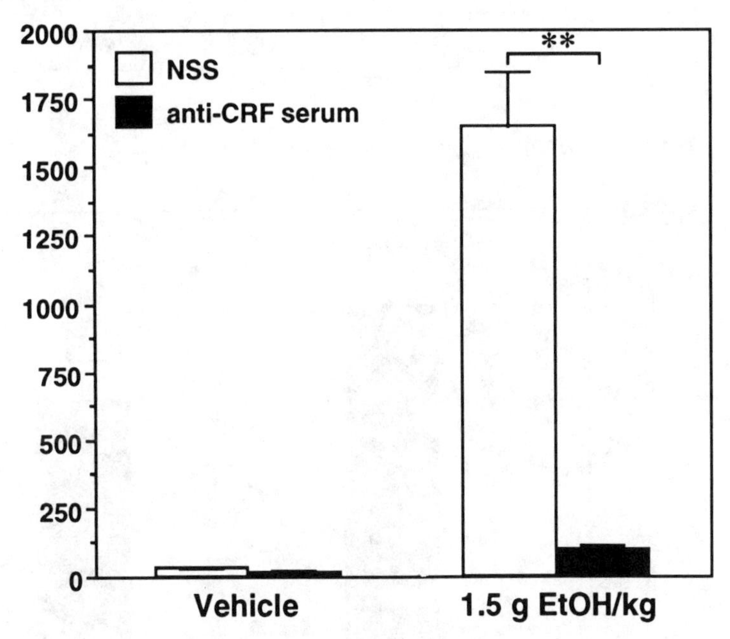

FIGURE 6. Effect of the prior (–2 minutes) i.v. injection of normal sheep serum (NSS, open bars) or anti-CRF serum (closed bars) on alcohol (EtOH)-induced ACTH secretion in intact male rats. Each bar represents the mean ± SEM of 5–6 animals. $^{**}p < 0.01$ from NSS. (Modified from Rivier *et al.*,[46] with the permission of Williams & Wilkins.)

between alcohol and physical stresses that result from a combination of increased steroid feedback, depletion of the readily releasable pool of ACTH in the corticotrophs, and changes in the activity of CRF cell bodies.[42]

Alcoholics and laboratory animals exposed to alcohol often show altered immune functions. While this drug's direct effects on immune cells have been proposed, we explored the possibility that at least a portion of these altered immune responses might be caused by pathological responses of the HPA axis to cytokines. We indeed observed that acute administration of alcohol 3 hours earlier, interfered with the stimulatory action of IL-1β on ACTH secretion.[38] However, although the net increase in ACTH levels was significantly diminished in rats injected with the higher dose of alcohol, then administered IL-1β, the absolute ACTH values were similar to those of control animals, or of animals injected with the smaller amount of alcohol (Figure 8). Rats exposed to alcohol vapors for 7 days (Figure 9), or rats ingesting an alcohol diet for the same period of time (Figure 10), also demonstrate a blunted release of ACTH in response to IL-1. Possible mechanisms responsible for this decreased pituitary activity include those described above, such as increased corticosteroid feedback, altered pituitary responsiveness to CRF, and changes in POMC

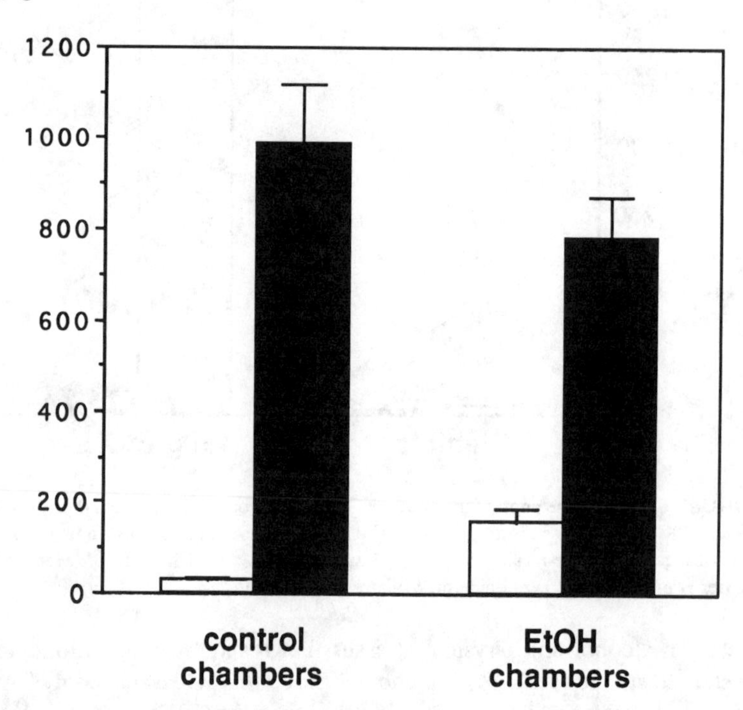

FIGURE 7. Effect of exposure to alcohol vapors (EtOH) on ACTH secretion in intact male rats exposed to mild electroshocks for 10 minutes. Blood samples were obtained at the end of the shock treatment. The difference between ACTH levels before and after the shock session was significantly ($p < 0.01$) smaller in rats exposed to alcohol. Each bar represents the mean ± SEM of 5–6 animals. (Reproduced from Rivier *et al., Brain Res.*, 520, 1, 1990, with permission from Elsevier Science Publishers.)

mRNA levels and/or in CRF receptors in the pituitary.[23,48] Because of the effects of POMC-related peptides on immune functions, as well as the effect of ACTH on corticosterone release, we propose that the ability of prior exposure to alcohol to alter the HPA axis' response to cytokines plays a role in modulating the altered immune activity observed during alcohol abuse.

Another intriguing model of alcohol abuse linked to abnormal immune responses is the paradigm of prenatal alcohol exposure. Clinical studies have shown that children born to alcoholic mothers, but not consuming alcohol themselves, often show an increased occurrence of infectious as well as

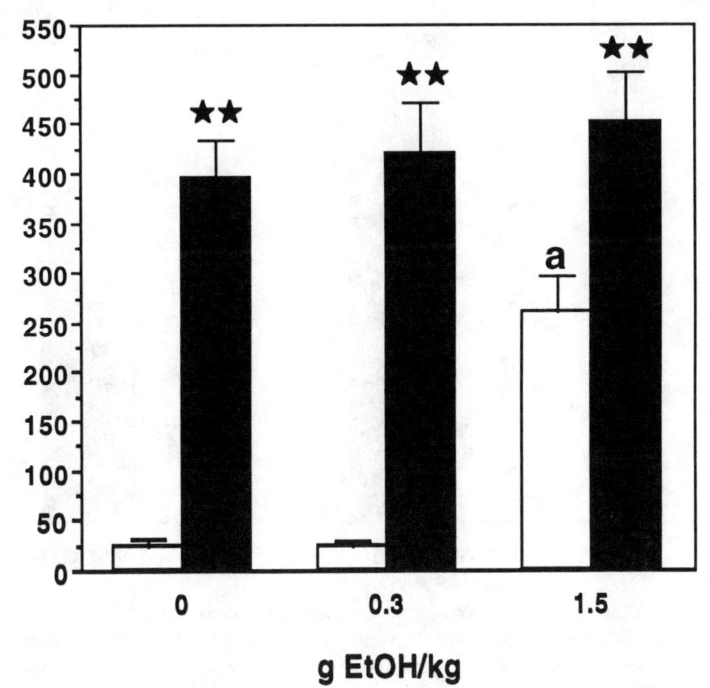

FIGURE 8. Effect of the prior (–3 hours) i.p. injection of alcohol (EtOH) on ACTH released in response to the i.v. injection of the vehicle (open bars) or IL-1β (300 ng/kg; closed bars). Each bar represents the mean ± SEM of 5 intact male rats. $^{**}p < 0.01$ from vehicle; $^{a}p < 0.01$ from pretreatment levels in rats preinjected with the vehicle or 0.3 g EtOH/kg.

inflammatory processes. Because alcohol influences many CNS structures during embryonic development, we became interested in the possibility that CRF-dependent pathways might be altered by prenatal alcohol exposure. We observed that, indeed, 21-day-old offspring of dams exposed to alcohol show increased ACTH release upon exposure to a short physical stress.[25] Because these rats have elevated CRF mRNA levels within the PVN, we proposed that this mechanism underlies, at least in part, the augmented responsiveness of the HPA axis to physical stress. Interestingly, however, peripheral administration of IL-1β causes a significantly blunted release of ACTH following prenatal alcohol treatment[24] (Figure 11). Because of our hypothesis that CRF terminals

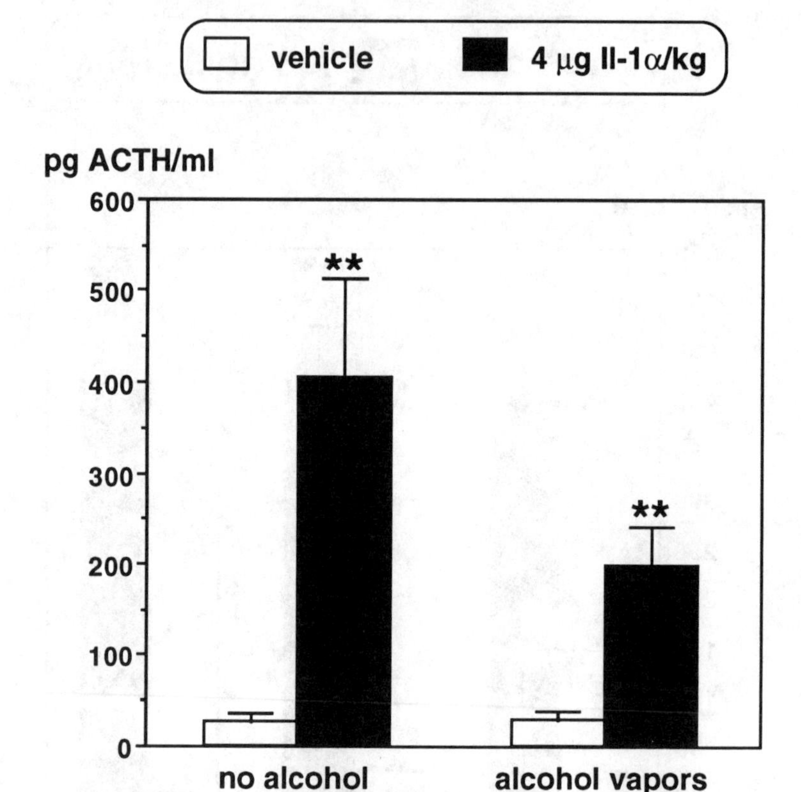

FIGURE 9. Effect of prior exposure to alcohol vapors for 7 days on ACTH released by the i.v. injection of the vehicle (open bars) or IL-1α (4 μg/kg; closed bars). Control intact male rats were kept in chambers without alcohol vapors. Blood samples were obtained 15 minutes after cytokine administration. Each bar represents the mean ± SEM of 6 animals. $^{**}p < 0.01$ from vehicle.

in the median eminence represent the primary target of IL-1 injected in the periphery,[33] these results suggest that prenatal alcohol treatment prevents the normal release of CRF into the portal circulation of immature rats acutely injected with IL-1β. Thus, it appears that exposure to alcohol during embryonic development exerts complex and opposite influences on the various elements of the HPA axis; in particular, CRF biosynthesis appears increased within the hypothalamus, resulting in augmented ACTH secretion in response to physical stresses. On the other hand, impaired release of CRF from nerve terminals blunts the activity of the corticotrophs in response to stimuli (such as increases in the levels of circulating cytokines) that act primarily at the level of the median eminence.

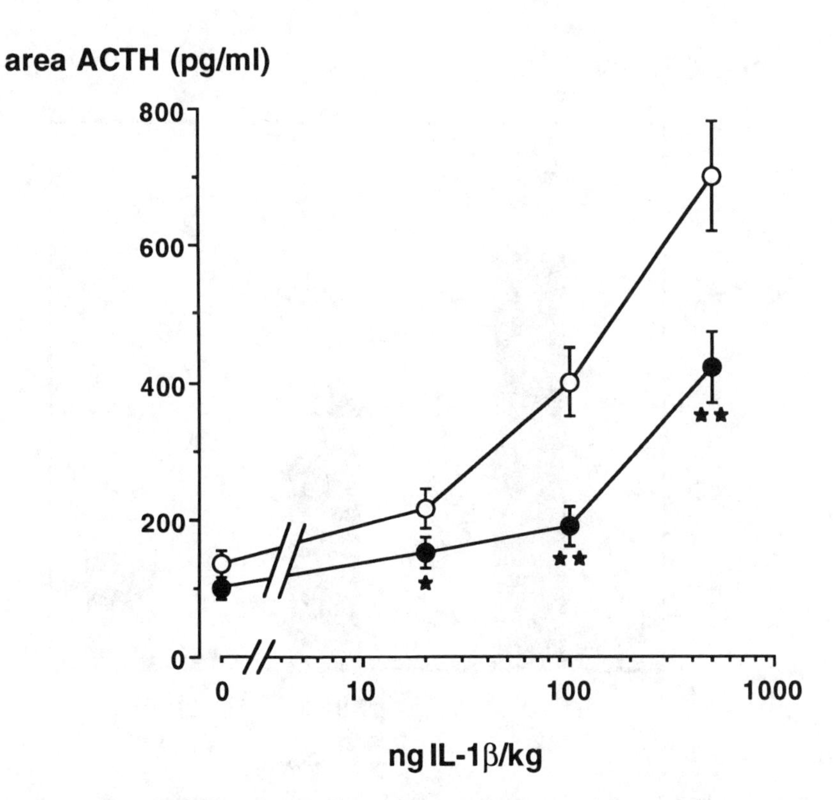

FIGURE 10. Effect of the vehicle or IL-1β on ACTH released by intact males exposed to an alcohol diet for 7 days (closed circles) or fed *ad libitum* (open circles). *$p < 0.05$ from *ad libitum*; **$p < 0.01$ from *ad libitum*. (Modified from Lee and Rivier.[23])

CONCLUSION

We have shown that endogenous CRF represents the primary mediator of the ability of physical stress, immune signals and drugs such as alcohol to activate the HPA axis (Figure 12). The stimulatory effect of various signals on the activity of CRF perikarya can be exerted through the release of neurotransmitters (indicated by the letters X, Y and Z) such as catecholamines, pros-

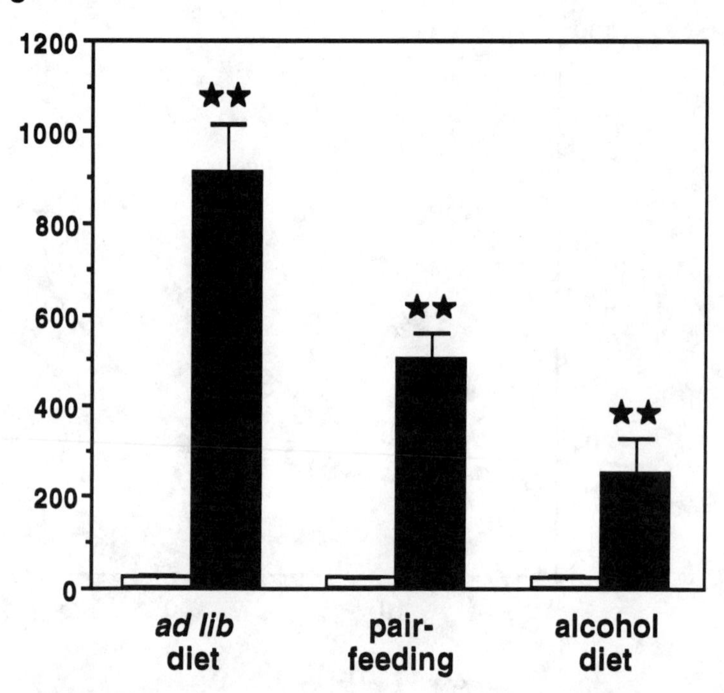

FIGURE 11. Effect of *ad libitum* feeding, pair-feeding or feeding with an alcohol diet during embryonic development (days 7–18 of gestation) on ACTH released by 22- to 24-day-old offspring injected with the vehicle (open bars) or IL-1β (2 μg/kg, i.p.; closed bars). The animals were rapidly decapitated 1 hour after i.p. treatment. $**p < 0.01$ from vehicle. (Modified from Lee and Rivier.[24])

taglandins, nitric oxide and peptides. Following exposure to one of these stimuli, the ability of the HPA axis to respond to another is altered. Changes in pituitary responsiveness to CRF (and possibly other secretagogues) and in corticosteroid feedback, increased CRF biosynthesis within the hypothalamus, and altered release of CRF from nerve terminals, all appear to play a role in these interactions. If, as presently believed, the integrity of the functional relationship between the immune and the HPA axes is essential to mount a

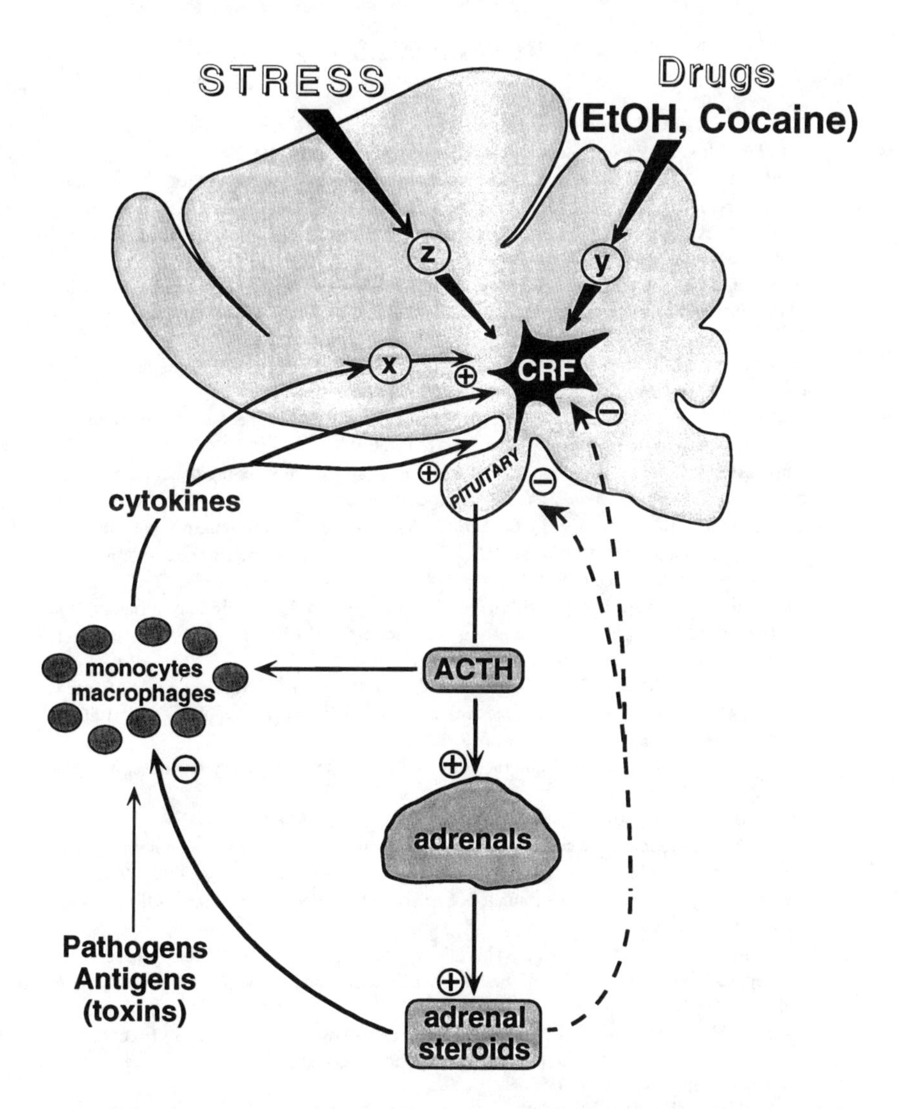

Figure 12. Schematic representation of the effects of, and possible interactions between, physical stress, cytokines (such as IL-1β) or drugs (such as EtOH) on the HPA axis of the rat. X, Y and Z indicate the possible involvement of neurotransmitters in mediating the stimulatory effect of the stresses.

proper response to pathogens, any alteration in the normal release of hormones from the HPA axis will cause deficiencies in the immune response. This may represent one of the mechanisms that underlies the increased occurrence of infectious or inflammatory processes in drug abusers.

REFERENCES

1. **Ban, E., Haour, F., and Lenstra, R.**, Brain interleukin 1 gene expression induced by peripheral lipopolysaccharide administration, *Cytokine*, 4, 48, 1992.
2. **Banks, W.A., and Kastin, A.J.**, Blood to brain transport of interleukin links the immune and central nervous systems, *Life Sci.*, 48, 117, 1991.
3. **Banks, W.A., Kastin, A.J., and Durham, D.A.**, Bidirectional transport of interleukin-1 alpha across the blood-brain barrier, *Brain Res. Bull.*, 23, 433, 1989.
4. **Bernton, E.W., Beach, J.E., Holaday, J.W., Smallridge, R.C., and Fein, H.G.**, Release of multiple hormones by a direct action of interleukin-1 on pituitary cells, *Science*, 238, 519, 1987.
5. **Besedovsky, H.O., and DelRey, A.**, Immuno-neuroendocrine circuits: Integrative role of cytokines, *Front. Neuroendocrinol.*, 13, 61, 1992.
6. **Blalock, J.E.**, A molecular basis for bidirectional communication between the immune and neuroendocrine systems, *Physiol. Rev.*, 69, 1, 1989.
7. **Busbridge, N.J., and Grossman, A.B.**, Stress and the single cytokine: Interleukin modulation of the pituitary-adrenal axis, *Mol. Cell Endocrinol.*, 82, C209, 1991.
8. **Cambronero, J.C., Rivas, F.J., Borrell, J., and Guaza, C.**, Interleukin-1-beta induces pituitary adrenocorticotropin secretion: Evidence for glucocorticoid modulation, *Neuroendocrinology*, 55, 648, 1992.
9. **Chuluyan, H.E., Saphier, D., Rohn, W.M., and Dunn, A.J.**, Noradrenergic innervation of the hypothalamus participates in adrenocortical responses to interleukin-1, *Neuroendocrinology*, 56, 106, 1992.
10. **Dallman, M.F., Akana, S.F., Scribner, K.A., Bradbury, M.J., Walker, C.-D., Strack, A.M., and Cascio, C.S.**, Stress, feedback and facilitation in the hypothalamo-pituitary-adrenal axis, *J. Neuroendocrinol.*, 4, 517, 1992.
11. **Dinarello, C.A.**, Interleukin-1 and the pathogenesis of the acute-phase response, *N. Engl. J. Med.*, 311, 1413, 1984.
12. **Dinarello, C.A.**, Biology of interleukin 1, *FASEB*, 2, 108, 1988.
13. **Dunn, A.J.**, Systemic interleukin-1 administration stimulates hypothalamic norepinephrine metabolism paralleling the increased plasma corticosterone, *Life Sci.*, 43, 429, 1988.
14. **Dunn, A.J.**, Interleukin-1 as a stimulator of hormone secretion, *Prog. Neurosci. Endocrinol. Immunol.*, 3, 26, 1990.
15. **Dunn, A.J., and Chuluyan, H.E.**, The role of cyclo-oxygtenase and lipoxygenase in the interleukin-1-induced activation of the HPA axis: Dependence on the route of injection, *Life Sci.*, 51, 219, 1992.
16. **Harbuz, M.S., Stephanou, A., Sarlis, N., and Lightman, S.L.**, The effects of recominant human interleukin (IL)-1α,IL-1β or IL-6 on hypothalamo-pituitary-adrenal axis activation, *J. Endocrinol.*, 133, 349, 1991.
17. **Hashimoto, M., Ishikawa, Y., Yokota, S., Goto, F., Bando, T., Sakakibara, Y., and Iriki, M.**, Action site of circulating interleukin-1 on the rabbit brain, *Brain Res.*, 540, 217, 1991.
18. **Imaki, F., Vale, W., and Sawchenko, P.E.**, Regulation of corticotropin-releasing factor mRNA in neuroendocrine and autonomic neurons by osmotic stimulation and volume loading, *Neuroendocrinology*, 56, 633, 1992.
19. **Imaki, T., Nahon, J.-L., Rivier, C., Sawchenko, P.E., and Vale, W.**, Differential regulation of corticotropin-releasing factor mRNA in rat brain cell types by glucocorticoids and stress, *J. Neurosci.*, 11, 585, 1991.
20. **Katsuura, G., Arimura, A., Koves, K., and Gottschall, P.E.**, Involvement of organum vasculosum of lamina terminalis and preoptic area in interleukin 1β-induced ACTH release, *Am. J. Physiol.*, 258, E163, 1990.
21. **Kehrer, P., Turnill, D., Dayer, J.-M., Muller, A.F., and Gaillard, R.C.**, Human recombinant interleukin-1beta and -alpha, but not recombinant tumor necrosis factor alpha stimulate ACTH release from rat anterior pituitary cells *in vitro* in a prostaglandin E_2 and cAMP independent manner, *Neuroendocrinology*, 48, 160, 1988.

22. **Koenig, J.I.**, Presence of cytokines in the hypothalamic-pituitary axis, *Prog. Neurosci. Endocrinol. Immunol.*, 4, 143, 1991.
23. **Lee, S., and Rivier, C.**, Effect of exposure to an alcohol diet for ten days on the ability of interleukin-1β to release ACTH and corticosterone in the adult ovariectomized female rat, *Alcoholism: Clin. Exp. Res.*, 17, 1009, 1993.
24. **Lee, S., and Rivier, C.**, Prenatal alcohol exposure blunts interleukin-1β-induced ACTH and β-endorphin secretion by immature rats, *Alcoholism: Clin. Exp. Res.*, 17, 940, 1993.
25. **Lee, S.Y., Imaki, T., Vale, W., and Rivier, C.L.**, Effect of prenatal exposure to ethanol on the activity of the hypothalamic-pituitary-adrenal axis of the offspring: Importance of the time of exposure to ethanol and possible modulating mechanisms, *Mol. Cell Neurosci.*, 1, 168, 1990.
26. **Martin, F.C., Anton, P.A., Gornbein, J.A., Shanahan, F., and Merrill, J.E.**, Production of interleukin-1 by microglia in response to substance P: Role for a non-classical NK-1 receptor, *J. Neuroimmunol.*, 42, 53, 1993.
27. **Parsadaniantz, S.M., Lenoir, V., Terlain, B., and Kerdelhué, B.**, Lack of effect of interleukins 1α and 1β, during *in vitro* perifusion, on anterior pituitary release of adrenocorticotropic hormone and β-endorphin in the male rat, *J. Neurosci. Res.*, 34, 315, 1993.
28. **Plotsky, P.M.**, Pathways to the secretion of adrenocorticotropin: A view from the portal, *J. Neuroendocrinol.*, 3, 1, 1991.
29. **Plotsky, P.M., and Vale, W.**, Hemorrhage-induced secretion of corticotropin-releasing factor-like immunoreactivity into the rat hypophysial portal circulation and its inhibition by glucocorticoids, *Endocrinology*, 114, 164, 1984.
30. **Plotsky, P., Cunningham, E.T.J., and Widmaier, E.P.**, Catecholaminergic modulation of corticotropin-releasing factor and adrenocorticotropin secretion, *Endocr. Rev.*, 10, 437, 1989.
31. **Rivest, S., and Rivier, C.**, Influence of the paraventricular nucleus of the hypothalamus in the alteration of neuroendocrine functions induced by physical stress or interleukin, *Endocrinology*, 129, 2049, 1991.
32. **Rivest, S., and Rivier, C.**, Interleukin-1β inhibits the endogenous expression of the early gene *c-fos* located within the nucleus of LHRH neurons and interferes with hypothalamic LHRH release during proestrus in the rat, *Brain Res.*, 613, 132, 1993.
33. **Rivest, S., Torres, G., and Rivier, C.**, Differential effects of central and peripheral injection of interleukin-1β on brain *c-fos* expression and neuroendocrine functions, *Brain Res.*, 587, 13, 1992.
34. **Rivier, C.**, Involvement of endogenous corticotropin-releasing factor (CRF) in modulating ACTH and LH secretion function during exposure to stress, alcohol or cocaine in the rat, in *Molecular Biology of Stress*, Zinder, O., and Bresnitz, S., Eds., Alan R. Liss, New York, 1989, 97.
35. **Rivier, C.**, Modulation of the rat pituitary and gonadal activity by factors from the endocrine and immune axes, in *Procedures of Serono Symposia, USA*, Yen, S.S.C., and Vale, W.W., Eds., Serono Symposia USA, Norwell, MA, 1990, 285.
36. **Rivier, C.**, Role of interleukins in the stress response, in *Methods and Achievements in Experimental Pathology (Series) Stress Revisited 1. Neuroendocrinology of Stress*, Jasmin, G., and Cantin, M., Eds., S. Karger, Basel, Switzerland, 1991, 63.
37. **Rivier, C.**, Neuroendocrine effects of cytokines in the rat, in *Reviews in the Neurosciences 4*, Huston, J.P., Ed., Freund Publishing House, London, 1993, 223.
38. **Rivier, C.**, Acute interactions between cytokines and alcohol on ACTH and corticosterone secretion in the rat, *Alcoholism: Clin. Exp. Res.*, 17, 946, 1993.
39. **Rivier, C.L., and Plotsky, P.M.**, Mediation by corticotropin-releasing factor (CRF) of adenohypophysial hormone secretion, *Ann. Rev. Physiol.*, 48, 475, 1986.
40. **Rivier, C., and Vale, W.**, Cytokines act within the brain to inhibit LH secretion and ovulation in the rat, *Endocrinology*, 127, 849, 1990.
41. **Rivier, C., and Vale, W.**, Influence of the frequency of ovine corticotropin-releasing factor (CRF) administration on ACTH and corticosterone secretion in the rat, *Endocrinology*, 113, 1422, 1983.

42. **Rivier, C., and Vale, W.,** Interaction between ethanol and stress on ACTH and β-endorphin secretion, *Alcoholism: Clin. Exp. Res.*, 12, 206, 1988.
43. **Rivier, C., and Vale, W.,** Interaction of corticotropin-releasing factor (CRF) and arginine vasopressin (AVP) on ACTH secretion *in vivo, Endocrinology*, 113, 939, 1983.
44. **Rivier, C., and Vale, W.,** Modulation of stress-induced ACTH release by corticotropin-releasing factor, catecholamines and vasopressin, *Nature*, 305, 325, 1983.
45. **Rivier, C., and Vale, W.,** Stimulatory effect of interleukin-1 on ACTH secretion in the rat: Is it modulated by prostaglandins? *Endocrinology*, 129, 384, 1991.
46. **Rivier, C., Bruhn, T., and Vale, W.,** Effect of ethanol on the hypothalamic-pituitary-adrenal axis in the rat: Role of corticotropin-releasing factor (CRF), *J. Pharmacol. Exp. Ther.*, 229, 127, 1984.
47. **Rivier, C., Chizzonite, R., and Vale, W.,** In the mouse, the activation of the hypothalamic-pituitary-adrenal axis by a lipopolysaccharide (endotoxin) is mediated through interleukin-1, *Endocrinology*, 125, 2800, 1989.
48. **Rivier, C., Imaki, T., and Vale, W.,** Prolonged exposure to alcohol: Effect on CRF mRNA levels, and CRF- and stress-induced ACTH secretion in the rat, *Brain Res.*, 520, 1, 1990.
49. **Rivier, C., Rivier, J., and Vale, W.,** Inhibition of adrenocorticotropic hormone secretion in the rat by immunoneutralization of corticotropin-releasing factor (CRF), *Science*, 218, 377, 1982.
50. **Rivier, C., Vale, W., and Brown, M.,** In the rat, interleukin-1α and -β stimulate adrenocorticotropin and catecholamine release, *Endocrinology*, 125, 3096, 1989.
51. **Sapolsky, R., Rivier, C., Yamamoto, G., Plotsky, P., and Vale, W.,** Interleukin-1 stimulates the secretion of hypothalamic corticotropin-releasing factor, *Science*, 238, 522, 1987.
52. **Stitt, J.T.,** Passage of immunomodulators across the blood-brain barrier, *Yale J. Biol. Med.*, 63, 121, 1990.
53. **Suda, T., Tozawa, F., Ushiyama, T., Sumitomo, T., Yamada, M., and Demura, H.,** Interleukin-1 stimulates corticotropin-releasing factor gene expression in rat hypothalamus, *Endocrinology*, 126, 1223, 1990.
54. **Suda, T., Tozawa, F., Ushiyama, T., Tomori, N., Sumitomo, T., Nakagami, Y., Yamada, M., Demura, H., and Shizume, K.,** Effects of protein kinase-C-related adrenocorticotropin secretagogues and interleukin-1 on proopiomelanocortin gene expression in rat anterior pituitary cells, *Endocrinology*, 124, 1444, 1989.
55. **Uehara, A., Gottschall, P.E., Dahl, R.R., and Arimura, A.,** Interleukin-1 stimulates ACTH release by an indirect action which requires endogenous corticotropin releasing factor, *Endocrinology*, 121, 1580, 1987.
56. **Vale, W.W., Rivier, C., Spiess, J., and Rivier, J.,** Corticotropin releasing factor, in *Brain Peptides*, Krieger, D., Brownstein, M., and Martin, J., Eds., John Wiley and Sons, New York, 1983, 961.
57. **Vale, W., Spiess, J., Rivier, C., and Rivier, J.,** Characterization of a 41 residue ovine hypothalamic peptide that stimulates the secretion of corticotropin and β-endorphin, *Science*, 213, 1394, 1981.
58. **Weidenfeld, J., Abramsky, O., and Ovadia, H.,** Effect of interleukin-1 on ACTH and corticosterone secretion in dexamethasone and adrenalectomized pretreated male rats, *Neuroendocrinology*, 50, 650, 1989.
59. **Weidenfeld, J., Abramsky, O., and Ovadia, H.,** Evidence for the involvement of the central adrenergic system in interleukin 1-induced adrenocortical response, *Neuropharmacology*, 28, 1411, 1989.
60. **Rivier, C.,** Neuroendocrine mechanisms of anterior pituitary regulation in the rat exposed to stress, in *Neurobiology and Neuroendocrinology of Stress*, Brown, M., Rivier, C., and Koob, G., Eds., Marcel Dekker, New York, 1991, 119.

Chapter 10

STRESS AND REPRODUCTION

D.A. Van Vugt and L.E. Heisler

TABLE OF CONTENTS

0-8493-2451-3/95/$0.00+.50

ABBREVIATIONS

ACTH–adrenocorticotropic hormone
CNS–central nervous system
CRF–corticotropin-releasing factor
CRH–corticotropin-releasing hormone
GABA–gamma-aminobutyric acid
GnRH–gonadotropin-releasing hormone
HPA–hypothalamic-pituitary-adrenal
HPO–hypothalamic-pituitary-ovarian
LH–luteinizing hormone

INTRODUCTION

Although the term stress did not appear in the biomedical literature until the 1930s when Selye applied this engineering term to medicine, the recognition that reproductive function can be compromised by a variety of conditions, which we now classify as stressors, long preceded the concept of stress. This is clearly exemplified by the following 17th century quotation: "Mr. Duke's daughter . . . in the eighteen year of her age . . . fell into a total suppression of her monthly courses from a multitude of cares and passions of her mind."[114] For the purposes of this chapter, stress will be defined as any stimulus that moves the animal away from homeostasis and evokes general responses that facilitate a return to the steady state.

As Selye described in his general adaptation syndrome, stress activates neural pathways, which increase arousal and attention, while inhibiting non-adaptive functions such as feeding, sexual behavior and reproduction.[150,151] Reduced reproductive function during times of stress is beneficial to both the individual and the species and is an important factor in population control.[35] For humans, it is viewed as a natural form of contraception that protects the woman from the physical and psychological demands of pregnancy during less than ideal conditions.[179] The birth of children is delayed until more hospitable conditions prevail, not unlike the situation in animals that give birth in a seasonal fashion as a result of environmental variables that regulate reproductive function.[162] Despite the beneficial effects of stress-induced amenorrhea, some negative aspects recently have been recognized. Because ovarian steroids have a protective effect on bone metabolism and the coronary vasculature, the opinion that stress-induced inhibition of reproduction function is a physiological response that need not be treated is being questioned.[21,55,100]

This chapter will summarize our current understanding of the effects of stress on reproductive function, with an emphasis on the literature that has appeared in the past 5 years. The chapter will focus on the central nervous system (CNS) mechanisms involved in stress-induced inhibition of the hypothalamic-pituitary-ovarian (HPO) axis. Because the effects of stress on the hypothalamic-pituitary-gonadal axis of the rat have been reviewed recently,[137] particular emphasis will be given to those studies performed in the female primate.

STRESS

Clinically Relevant Stressors

Amenorrhea may result from a variety of stressors. An association between exercise and amenorrhea in women or delayed puberty in premenarchial girls has been described.[52,167] Dietary restriction, either voluntary or involuntary, is similarly associated with amenorrhea and reduced fertility.[168] Psychosocial stresses resulting from changes in one's environment, such as occurs upon leaving home to attend school, joining the military, or being incarcerated, result in a significant incidence of amenorrhea.[110,171,179] Although the various stressors may be categorized as physical, nutritional, or psychosocial, we will not attempt to detail how they differ from one another and how they impact differently on the reproductive system. Rather, we will assume that the mechanisms whereby the various stressors exert their effects are more similar than different. The overlap between the various categories of stressors is probable justification for this viewpoint. For example, exercise is not purely a physical stress. Nutrition is often a factor, as is psychosocial stress associated with either competition or one's motivation for exercising.

Endocrine Features

Amenorrhea associated with stress is usually categorized as hypothalamic amenorrhea.[181] It is a diagnosis of exclusion that, as the name implies, has an etiology of primarily CNS origin. Gonadotropin levels range from low to normal with reduced pulse frequency, while estradiol levels are usually in the early follicular phase range or lower.[90,131] Pituitary responsiveness to gonadotropin-releasing hormone (GnRH) stimulation may be blunted, normal or accentuated.[131,181] This variability in response is most likely due to the degree of endogenous GnRH priming at the time of the exogenous GnRH challenge.[2] Similar gonadotropin levels and responsiveness have been described for women who exhibit exercise-induced amenorrhea.[43,44,101] Hypogonadotropic-hypogonadism is most extreme in amenorrhea associated with nutritional deficits. The pattern of luteinizing hormone (LH) secretion may be prepubertal with nocturnal pulsations.[181]

It has been inferred from the pattern of gonadotropin secretion and gonadotroph sensitivity to GnRH that hypothalamic amenorrhea is due to reduced GnRH secretion. In those women who desire pregnancy and in whom counseling has not achieved initiation of menstrual cyclicity, pulsatile GnRH therapy is very effective.[19,97,132] This observation supports the theory that the GnRH secretion is reduced, although a reduction in gonadotroph responsiveness to GnRH cannot be entirely excluded.

A second endocrine feature common to women with stress-induced amenorrhea is hypercortisolemia. Regardless of whether the cause of the stress-induced amenorrhea is psychogenic, physical, or nutritional, it is apparent that the hypothalamic-pituitary-adrenal (HPA) axis is activated.[16,22,68,101,158,164] A good example of this correlation between HPA axis activation and HPO axis inhibition is seen with exercise. Although exercise intensity is correlated with the incidence of amenorrhea, not all women who exercise become amenorrheic. Some women will continue to have regular menstrual cycles while others will become amenorrheic even though the amount of exercise and caloric intake is indistinguishable between the two groups. However, activation of the HPA axis, as reflected by increased cortisol, was shown to be a positive predictor of whether or not amenorrhea developed.[53]

This correlation between HPA activation and HPO inactivation has led to a detailed examination of possible interactions between these two axes. Therefore, when discussing stress and reproduction it is important to include a discussion of stress and the HPA axis. For a more detailed description of how the HPA axis responds to stress, the reader is referred to Rivier's discussion (chapter 9 in this volume), as well as to other reviews.[1,5,37]

Corticotropin-Releasing Hormone (CRH) and the HPA Axis

Evidence of HPA axis activation is based clinically on increased cortisol concentration (Figure 1). HPA activation is thought to result from increased secretion of CRH and/or other corticotropin-releasing factors (CRFs). This conclusion is supported by several lines of evidence. Even though cortisol levels are increased, adrenocorticotropin (ACTH) levels are in the normal range. Increased CRH secretion would explain the occurrence of normal ACTH levels in face of increased cortisol negative feedback. A reduced ACTH responsiveness to CRH also is in keeping with increased endogenous CRH release.[13,82,84,101] The conclusion that CRH secretion is increased during stress is supported by studies that measured CRH. CRH concentrations are elevated in cerebrospinal fluid of women with anorexia nervosa.[68,82] A variety of stresses have been shown to increase CRH concentrations in hypophyseal portal blood of rats and sheep and to increase synthesis of CRH as determined by *in situ* hybridization studies.[14,32,33,70,83,117,129,157] A third line of evidence for increased CRH secretion comes from ablation experiments. Ablation of the paraventricular nucleus of rats reduces stress-induced activation of the HPA axis.[134] Although this experiment is not specific for CRH, because other CRFs also are eliminated by lesioning the paraventricular nucleus, the ACTH response to acute stress is reduced in rats by antagonism of CRH.[128,140] Therefore, the data as a whole support CRH as playing a pivotal role in stress-induced activation of the HPA axis.

STRESS AND THE HYPOTHALAMIC-PITUITARY-GONADAL AXIS

CRH Effects on LH

Increased CRH secretion during stress can explain how an increase in HPA activity is coordinated with a decrease in HPO activity. Administration of CRH to rats, sheep, monkeys, and humans acutely inhibits LH secretion.[4,8,9,124,140] This effect is the result of CRH inhibition of GnRH release as demonstrated by the measurement of GnRH in hypophyseal portal blood of rats[125] and by electrophysiologic correlates of the GnRH pulse generator of the rhesus monkey.[175] Inhibition of GnRH release by CRH also has been demonstrated by *in vitro* studies using rat and human hypothalami.[63,118] The observation that CRH neurons synapse on GnRH cell bodies in the medial preoptic area further supports the conclusion that CRH inhibits GnRH.[104] An alternative site of CRH effects on GnRH secretion may be the midbrain where CRH

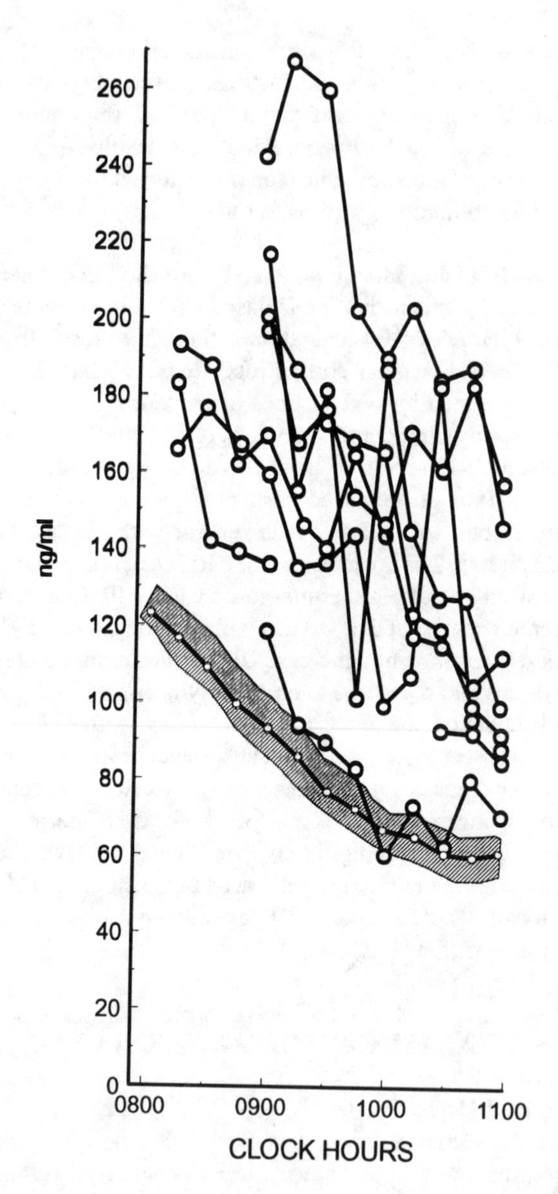

FIGURE 1. Serum cortisol levels in 8 hypothalamic amenorrhea patients during the morning hours compared to mean (± SEM) cortisol values in 10 women in the early follicular phase. (Used with permission from Suh, B.Y., Liu, J.H., Berga, S.L., Quigley, M.E., Laughlin, G.A., and Yen, S.S.C., Hypercortisolism in patients with functional hypothalamic-amenorrhea, *J. Clin. Endocrinol. Metab.*, 66, 733, 1988. © The Endocrine Society.)

innervates noradrenergic cell bodies.[161] Innervation of the medial preoptic area by noradrenergic neurons via the medial forebrain bundle could explain how activation of CRH neurons within the hypothalamus and/or other brain regions results in inhibition of LH secretion.[11] In addition, CRH may inhibit

GnRH through an opioid mechanism (to be discussed). In contrast to the large body of evidence for a hypothalamic site of action, there is limited evidence in support of a direct effect of CRH on the gonadotroph.[17] The majority of studies found no direct effect of CRH.[18,139]

The observations that CRH secretion is stimulated by stressors that also inhibit gonadotropin secretion, together with reports that CRH administration inhibits LH release, implicate CRH in stress-induced inhibition of gonadotropin secretion and reproductive function. However, a cause-and-effect relationship is not proven by this association nor does it exclude the possibility of other mechanisms. There has been only one study that showed CRH antagonism could block stress-induced inhibition of LH secretion.[138] This research direction has been hampered by the unavailability of potent CRH antagonists.

Stress and Endogenous Opiates

Endogenous opiates, particularly β-endorphin and dynorphin, have been implicated as inhibitory neuromodulators of LH secretion.[60,127,180] The role of endogenous opiates in stress-induced inhibition of reproductive function has been extensively studied, due in part to the availability of opiate antagonists such as naloxone and naltrexone. Naloxone was reported to stimulate LH secretion in a subgroup of women with hypothalamic amenorrhea,[6,90,130] and to both stimulate and have no effect on LH secretion in anorexia nervosa.[6,7,64] In a very small uncontrolled series, tonic naltrexone administration was reported to stimulate gonadotropin secretion and stimulate follicular maturation and ovulation in women with hypothalamic amenorrhea.[173] Recently, this result was confirmed in a much larger series of patients, although this study lacked placebo controls.[174]

An opioid mechanism of stress-induced hypogonadotropism has been examined in a variety of animal models. In nonhuman primates, inhibition of LH secretion by physical or chemical restraint,[121,143] or restraint in combination with receipt of aggression from a conspecific,[122] was reversed by naloxone. In other species, such as rats and sheep, inhibition of LH secretion by a variety of stressors including fasting,[29] intermittent foot shock,[126] restraint,[24,25] insulin-induced hypoglycemia,[38] and swimming,[25] was reversed or prevented by opiate antagonism. These results have led to the conclusion that stress inhibits LH secretion by stimulating endogenous opiates. An alternative interpretation is that naloxone is simply offsetting stress-induced suppression of LH as a result of stimulating LH secretion. This explanation is particularly viable when applied to those situations (endocrine conditions and animal models) where naloxone stimulates LH secretion in the absence of stress.

There also have been reports that opiate antagonism does not reverse stress-induced suppression of LH secretion. As already mentioned, naloxone does not stimulate LH secretion in all cases of hypothalamic amenorrhea.[130] Reports of negative LH responses to naloxone in exercise-induced amenorrhea[109,142] or anorexia[6,64] also have been published. Whereas insulin-induced hypoglycemia was reported to inhibit LH secretion in a naloxone

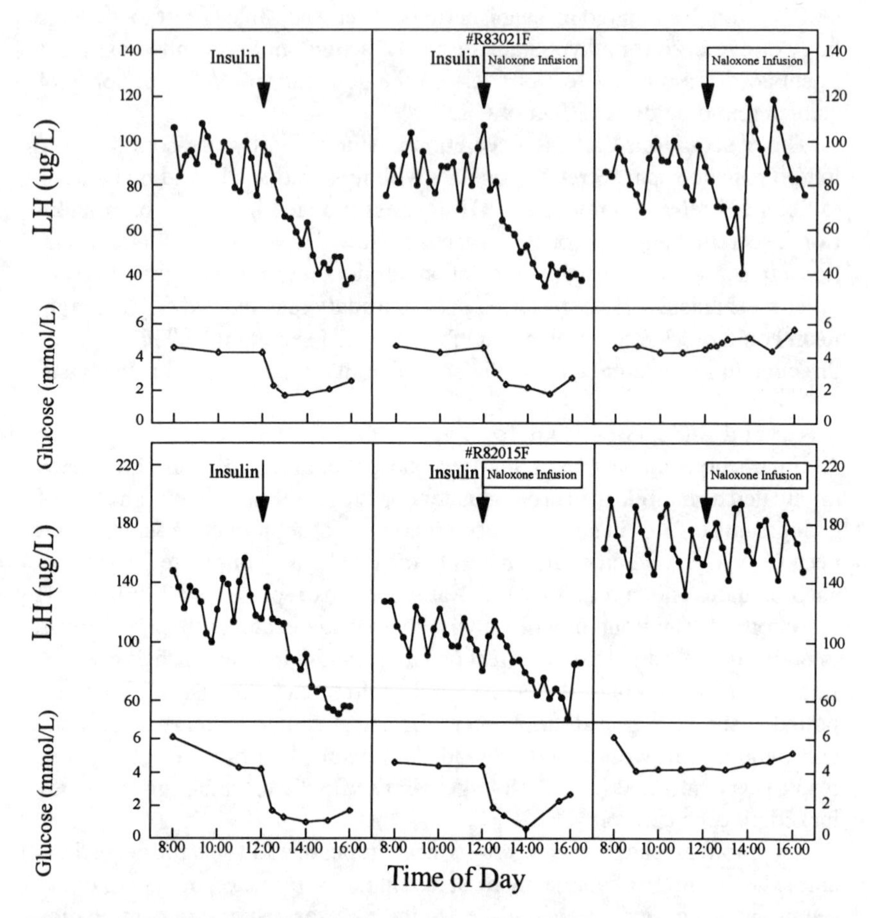

FIGURE 2. The effects of insulin, insulin plus naloxone, or naloxone alone on LH and glucose concentrations in 3 ovariectomized monkeys. (Used with permission from Heisler, L.E., Pallotta, C.M., Reid, R.L., and Van Vugt, D.A., Hypoglycemia-induced inhibition of luteinizing hormone secretion in the rhesus monkey is not mediated by endogenous opioid peptides, *J. Clin. Endocrinol. Metab.*, 76, 1280, 1993. © The Endocrine Society.)

reversible manner in ovariectomized sheep,[38] Heisler *et al.*[74] and Chen *et al.*[34] reported that naloxone did not reverse hypoglycemia-induced inhibition of LH secretion in ovariectomized monkeys (Figure 2). In addition, naloxone did not reverse insulin-induced inhibition of multiunit activity.[34] Similarly, bacteremia-induced suppression of LH secretion in sheep was not reversed by naloxone.[96] Helmreich *et al.* reported that acute fasting-induced inhibition of LH secretion in male rhesus monkeys was not blocked by naloxone administration[76] (Figure 3). These results indicate that endogenous opioid peptides do not mediate the suppressive effects of all stressors on LH secretion and highlight the fact that not all stresses work through identical mechanisms to suppress GnRH/LH secretion. Similar or identical stresses may work through different mechanisms in different species.

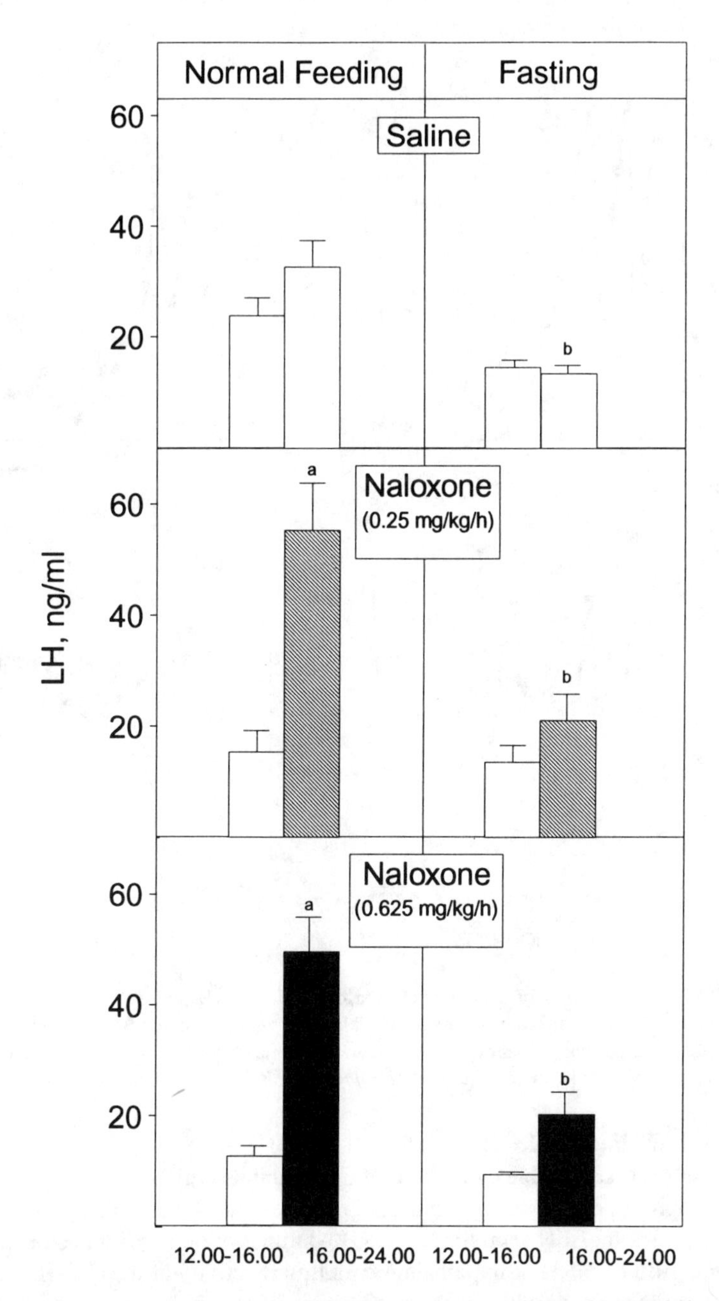

FIGURE 3. Mean LH concentrations on days of normal feeding and days of fasting during saline (*top panel*), 0.25 mg naloxone/kg/h (*middle panel*), or 0.625 mg naloxone/kg/h (*bottom panel*) infusion. [a]Significantly different from preceding 4-hour period; [b]significantly different from same period on day of normal feeding. (Used with permission from Helmreich, D.L., and Cameron, J.L., Suppression of luteinizing hormone secretion during food restriction in male rhesus monkeys (*Macaca mulatta*): Failure of naloxone to restore normal pulsatility, *Neuroendocrinology*, 56, 464, 1992. © S. Karger AG, Basel.)

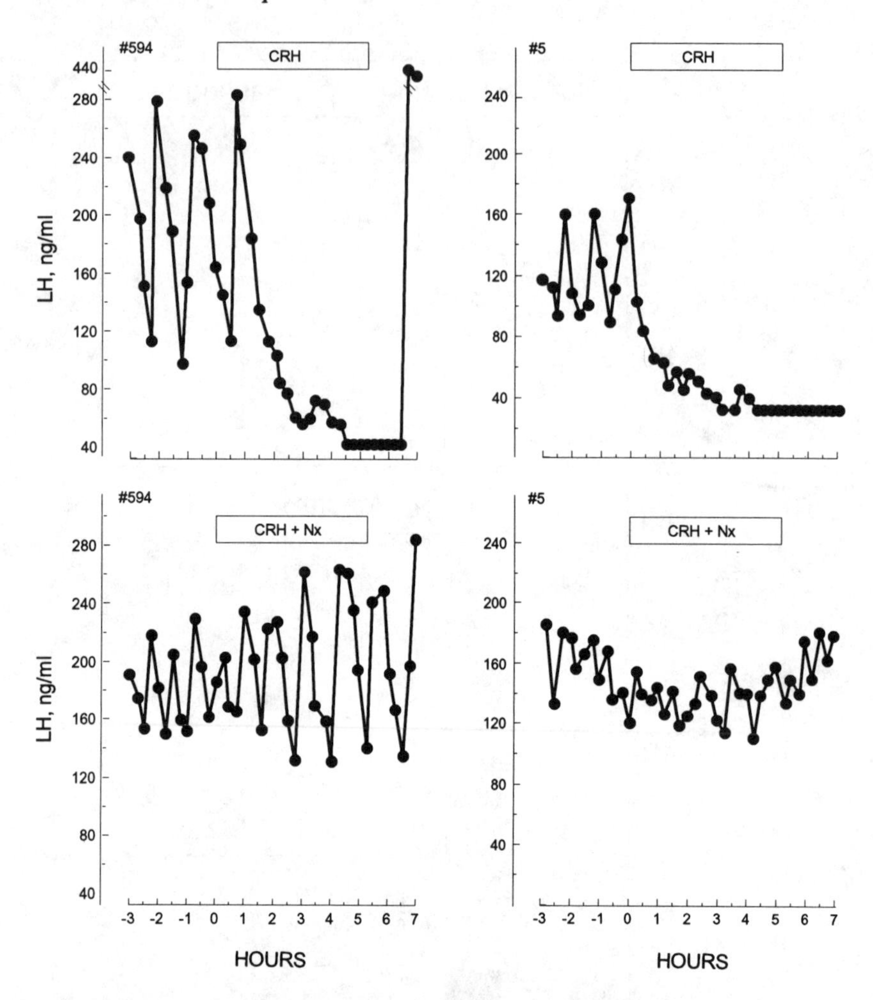

FIGURE 4. The effects of a 5-hour infusion of CRH or CRH plus naloxone (Nx) on LH concentrations in 2 ovariectomized monkeys. (Modified with permission from Gindoff, P., and Ferin, M., Endogenous opioid peptides modulate the effect of corticotropin-releasing factor on gonadotropin release in the primate, *Endocrinology*, 121, 837, 1987. © The Endocrine Society.)

Opioid Mediation of CRH Effect

There is considerable evidence that CRH inhibits GnRH through an endogenous opioid peptide mechanism. The primary evidence for this hypothesis is that inhibition of LH secretion by CRH administration in primates or rats or suppression by CRH of hypothalamic multiunit activity of the GnRH oscillator in monkeys was blocked by administration of the opiate antagonist naloxone[4,9,67,126,175] (Figure 4). Furthermore, both *in vivo* and *in vitro* release of β-endorphin and dynorphin from the hypothalamus were stimulated by CRH.[10,26,120] Conversely, a CRH antagonist inhibited β-endorphin release and increased GnRH release.[119] The site of action would appear to be other than the arcuate nucleus, the region where the majority of β-endorphin neurons are located. This region of the medial basal hypothalamus is not innervated by CRH and lacks receptors for CRH.[51] It has been suggested that the effects of CRH on β-endorphin release may be exerted at the level of the median eminence where both CRH and opioid neurons terminate. Alternatively, the effects of CRH on β-endorphin release may be mediated by other neuromodulators, a good candidate being vasopressin. Vasopressin has been shown to be as potent as CRH in stimulating β-endorphin release.[10,26] Furthermore, the effects of CRH on β-endorphin release are blocked by a vasopressin antagonist.[28]

Vasopressin and Stress-Induced Inhibition of LH Secretion

Vasopressin is colocalized with CRH in parvocellular neurons of the paraventricular nucleus[91,115,147,148,160] and synergizes with CRH to stimulate ACTH secretion.[50,66,94,99] A variety of stresses have been shown to increase vasopressin release from both magnocellular and parvocellular neurons.[46,47] Insulin-induced hypoglycemia increased the proportion of CRH-containing parvocellular neurons, which synthesize vasopressin, and also resulted in a depletion of vasopressin-containing granules.[48,172] These observations agree with the report that insulin-induced hypoglycemia significantly increased vasopressin concentrations in hypophyseal portal blood.[32,58,128]

Recent evidence suggests that vasopressin may play a role in the regulation of LH secretion. Vasopressin-containing neurons were shown to synapse with GnRH neurons in the medial preoptic area of the monkey.[159] Vasopressin inhibited *in vitro* release of GnRH from cultured GT_1 cells, suggesting a direct effect on the GnRH neuron.[170] Intraventricular administration of vasopressin inhibited LH concentrations in rats and ovariectomized rhesus monkeys[140,153] and inhibited lordosis in the rat.[156]

We recently addressed the role of vasopressin in stress-induced inhibition of LH secretion in the rhesus monkey by examining the effect of vasopressin antagonism on hypoglycemia-induced suppression of LH secretion. Whereas naloxone was totally ineffective in this regard,[34,74] administration of a vasopressin antagonist blocked hypoglycemia-induced inhibition of LH[75] (Figure 5). While these results suggest that hypoglycemia inhibits LH secretion through a mechanism involving vasopressin, vasopressin may also be involved in the suppression of LH by other types of stressors. Whereas histamine resulted in reduced LH concentrations in control rats, no inhibition was observed in the vasopressin-deficient Brattleboro rat.[39]

Locus of Inhibitory CRH and Vasopressin Neurons

The largest source of CRH and vasopressin in the hypothalamus is the paraventricular nucleus.[81,146-148] Although it would appear that parvocellular CRH and vasopressin neurons activate the HPA axis in response to stressful stimuli, it does not appear that these same neurons inhibit the HPO axis. Rivest and Rivier recently reported that lesions of the paraventricular nucleus, including the parvocellular neurons, significantly blunted the ACTH/corticosterone

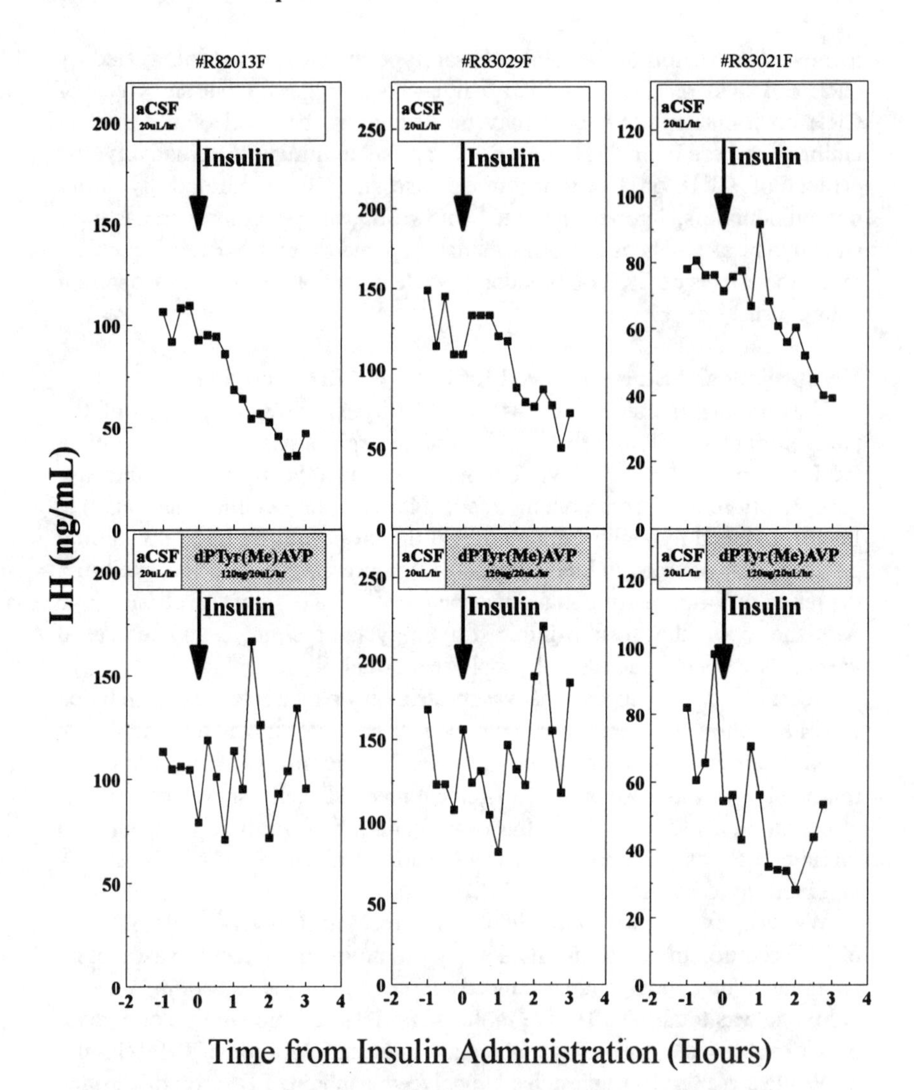

FIGURE 5. Effect of a vasopressin antagonist infusion (dPTyr(Me)AVP) on LH responses to insulin-induced hypoglycemia in 3 ovariectomized monkeys. (Heisler, L.E., Tumber, A.J., Reid, R.L., and Van Vugt, D.A.; unpublished results.)

response to stress or interleukin injection.[134] However, inhibition of LH secretion by either challenge was undiminished by lesions of the paraventricular nucleus. Furthermore, osmotic stimuli that increased vasopressin secretion and activated the HPA axis in male rhesus monkeys did not inhibit LH secretion.[78] Although both results suggest that neuroendocrine factors originating in the paraventricular nucleus are not involved in stress-induced suppression of LH secretion, extrapolation of experimental results from the male rat with paraventricular nucleus lesions to the female primate may not be warranted. With regard to osmotic stimuli, it is conceivable that a subset of vasopressin neurons in the paraventricular nucleus that influences LH secretion is not activated by this stimuli.

Regardless of the validity of such explanations, the conclusion that stress-induced inhibition of LH secretion can occur independent of the paraventricular nucleus does not preclude the involvement of CRH or vasopressin. Rather, it suggests that CRH and vasopressin-containing neurons originating outside the paraventricular nucleus are involved. One such locus may be the central amygdaloid nucleus. CRH neurons project from the central amygdaloid nucleus to the preoptic area and the bed nucleus of the stria terminalus.[69] Lesions of the central amygdaloid nucleus result in a reduction in CRH concentrations in the median eminence.[12]

ACTH and Glucocorticoid Effects

In addition to the CNS effects just reviewed, it is conceivable that the inhibitory effects of CRH and other CRFs on the reproductive axis are mediated by ACTH and glucocorticoids. Chronic ACTH administration inhibits gonadotropin secretion and interferes with sexual maturation and reproduction.[35,36,123] Although some reports have suggested a pituitary effect of ACTH, the majority of studies indicate that the anti-reproductive effects of ACTH are due to stimulation of glucocorticoids.[103,106,107] Glucocorticoids have been shown to inhibit gonadotropin and gonadal steroid secretion, block ovulation, and decrease gonadotroph responsiveness to GnRH.[87,102,133] In contrast to a pituitary effect, Dubey and Plant reported that chronic hydrocortisone administration to male monkeys inhibited gonadotropin secretion without altering pituitary responsiveness to GnRH.[56] They concluded that chronic glucocorticoid administration inhibits GnRH secretion.

Neither ACTH nor glucocorticoids appear to influence LH secretion acutely in the primate nor are they necessary for CRH to acutely inhibit LH secretion. CRH was equally effective in suppressing gonadotropin secretion or multiunit activity in monkeys when cortisol secretion was impaired either through adrenalectomy or blockade of adrenal steroid biosynthesis.[175,178] Whereas CRH inhibited LH secretion in the rhesus monkey, acute ACTH infusion had no effect, even though cortisol levels were increased to the same levels as those achieved with CRH administration.[177] Although these results demonstrate the importance of the hypothalamic component of the HPA axis, it should not be concluded from these results that glucocorticoids are unimportant in stress-induced inhibition of LH. Potential differences between acute and chronic stress should be kept in mind. Glucocorticoids may be very important in situations of chronic stress because they clearly inhibit GnRH and LH secretion.[56] It is possible that GnRH/LH are acutely inhibited by CRH whereas both CRH and glucocorticoids are involved chronically.

Cytokines and the Reproductive Axis

Recent evidence suggests that the inhibitory effects of stress on the reproductive axis may involve cytokines. Cytokine synthesis is increased by infectious states, exercise, trauma, and endotoxemia.[30,31,112] These same conditions have numerous neuroendocrine effects, including inhibition of the reproductive axis and stimulation of the HPA axis.[42,113,166] Administration of cytokines inhibited the hypothalamic-pituitary-gonadal axis while stimulating the HPA axis.[15,85,86,136,141] Conversely, antagonism of interleukin-1 by either receptor antagonists or anti-interleukin antiserum attenuated both interleukin and lipopolysaccharide-induced inhibition of LH,[57] suggesting that endotoxin-induced suppression of LH secretion is mediated by interleukin. Lipopolysaccharide injection has been reported to increase interleukin-1 synthesis in the mouse brain.[61,80]

Cytokines are synthesized in several different brain tissues.[23,73,80,112] These CNS cytokines are most likely responsible for GnRH/LH inhibition rather than peripheral cytokines. Intracisternal administration of cytokines and endotoxins consistently has been shown to be more effective than systemic administration.[57,141] Cytokine administration inhibited GnRH release from the hypothalamus and reduced c-fos expression in GnRH neurons of proestrous rats.[85,86,135] These observations are consistent with a CNS site of action.

Interleukin inhibition of GnRH release appears to be mediated by several different neuropeptides. Interleukin 1-α inhibition of LH secretion in ovariectomized rhesus monkeys was blocked by pretreatment with either a CRH antagonist or a vasopressin antagonist.[59,152] An effect of interleukin on CRH synthesis and release supports the hypothesis that interleukin stimulation of the HPA axis and inhibition of the reproductive axis is mediated by CRH and possibly vasopressin.[15,136,144] Naloxone also was reported to prevent interleukin-induced inhibition of GnRH/LH secretion and ovulation, suggesting that interleukins may activate endogenous opioid neurons.[85,141] Interleukin-1 has been shown to stimulate β-endorphin in pituitary and lymphocytes.[27,88] Interleukin may have a similar effect on opioid neurons. It is conceivable that the effect of interleukin on endogenous opiates is mediated by CRH.

Other CNS Mechanisms

Regulation of gonadotropin secretion may be influenced by metabolic signals. Consideration of such signals may be particularly important when nutrition is a suspected factor in the etiology of hypogonadotropism such as anorexia, delayed puberty, or exercise-induced amenorrhea. There is some evidence that such a metabolic signal may work independently of the HPA axis. Using short-term food deprivation to inhibit LH secretion in male monkeys, Helmreich *et al.* showed that pretreatment with dexamethasone, which exerts a glucocorticoid-like negative feedback effect on CRH and vasopressin synthesis and secretion,[45,91,92,145] blocked the increase in cortisol secretion associated with food deprivation but did not block the suppression of LH secretion.[77] Conversely, a high caloric meal given the day before acute fasting resulted in HPA axis activation (based on behavioral indices such as overt signs of agitation) without inhibiting the hypothalamic-pituitary-gonadal axis.[149]

The dissociation of HPA axis activation and LH inhibition demonstrated in these two experiments suggests that LH inhibition occurred independently of CRH or vasopressin activation of the pituitary-adrenal axis or direct inhibition of GnRH secretion by these two peptides. Rather, it was concluded that a metabolic signal is responsible for inhibiting the hypothalamic-pituitary-gonadal axis during acute food deprivation. This metabolic signal would presumably inhibit GnRH/LH secretion independently of a CRH, vasopressin, or opioid mechanism.

OVARIAN STEROID EFFECTS ON STRESS RESPONSE

Effect on HPO Axis

Ovarian steroids appear to sensitize the HPO axis to the inhibitory effects of stress. Ovariectomized rats exhibited reduced LH pulse frequency to fasting only when given estrogen replacement.[29] Ovariectomized rhesus monkeys exhibited greater sensitivity to hypoglycemia-induced GnRH and LH suppression when given estrogen replacement[34] (Figure 6). Increased responsiveness of the HPO axis to inhibition by restraint also was observed. Multiunit activity was abolished and LH secretion reduced when ovariectomized monkeys treated with estrogen were restrained, whereas these parameters were unaffected in untreated ovariectomized monkeys.[89]

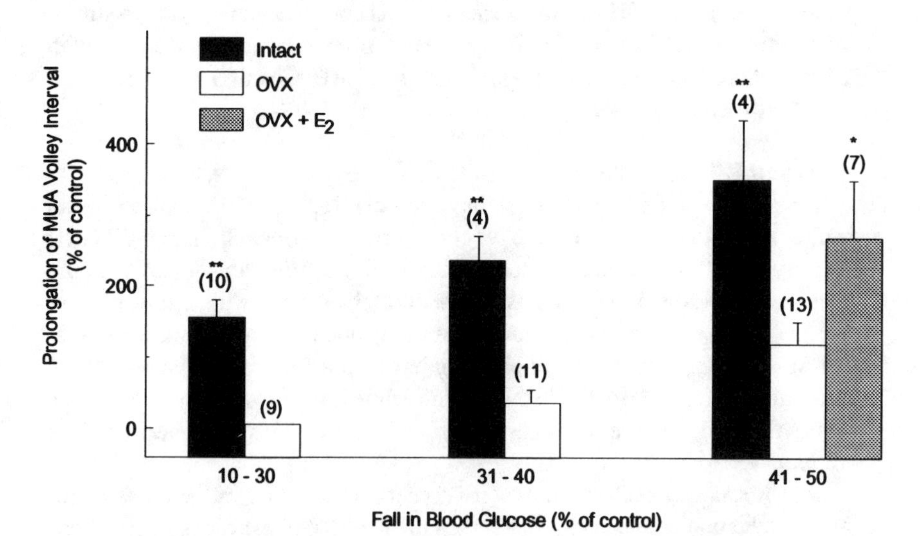

FIGURE 6. GnRH pulse generator activity in intact, ovariectomized and estradiol-treated ovariectomized monkeys in response to insulin-induced reductions in blood glucose concentrations of increasing magnitude. (Used with permission from Chen, M.-D., O'Byrne, K., Chiappini, S., Hotchkiss, J., and Knobil, E., Hypoglycemic 'stress' and gonadotropin-releasing hormone pulse generator activity in the rhesus monkey: Role of the ovary, *Neuroendocrinology*, 56(5), 666, 1992. © S. Karger AG, Basel.)

Effect on HPA Axis

While ovarian steroids appear to sensitize the HPO axis to the inhibitory effects of stress, the HPA axis is sensitized in an opposite manner (i.e., ovarian steroids appear to increase its activity). ACTH and corticosterone levels are greater in female compared to male rats, both basally and in response to stress.[41,95,163] These differences are eliminated by ovariectomy and restored by estrogen replacement.[163] Similar effects have been reported for primates. In women, ACTH and cortisol levels increased in the preovulatory phase,[65] whereas in monkeys, orchidectomy reduced cortisol levels and eliminated the diurnal pattern of secretion.[154] Ovariectomy similarly reduced cortisol levels but diurnal rhythmicity was maintained.[155] Contrary to the majority of results suggesting that ovarian steroids stimulate the HPA axis, an opposite effect of estrogen in postmenopausal women was reported. Estrogen replacement for 6 weeks to postmenopausal women reduced the responses of ACTH, cortisol, norepinephrine, and androstenedione to psychological stresses when compared to the response observed in postmenopausal women given a placebo.[98]

The effect of estrogen on HPA axis activity may result from a variety of actions. Ovariectomy decreased CRH synthesis, although an effect of estrogen has not been demonstrated.[71,72] CRH content was decreased whereas CRH mRNA was increased on proestrus, suggesting that estrogen increases both synthesis and release.[20] An effect of estrogen at the pituitary level also has been suggested. ACTH responsiveness to CRH may be increased by a stimulatory effect of estrogen on ACTH synthesis[40] or by estrogen interference with glucocorticoid feedback resulting in increased CRH secretion or an increased ACTH response to CRH.

Estrogen Effects on Neurotransmitters and Synapses

Regardless of whether estrogen has a direct effect on CRH synthesis or an indirect effect via interference with glucocorticoid feedback, increased CRH activity could explain the greater sensitivity of the HPO axis to stress in the presence of estrogen. Neurotransmitters other than CRH, which may account for this effect of estrogen, include endogenous opiates, GABA, and vasopressin since stimulatory effects of estrogen have been reported for each of these neuromodulators.[3,49,79,105,111,165,169] Although less is known about the effects of GABA and vasopressin on LH secretion, like opiates, they appear to be inhibitory.[54,93,108]

An alternative mechanism may involve the stimulatory influence of estrogen on synaptic connectivity.[62,116,176] In the absence of estrogen, connections between inhibitory neuromodulators and GnRH may be reduced below a threshold necessary to achieve suppression of GnRH and LH. It is conceivable that chronic stress suppresses ovarian function and reduces estrogen below such a threshold. This could constitute a feedback mechanism for tempering the effects of stress on both the HPO and HPA axes.

SUMMARY

It has been recognized that stress has a profound influence on reproductive function. Significant advances have been made in the past decade toward understanding the mechanisms involved. Of paramount importance has been the isolation and sequencing of CRH and endogenous opioid peptides. The influence of cytokines on both the HPO and HPA axes is sure to shed further light on how stress disrupts reproductive function.

REFERENCES

1. **Aguilera, G., Kiss, A., Hauger, R., and Tizabi, Y.,** Regulation of the hypothalamic-pituitary-adrenal axis during stress: Role of neuropeptides and neurotransmitters, in *Stress: Neuroendocrine and Molecular Approaches*, Kvetnansky, R., McCarty, R., and Axelrod, J., Eds., Gordon and Breach Science Publishers S.A., New York, 1992, 365.
2. **Aiyer, M., Chiappa, S., and Fink, G.,** A priming effect of luteinizing hormone releasing factor on the anterior pituitary gland in the female rat, *J. Endocrinol.*, 62, 573, 1974.
3. **Almeida, O., Hassan, A., Harbuz, M., Linton, E., and Lightman, S.,** Hypothalamic corticotropin-releasing hormone and opioid peptide neurons: Functional changes after adrenalectomy and/or castration, *Brain Res.*, 571, 189, 1992.
4. **Almeida, O.F.X., Nikolarakis, K.E., and Herz, A.,** Evidence for the involvement of endogenous opioids in the inhibition of luteinizing hormone by corticotropin-releasing factor, *Endocrinology*, 122, 1034, 1988.
5. **Antoni, F.,** Hypothalamic control of adrenocorticotropin secretion: Advances since the discovery of 41-residue corticotropin-releasing factor, *Endocr. Rev.*, 7, 351, 1986.
6. **Armeanu, M.C., Berkhout, G.M., and Schoemaker, J.,** Pulsatile luteinizing hormone secretion in hypothalamic amenorrhea, anorexia nervosa, and polycystic ovarian disease during naltrexone treatment, *Fertil. Steril.*, 57, 762, 1992.
7. **Baranowska, B.,** Are disturbances in opioid and adrenergic systems involved in the hormonal dysfunction of anorexia nervosa? *Psychoneuroendocrinology*, 15, 371, 1990.
8. **Barbarino, A., De Marinis, L., Folli, G., Tofani, A., Della Casa, S., D'Amico, C., Mancini, A., Corsello, S., Sambo, P., and Barini, A.,** Corticotropin-releasing hormone inhibition of gonadotropin secretion during the menstrual cycle, *Metabolism*, 38, 504, 1989.
9. **Barbarino, A., De Marinis, L., Tofani, G., Della Casa, S.D, Mancini, A., Corsello, S.M., Sciuto, R., and Barini, A.,** Corticotropin-releasing hormone inhibition of gonadotropin release and the effect of opioid blockade, *J. Clin. Endocrinol. Metab.*, 68, 523, 1989.
10. **Barna, I., Sweep, C.G.J., Veldhuis, H.D., Wiegant, V.M., and De Wied, D.,** Effects of pituitary beta-endorphin secretagogues on the concentration of beta-endorphin in rat cerebrospinal fluid: Evidence for a role of vasopressin in the regulation of brain beta-endorphin release, *Neuroendocrinology*, 51, 104, 1990.
11. **Barraclough, C.A., and Wise, P.M.,** The role of catecholamines in the regulation of pituitary luteinizing hormone and follicle-stimulating hormone secretion, *Endocr. Rev.*, 3, 91, 1982.
12. **Beaulieu, S., Pelletier, G., Vaudry, H., and Barden, N.,** Influence of the central nucleus of the amygdala on the content of corticotropin-releasing factor in the median eminence, *Neuroendocrinology*, 19, 115, 1975.
13. **Berga, S., Mortola, J., Girton, L., Suh, B., Laughlin, G., Pham, P., and Yen, S.,** Neuroendocrine aberrations in women with functional hypothalamic amenorrhea, *J. Clin. Endocrinol. Metab.*, 68, 301, 1989.
14. **Berkenbosch, F., De Goeij, D.C.E., and Tilders, F.J.H.,** Hypoglycemia enhances turnover of corticotropin-releasing factor and of vasopressin in the zona externa of the rat median eminence, *Endocrinology*, 125, 28, 1989.

15. **Berkenbosch, F., de Goeij, D.E., Rey, A.D., and Besedovsky, H.O.**, Neuroendocrine, sympathetic and metabolic responses induced by interleukin-1, *Neuroendocrinology*, 50, 570, 1989.
16. **Biller, B., Federoff, H., Koenig, J., and Klibanski, A.**, Abnormal cortisol secretion and responses to corticotropin-releasing hormone in women with hypothalamic amenorrhea, *J. Clin. Endocrinol. Metab.*, 70, 311, 1990.
17. **Blank, M.S., Fabbri, A., Catt, K.J., and DuFau, M.L.**, Inhibition of luteinizing hormone release by morphine and endogenous opioids in cultured pituitary cells, *Endocrinology*, 118, 2097, 1986.
18. **Blumenfield, Z., Kuhn, R.W., and Jaffe, R.B.**, Corticotropin-releasing factor can stimulate gonadotropin secretion by human fetal pituitaries in superfusion, *Am. J. Obstet. Gynecol.*, 154, 606, 1986.
19. **Blunt, S., and Butt, W.**, Pulsatile GnRH therapy for the induction of ovulation in hypogonadotropic hypogonadism, *Acta Endocrinol.*, 288, 58, 1988.
20. **Bohler, H.C.L.J., Zoeller, R.T., King, J.C., Rubin, B.S., Weber, R., and Merriam, G.R.**, Corticotropin-releasing hormone mRNA is elevated on the afternoon of proestrus in the parvocellular paraventricular nuclei of the female rat, *Mol. Brain Res.*, 8, 259, 1990.
21. **Bourne, T., Hillard, T.C., Whitehead, M.I., Crook, D., and Campbell, S.**, Oestrogens, arterial status, and postmenopausal women, *Lancet*, 335, 1470, 1990.
22. **Boyar, R.M., Hellman, L.D., Roffwang, H., Karz, J., Zumoff, B., O'Connor, J., Bradlow, H.L., and Fukushima, D.K.**, Cortisol secretion and metabolism in anorexia nervosa, *N. Engl. J. Med.*, 296, 190, 1977.
23. **Breder, C.C., Dinarello, C.A., and Saper, C.B.**, Interleukin-1 immunoreactive innervation of the human hypothalamus, *Science*, 240, 321, 1988.
24. **Briski, K., Quigley, K., and Meites, J.**, Endogenous opiate involvement in acute and chronic stress-induced changes in plasma LH concentrations in the male rat, *Life Sci.*, 34, 2485, 1984.
25. **Briski, K., and Sylvester, P.**, Effect of specific acute stressors on luteinizing hormone release in ovariectomized and ovariectomized estrogen-treated female rats, *Neuroendocrinology*, 47, 194, 1988.
26. **Bronstein, D.M., and Akil, H.**, *In vitro* release of hypothalamic β-endorphin (βE) by arginine vasopressin, corticotropin-releasing hormone and 5-hydroxytryptamine: Evidence for release of opioid active and inactive βE forms, *Neuropeptides*, 16, 33, 1990.
27. **Brown, S.L., Smith, L.R., and Blalock, J.E.**, Interleukin-1 and interleukin-2 enhance proopiomelanocortin gene expression in pituitary cells, *J. Immunol.*, 139, 3181, 1987.
28. **Burns, G., Almeida, O., Passarelli, F., and Herz, A.**, A two-step mechanism by which corticotropin-releasing hormone releases hypothalamic β-endorphin: The role of vasopressin and G-proteins, *Endocrinology*, 125, 1365, 1989.
29. **Cagampang, F., Maeda, K.-I., Tsukamura, H., Ohkura, S., and Ota, K.**, Involvement of ovarian steroids and endogenous opioids in the fasting-induced suppression of pulsatile LH release in ovariectomized rats, *J. Endocrinol.*, 129, 321, 1991.
30. **Cannon, J.G., Evans, W.J., Hughes, V.A., Meredith, C.N., and Dinarello, C.A.**, Physiological mechanisms contributing to increased interleukin-1 secretion, *J. Appl. Physiol.*, 61, 1869, 1986.
31. **Cannon, J.G., Fielding, R.A., Fiatarone, M.A., Ornencole, S.F., Dinarello, C.A., and Evans, W.J.**, Increased interleukin-1 beta in human skeletal muscle after exercise, *Am. J. Physiol.*, 257, R451, 1989.
32. **Caraty, A., Grino, M., Locatelli, A., Guillaume, V., Boudouresque, F., Conte-Devoix, B., and Oliver, C.**, Insulin-induced hypoglycemia stimulates corticotropin-releasing factor and arginine vasopressin secretion into hypophysial portal blood of conscious, unrestrained rams, *J. Clin. Invest.*, 85, 1716, 1990.
33. **Chappell, P.B., Smith, M.A., Kits, C., Bassette, G., Ritchie, J., Anderson, C., and Nemeroff, C.B.**, Alterations in corticotropin-releasing factor-like immunoreactivity in discrete rat brain regions after acute and chronic stress, *J. Neurosci.*, 6, 2908, 1986.

34. **Chen, M.-D., O'Byrne, K., Chiappini, S., Hotchkiss, J., and Knobil, E.**, Hypoglycemic 'stress' and gonadotropin-releasing hormone pulse generator activity in the rhesus monkey: Role of the ovary, *Neuroendocrinology*, 56 (5), 666, 1992.
35. **Christian, J.**, Population density and reproductive efficiency, *Biol. Reprod.*, 4, 248, 1971.
36. **Christian, J.J.**, Actions of ACTH in intact and corticoid-maintained adrenalectomized female mice with emphasis on the reproductive tract, *Endocrinology*, 75, 653, 1964.
37. **Chrousos, G.P., and Gold, P.W.**, The concepts of stress and stress system disorders. Overview of physical and behavioral homeostasis, *J. Am. Med. Assoc.*, 267, 1244, 1992.
38. **Clarke, I., Horton, R., and Doughton, B.**, Investigation of the mechanism by which insulin-induced hypoglycemia decreases luteinizing hormone-releasing hormone in ovariectomized ewes, *Endocrinology*, 127, 1470, 1990.
39. **Cover, P.O., Laycock, J.F., Artside, I.B., and Buckingham, J.C.**, A role for vasopressin in the stress-induced inhibition of gonadotropin secretion. Studies in the Brattleboro rat, *J. Neuroendocrinol.*, 3, 413, 1991.
40. **Coyne, M.D., and Kitay, J.I.**, Effect of ovariectomy on pituitary secretion of ACTH, *Endocrinology*, 85, 1097, 1969.
41. **Critchlow, V., Liebelt, A., Bar-Sela, M., Mountcastle, W., and Lipscomb, H.S.**, Sex differences in resting pituitary-adrenal function in the rat, *Am. J. Physiol.*, 205, 807, 1963.
42. **Croxson, T.S., Chapman, W.E., Miller, L.K., Levit, C.D., Senie, R., and Sumoff, B.**, Changes in hypothalamic-pituitary-gonadal axis in human immunodeficiency virus-infected homosexual men, *J. Clin. Endocrinol. Metab.*, 68, 317, 1989.
43. **Cumming, D., Vickovic, M., Wall, S., Fluker, M., and Belcastro, A.**, The effect of acute exercise on pulsatile release of luteinizing hormone in women runners, *Am. J. Obstet. Gynecol.*, 153, 482, 1985.
44. **Cumming, D.C., and Rebar, R.W.**, Exercise and reproductive function in women, *Am. J. Ind. Med.*, 4, 113, 1983.
45. **Davis, L.G., Arentzen, R., Reid, J.M., Manning, R.W., Wolfson, B., Lawrence, K.L., and Baldino, F.J.**, Glucocorticoid sensitivity of vasopressin mRNA levels in the paraventricular nucleus of the rat, *Proc. Natl. Acad. Sci. USA*, 83, 1145, 1986.
46. **de Goeij, D., Binnekade, R., and Tilders, F.**, Chronic stress enhances vasopressin but not corticotropin-releasing factor secretion during hypoglycemia, *Am. J. Physiol.*, 263, E394, 1992.
47. **de Goeij, D., Jezova, D., and Tilders, F.**, Repeated stress enhances vasopressin synthesis in corticotropin-releasing factor neurons in the paraventricular nucleus, *Brain Res.*, 577, 165, 1992.
48. **de Goeij, D., Kvetnansky, R., Whitnall, M., Jezova, D., Berkenbosch, F., and Tilders, F.**, Repeated stress-induced activation of corticotropin-releasing factor neurons enhances vasopressin stores and colocalization with corticotropin-releasing factor in the median eminence of rats, *Neuroendocrinology*, 53, 150, 1991.
49. **De Vries, G.J., Duetz, W., Buijs, R.M., and Van Heerkhuize, J.**, Effects of androgens and estrogens on the vasopressin and oxytocin innervation of the adult rat brain, *Brain Res.*, 399, 296, 1986.
50. **DeBold, C.R., Sheldon, W.R., DeCherney, G.S., Jackson, R.V., Alexander, A.N., Vale, W., Rivier, J., and Orth, D.N.**, Arginine vasopressin potentiates adrenocorticotropin release induced by ovine corticotropin-releasing factor, *J. Clin. Invest.*, 73, 533, 1984.
51. **DeSouza, E.B., and Insel, T.R.**, Corticotropin-Releasing factor (CRF) receptors in the rat central nervous system: Autoradiographic localization studies, in *Corticotropin-Releasing Factor: Basic and Clinical Studies of a Neuropeptide*, DeSouza, E.B., and Nemeroff, C.B., Eds., CRC Press, Boca Raton, FL, 1990, 69.
52. **DeSouza, M.J., and Metzger, D.A.**, Reproductive dysfunction in amenorrheic athletes and anorexic patients: A review, *Med. Sci. Sports Exercise*, 23, 995, 1991.
53. **Ding, J.H., Sheckter, C.B., Drinkwater, B.L., Soules, M.R., and Bremner, W.J.**, High serum cortisol levels in exercise-associated amenorrhea, *Ann. Intern. Med.*, 108, 530, 1988.

54. **Donoso, A.O., and Banzan, A.M.**, Effects of increase of brain GABA levels on hypothalamic-pituitary-luteinizing hormone axis in rats, *Acta Endocrinol.*, 106, 298, 1984.
55. **Drinkwater, B.L., Nilson, K., Chestnut, C.H. III, Bremner, W.J., Shainholtz, S., and Southworth, M.B.**, Bone mineral content of amenorrheic and eumenorrheic athletes, *N. Engl. J. Med.*, 311, 277, 1984.
56. **Dubey, A., and Plant, T.**, A suppression of gonadotropin secretion by cortisol in castrated male rhesus monkeys (*Macaca mulatta*) mediated by the interruption of hypothalamic gonadotropin-releasing hormone release, *Biol. Reprod.*, 33, 423, 1985.
57. **Ebisui, O., Fukata, J., Tominaga, T., Murakami, N., Kobayashi, H., Segawa, H., Muro, S., Naito, Y., Nakai, Y., Masui, Y., Nishida, T., and Imura, H.**, Roles of interleukin-1α and -1β in endotoxin-induced suppression of plasma gonadotropin levels in rats, *Endocrinology*, 130, 3307, 1992.
58. **Engler, D., Pham, T., Fullerton, M.J., Ooi, G., Funder, J.W., and Clarke, I.J.**, Studies of the secretion of corticotropin-releasing factor and arginine vasopressin into the hypophysial-portal circulation of the conscious sheep. I. Effect of an audiovisual stimulus and insulin-induced hypoglycemia, *Neuroendocrinology*, 49, 367, 1989.
59. **Feng, Y.J., Shalts, E., Xia, L.N., Rivier, J., Rivier, C., Vale, W., and Ferin, M.**, An inhibitory effects of interleukin-1α on basal gonadotropin release in the ovariectomized rhesus monkey: Reversal by a corticotropin-releasing factor antagonist, *Endocrinology*, 128, 2077, 1991.
60. **Ferin, M., Van Vugt, D., and Wardlaw, S.**, The hypothalamic control of the menstrual cycle and the role of endogenous opioid peptides, *Recent Prog. Horm. Res.*, 40, 441, 1984.
61. **Fontana, A., Weber, E., and Dayer, J.M.**, Synthesis of interleukin-1/endogenous pyrogen in the brain of endotoxin-treated mice: A step in fever induction? *J. Immunol.*, 133, 1696, 1984.
62. **Frankfurt, M., Gould, E., Woolley, C.S., and McEwen, B.S.**, Gonadal steroids modify dendritic spine density in ventromedial hypothalamic neurons: A golgi study in the adult rat, *Neuroendocrinology*, 51, 530, 1990.
63. **Gambacciani, M., Yen, S., and Rasmussen, D.**, GnRH release from the mediobasal hypothalamus: *In vitro* inhibition by corticotropin-releasing factor, *Neuroendocrinology*, 43, 533, 1986.
64. **Garcia-Rubi, E., Vazquez-Aleman, D., Mendez, J.P., Salinas, J.L., Garza-Flores, J., Ponce-de-Leon, S., Perez-Palacios, G., and Ulloa-Aguirre, A.**, The effects of opioid blockade and GnRH administration upon luteinizing hormone secretion in patients with anorexia nervosa during the stages of weight loss and weight recovery, *Clin. Endocrinol.*, 37, 520, 1992.
65. **Genazzani, A.R., Lemarchand-Berand, T.H., Aubert, M.L., and Felber, J.P.**, Patterns of plasma ACTH, GH, and cortisol during menstrual cycle, *J. Clin. Endocrinol. Metab.*, 41, 431, 1975.
66. **Gillies, G.E., Linton, E.A., and Lowry, P.J.**, Corticotropin-releasing activity of the new CRF is potentiated several times by vasopressin, *Nature*, 299, 355, 1982.
67. **Gindoff, P., and Ferin, M.**, Endogenous opioid peptides modulate the effect of corticotropin-releasing factor on gonadotropin release in the primate, *Endocrinology*, 121, 837, 1987.
68. **Gold, P.W., Gwirtsman, H., and Augerinos, C.**, Abnormal hypothalamic-pituitary-adrenal function in anorexia nervosa, *N. Engl. J. Med.*, 314, 1335, 1986.
69. **Gray, T.S.**, The organization and possible function of amygdaloid corticotropin-releasing factor pathways, in *Corticotropin-Releasing Factor: Basic and Clinical Studies of a Neuropeptide*, DeSouza, E.B., and Nemeroff, C.B., Eds., CRC Press, Boca Raton, FL, 1990, 53.
70. **Guillaume, V., Grino, M., Conte-Devolx, B., Boudouresque, F., and Oliver, C.**, Corticotropin-releasing factor secretion increases in rat hypophysial portal blood during insulin-induced hypoglycemia, *Neuroendocrinology*, 49, 676, 1989.
71. **Haas, D.A., and George, S.R.**, Gonadal regulation of corticotropin-releasing factor immunoreactivity in hypothalamus, *Brain Res. Bull.*, 20, 361, 1988.

72. **Haas, D.A., and George, S.R.,** Estradiol and ovariectomy decrease CRF synthesis in hypothalamus, *Brain Res. Bull.*, 23, 215, 1989.
73. **Hashimoto, M., Ishikawa, Y., Yokota, S., Goto, F., Bando, T., Sakakibara, Y., and Iriki, M.,** Action site of circulating interleukin-1 on the rabbit brain, *Brain Res.*, 540, 217, 1991.
74. **Heisler, L.E., Pallotta, C.M., Reid, R.L., and Van Vugt, D.A.,** Hypoglycemia-induced inhibition of luteinizing hormone secretion in the rhesus monkey is not mediated by endogenous opioid peptides, *J. Clin. Endocrinol. Metab.*, 76, 1280, 1993.
75. **Heisler, L.E., Tumber, A.J., Reid, R.L., and Van Vugt, D.A.,** Vasopressin mediates the suppression of luteinizing hormone in response to insulin-induced hypoglycemic stress in the rhesus monkey, *75th Ann. Meet. Endocr. Soc.*, Abstr. 1993, 341.
76. **Helmreich, D.L., and Cameron, J.L.,** Suppression of luteinizing hormone secretion during food restriction in male rhesus monkeys (*Macaca mulatta*): Failure of naloxone to restore normal pulsatility, *Neuroendocrinology*, 56, 464, 1992.
77. **Helmreich, D.L., Maturn, L.G., and Cameron, J.L.,** Lack of a role of the hypothalamic pituitary adrenal axis in the fasting induced suppression of luteinizing hormone secretion in adult male rhesus monkeys (*Macaca mulatta*), *Endocrinology*, 132, 2427, 1993.
78. **Helmreich, D.L., Verbalis, J.G., and Cameron, J.L.,** Lack of a role for AVP from osmotically sensitive neurons in the modulation of LH secretion in primates, *75th Ann. Meet. Endocr. Soc.*, Abstr., 1993, 341.
79. **Herbison, A.E., Heavens, R.P., and Dyer, R.G.,** Oestrogen and noradrenaline modulate endogenous GABA release from slices of the rat medial preoptic area, *Brain Res.*, 486, 195, 1989.
80. **Higgins, G.A., and Olschowka, J.A.,** Induction of interleukin-1 beta mRNA in adult rat brain, *Mol. Brain Res.*, 9, 143, 1991.
81. **Holmes, M., Antoni, F., Aguilera, G., and Catt, K.,** Magnocellular axons in passage through the median eminence release vasopressin, *Nature*, 319, 326, 1986.
82. **Hotta, M., Shibasaki, T., Masuda, A., Imaki, T., Demuna, H., Ling, N., and Shizume, K.,** The responses of plasma adrenocorticotropin and cortisol to corticotropin-releasing hormone and cerebrospinal fluid immunoreactive CRH in anorexia nervosa patients, *J. Clin. Endocrinol. Metab.*, 62, 319, 1986.
83. **Imaki, T., Nahan, J., Rivier, C., Sawchenko, P., and Vale, W.,** Differential regulation of corticotropin-releasing factor mRNA in rat brain regions by glucocorticoids and stress, *J. Neurosci.*, 11, 585, 1991.
84. **Ixart, G., Barbanel, G., Conte-Devolx, B., Grino, M., Oliver, C., and Assenmacher, I.,** Evidence for basal and stress-induced release of corticotropin-releasing factor in the push-pull cannulated median eminence of conscious free-moving rats, *Neurosci. Lett.*, 74, 85, 1987.
85. **Kalra, P.S., Fuentes, M., Sahu, A., and Kalra, S.P.,** Endogenous opioid peptides mediate the interleukin-1-induced inhibition of the release of luteinizing hormone (LH)-releasing hormone and LH, *Endocrinology*, 127, 2381, 1990.
86. **Kalra, P.S., Sahu, A., and Kalra, S.P.,** Interleukin-1 inhibits the ovarian steroid-induced luteinizing hormone surge and release of hypothalamic luteinizing hormone-releasing hormone in rats, *Endocrinology*, 126, 2145, 1990.
87. **Kamel, F.A., and Kubajak, C.L.,** Modulation of gonadotropin secretion by corticosterone: Interaction with gonadal steroids and mechanism of action, *Endocrinology*, 121, 561, 1987.
88. **Kavelaars, A., Ballieux, R.E., and Heijnen, C.J.,** The role of IL-1 in the corticotropin-releasing factor and arginine vasopressin-induced secretion of immunoreactive beta-endorphin by human peripheral blood mononuclear cells, *J. Immunol.*, 142, 2338, 1989.
89. **Kesner, J.S., Wilson, R.C., Kaufman, J.-M., Hotchkiss, J., Chen, Y., Yamamoto, H., Pardo, R.R., and Knobil, E.,** Unexpected responses of the hypothalamic gonadotropin-releasing hormone 'pulse generator' to physiological estradiol inputs in the absence of the ovary, *Proc. Natl. Acad. Sci. USA*, 84, 8745, 1987.

90. **Khoury, S., Reame, N., Kelch, R., and Marshall, J.,** Diurnal patterns of pulsatile luteinizing hormone secretion in hypothalamic amenorrhea: Reproducibility and responses to opiate blockade and an α2-adrenergic agonist, *J. Clin. Endocrinol. Metab.*, 64, 755, 1987.
91. **Kiss, J.Z., Mezey, E., and Skirboll, L.,** Corticotropin-releasing factor-immunoreactive neurons of the paraventricular nucleus become vasopressin positive after adrenalectomy, *Proc. Natl. Acad. Sci. USA*, 81, 1854, 1984.
92. **Kovacs, K., and Mezey, E.,** Dexamethasone inhibits corticotropin-releasing factor gene expression in the rat paraventricular nucleus, *Neuroendocrinology*, 46, 365, 1987.
93. **Lamberts, R., Vijayan, E., Graf, M., Mansky, H., and Wuttke, W.,** Involvement of preoptic-anterior hypothalamic GABA neurons in the regulation of pituitary LH and prolactin release, *Exp. Brain Res.*, 52, 356, 1983.
94. **Lamberts, S.W.J., Verleun, T., Oosterom, R., De Jong, F., and Hackeng, W.H.L.,** Corticotropin-releasing factor (ovine) and vasopressin exert a synergistic effect on adrenocorticotropin release in man, *J. Clin. Endocrinol. Metab.*, 58, 298, 1984.
95. **Le Mevel, J.C., Abitbol, S., Beraud, G., and Mainey, J.,** Temporal changes in plasma adrenocorticotropin concentration after repeated neurotropic stress in male and female rats, *Endocrinology*, 105, 812, 1979.
96. **Leshin, L.S., and Malven, P.V.,** Bacteremia-induced changes in pituitary hormone release and effect of naloxone, *Am. J. Phsyiol.*, 247, E-585, 1984.
97. **Leyendecker, G., Wildt, L., and Hansmann, M.,** Pregnancies following chronic intermittent (pulsatile) administration of GnRH by means of a portable pump (Zyklomat): A new approach in the treatment of infertility in hypothalamic amenorrrhea, *J. Clin. Endocrinol. Metab.*, 51, 1214, 1980.
98. **Lindheim, S.R., Legro, R.S., Bernstein, L., Stanczyk, F.Z., Vijod, M.A., Presser, S.C., and Lobo, R.A.,** Behavioral stress responses in premenopausal and postmenopausal women and the effects of estrogen, *Am. J. Obstet. Gynecol.*, 167, 1831, 1992.
99. **Liu, J.H., Muse, K., Contreras, P., Gibbs, D., Vale, W., Rivier, J., and Yen, S.S.C.,** Augmentation of ACTH-releasing activity of synthetic corticotropin-releasing factor (CRF) by vasopressin in women, *J. Clin. Endocrinol. Metab.*, 57, 1087, 1983.
100. **Lobo, R.A.,** Effects of hormonal replacement on lipids and lipoproteins in postmenopausal women, *J. Clin. Endocrinol. Metab.*, 73, 925, 1991.
101. **Loucks, A., Mortola, J., Girton, L., and Yen, S.,** Alterations in the hypothalamic-pituitary-ovarian and the hypothalamic-pituitary-adrenal axes in athletic women, *J. Clin. Endocrinol. Metab.*, 68, 402, 1989.
102. **Luton, J., Thiebolt, P., Valcke, J., Mahoudeau, J.A., and Bricaire, H.,** Reversible gonadotropin deficiency in male Cushing's disease, *J. Clin. Endocrinol. Metab.*, 45, 488, 1977.
103. **MacFarland, L.A., and Mann, D.R.,** The inhibitory effects of ACTH and adrenalectomy on reproductive maturation in female rats, *Biol. Reprod.*, 166, 306, 1977.
104. **MacLusky, N., Naftolin, F., and Leranth, C.,** Immunocytochemical evidence for direct synaptic connections between corticotrophin-releasing factor (CRF) and gonadotrophin-releasing hormone (GnRH)-containing neurons in the preoptic area of the rat, *Brain Res.*, 439, 391, 1988.
105. **Maggi, A., and Perez, J.,** Progesterone and estrogens in rat brain: Modulation of GABA (γ-aminobutyric acid) receptor activity, *Eur. J. Pharmacol.*, 103, 165, 1984.
106. **Mann, D.R., Evans, D., Edoimioy, F., Kamel, F., and Butterstein, G.M.,** A detailed examination of the *in vivo* and *in vitro* effects of ACTH on gonadotropin secretion in the adult rat, *Neuroendocrinology*, 40, 297, 1985.
107. **Mann, D.R., Jackson, G.G., and Blank, M.S.,** Influence of adrenocorticotropin and adrenalectomy on gonadotropin secretion in immature rats, *Neuroendocrinology*, 34, 20, 1982.
108. **Mansky, T., Mestres-Ventura, P., and Wuttke, W.,** Involvement of GABA in the feedback action of estradiol on gonadotropin and prolactin release: Hypothalamic GABA and catecholamine turnover rates, *Brain Res.*, 231, 353, 1982.

109. **McArthur, J.W., Turnbull, B.A., Pehrson, J., Bauman, M., Henley, K., Turner, A., Evans, W.J., Bullen, B.A., and Skrinar, G.S.**, Nalmefene enhances LH secretion in a proportion of oligo-amenorrheic athletes, *Acta Endocrinol.*, 128, 325, 1993.
110. **McCormick, W.O.**, Amenorrhea and other menstrual symptoms in student nurses, *J. Psychosom. Res.*, 19, 131, 1975.
111. **Miller, M.A., Urban, J.H., and Dorsa, D.M.**, Steroid dependency of vasopressin neurons in the bed nucleus terminalis by *in situ* hybridization, *Endocrinology*, 125, 2335, 1989.
112. **Minami, M., Kuraishi, K., Yamaguchi, T., Nakai, S., Hirai, Y., and Satoh, M.**, Immobilization stress induces interleukin-1 beta mRNA in the rat hypothalamus, *Neurosci. Lett.*, 123, 254, 1991.
113. **Molitch, M.E., and Hou, S.H.**, Neuroendocrine alterations in systemic disease, *J. Clin. Endocrinol. Metab.*, 12, 825, 1983.
114. **Morton, R.**, Opera medica liber primus pathisologiae, donatum donati amstelodami 1696. Cited in Nehimiah, J.C.: Anorexia nervosa, *Medicine*, 29, 225, 1950.
115. **Mouri, T., Itoi, K., Takahashi, K., Suda, T., Murakami, O., Yoshinaga, K., Andoh, N., Ohtani, H., Masuda, T., and Sasano, N.**, Colocalization of corticotropin-releasing factor and vasopressin in the paraventricular nucleus of the human hypothalamus, *Neuroendocrinology*, 57, 34, 1993.
116. **Naftolin, F., Garcia-Segura, L.M., Keefe, D., Leranth C., MacLusky, N.J., and Brawer, J.R.**, Estrogen effects of the synaptology and neuronal membranes of the rat hypothalamic arcuate nucleus, *Biol. Reprod.*, 42, 21, 1990.
117. **Nakane, T., Audhya, T., Kanie, N., and Hollander, C.S.**, Evidence for a role of endogenous corticotropin-releasing factor in cold, ether, immobilization and traumatic stress, *Proc. Natl. Acad. Sci. USA*, 82, 1247, 1985.
118. **Nikolarakis, K., Almeida, O., and Herz, A.**, Corticotropin-releasing factor (CRF) inhibits gonadotropin-releasing hormone (GnRH) release from superfused rat hypothalami *in vitro*, *Brain Res.*, 377, 388, 1986.
119. **Nikolarakis, K., Almeida, O., Sirinathsinghji, D., and Herz, A.**, Concomitant changes in the *in vitro* and *in vivo* release of opioid peptides and luteinizing hormone-releasing hormone from the hypothalamus following blockade of receptors for corticotropin-releasing factor, *Neuroendocrinology*, 47, 545, 1988.
120. **Nikolarakis, K.E., Almeida, O.F.X., and Herz, A.**, Stimulation of hypothalamic β-endorphin and dynorphin release by corticotropin-releasing factor (*in vitro*), *Brain Res.*, 399, 152, 1986.
121. **Norman, R., and Smith, C.**, Restraint inhibits luteinizing hormone and testosterone secretion in intact male rhesus macaques: Effects of concurrent naloxone administration, *Neuroendocrinology*, 55, 405, 1992.
122. **O'Byrne, K., Lunn, S., and Dixson, A.**, Naloxone reversal of stress-induced suppression of LH release in the common marmoset, *Physiol. Behav.*, 45, 1077, 1989.
123. **Ogle, T.F.**, Modification of serum luteinizing hormone and prolactin concentration by corticotropin and adrenalectomy in ovariectomized rats, *Endocrinology*, 101, 494, 1977.
124. **Olster, D., and Ferin, M.**, Corticotropin-releasing hormone inhibits gonadotropin secretion in the ovariectomized rhesus monkey, *J. Clin. Endocrinol. Metab.*, 65, 262, 1987.
125. **Petraglia, F., Sutton, S., Vale, W., and Plotsky, P.**, Corticotrophin-releasing factor decreases plasma luteinizing hormone levels in female rats by inhibiting gonadotropin-releasing hormone release into hypophysial-portal circulation, *Endocrinology*, 120, 1083, 1987.
126. **Petraglia, F., Vale, W., and Rivier, C.**, Opioids act centrally to modulate stress-induced decrease in luteinizing hormone in the rat, *Endocrinology*, 119, 2445, 1986.
127. **Pfeiffer, D.G., Pfeiffer, A., Shimohigashi, Y., Merriam, G.R., and Loriaux, D.L.**, Predominant involvement of mu- rather than delta- or kappa-opiate receptors in LH secretion, *Peptides*, 4, 647, 1983.
128. **Plotsky, P.M., Bruhn, T.O., and Vale, W.**, Hypophysiotropic regulation of adrenocorticotropin secretion in response to insulin-induced hypoglycemia, *Endocrinology*, 117, 323, 1985.

129. **Plotsky, P.M., and Vale, W.,** Hemorrhage-induced secretion of corticotropin-releasing factor-like immunoreactivity into the rat hypophysial portal circulation and its inhibition by glucocorticoids, *Endocrinology*, 114, 164, 1984.
130. **Quigley, M., Sheehan, K., Casper, R., and Yen, S.,** Evidence for increased dopaminergic and opioid activity in patients with hypothalamic hypogonadotropic amenorrhea, *J. Clin. Endocrinol. Metab.*, 50, 949, 1980.
131. **Reame, N.E., Sauder, S.E., Case, G.D., Kelch, R.P., and Marshall, J.C.,** Pulsatile gonadotropin secretion in women with hypothalamic amenorrhea: Evidence that reduced frequency of gonadotropin secretion is the mechanism of persistent anovulation, *J. Clin. Endocrinol. Metab.*, 61, 851, 1985.
132. **Reid, R.L., Leopold, G.R., and Yen, S.S.C.,** Induction of ovulation and pregnancy with pulsatile luteinizing hormone releasing factor; dosage and mode of delivery, *Fertil. Steril.*, 36, 553, 1981.
133. **Ringstrom, S.J., and Schwartz, N.B.,** Cortisol suppresses the LH release and ovulation in the cyclic rat, *Endocrinology*, 116, 472, 1985.
134. **Rivest, S., and Rivier, C.,** Influence of the paraventricular nucleus of the hypothalamus in the alteration of neuroendocrine functions induced by intermittent footshock or interleukin, *Endocrinology*, 129, 2049, 1991.
135. **Rivest, S., and Rivier, C.,** Interleukin-1 beta inhibits the endogenous expression of the early gene c-fos located within the nucleus of LH-RH neurons and interferes with hypothalamic LH-RH release during proestrus in the rat, *Brain Res.*, 613, 132, 1993.
136. **Rivier, C.,** Role of endotoxin and interleukin-1 in modulating ACTH, LH and sex steroid secretion, *Adv. Exp. Med. Biol.*, 274, 295, 1990.
137. **Rivier, C., and Rivest, S.,** Effect of stress on the activity of the hypothalamic-pituitary-gonadal axis: Peripheral and central mechanisms, *Biol. Reprod.*, 45, 523, 1991.
138. **Rivier, C., Rivier, J., and Vale, W.,** Stress-induced inhibition of reproductive functions: Role of endogenous corticotropin-releasing factor, *Science*, 231, 607, 1986.
139. **Rivier, C., and Vale, W.,** Influence of corticotropin-releasing factor on reproductive functions in the rat, *Endocrinology*, 114, 914, 1984.
140. **Rivier, C., and Vale, W.,** Effects of corticotropin-releasing factor, neurohypophyseal peptides, and catecholamines on pituitary function, *Fed. Proc.*, 44, 189, 1985.
141. **Rivier, C., and Vale, W.,** Cytokines act within the brain to inhibit luteinizing hormone secretion and ovulation in the rat, *Endocrinology*, 127, 849, 1990.
142. **Samuels, M.H., Sanborn, C.F., Hofeldt, F., and Robins, R.,** The role of endogenous opiates in athletic amenorrhea, *Fertil. Steril.*, 55, 507, 1991.
143. **Sapolsky, R., and Krey, L.,** Stress-induced suppression of luteinizing hormone concentrations in wild baboons: Role of opiates, *J. Clin. Endocrinol. Metab.*, 66, 722, 1988.
144. **Sapolsky, R., Rivier, C., Yamamoto, G., Plotsky, P., and Vale, W.,** Interleukin-1 stimulates the secretion of hypothalamic corticotropin-releasing factor, *Science*, 238, 522, 1987.
145. **Sawchenko, P.,** Evidence for a local site of action for glucocorticoids in inhibiting CRF and vasopressin expression in the paraventricular nucleus, *Brain Res.*, 403, 213, 1987.
146. **Sawchenko, P.E., and Swanson, L.W.,** Organization of CRF immunoreactive cells and fibers in the rat brain: Immunohistochemical studies, in *Corticotropin-Releasing Factor: Basic and Clinical Studies of a Neuropeptide*, DeSouza, E.B., and Nemeroff, C.B., Eds., CRC Press, Boca Raton, FL, 1990, 29.
147. **Sawchenko, P.E., Swanson, L.W., and Vale, W.W.,** Coexpression of corticotropin-releasing factor and vasopressin immunoreactivity in parvocellular neurosecretory neurons of the adrenalectomized rat, *Proc. Natl. Acad. Sci. USA*, 81, 1883, 1984.
148. **Swanson, L.W., Sawchenko, P.E., and Lind, R.W.,** Regulation of multiple peptides in CRF parvocellular neurosecretory neurons: Implications for the stress response, *Prog. Brain Res.*, 68, 169, 1986.
149. **Schreihofer, D.A., Parfeitt, D.B., and Cameron, J.L.,** Suppression of luteinizing hormone secretion during short-term fasting in male rhesus monkeys: The role of metabolic versus stress signals, *Endocrinology*, 132, 1881, 1993.

150. **Selye, H.,** Effect of adaptation to various damaging agents on the female sex organs in the rat, *Endocrinology*, 25, 615, 1939.
151. **Selye, H.,** *The Stress of Life*, McGraw-Hill, New York, 1976.
152. **Shalts, E., Feng, Y.-J., and Ferin, M.,** Vasopressin mediates the interleukin-1-induced decrease in luteinizing hormone secretion in the ovariectomized rhesus monkey, *Endocrinology*, 131, 153, 1992.
153. **Shalts, E., Xiao, E., Xia, L., and Ferin, M.,** An inhibitory role of vasopressin on luteinizing hormone secretion in the rhesus monkey, *75th Ann. Meet. Endocr. Soc.*, Abstr., 1993, 341.
154. **Smith, C.J., and Norman, R.L.,** Circadian periodicity in circulating cortisol is absent after orchidectomy in rhesus macaques, *Endocrinology*, 121, 2186, 1987.
155. **Smith, C.J., and Norman, R.L.,** Influence of the gonads on cortisol secretion in female rhesus macaques, *Endocrinology*, 121, 2192, 1987.
156. **Södersten, P., Henning, M., Melin, P., and Ludin, S.,** Vasopressin alters female sexual behaviour by acting on the brain independently of alterations in blood pressure, *Nature*, 301, 608, 1983.
157. **Suda, T., Tozawa, F., Yamada, M., Ushiyama, T., Tomori, N., Sumitomo, T., Nakagami, Y., Demura, H., and Shizume, K.,** Insulin-induced hypoglycemia increases corticotropin-releasing factor messenger ribonucleic acid levels in rat hypothalamus, *Endocrinology*, 123, 1371, 1988.
158. **Suh, B.Y., Liu, J.H., Berga, S.L., Quigley, M.E., Laughlin, G.A., and Yen, S.S.C.,** Hypercortisolism in patients with functional hypothalamic-amenorrhea, *J. Clin. Endocrinol. Metab.*, 66, 733, 1988.
159. **Thind, K.K., Boggan, J.E., and Goldsmith, P.C.,** Interactions between vasopressin- and gonadotropin-releasing hormone-containing neuroendocrine neurons in the monkey supraoptic nucleus, *Neuroendocrinology*, 53, 287, 1991.
160. **Tramu, G., Croix, C., and Pillez, A.,** Ability of the CRF immunoreactive neurons of the paraventricular nucleus to produce a vasopressin-like material, *Neuroendocrinology*, 37, 467, 1983.
161. **Valentino, R.J.,** Corticotropin-releasing factor: Putative neurotransmitter in the noradrenergic nucleus locus coeruleus, *Psychopharmacol. Bull.*, 25, 306, 1989.
162. **Van Vugt, D.A.,** Influences of the visual and olfactory systems on reproduction, *Semin. Reprod. Endocrinol.*, 8, 1, 1990.
163. **Viau, V., and Meaney, M.J.,** Variations in the hypothalamic-pituitary-adrenal response to stress during the estrous cycle in the rat, *Endocrinology*, 129, 2503, 1991.
164. **Walsh, T., Katz, J.L., Levin, J., Kream, J., Fukushima, D., Weiner, H., and Zumoff, B.,** The production rate of cortisol declines during recovery from anorexia nervosa, *J. Clin. Endocrinol. Metab.*, 53, 203, 1981.
165. **Wardlaw, S., Wehrenberg, W., Ferin, M., Antunes, J., and Frantz, A.,** Effect of sex steroids on β-endorphin in hypophyseal portal blood, *J. Clin. Endocrinol. Metab.*, 55, 877, 1982.
166. **Warner, B.A., DuFau, M.L., and Santen, R.J.,** Effects of aging and illness on the pituitary testicular axis in men, *J. Clin. Endocrinol. Metab.*, 60, 263, 1985.
167. **Warren, M.,** Amenorrhea in endurance runners, *J. Clin. Endocrinol. Metab.*, 75, 1393, 1992.
168. **Warren, M.P., and Vande de Wiele, R.L.,** Clinical and metabolic features of anorexia nervosa, *Am. J. Obstet. Gynecol.*, 117, 435, 1973.
169. **Wehrenberg, W., Wardlaw, S., Frantz, A., and Ferin, M.,** β-Endorphin in the hypophyseal portal blood: Variations throughout the menstrual cycle, *Endocrinology*, 111, 879, 1982.
170. **Weiner, R.I., and Martinez de la Escalera, G.,** Inhibitory regulation of pulsatile GnRH release from GT1 GnRH cell lines by vasopressin and GABA, *Soc. Neurosci. 22nd Ann. Meet.*, Abstr., 1992, 928.
171. **Whitacre, F.E., and Barrera, B.,** War amenorrhea. A clinical and laboratory study, *J. Am. Med. Assoc.*, 124, 399, 1944.

172. **Whitnall, M.,** Stress selectively activates the vasopressin-containing subset of corticotropin-releasing hormone neurons, *Neuroendocrinology*, 50, 702, 1989.

173. **Wildt, L., and Leyendecker, G.,** Induction of ovulation by the chronic administration of naltrexone in hypothalamic amenorrhea, *J. Clin. Endocrinol. Metab.*, 64, 1334, 1987.

174. **Wildt, L., Leyendecker, G., Sir-Petermann, T., and Waibel-Treber, S.,** Treatment with naltrexone in hypothalamic ovarian failure: Induction of ovulation and pregnancy, *Hum. Reprod.*, 8, 350, 1993.

175. **Williams, C.L., Nishihara, M., Thalabard, J.-C., Grosser, P.M., Hotchkiss, J., and Knobil, E.,** Corticotropin-releasing factor and gonadotropin-releasing hormone pulse generator activity in the rhesus monkey, *Neuroendocrinology*, 52, 133, 1990.

176. **Witkin, J.W., Ferin, M., Popilskis, S.J., and Silverman, A.J.,** Effects of gonadal steroids on the ultrastructure of GnRH neurons in the rhesus monkey: Synaptic input and glial apposition, *Endocrinology*, 129, 1083, 1991.

177. **Xiao, E., and Ferin, M.,** The inhibitory action of corticotropin-releasing hormone on gonadotropin secretion in the ovariectomized rhesus monkey is not mediated by adrenocorticotropic hormone, *Biol. Reprod.*, 38, 763, 1988.

178. **Xiao, E., Luckhaus, J., Niemann, W., and Ferin, M.,** Acute inhibition of gonadotropin secretion by corticotropin-releasing hormone in the primate: Are the adrenal glands involved? *Endocrinology*, 124, 1632, 1989.

179. **Yamamora, D., and Reid, R.,** Psychological stress and the reproductive system, *Semin. Reprod. Endocrinol.*, 8, 65, 1990.

180. **Yen, S.S., Quigley, M.E., Reid, R.L., Ropert, J.F., and Cetel, N.S.,** Neuroendocrinology of opioid peptides and their role in the control of gonadotropin and prolactin secretion, *Am. J. Obstet. Gynecol.*, 152, 485, 1985.

181. **Yen, S.S.C.,** Chronic anovulation due to CNS-hypothalamic-pituitary dysfunction, in *Reproductive Endocrinology*, Yen, S.S.C., and Jaffe, R.B., Eds., W.B. Saunders, Philadelphia, 1991, 631.

Chapter 11

SEXUAL DIFFERENTIATION: EFFECTS OF PERINATAL EXPOSURE TO ALCOHOL, COCAINE, MORPHINE OR NICOTINE

R.F. McGivern, E. Redei and W.J. Raum

TABLE OF CONTENTS

0-8493-2451-3/95/$0.00+.50

ABBREVIATIONS

ACTH–adrenocorticotropic hormone
AG–anogenital
CNS–central nervous system
DHT–dihydrotestosterone
FAE–fetal alcohol exposed
GnRH–gonadotropin hormone-releasing hormone
HPA–hypothalamic-pituitary-adrenal
HPG–hypothalamic-pituitary-gonadal
LH–luteinizing hormone
LHRH–luteinizing hormone-releasing hormone
MIS–Müllerian inhibitory substance
MPOA–medial preoptic area
NE–norepinephrine
POA–preoptic area
SDN–sexually dimorphic nucleus
SHBG–sex hormone binding globulin

INTRODUCTION

Sexually dimorphic behaviors, or sex-related behaviors, are perhaps the most important category of social behavior in mammals primarily because of their central role in reproduction. While studies over the past 100 years have established the basic mechanisms in mammalian reproduction, more recent work has demonstrated a pervasiveness to the sexual differentiation process that extends well beyond the act of reproduction. The range of this sexual differentiation extends to include differences in cognition and affect,[64] as well as biological differences in the central mechanisms involved in their expression.[9] Sex differences in humans have been observed in communication style, verbal and spatial skills, mathematical reasoning ability, and even play behavior in children.[17,20,68,117,129] Many of the nonlinguistic sex differences in humans can be observed in lower species, attesting to the primary role of biology over cultural factors in the determination of such differences. For instance, in the rat, nonreproductive, sex-related differences have been observed in a wide variety of behaviors including pain sensitivity,[111,118] social interaction,[80] taste preferences,[156] and spatial skills.[95,152]

The process of neurobehavioral sexual differentiation is strongly influenced by environmental factors acting during early development. One of the primary environmental influences that alters the normal development and expression of sexually dimorphic behaviors is prenatal exposure to drugs or other substances that disrupt gonadal development or neuronal and/or glial development in sexually dimorphic brain regions, leading to enduring alterations in the expression of sexually dimorphic behaviors. Before considering the effects of selected drugs with addictive properties on the mechanisms associated with neurobehavioral sexual differentiation, we will briefly review the physiology related to this process.

SEXUAL DIFFERENTIATION

Differentiation between the sexes in most mammals is easily discerned on the basis of observable physical differences. However, data accumulated in recent years have demonstrated marked anatomical sex differences in several areas of the central nervous system (CNS) as well, such as the hypothalamus, corpus callosum, and spinal cord.[9,27] Such differences have been identified in a number of mammalian species, including rodents and humans,[5,6,28,41] and are presumed to underlie the sex differences observed not only in copulatory behaviors, but also in nonreproductive behaviors such as aggression, play behavior, maternal behavior and spatial skills.[15,16] The result is a normal sexual dimorphism in mammals that is evident in brain, body and behavior.

Normal sexual differentiation is determined by genetic and phenotypic factors acting at different stages of development (Table 1). Genetic or chromosomal sex is determined at the time of fertilization by unidentified factors associated with the X and Y sex chromosomes. Subsequent genetic activity

TABLE 1
Human Sexual Development

	Male	**Female**
	Genetic Sex	
Fertilization	XY (sperm)	XX (ovum)
	Phenotypic Sex	
Fetus	Y chromosome causes undifferentiated gonad to develop into testes, which subsequently secrete testosterone.	In the absence of Y chromosome, undifferentiated gonad develops into ovaries.
	Testosterone stimulates the development of internal sex organs (i.e., prostate, seminal vesicles) from the Wollfian structures. Müllerian Inhibitory Substance (MIS), another hormone secreted by testes, causes the Müllerian structures to degenerate.	In the absence of testosterone, the Wollfian structures degenerate. The Müllerian structures develop into female internal sex organs (i.e., uterus, fallopian tubes) in the absence of MIS.
	DHT, a metabolite of testosterone, stimulates development of external genitalia into penis and scrotal sacs.	In the absence of DHT, external genitalia develop into clitoris, vagina and labia.
Fetus/child	Testosterone/estrogen induce structural changes in brain areas such as hypothalamus and corpus callosum. Such changes are considered to mediate masculinization and defeminization of behavior.	Absence of testosterone leads to feminization and demasculinization of the brain and behavior.
Puberty	Testosterone and DHT stimulate development of secondary sex characteristics (i.e., facial hair, muscles, penile growth, voice change) caused by testosterone.	Estrogens stimulate development of female secondary sex characteristics (i.e., breast enlargement, skin texture, fat deposition in hips).

related to the sex chromosomes differentiates an indifferent embryonic gonad into a testis or an ovary from the Wolffian or Müllerian duct, respectively. Phenotypic sex is under the direct control of gonadal hormone secretion, and other biochemical factors.[57] Phenotypical expression relates to development of sex organs such as the penis or vagina, as well as sex-specific regional differentiation of the CNS.

It is the phenotypic aspect of the sexual differentiation process that is most susceptible to environmental influences such as prenatal drug exposure. Drugs that influence biochemical factors critical to the sexual differentiation process, such as sex steroid hormones[92] or catecholamines,[110] are those of most concern. In some cases, the influences of such drugs are not obvious from their intended use, as is the case with cimetidine. Cimetidine is an H^2 receptor antagonist that is a widely prescribed ulcer medication. However, it was later identified as a weak antiandrogen that competes with testosterone and dihydrotestosterone at the site of the androgen receptor. Not surprisingly, it has been shown to also disrupt the masculinization process in rats exposed to the drug perinatally.[7,94]

Gonadal hormones serve to 'organize' the brain during critical periods of development to respond with a masculine or feminine pattern to later 'activational' hormonal stimulation,[93] although this organizational/activational distinction is not as clear as originally formulated.[8] Organizational actions of hormones are restricted by definition to critical periods in development to induce their long-lasting consequences. Regardless of the developmental time frame of a given species to reach maturity, there is a good correlation with the timing of the critical period for sexual differentiation relative to the stage of biological development.[122] Organizational effects of hormones are long lasting and occur in both brain and periphery through their effects on gene activity in the developing organism.[93] Thus, prenatal exposure of males or females to adequate levels of testosterone at the proper point in fetal development will lead to later masculine sex behavior patterns (i.e., mounting or intromission) in castrated adult animals following an injection of testosterone in the presence of a receptive female. The lack of adequate hormonal exposure during a critical organizational period will produce an adult that fails to respond to the same testosterone injection with normal sex behavior. Activational effects of a hormone are those that occur only in the presence of the hormone. Such effects can be determined by the prior exposure to hormones during critical organizational periods. However, activational effects of hormones can also be influenced in the presence of environmentally induced alterations in the biochemical milieu at the time the hormone is present.[18]

Some aspects of masculine sex behavior are organized by genomic actions of androgens such as testosterone or dihydrotestosterone acting directly through the androgen receptor, while other aspects of male sex behavior are organized by estrogens such as estradiol acting in specific brain regions such as hypothalamus, preoptic area (POA) and amygdala. The role of estrogen in masculine sexual differentiation of the brain is intimately tied to testosterone through the action of aromatase, an intracellular P-450 enzyme that converts testosterone or androstenedione intracellularly to estrogen. In this manner, a male can be influenced by estrogen in the absence of ovaries. Thus, part of testosterone's role in the masculinization process is to serve as a substrate for estrogen. Females are protected from the masculinizing effects of estrogen derived from their ovaries by a plasma-binding protein that prevents estrogen from crossing the cell membrane. In the rat, this binding protein is α-fetoprotein, whereas in the human it is sex hormone binding globulin (SHBG). (See McEwen[92] and Goy and McEwen[174] for excellent reviews of the mechanisms involved in neurobehavioral sexual differentiation.) Masculinized behavior patterns have been observed in women prenatally exposed to high levels of adrenal androgens such as androstenedione[53] or to estrogens such as diethylstilbestrol,[71] which are not bound by SHBG. Limited evidence also exists indicating that ovarian steroids may be important for complete feminization of the brain.[45,50,58]

Estrogen or androgen receptor occupation leads to a binding of the complex to DNA response elements in the nucleus, which subsequently inhibits or triggers protein production within the cell. Such genomic actions appear to be intimately connected with neuronal survival and growth.[93] In addition to its effects on neurogenesis, this process is thought to permanently define a neuron's subsequent responsiveness to stimulation. Present evidence indicates that sex steroids act in brain regions such as the hypothalamus to produce long-term changes in synapses and postsynaptic membranes to estrogen stimulation.[69,113]

The timing of any disruption in the steroid milieu during development is of particular importance in determining the long-term consequences on brain and behavior. The normal masculinization process in males is dependent in part on surges of testosterone levels occurring during critical periods of development. In the rat, the critical period for sexual differentiation is considered to extend from approximately the last week of gestation through the first few days postnatally. During this period, there is a normal surge of testosterone on days 18 and 19 of gestation, as well as another surge that occurs at birth. If the amplitude or time course of either surge is altered, sexual differentiation of the male will be incomplete.[34,97,100,168] For example, exposing a pregnant rat to prenatal stress will accelerate the appearance of the surge in the male fetus to day 17, subsequently resulting in incomplete masculinization of sex behavior in the adult male offspring.[165] Similar results have been found in adult animals castrated at birth to prevent the effects of the postnatal testosterone surge.[134]

Similar surges of testosterone have been identified in the human male fetus and neonate. A prenatal surge of testosterone occurs around 16–20 weeks of gestation.[119] A postnatal testosterone surge has also been observed during the first few hours after birth.[35,151] It is presumed that interference with either surge will lead to incomplete masculinization as occurs in the rat.[36,119]

Studies in recent years have also revealed notable effects of monoamines on the process of sexual differentiation in the CNS. In addition to their neurotrophic actions early in brain development,[86] actions of serotonin and NE have been identified as playing a significant modulatory role on sex steroid actions during development.[21,73,125,127] Such findings have led to an increased appreciation of alternative mechanisms to explain alterations in sexual differentiation in the absence of any apparent change in steroid secretion or action.[22,128]

PRENATAL DRUG EXPOSURE

Prenatal exposure to drugs that alter sex steroid production or secretion, or influence monoaminergic activity, pose a significant risk of disrupting the sexual differentiation process in the fetus. At present, with the exception of ethanol, limited information is available regarding the extent or degree to which prenatal exposure to commonly used legal and illicit drugs influences the sexual differentiation process in animal models. Such drugs include ethanol, cocaine, morphine, and nicotine. Below is a review of current information regarding each of these substances on this process in rodents. Effects of additional drugs on this process have been recently reviewed elsewhere.[141,164]

Alcohol

Jones and Smith[75] and Lemoine *et al.*[87] published the first studies describing the developmental consequences of fetal alcohol exposure in humans nearly 25 years ago. Since that time, the rodent model of fetal alcohol exposure has yielded considerable preclinical data, which has essentially paralleled results observed in children with Fetal Alcohol Syndrome.[46] Although no studies in humans have been conducted to date to examine the consequences of fetal alcohol exposure on neurobehavioral sexual differentiation in humans, results from animal studies indicate a pervasive influence of ethanol on this process. Typically, alcohol is administered in a fortified liquid diet prenatally, neonatally or perinatally. Postnatal administration employs a gastric intubation and artificial rearing procedures for the first 7–10 days after birth.[138] The rationale for exposing the newborn rat to alcohol is based on the temporal differences in development of the CNS across species. Although the time course of brain development is similar across mammalian species, one marked difference between species is when birth occurs relative to CNS development. In humans, the third trimester of pregnancy is characterized by a period of rapid neuronal growth and proliferation that has been referred to as the "brain growth spurt." In contrast, this period of CNS growth occurs in the rat during the first weeks after birth.[44] Therefore, neonatal alcohol exposure with the rodent model is used to approximate ethanol exposure during the third trimester equivalent period of CNS development in the human.

Reproductive and Maternal Behaviors

Several aspects of sexual behavior have been found to be altered in adult rodents prenatally exposed to alcohol. Chen and Smith[30] reported the first study of sexual behavior in the fetal alcohol exposed (FAE) male rat in which they observed poorer penile reflexes in FAE males compared to controls. Udani *et al.*[154] later reported a decrease in intromission behavior in FAE males in the presence of receptive females, suggesting incomplete masculinization. Hard *et al.*[67] reported increased lordosis in FAE males primed with estrogen and progesterone, indicating incomplete defeminization in FAE males. However, others have not observed differences in FAE males with respect to masculinization[38,67,96] or feminization[38,96] of sexual behavior.

This inconsistency between studies may reflect the effect of prenatal alcohol exposure on the sensitivity of the hypothalamic-pituitary-adrenal (HPA) axis. Prenatal ethanol exposure has also been found to have long-term effects on the developing HPA axis. FAE female rats exhibit a greater response to stress in adulthood as measured by the release of corticosterone from the adrenal gland. This effect has been found with both prenatal[153,166] and postnatal alcohol exposure.[82] Stress responsiveness of FAE males was not affected in these studies, but a recent study by Weinberg indicates that increased HPA activation in response to stress can also be observed in adult FAE males.[167] Stress-induced suppression of reproductive behavior is well known,[47,128] an effect that is mediated by glucocorticoids.[26,84] Thus it is possible that the decreases in sexual behavior observed by some investigators reflect a lack of habituation to the open field testing situation, a situation well known to increase HPA activation in the rat.[51] Alternatively, olfactory cues, important to initiation of sexual behavior in the rat, may be compromised due to the loss of mitral cells in the olfactory bulb following early alcohol exposure.[25]

Fewer sex-related effects of prenatal alcohol exposure have been observed in exposed female offspring. However, a delay in the onset of sexual maturation in mice and rats, as measured by date of vaginal opening, has been a consistently reported effect of fetal alcohol exposure in females.[24,47,49,99] In a recent study, the date of vaginal opening in females exposed to ethanol during days 7–21 of gestation was compared with that of females exposed during days 14–21 of gestation.[99] An equal amount of delay was observed in both groups. These results suggest that exposing the developing hypothalamus to ethanol during the last week of gestation is a more important factor in causing this delay than alcohol exposure to the developing ovary, which differentiates around day 12 of gestation. This suggestion is supported by the findings of Sonderegger *et al.*,[149] who observed no effect of prenatal ethanol exposure on days 1–7 or 8–14 of gestation on female reproductive function. However, females exposed to a much higher level of ethanol only on day 8 of gestation have been reported to be more sexually responsive to estrogen in adulthood.[109] Neither prenatal nor postnatal alcohol exposure has been found to alter fertility.[149,176]

However, female FAE rats do display deficits in maternal behavior, taking longer to retrieve their pups and making poorer nests.[68] Retrieval deficits can also be observed in virgin FAE rats presented repeatedly with pups from another mother,[12] indicating that alcohol has a disruptive effect on the organization of this behavior. However, the role played by increased stress responsiveness of FAE females needs to be determined to better assess whether prenatal ethanol exposure is acting to directly influence the organization of maternal behavior.

Nonreproductive Behaviors

Several sex-related behaviors unrelated to reproduction are feminized in adult FAE males. Such behaviors include saccharin preference, maze performance and juvenile play behavior, all of which are organizationally dependent on testosterone for their sexually dimorphic expression. Normally, adult female rats consume greater quantities of sweetened solutions such as saccharin than do males.[156] This behavior, like most sexually dimorphic behaviors, can be altered by changes in the steroid hormonal environment during either late prenatal or early postnatal life.[15] Following both prenatal[95] and neonatal[10] alcohol exposure, the normal sex difference in saccharin preference is eliminated. Alcohol-exposed males consume more saccharin than control males, suggesting some demasculinization of this behavior, whereas females consume less saccharin, suggesting masculinization.

The literature on the effects of prenatal alcohol exposure on learning demonstrates consistent decrements in learning abilities of alcohol-exposed animals.[108] However, in studies involving spatial mapping where both sexes have been tested, sex-dependent effects have also been observed in complex learning paradigms that involve spatial learning or spatial mapping. Male learning of spatial types of tasks is generally found to be significantly better than that of females.[15] Prenatal alcohol exposure appears to influence adult performance of a spatial task in a sex-dependent manner. In the Lashley III maze, FAE males required more trials to learn the maze than controls while the performance of FAE females improved to the level of control males.[95] Performance of FAE males in other complex mazes is also impaired.[23,173]

Sex-dependent effects of prenatal alcohol exposure have also been observed in play behavior. Juvenile rats engage in a type of rough and tumble play that resembles wrestling. The degree to which this behavior is expressed is dependent on perinatal testosterone levels; males typically engage in more of this play behavior than females.[103] FAE males have been found to display less of this aggressive behavior than controls, while females display more,[107] again suggesting a partial masculinization of females and a demasculinization of males. Such effects of alcohol on nonreproductive sexually dimorphic behaviors are not restricted to rodents. Fadem[177] has recently observed masculinization of sexually dimorphic behaviors in adult male opossums exposed perinatally to alcohol. The nonreproductive behavior of male littermates was feminized.

Originally, we proposed that masculinization of FAE females might be due to ethanol-induced activation of the HPA axis, resulting in excessive release of adrenal androgens such as androstenedione.[95] However, recent evidence indicates that androstenedione is not released in response to stress in the rat,[51] unlike the human, because it is not synthesized in significant quantities in the adrenal gland.[157] Thus it seems unlikely that the masculinization observed in FAE females is related to excessive androgen production.

Other learning paradigms also reveal sex differences in sensitivity to alcohol's effects, but these differences also appear to depend on the period of exposure. While males appear more affected in some paradigms following prenatal alcohol exposure, the reverse appears to be true following postnatal alcohol exposure. When mice exposed to alcohol prenatally were trained to press a bar for reward in an operant learning task, the normal sex differences were eliminated and males appeared more affected than females.[56] In a passive avoidance paradigm, a simple learning task in which the subject must learn to inhibit its normally preferred response to avoid punishment, both sexes have difficulty with learning following prenatal alcohol exposure.[131] However, following neonatal alcohol exposure only females are impaired.[11] When spatial navigation was examined in a Morris water maze, the performance of adult females was impaired while adult male performance was unaffected following postnatal alcohol exposure.[81]

Alterations in Neuroendocrine Function

Ethanol is well known to depress hypothalamic-pituitary-gonadal (HPG) function in both males and females,[123] resulting in reduced testosterone levels in males. Ethanol also has been found to produce a marked depression in fetal and neonatal production of testosterone in males, an effect that is quite consistent with its behavioral effects. The prenatal testosterone surge on days 18 and 19 of gestation are greatly attenuated in FAE male fetuses.[100] A less marked but significant attenuation has also been observed in the postnatal surge of males from dams consuming approximately 14 g/kg/day of ethanol during the last week of gestation.[97] These data are shown in Figure 1. A decrease in testosterone levels of FAE males around the time of birth has been reported by others,[79,136] which is consistent with an attenuation of the postnatal testosterone surge. This depression of testosterone production appears to relate to a depression in 17-α-hydroxylase activity in neonatal testes from FAE males,[78,79] although changes in luteinizing hormone (LH) secretion or sensitivity to LH cannot be ruled out.[100] Exposures to lower amounts of

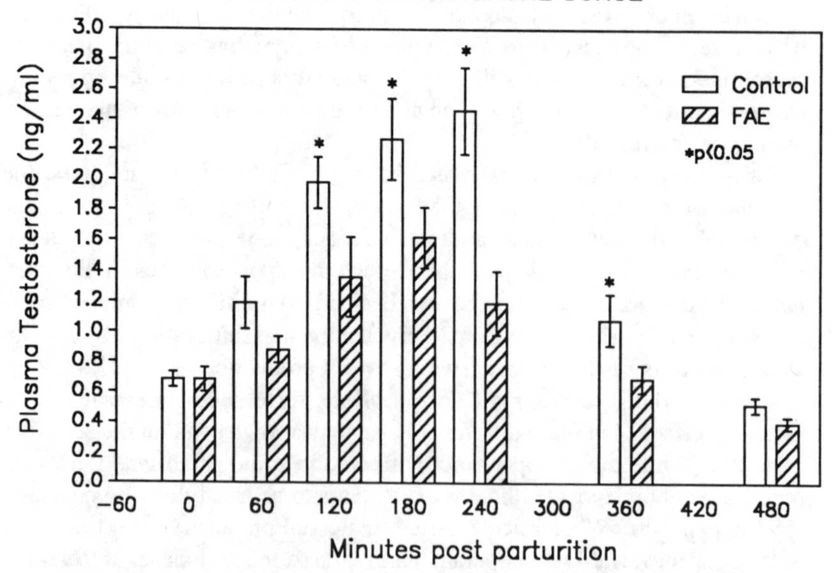

FIGURE 1. Plasma testosterone levels in male rats during the first 8 hours after birth. Animals were delivered by cesarean section. Each time point represents mean and standard error of the level in 10–12 males. (See McGivern *et al.*[97] for details.) (Reproduced from McGivern, R.F., Handa, R.J., and Redei, E., Decreased postnatal testosterone surge in male rats exposed to ethanol during the last week of gestation, *Alcoholism: Clin. Exp. Res.*, 17, 1215, 1993, with permission from Williams & Wilkins, ©Research Society on Alcoholism.)

ethanol during this period do not appear to influence the postnatal rise in testosterone levels.[38]

Present evidence indicates that LH secretion is decreased in older adult FAE animals of both sexes. Basal LH secretion in adult castrated FAE males and females at 5–6 months of age was found to be reduced to nearly half the level of pair-fed controls.[66] In addition, alterations in the amplitude and duration of pulsatile LH release were observed. Both sexes have been reported to exhibit reduced sensitivity to sex steroid feedback in the brain.[66,76] Such effects may contribute significantly to the delay in puberty onset in FAE females, as well as to the demasculinized sex behavior of adult FAE males. However, corticosteroid levels were not measured in the studies cited above, leaving open the possibility that decreases in basal LH or in response to estrogen-induced positive feedback result in part from the increased stress responsiveness of FAE females.

Data concerning the effects of prenatal ethanol exposure on the adult male HPG axis are inconsistent. Reduced sex organ weights in FAE adult males, including testes, prostate and seminal vesicles, have been reported by Udani *et al.*[154] in animals exposed to ethanol from day 12 through parturition. However, similar reductions were not observed in FAE males of the same strain exposed to ethanol during either the last 2 weeks of gestation or the last week alone.[99] In addition, males in this study were observed to have normal testosterone levels and normal sperm counts. Other studies have reported reduced plasma testosterone levels in adult FAE males,[38,154] although the measured plasma values were still in the normal male range. Given this fact, as well as the inconsistency between studies with respect to male sex behavior, the significance of these reductions is not clear.

A number of studies have reported that uncorrected anogenital (AG) distance in FAE males is reduced at birth.[30,136,154] Such results have been interpreted to indicate either reduced levels of testosterone or a decrease in 5-α reductase activity to convert testosterone to dihydrotestosterone, and they appear to be consistent with the attenuated prenatal surge of testosterone in FAE males.[99] However, AG distance is positively correlated with both body weight and length. Graham and Gandelman[63] have shown that accurate assessment of the steroid influence on AG distance can best be measured by indexing AG distance to body weight. We also observed significant reductions in AG distance of FAE males at birth,[94,100] but the results were not significant when indexed to body weight. Thus, we believe that the reduction in AG distance primarily reflects an effect of prenatal ethanol exposure on somatic growth rather than a specific effect of the drug on peripheral androgen metabolism or sensitivity.

Neurotransmitter Function

Animal studies suggest that the activity of monoamine neurotransmitters such as NE and serotonin during prenatal development can act as modulators of neuroanatomical and behavioral sexual differentiation.[66,73] Excessive activity of NE is known to inhibit the actions of sex steroid hormones in areas of the brain such as the hypothalamus in the neonatal rat.[127] Data from other studies indicate that both NE and serotonin play an important role in the structural development of the brain.[86,110] Finally, prenatal alcohol has been shown in several studies to have long-term effects on neurotransmitters in the developing brain.[33,42,124] Taken together, these data indicate that an interaction between monoamines and sex steroid hormones on brain development may be an important variable when considering the effects of prenatal alcohol exposure on sexual differentiation.

Neuroanatomical Changes

Since prenatal alcohol exposure alters sex steroid levels during development, resulting in altered sexually dimorphic behaviors, one would expect that structures of the brain influenced by neonatal steroid levels should also be affected. Indeed, there are data showing that this is the case, and it is likely that these structural changes mediate some of the behavioral changes that have been reported. Neuroanatomically, there are a number of areas in the rat brain that are sexually dimorphic in size, such as regions of the cortex, corpus callosum and the hypothalamus. The size of these areas is determined by levels of circulating testosterone during development and has been shown to be affected by perinatal alcohol exposure.

The POA of the hypothalamus is known to play an important role in sex and maternal behaviors in rats.[116] Within this area is a sexually dimorphic nucleus termed the SDN-POA, which is severalfold larger in males than in females.[60] Following either prenatal or perinatal alcohol exposure, this nucleus was smaller in adult males, indicating a demasculinizing effect of alcohol during perinatal development.[13,136]

The corpus callosum in rats has also been shown to be influenced by prenatal alcohol exposure in a sexually dimorphic manner. Typically, the corpus callosum is larger in males than females. However, data from a recent study suggest that prenatal alcohol exposure reduces or eliminates this sex difference.[172] A similar effect of prenatal alcohol exposure on cortical asymmetry has also been reported.[171] In normal males, the right hemisphere of the cerebral cortex is thicker than the left cortex, whereas females show no such asymmetry. Following prenatal alcohol exposure, this cortical asymmetry appears reduced in males, again suggesting a demasculinizing influence of ethanol.

An important question regarding a differential sensitivity between the sexes to perinatal alcohol exposure has been addressed indirectly by several studies. Behavioral data indicate that female rats tend to be more adversely affected by neonatal alcohol exposure than males in certain learning paradigms, such as Morris water maze performance[81] or passive avoidance learning.[11] Similar findings emerge from neuroanatomical studies. Female rats exposed to alcohol during this third trimester equivalent tend to show greater brain weight deficits relative to males.[61,170]

In the mouse, ethanol reduced the number of immunoreactive gonadotropin hormone-releasing hormone (GnRH) neurons detectable at 18 days of gestation following exposure to a high dose of ethanol administered on day 8 of pregnancy.[140] However, no effect of prenatal ethanol exposure was observed in the rat in the number of immunoreactive GnRH neurons in 44-day-old female FAE animals with delayed onset of puberty.[101] Subtle alterations were detected in the morphological characteristics of the neuronal processes of GnRH cells in these animals, but their significance is unclear at the present time. Differences in species, as well as prenatal timing and amount of exposure to alcohol, make comparisons to the results found in the mouse difficult. It remains to be determined whether the decreases in LH secretion of FAE animals reflect a functional deficit related to the GnRH neuron.

Cocaine

Little is known regarding the effects of prenatal cocaine exposure on neurobehavioral sexual differentiation. Cocaine is well known to block reuptake of monoamines in adult animals, as well as to release dopamine.[70,83,132,135] Because monoamines have been demonstrated to modulate steroid actions as well as to have neurotrophic actions early in development,[86] prenatal exposure might be expected to have significant effects on the sexual differentiation process. Fetal brain tyrosine hydroxylase activity is significantly increased by cocaine exposure,[4,106] which is consistent with the marked hyperactivity of the noradrenergic system observed in neonatal brains of males and females exposed to cocaine from days 8–20 of gestation.[144] Little or no effect of prenatal cocaine exposure was observed on dopamine activity.

Raum *et al.*[125] found that adrenergic stimulation of the neonatal brain will inhibit hypothalamic nuclear incorporation of estrogen. It was subsequently demonstrated that cocaine administered intracerebroventricularly to 4-day-old female rats also significantly inhibited the incorporation of estradiol in the hypothalamus.[126] The effectiveness of cocaine and adrenergic stimulation on this process is shown in Figure 2. Because estradiol is a critical factor in the masculinization of the brain, such results would suggest that prenatal cocaine exposure might interfere with neurobehavioral masculinization in males. To date two published studies have addressed this question directly with conflicting behavioral results.

Raum *et al.*[126] treated pregnant Sprague-Dawley dams with cocaine (10 mg/kg, s.c.) twice a day during the last week of gestation and studied only male offspring. No effects of cocaine were observed on AG distance at birth. In adulthood, males prenatally exposed to cocaine were found to have increased latencies to initiate sexual behavior, but other aspects of masculine sex behavior were similar to controls. However, another testosterone-dependent behavior, territorial scent marking, was significantly reduced in these animals.

Some evidence for alterations of adult endocrine function was also detected. Plasma LH in cocaine-exposed males was significantly higher than controls, whereas testosterone levels were the same as controls, suggesting

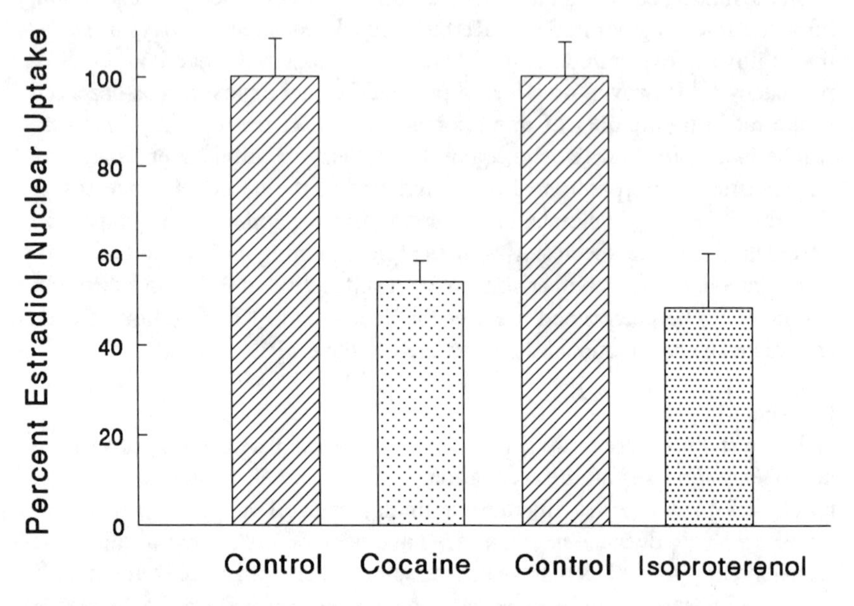

FIGURE 2. Inhibition of estradiol incorporation into hypothalamic nuclei of 4-day-old females. Data shown are mean and SEM of 5–11 animals per group. (Data are adapted from results reported in Raum *et al.*[125,126])

some measure of insensitivity to negative feedback by testosterone in these animals. Other endocrine measures, as well as sex organ weights, were normal. However, sperm counts were significantly reduced. The reduction in scent marking in the face of normal circulating levels of testosterone and elevated LH suggests a relative CNS insensitivity to androgens in these animals compared with controls.

Vathy *et al.*[161] treated dams of the same strain with the same dose regimen of cocaine and observed a different pattern of results. Adult cocaine-exposed males were observed to have facilitated sexual activity patterns, as well as markedly reduced postejaculatory intervals. Conversely, sexual behavior of cocaine-exposed females was significantly inhibited. Catecholamine levels in the POA of cocaine-exposed males, but not females, were significantly higher than controls. Reasons for the differences between the two studies are not immediately apparent. However, one possibility is that animals in the Raum *et al.*[126] study were group housed during the several weeks of sex behavior testing, whereas the animals in the study by Vathy *et al.*[161] were singly housed (personal communication). Extended group housing is known to be related to a number of biochemical changes in the brain,[2,3,62,169] as well as result in decreased sexual behavior in males.[59] Thus, housing conditions may have influenced the pattern of results observed. If so, the results imply that prenatal exposure to cocaine may significantly influence the way males and females respond to long-term environmental stress in adulthood. Clearly, additional studies will be needed to assess the long-term effects of cocaine on sexual differentiation.

Peris *et al.*[120] found that prenatal cocaine exposure appears to influence dopamine release from nigrostriatal terminals in a sex-dependent manner. They studied sex differences in locomotor and stereotypic behaviors following cocaine or saline administration in adult males and females exposed prenatally to cocaine. Both males and females exhibited an increased sensitivity to cocaine compared with controls. In addition, females, but not males, exposed prenatally to cocaine exhibited increased locomotor activity following saline injection. However, the stage of estrus was not reported, leaving open the interpretation of this finding. Sex differences were also detected in amphetamine-induced release of ^{3}[H]dopamine from striatal slices. *In utero* cocaine exposure increased amphetamine-stimulated release in females, but decreased release in males.

Morphine

Data concerning the long-term effects of morphine on sexual differentiation of the brain are limited to a few studies. Exposure to morphine from days 5–14[159] or from days 11–18[158,160] reduces lordosis frequency in females. This effect is more pronounced in animals exposed from days 11–18 of gestation (40–57%) compared to those exposed from days 5–14 (20%), indicating the importance of morphine exposure during the entire period of hypothalamic differentiation in producing this effect. Sex behavior of males was unaltered, with the exception that morphine-exposed males had significantly shorter postejaculatory intervals.[158,160] Lordosis behavior was not examined in males. Hypothalamic norepinephrine (NE) content was dramatically altered in these animals. Relative to controls, NE content of morphine-exposed females was elevated by 95%, while content of males was reduced by 57%.[160] It should be noted that animals from these studies were single housed from weaning on, an environmental variable that may have contributed to eliciting the differences in drug-exposed animals. No differences were observed in estrogen receptor regulation in the hypothalamic-preoptic region of morphine-exposed animals.[158] In golden hamsters, perinatal morphine exposure had no effect on the ability of the female to display hormone-induced male or female sexual behavior patterns.[74] Morphine-exposed males exhibited normal masculine sex behavior patterns but significantly more lordosis behavior than controls, indicating incomplete defeminization.

Increased AG distance was reported by Lapointe and Nosal[85] in females at weaning that were exposed to morphine from conception through day 16, suggestive of some masculinization. Such results are consistent with the effects of morphine on female sex behavior, as well as the report of delayed vaginal opening in females exposed to morphine from days 5–12 of gestation.[91] Postnatal administration of morphine has been found to induce precocious puberty in females, as evidenced by an accelerated date of vaginal opening.[148] Davis and Lin[40] examined offspring of both sexes from dams administered morphine from days 5–18 of gestation. They found increased activity levels in drug-exposed offspring of both sexes exposed at 35 and 75 days of age, but no sex-related effect of the drug.

Singh *et al.*[146] reported that plasma testosterone and androstenedione on day 20 of gestation were significantly reduced in males but not females exposed to methadone from days 14–19 of gestation. They found no evidence for a direct effect of the drug at the site of the testis, nor any effect of the drug on aromatization of testosterone to estrogen. However, the results of this study, as well as many of the above-cited studies on the prenatal effects of morphine must be interpreted with caution. Cessation of morphine treatment causes severe withdrawal in rodents, which itself is known to induce long-term effects on growth and differentiation. Sparber has provided strong evidence that many of the effects attributed to prenatal morphine exposure, both biochemical and behavioral, are a result of withdrawal rather than direct effects of the drug.[88,150]

However, although there is limited and somewhat conflicting evidence that morphine influences sex differentiation, circumstantial evidence strongly justifies further study. The μ opioid receptors, which bind opiates such as morphine, heroin or methadone, are distributed throughout brain regions integral to reward and the expression of sex behavior, including the medial preoptic area (MPOA) and the ventral tegmental area. These receptors are present in brain as early as day 14 of gestation.[14,32] Stimulation of μ receptors in adult animals inhibits luteinizing hormone-releasing hormone (LHRH) release through an inhibitory influence on excitatory NE projection to the MPOA[77] and inhibits female sexual behavior through an inhibition of NE release in the ventromedial hypothalamus of females.[162] Injection of β-endorphin directly into the MPOA produces a cessation of copulation in male rats[72] and inhibits LHRH release in sheep.[175]

Chronic morphine treatment during the last 2 weeks of gestation did not alter the number or functional efficacy of the μ receptors in striatal and cortical slices from fetal brain at 21 days of gestation.[43] However, electrically stimulated, calcium-dependent release of both dopamine and NE in morphine-exposed tissue was dramatically enhanced, indicating an excessive activation of signal transduction mechanisms regulating catecholamine release. These opioid-related actions on catecholamines, combined with evidence that opiate receptor systems modulate development of catecholaminergic systems,[145] suggest that morphine has a strong potential to alter the normal development of neurobehavioral sexual differentiation. Future studies designed to examine a spectrum of sex-related behaviors will be helpful in ascertaining such potential.

Nicotine

Male offspring exposed to nicotine prenatally have been found to exhibit demasculinized behavior patterns in adulthood. Decreased mounting and intromission behavior in adult nicotine-exposed males has been reported,[142] indicating incomplete masculinization of the brain in these animals. Bernardi *et al.*[19] reported a decrease in the postejaculatory interval of males exposed prenatally to cigarette smoke, suggesting an increase in sexual drive. Evidence for feminization of the brain in males prenatally exposed to nicotine is provided by data indicating an increase in saccharin preference in these animals.[90]

Female sex behavior has not been found to be altered, but an increase in ovarian weight in females exposed to nicotine prenatally was observed by Segarra and McEwen.[141] Other sex-related nonreproductive behaviors have generally been found to be unaffected in females following perinatal nicotine exposure. These include saccharin preference,[90] salt preference,[141] and open-field behavior.[121] An exception is active avoidance behavior, which was found to be improved in adult females exposed to nicotine throughout gestation.[54] Meyer and Carr[105] found a consistent delay in vaginal opening in females exposed to a low or high dose of nicotine either prenatally or postnatally. In addition, elevations in peripubertal LH values were observed in nicotine-exposed males and females, suggestive of a relative insensitivity to steroid negative feedback.

In males exposed prenatally to nicotine, plasma testosterone levels have been reported to be significantly reduced in adulthood[142] or when measured on day 18 of gestation in male fetuses,[90] which appears to be consistent with the effects of the drug on masculine sex behavior. However, since the prenatal surge peaks on days 18–19,[100,168] additional time points during this period of gestation will need to be measured to reach a more definitive assessment of nicotine's effect on fetal testosterone levels.

AG distance at birth was also reported to be significantly smaller in nicotine-exposed males, but birth weight was significantly lower in these animals compared to controls, which likely accounted for the decrease in AG distance. Peters and Tang[121] observed decreased birth weight in male, but not female, offspring from dams treated with 6 mg/kg of nicotine prior to and during gestation, which is in the same range as the animals from the study of Lichtensteiger and Schlumpf.[90] Segarra and Strand[142] failed to see a difference in birth weights of either sex from dams treated with 0.25 mg/kg of nicotine twice daily from days 3–21 of gestation.

The mechanisms whereby nicotine might alter the sexual differentiation process are multiple. Nicotine receptors in brain are most dense in the hypothalamus and POA[31] and cholinergic regulation in the POA is modulated by gonadal steroids.[37] Prenatal nicotine exposure has been reported to induce transient increases in nicotinic receptors in fetal and postnatal brains of rats;[147] however, long-lasting effects of prenatal nicotine exposure on the cholinergic system have not been studied. Nicotine treatment of the fetus is known to have long-term effects on catecholaminergic function. Deficiencies in postnatal catecholamine activity have been observed,[114,115,130,143] in contrast to significant increases in catecholamine turnover observed in fetal brain.[89] Such increases could hypothetically result in a decreased hypothalamic nuclear binding of estradiol, and subsequent demasculinization of the male.[125,126]

Nicotine administration elevates several pituitary hormones, including adrenocorticotropin (ACTH), LH, vasopressin, endorphins and prolactin.[52]

Fetal adrenal function is affected by prenatal nicotine exposure, as well as aromatase activity in fetal forebrain.[163] Activation of the HPA axis in pregnant animals by stress or ACTH injection is known to demasculinize and/or feminize reproductive behavior of male offspring,[141] and this mechanism may account for a significant portion of the long-term effects of prenatal nicotine exposure on sex-related behaviors.

CONCLUSIONS

Overall, the data from animal studies provide strong support that prenatal alcohol exposure can alter the sexual differentiation process by interfering with the secretion, metabolism or actions of sex steroid hormones during critical developmental periods. The consequences are to feminize sexually dimorphic behaviors of males and, to a lesser extent, masculinize these behaviors in females. The size of sexually dimorphic regions of the brain, as well as adult endocrine function, are also influenced. Case reports in humans exposed to alcohol *in utero* have been reviewed by Abel;[1] these cases show a number of genital anomalies in children with fetal alcohol syndrome, suggesting the possibility that prenatal alcohol exposure can disrupt the sexual differentiation process in humans as well.

To a lesser extent, a growing literature supports the conclusion that morphine, cocaine and nicotine can also influence the sexual differentiation process. However, the extent of this disruption to endocrine, behavioral, and cognitive sex differences is largely undetermined. Further studies are needed to define the anatomical, behavioral and endocrinological parameters influenced by each drug. A small amount of evidence hints at some teratogenic potential for cocaine and morphine on the sexual differentiation process in humans. Chasnoff *et al.*[29] noted an abnormally high incidence of hypospadia (4/50) in male infants exposed to cocaine *in utero*, while there is a single report that boys exposed to opiates *in utero* tended to exhibit a feminized gender orientation.[139]

It is increasingly clear that the neurobehavioral development of reproductive and nonreproductive behaviors is not influenced to the same degree by alterations in the perinatal hormonal or monoaminergic environment, probably reflecting a fundamental underlying difference in the relative contributions of different brain areas to each behavior.[102] This fact points to the necessity of greater inclusion of sex-related behaviors in animal models used to assess the teratogenic potential of a given drug on the sexual differentiation process. However, one major problem with the animal model should be noted with respect to its relevance to human sexual behavior. Some reports from animal studies of prenatal drug exposure have indicated a relevance of the data to the causes of homosexuality in humans.[39,67] While data from the animal models reviewed above can provide invaluable preclinical evidence to help understand the effects of perinatal drug exposure on brain development and the process of sexual differentiation, the results from these studies probably provide minimal information with respect to the prenatal influence of these drugs on homosexual behavior in humans. This appears likely for several reasons. First, no adequate model exists for homosexual behavior in the rodent in the absence of pharmacological administration of steroids. Normal male rats that show low levels of masculine sex behavior in the presence of estrous females do not exhibit increased tendencies to mount other males nor to lordosis when mounted by another male. In fact, male preference behavior for an estrous female rat does not appear to be influenced by perinatal androgen exposure.[104] Second, sexual orientation in humans is determined by hormonal, environmental and cultural factors,[112] which are not present in animal models.

These and other problems with a developmental animal model of human homosexuality have been extensively discussed by Sachs and Meisel.[137] Finally, in the human, there is also the area of gender identity, which refers to traits or conditions of maleness or femaleness. The degree to which gender identity in humans is causally linked to cultural or biological influences is an area of current debate,[55,155] but such identity is clearly beyond the scope of animal modeling at the present time. Therefore, issues related to sexual orientation of humans and prenatal drug exposure likely await data from future human studies for further resolution.

ACKNOWLEDGMENTS

Published and unpublished work was supported by grants from the National Institute on Alcohol Abuse and Alcoholism (AA06478) and the National Institute of Drug Abuse (DA06478).

REFERENCES

1. **Abel, E.L.,** *Fetal Alcohol Syndrome,* Medical Economic Books, Oradell, NJ, 1990.
2. **Adler, M.W., Bendotti, C., Ghezzi, D., Samanin, R., and Valzelli, L.,** Dependence to morphine in differentially housed rats, *Psychopharmacologica,* 41, 15, 1975.
3. **Adler, M.W., Mauron, C., Samanin, R., and Valzelli, L.,** Morphine analgesia in grouped and isolated rats, *Psychopharmacologica,* 41, 11, 1975.
4. **Akbari, H.M., and Azmitia, E.C.,** Increased tyrosine hydroxylase immunoreactivity in the rat cortex following prenatal cocaine exposure, *Dev. Brain Res.,* 66, 277, 1992.
5. **Allen, L.S., Hines, M., Shyrne, J.E., and Gorski, R.A.,** Two sexually dimorphic cell groups in the human brain, *J. Neurosci.,* 9, 497, 1989.
6. **Allen, L.S., Richey, M.S., Chai, Y.M., and Gorski, R.A.,** Sexual differences in the corpus callosum of the living human being, *J. Neurosci.,* 11, 933, 1991.
7. **Anand, S., and Van Thiel, D.H.,** Prenatal and neonatal exposure to cimetidine results in gonadal and sexual dysfunction in adult males, *Science,* 218, 493, 1982
8. **Arnold, A.P., and Breedlove, S.M.,** Organizational and activational effects of sex steroids on brain and behavior: A re-analysis, *Horm. Behav.,* 19, 469, 1985.
9. **Arnold, A.P., and Gorski, R.A.,** Gonadal steroid induction of structural sex differences in the central nervous system, *Ann. Rev. Neurosci.,* 7, 413, 1984.
10. **Barron, S., Razani, L.J., Gallegos, R.A., and Riley, E.P.,** The effects of neonatal alcohol exposure on saccharin preference, *Alcoholism: Clin. Exp. Res.,* in press.
11. **Barron, S., and Riley, E.P.,** Passive avoidance performance following neonatal alcohol exposure, *Neurotoxicol. Teratol.,* 12, 135, 1990.

12. **Barron, S., and Riley, E.P.**, Pup-induced maternal behavior in adult and juvenile rats exposed to alcohol prenatally, *Alcoholism: Clin. Exp. Res.*, 9, 360, 1985.
13. **Barron, S., Tieman, S.B., and Riley, E.P.**, Effects of prenatal alcohol exposure on the sexually dimorphic nucleus of the preoptic area in male rats, *Alcoholism: Clin. Exp. Res.*, 12, 59, 1988.
14. **Bayon, A., Shoemaker, W.J., Bloom, F.E., Mauss, A., and Guillemin, R.**, Perinatal development of the endorphin and enkephalin-containing systems in the rat brain, *Brain Res.*, 179, 93, 1979.
15. **Beatty, W.W.**, Gonadal hormones and sex differences in non-reproductive behavior in rodents: Organizational and activational effects, *Horm. Behav.*, 12, 112, 1979
16. **Beatty, W.W.**, Gonadal hormones and sex differences in nonreproductive behaviors, in *Handbook of Behavioral Neurobiology, Vol. 11*, Gerall, A.A., Moltz, H., and Ward, I.L., Eds., Plenum Press, New York, 1992, 85.
17. **Beatty, W.W.**, Hormonal organization of sex differences in play fighting and spatial behavior, *Prog. Brain Res.*, 61, 320, 1984.
18. **Becker, J.B., and Cha, J.**, Estrous cycle-dependent variation in amphetamine-induced behaviors and striatal dopamine release assessed with microdialysis, *Behav. Brain Res.*, 35, 117, 1989.
19. **Bernardi, M., Genedani, S., and Bertolini, A.**, Sexual behavior in the offspring of rats exposed to cigarette smoke or treated with nicotine during pregnancy, *Riv. Farmacol. Ter. XII*, 3, 197, 1981.
20. **Benbow, C.P.**, Sex differences in mathematical reasoning ability in intellectually talented preadolescents: Their nature, effects, and possible causes, *Behav. Brain Sci.*, 11, 169, 1988.
21. **Beyer, C., and Feder, H.H.**, Sex steroids and afferent input: Their roles in brain sexual differentiation, *Ann. Rev. Physiol.*, 49, 349, 1987.
22. **Beyer, C., Kolbinger, W., Froehlich, U., Pilgrim, C., and Reisert, I.**, Sex differences of hypothalamic prolactin cells develop independently of the presence of sex steroids, *Brain Res.*, 593, 253, 1992.
23. **Blanchard, B.A., Riley, E.P., and Hannigan, J.H.**, Deficits on a spatial navigation task following prenatal exposure to ethanol, *Neurotoxicol. Teratol.*, 9, 253, 1987.
24. **Boggan, W.O., Randall, C.L., and Dodds, H.M.**, Delayed sexual maturation in female C57BL/6J mice prenatally exposed to alcohol, *Res. Commun. Chem. Pathol. Pharmacol.*, 23, 117, 1979.
25. **Benthos, DB., Benthos, N.E., Napery, R.M.A., and West, J.R.**, Early postnatal alcohol exposure acutely and permanently reduces the number of granule cells and mitral cells in the rat olfactory bulb: A sterological study, *J. Comp. Neurol.*, 324, 557, 1992.
26. **Brann, D.W., Putnam, C.D., and Mahesh, V.B.**, Corticosteroid regulation of gonadotropin and prolactin secretion in the rat, *Endocrinology*, 126, 159, 1990.
27. **Breedlove, S.M.**, Sexual dimorphism in the vertebrate nervous system, *J. Neurosci.*, 12, 4133, 1992.
28. **Breedlove, S.M.**, Steroid influences on the development and function of a neuromuscular system, *Prog. Brain Res.*, 61, 147, 1984.
29. **Chasnoff, I.J., Chisum, G.M., and Kaplan, W.E.**, Maternal cocaine use and genitourinary tract malformations, *Teratology*, 37, 201, 1988.
30. **Chen, J.J., and Smith, E.R.**, Effects of perinatal alcohol on sexual differentiation and open-field behavior in rats, *Horm. Behav.*, 13, 219, 1979.
31. **Clarke, P.B.S., Schwartz, R.D., Paul, S.M., Pert, C.B., and Pert, A.**, Nicotinic binding in rat brain: Autoradiographic comparison of ^{3}[H]acetyl-choline, ^{3}[H]nicotine, and [^{125}]a-bungarotoxin, *J. Neurosci.*, 5, 1307, 1988.
32. **Clendeninn, N.J., Petraitis, M., and Simon, E.J.**, Ontological development of opiate receptors in rodent brain, *Brain Res.*, 118, 157, 1976.
33. **Cooper, J.D., and Rudeen, P.K.**, Alterations in regional catecholamine content and turnover in the male rat brain in response to *in utero* ethanol exposure, *Alcoholism: Clin. Exp. Res.*, 12, 282, 1988.

34. **Corbier, P., Kerdelhué, B., Picon, R., and Roffi, J.,** Changes in testicular weight and serum gonadotropin and testosterone levels before, during, and after birth in the perinatal rat, *Endocrinology,* 103, 1985, 1978.
35. **Corbier, P., Dehennin, M., Castanier, A., Mebazaa, D.A., and Roffi, J.,** Sex differences in serum luteinizing hormone and testosterone in the human neonate during the first few hours of life, *J. Clin. Endocrinol. Metab.,* 71, 1344, 1990.
36. **Corbier, P., Edwards, D.A., and Roffi, J.,** The neonatal testosterone surge: A comparative study, *Arch. Int. Physiol. Biophys.,* 100, 127, 1992.
37. **Commins, D., and Yahr, P.,** Acetylcholinesterase activity in the sexually dimorphic area of the gerbil brain: Sex differences and influences of adult gonadal steroids, *J. Comp. Neurol.,* 224, 123, 1984.
38. **Dahlgren, I.L., Eriksson, C.J.P., Gustafsson, B., Harthon, C., Hard, E., and Larsson, K.,** Effects of chronic and acute ethanol treatment during prenatal and early postnatal ages on testosterone levels and sexual behaviors in rats, *Pharmacol. Biochem. Behav.,* 33, 867, 1989.
39. **Dahlgren, I.L., Matuszczyk, J.V., and Hard, E.,** Sexual orientation in male rats prenatally exposed to ethanol, *Neurotoxicol. Teratol.,* 13, 267, 1991.
40. **Davis, W.M., and Lin, C.H.,** Prenatal morphine effects on survival and behavior of rat offspring, *Res. Commun. Chem. Pathol. Pharmacol.,* 3, 205, 1972.
41. **deLacoste-Utamsing, C., and Holloway, R.L.,** Sexual dimorphism in the human corpus callosum, *Science,* 216, 1431, 1982.
42. **Detering, N., Collins, R.M., Hawkins, R.L., Ozland, P.T., and Karahasan, A.M.,** Comparative effects of ethanol and malnutrition on the development of catecholamine neurons: Changes in norepinephrine turnover, *J. Neurochem.,* 34, 1788, 1980.
43. **De Vries, T.J., Van Vliet, B.J., Hogenboom, F., Wardeh, G., Van der Laan, J.W., Mulder, A.H., and Schoffelmeer, A.N.M.,** Effect of chronic prenatal morphine treatment on μ-opioid receptor-regulated adenylate cyclase activity and neurotransmitter release in rat brain slices, *Eur. J. Pharmacol.,* 208, 97, 1991.
44. **Dobbing, J., and Sands, J.,** Comparative aspects of the brain growth spurt, *Early Hum. Dev.,* 3, 79, 1979.
45. **Dohler, K.D., Hancke, J.L., Srivastava, S.S., Hofmann, C., Shryne, J.E., and Gorski, R.A.,** Participation of estrogens in female sexual differentiation of the brain; neuroanatomical, neuroendocrine and behavioral evidence, *Prog. Brain Res.,* 61, 99, 1984.
46. **Driscoll, C.D., Streissguth, A.P., and Riley, E.P.,** Prenatal alcohol exposure: Comparability of effects in humans and animal models, *Neurotoxicol. Teratol.,* 12, 231, 1990.
47. **Esquifino, A.I., Sanchis, R., and Guerri, C.,** Effect of prenatal alcohol exposure on sexual maturation of female rat offspring, *Neuroendocrinology,* 44, 483, 1986.
48. **Euker, J.S., Meites, J., and Reigle, G.D.,** Effects of acute stress on serum LH and prolactin in intact, castrated and dexamethasone-treated male rats, *Endocrinology,* 96, 85, 1975.
49. **Farry, K., and Tittmar, H.G.,** Alcohol as a teratogen: Effects of maternal administration in rats on sexual development in female offspring, *IRCS Med. Sci. Pharmacol. Reprod. Obstet. Gynecol.,* 3, 620, 1975.
50. **Fitch, R.H., Cowell, P.E., Schrott, L.M., and Denenberg, V.H.,** Corpus callosum: Ovarian hormones and feminization, *Brain Res.,* 542, 313, 1991.
51. **Fitch, R.H., McGivern, R.F., Redei, E., Schrott, L.M., Cowell, P.E., and Denenberg, V.H.,** Neonatal ovariectomy and pituitary adrenal responsiveness in the adult rat, *Acta Endocrinol.,* 126, 44, 1992.
52. **Fuxe, K., Anderson, K., Eneroth, P., Harfstrand, A., and Agnati, L.F.,** Neuroendocrine actions of nicotine and of exposure to cigarette smoke: Medical implications, *Psychoneuroendocrinology,* 14, 19, 1989.
53. **Galatzer, A., and Laron, Z.,** The effects of prenatal androgens on behavior and cognitive functions, *Pediatr. Adolesc. Endocrinol.,* 19, 98, 1989.
54. **Genedani, S., Bernardi, M., and Bertolini, A.,** Sex-linked differences in avoidance learning in the offspring of rats treated with nicotine during pregnancy, *Psychopharmacology,* 80, 93, 1983.

55. **Gentile, D.A.**, Just what are sex and gender, anyway? *Psychol. Sci.*, 4, 120, 1993.
56. **Gentry, G.D., and Middaugh, L.D.**, Prenatal ethanol weakens the efficacy of reinforcers for adult mice, *Teratology*, 37, 135, 1988.
57. **George, F.W., and Wilson, J.D.**, Sex determination and differentiation, in *The Physiology of Reproduction, Vol. 1*, Knobil, E., and Neill, J.D., Eds., Raven Press, New York, 1988, 3.
58. **Gerall, A.A., and Dunlap, J.L.**, Evidence that the ovaries of the neonatal rat secrete active substances, *J. Endocrinol.*, 50, 529, 1971.
59. **Gerall, H.D., Ward, I.L., and Gerall, A.A.**, Disruption of the male's sexual behavior induced by social isolation, *Anim. Behav.*, 15, 54, 1967.
60. **Gorski, R.A., Gordon, J.H., Shryne, J.E., and Southam, A.M.**, Evidence for a morphological sex difference within the medial preoptic area of the rat brain, *Brain Res.*, 148, 33, 1978.
61. **Grant, K.A., and Samson, H.H.**, Neonatal ethanol exposure: Effects on adult behavior and growth parameters, *Pharmacol. Biochem. Behav.*, 18, 331, 1983.
62. **Greco, A.M., Gambardella, P., Sticchi, R., D'Aponte, D., and De Franciscis, P.**, Circadian rhythms of hypothalamic norepinephrine and of some circulating substances in individually housed adult rats, *Physiol. Behav.*, 52, 1167, 1992.
63. **Graham, S., and Gandelman, R.**, The expression of ano-genital distance data in the mouse, *Physiol. Behav.*, 36, 103, 1986.
64. **Halpern, D.F.**, *Sex Differences in Cognitive Abilities, 2nd edition*, Lawrence Erlbaum, Hillsdale, NJ, 1992.
65. **Handa, R.J., Hines, M., Schoonmaker, J.N., Shryne, J.E., and Gorski, R.A.**, Evidence that serotonin is involved in the sexually dimorphic development of the preoptic area in the rat brain, *Dev. Brain Res.*, 30, 278, 1986.
66. **Handa, R.J., McGivern, R.F., Noble, E.P., and Gorski, R.A.**, Exposure to alcohol *in utero* alters the adult pattern of luteinizing hormone secretion in male and female rats, *Life Sci.*, 37, 1683, 1985.
67. **Hard, E., Dahlgren, I.L., Engel, J., Larsson, K., Liljequist, S., Lindh, A.S., and Musi, B.**, Impairment of reproductive behavior in prenatally ethanol-exposed rats, *Drug Alcohol Depend.*, 14, 51, 1984.
68. **Hard, E., Engel, J., Larsson, K., Liljequist, S., and Musi, B.**, Effects of maternal ethanol consumption on the offspring sensory-motor development, ultrasonic vocalization, audiogenic immobility reaction and brain monoamine synthesis, *Acta Pharmacal. Toxicol.*, 54, 354, 1985.
69. **Hasegawa, T., and Sakuma, Y.**, Developmental effect of testosterone on estrogen sensitivity of the rat preoptic neurons with axons to the ventral tegmental area, *Brain Res.*, 611, 1, 1993.
70. **Heikkila, R.E., Orlansky, H., and Cohen, G.**, Studies on the distinction between uptake inhibition and release of ^{3}[H]dopamine in rat brain tissue slices, *Biochem. Pharmacol.*, 24, 847, 1975.
71. **Hines, M., and Shipley, C.**, Prenatal exposure to diethylstilbestrol (DES) and the development of sexually dimorphic cognitive abilities and cerebral lateralization, *Dev. Psychol.*, 20, 81, 1984.
72. **Hughes, A.M., Everitt, B.J., and Herbert, J.**, Selective effects of beta-endorphin infused into the hypothalamus, preoptic area and bed nucleus of the stria terminalis on the sexual and ingestive behavior of the male rat, *Neuroscience*, 23, 1063, 1987.
73. **Jarzab, B., Lindner, G., Lindner, T., Sickmoller, P.M., Geerling, H., and Dohler, K.D.**, Adrenergic influences on sexual differentiation of the rat brain, *Monogr. Neural Sci.*, 12, 191, 1986.
74. **Johnston, H.M., Payne, A.P., and Gilmore, D.P.**, Perinatal exposure to morphine affects adult sexual behavior of the male golden hamster, *Pharmacol. Biochem. Behav.*, 42, 41, 1992.
75. **Jones, K.L., and Smith, D.W.**, Recognition of the fetal alcohol syndrome in early infancy, *Lancet*, 2, 999, 1973.

76. **Jungkuntz-Burgett, L., Paredez, S., and Rudeen, P.K.,** Reduced sensitivity of hypothalamic-preoptic area norepinephrine and dopamine to testosterone feedback in adult fetal ethanol-exposed male rats, *Alcohol*, 7, 513, 1990.
77. **Kalra, S.P., and Kalra, P.S.,** Opioid-adrenergic-steroid connection in the regulation of luteinizing hormone secretion in the rat, *Neuroendocrinology*, 38, 418, 1984.
78. **Kelce, W.R., Ganjam, V.K., and Rudeen, P.K.,** Inhibition of testicular steroidogenesis in the neonatal rat following acute ethanol exposure, *Alcohol*, 7, 75, 1990.
79. **Kelce, W.R., Rudeen, P.K., and Ganjam, V.K.,** Prenatal ethanol exposure alters steroidogenic enzyme activity in newborn rat testes, *Alcoholism: Clin. Exp. Res.*, 13, 617, 1989.
80. **Kellogg, C.K., Primus, R.J., and Bitran, D.,** Sexually dimorphic influence of prenatal exposure to diazepam on behavioral responses to environmental challenge and on GABA-stimulated chloride uptake in the brain, *J. Pharmacol. Exp. Ther.*, 256, 259, 1991.
81. **Kelly, S.J., Goodlett, C.R., Hulsether, S.A., and West, J.R.,** Impaired spatial navigation in adult female but not adult male rats exposed to alcohol during the brain growth spurt, *Behav. Brain Res.*, 27, 247, 1988.
82. **Kelly, S.J., Mahoney, J.C., Randich, A., and West, J.R.,** Indices of stress in rats: Effects of sex, perinatal alcohol and artificial rearing, *Physiol. Behav.*, 49, 751, 1991.
83. **Komiskey, H.L., Miller, D.D., Lapidus, J.B., and Patil, P.N.,** The isomers of cocaine and tropococaine: Effect on ^{3}H-catecholamine uptake by rat brain synaptosomes, *Life Sci.*, 21, 1117, 1977.
84. **Kononen, J., Honkaniemi, J., Gustafsson, J.A., and Pelto-Huikko, M.,** Glucocorticoid receptor colocalization with pituitary hormones in the rat pituitary gland, *Mol. Cell. Endocrinol.*, 93, 97, 1993.
85. **Lapointe, G., and Nosal, G.,** Morphine treatment during pregnancy: Neonatal and preweaning consequences, *Biol. Neonate*, 42, 22, 1982.
86. **Lauder, J.M., and Krebs, H.,** Do neurotransmitters, neurohumors and hormones specify critical periods? in *Developmental Psychobiology*, Greenough, W.T., and Juraska, J.M., Eds., Academic Press, New York, 1986, 119.
87. **Lemoine, P., Haronsseau, H., Borteyru, J.P., and Menuet, J.C.,** Les enfants de parents alcoholiques: Anomalies observed a propos de 127 cas, *Ouest Med.*, 125, 476, 1968.
88. **Lichtblau, L., and Sparber, S. B.,** Opioids and development: A perspective on experimental models and methods, *Neurobehav. Toxicol. Teratol.*, 6, 3, 1984.
89. **Lichtensteiger, W., Ribary, U., Schlumpf, M., Odermatt, B., and Widmar, H.R.,** Prenatal adverse effects of nicotine on the developing brain, *Prog. Brain Res.*, 73, 137, 1988.
90. **Lichtensteiger, W., and Schlumpf, M.,** Prenatal nicotine affects fetal testosterone and sexual dimorphism of saccharin preference, *Pharmacol. Biochem. Behav.*, 23, 439, 1985.
91. **Litto, W.J., Griffin, J.P., and Rabii, J.,** Influence of morphine during pregnancy on neuroendocrine regulation of pituitary hormone secretion, *J. Endocrinol.*, 98, 289, 1983.
92. **McEwen, B.S.,** Actions of sex hormones on the brain: 'Organization' and 'activation' in relation to functional teratology, *Prog. Brain Res.*, 73, 121, 1988.
93. **McEwen, B.S.,** Steroid hormones: Effect on brain development and function, *Horm. Res.*, 37 (Suppl. 3), 1, 1992.
94. **McGivern, R.F.,** Influence of prenatal exposure to cimetidine and alcohol on selected morphological parameters of sexual differentiation: A preliminary report, *Neurotoxicol. Teratol.*, 9, 23, 1987.
95. **McGivern, R.F., Clancy, A.N., Hill, M.A., and Noble, E.P.,** Prenatal alcohol exposure alters adult expression of sexually dimorphic behavior in the rat, *Science*, 224, 896, 1984.
96. **McGivern, R.F., and Handa, R.J.,** unpublished observations.
97. **McGivern, R.F., Handa, R.J., and Redei, E.,** Decreased postnatal testosterone surge in male rats exposed to ethanol during the last week of gestation, *Alcoholism: Clin. Exp. Res.*, 17, 1215, 1993.
98. **McGivern, R.F., Holcomb, C., and Poland, R.E.,** Effects of prenatal testosterone propionate treatment on saccharin preference of adult rats exposed to ethanol *in utero*, *Physiol. Behav.*, 39, 241, 1987.

99. **McGivern, R.F., Raum, W.J., Handa, R.J., and Sokol, R.Z.**, Comparison of two weeks versus one week of prenatal ethanol exposure in the rat on gonadal organ weights, sperm count and onset of puberty, *Neurotoxicol. Teratol.*, 14. 351, 1992.
100. **McGivern, R.F., Raum, W.J., Salido, E., and Redei, E.**, Lack of prenatal testosterone surge in fetal rats exposed to alcohol: Alterations in testicular morphology and physiology, *Alcoholism: Clin. Exp. Res.*, 12, 243, 1988.
101. **McGivern, R.F., and Yellon, S.M.**, Delayed onset of puberty and subtle alterations in GnRH neuronal morphology, *Alcohol*, 9, 335, 1992.
102. **Meaney, M.J., and McEwen, B.S.**, Testosterone implants into the amygdala during the neonatal period masculinizes the social play of juvenile female rats, *Brain Res.*, 398, 324, 1986.
103. **Meaney, M.J., and Stewart, J.**, Neonatal androgens influence the social play of prepubescent rats, *Horm. Behav.*, 15, 197, 1981.
104. **Merkx, J.**, Effects of neonatal testicular hormones on preference behavior in the rat, *Behav. Brain Res.*, 12, 1, 1984.
105. **Meyer, D.C., and Carr, L.A.**, The effects of perinatal exposure to nicotine on plasma LH levels in prepubertal rats, *Neurotoxicol. Teratol.* 9, 95, 1987.
106. **Meyer, J.S., and Dupont, S.A.**, Prenatal cocaine administration stimulates fetal brain tyrosine hydroxylase activity, *Brain Res.*, 608, 129, 1993.
107. **Meyer, L.S., and Riley, E.P.**, Social play in juvenile rats prenatally exposed to alcohol, *Teratology*, 34, 1, 1986.
108. **Meyer, L.S., and Riley, E.P.**, Behavioral teratology of alcohol, in *Handbook of Behavioral Teratology*, Riley, E.P., and Vorhees, C., Eds., Plenum Press, New York, 1986, 101.
109. **Minetti, S.A., and Fulginiti, S.**, Sexual receptivity of adult female rats prenatally intoxicated with alcohol on gestational day 8, *Neurotoxicol. Teratol.*, 13, 531, 1991.
110. **Mirmiran, M., Feenstra, M.G.P., Dijcks, F.A., Bos, N.P.A., and Van Haaren, F.**, Functional deprivation of noradrenaline neurotransmission: Effects of clonidine on brain development, *Prog. Brain Res.*, 73, 159, 1988.
111. **Mogil, J.S., Sternberg, W.F., Kest, B., Marek, M., and Liebeskind, J.C.**, Sex differences in the antagonism of swim stress-induced analgesia: Effects of gonadectomy and estrogen replacement, *Pain*, 53, 17, 1993.
112. **Money, J.**, Sin, sickness, or status? *Am. Psychol.*, 42, 384, 1987.
113. **Naftolin, F., Garcia-Segura, L.M., Keefe, D., Leranth, C., MacLusky, N.J., and Brawer, J.R.**, Estrogen effects on the synaptology and neural membranes of the rat hypothalamic arcuate nucleus, *Biol. Reprod.*, 42, 21, 1990.
114. **Navarro, H.A., Mills, E., Seidler, F.J., Baker, F.E., Lappi, S.E., Tayyeb, M.I., Spencer, J.R., and Slotkin, T.A.**, Prenatal nicotine exposure impairs β-adrenergic function: Persistent chronotropic subsensitivity despite recovery from deficits in receptor binding, *Brain Res. Bull.*, 25, 233, 1990.
115. **Navarro, H.A., Seidler, F.J., Whitmore, W.L., and Slotkin, T.A.**, Prenatal exposure to nicotine via maternal infusions: Effects on development of catecholaminergic systems, *J. Pharmacol. Exp. Ther.*, 244, 940, 1988.
116. **Numan, M.**, Maternal behavior, in *The Physiology of Reproduction, Vol.* 2, Knobil, E., and Neill, J.D., Eds., Raven Press, New York, 1988, 1569.
117. **Nyborg, H.**, Performance and intelligence in hormonally different groups, *Prog. Brain Res.*, 61, 491, 1984.
118. **Pare, W.P.**, Age, sex, and strain differences in the aversive threshold to grid shock in the rat, *J. Comp. Physiol. Psychol.*, 69, 214, 1969.
119. **Parker, C.R.**, The endocrinology of pregnancy, in *Textbook of Reproductive Medicine*, Carr, B.R., and Blackwell, R.E., Eds., Appleton and Lange, Norwalk, CT, 1993, 17.
120. **Peris, J., Coleman-Hardee, M., and Millard, W.J.**, Cocaine *in utero* enhances the behavioral response to cocaine in adult rats, *Pharmacol. Biochem. Behav.*, 42, 509, 1992.
121. **Peters, D.A.V., and Tang, S.**, Sex-dependent biological changes following prenatal nicotine exposure in the rat, *Pharmacol. Biochem. Behav.*, 17, 1077, 1982.

122. **Phoenix, C.H., Goy, R.W., Gerall, A.A., and Young, W.C.,** Organizing action of prenatally administered testosterone propionate on the tissues mediating mating behavior in the female guinea pig, *Endocrinology,* 65, 369, 1959.
123. **Purohit, V.,** Effects of alcohol on the hypothalamic-pituitary-gonadal axis, in *Alcohol and the Endocrine System,* Zakhari, S., Ed., Nat. Inst. Alcohol Abuse and Alcoholism, Monograph 23, NIH Publication No. 93-3533, 1993, 189.
124. **Rathbun, W., and Druse, M.J.,** Dopamine, serotonin and acid metabolites in brain regions from the developing offspring of ethanol-treated rats, *J. Neurochem.,* 44, 57, 1985.
125. **Raum, W.J., Marcano, M., and Swerdloff, R.S.,** Nuclear accumulation of estradiol derived from aromatization of testosterone is inhibited by hypothalamic beta-receptor stimulation in the neonatal female rat brain, *Biol. Reprod.,* 30, 388, 1984.
126. **Raum, W.J., McGivern, R.F., Peterson, M.A., Shyrne, J.H., and Gorski, R.A.,** Prenatal inhibition of hypothalamic sex steroid uptake by cocaine: Effects on neurobehavioral sexual differentiation in male rats, *Dev. Brain Res.,* 53, 230, 1990.
127. **Raum, W.J., and Swerdloff, R.S.,** The role of hypothalamic adrenergic receptors in preventing testosterone-induced androgenization in the female, *Endocrinology,* 109, 273, 1981.
128. **Reisert, I., and Pilgrim, C.,** Sexual differentiation of monoaminergic neurons-genetic or epigenetic? *TINS,* 10, 468, 1991.
129. **Reinisch, J.M., and Sanders, S.A.,** Prenatal hormonal contributions to sex differences in human cognitive and personality development, in *Handbook of Behavioral Neurobiology, Vol. 11,* Gerall, A.A., Moltz, H., and Ward, I.L., Eds., Plenum Press, New York, 1992, 221.
130. **Ribary, U., and Lichtensteiger, W.,** Effects of acute and chronic prenatal nicotine treatment on central catecholamine systems of male and female rat fetuses and offspring, *J. Pharmacol. Exp. Ther.,* 248, 786, 1989.
131. **Riley, E.P., Lochry, E.A., Baldwin, J., and Shapiro, N.R.,** Lack of response inhibition in rats prenatally exposed to alcohol, *Psychopharmacology,* 62, 47, 1979.
132. **Ritz, M.C., Lamb, R.J., Goldberg, S.R., and Kuhar, M.J.,** Cocaine receptors on dopamine transporters are related to self administration of cocaine, *Science,* 237, 1219, 1987.
133. **Rivier, C., Rivier, J., and Vale, W.,** Stress-induced inhibition of reproductive functions: Role of endogenous corticotropin-releasing factor, *Science,* 231, 607, 1986.
134. **Roffi, J., Chami, F., Corbier, P., and Edwards, D.A.,** Testicular hormones during the first few hours after birth augment the tendency of adult males to mount receptive females, *Physiol. Behav.,* 39, 625, 1987.
135. **Ross, S.B., and Renyi., A.L.,** Inhibition of the uptake of tritiated 5-hydroxytryptamine in brain tissue, *Eur. J. Pharmacol.,* 7, 270, 1969.
136. **Rudeen, P.K., Kappel, C.A., and Lear, K.,** Postnatal or *in utero* ethanol exposure reduction of the sexually dimorphic nucleus of the preoptic area in male rats, *Drug Alcohol Depend.,* 18, 247, 1986.
137. **Sachs, B.D., and Meisel, R.L.,** The physiology of male sexual behavior, in *The Physiology of Reproduction, Vol. 2,* Knobil, E., and Neill, J.D., Eds., Raven Press, New York, 1988, 1393.
138. **Samson, H.H., and Grant, K.A.,** Ethanol-induced microcephaly in neonatal rats: Relation to dose, *Alcoholism: Clin. Exp. Res.,* 8, 201, 1984.
139. **Sandberg, D.E., Meyer-Bahlburg, H.F.L., Rosen, T.S., and Johnson, H.L.,** Effects of prenatal methadone exposure on sex dimorphic behavior in early school age children, *Psychoneuroendocrinology,* 15, 77, 1990.
140. **Scott, H.C., Westling, E., Paull, W.K., and Rudeen, P.K.,** LHRH neuron migration in mice exposed to ethanol *in utero, Soc. Neurosci. Abstr.,* 16, 32, 1990.
141. **Segarra, A.C., and McEwen, B.S.,** Drug effects on sexual differentiation of the brain: Role of stress and hormones in drug actions, in *Maternal Substance Abuse and the Developing Nervous System,* Zagon, I.S., and Slotkin, T.A., Eds., Academic Press, San Diego, CA, 1992, 323.

142. **Segarra, A.C., and Strand, F.L.,** Prenatal administration of nicotine alters subsequent sexual behavior and testosterone levels of male rats, *Brain Res.*, 480, 151, 1989.
143. **Seidler, F.J., Levin, E.D., Lappi, S.E., and Slotkin, T.A.,** Fetal nicotine exposure ablates the ability of postnatal nicotine challenge to release norepinephrine from rat brain regions, *Dev. Brain Res.*, 69, 288, 1992.
144. **Seidler, F.J., and Slotkin, T.A.,** Fetal cocaine exposure causes persistent noradrenergic hyperactivity in rat brain regions: Effects on neurotransmitter turnover and receptors, *J. Pharmacol. Exp. Ther.*, 263, 413, 1992.
145. **Seidler, F.J., Whitmore, W.L., and Slotkin, T.A.,** Delays in growth and biochemical development of rat brain caused by maternal methadone administration: Are the alterations in synaptogenesis and cellular maturation independent of reduced maternal food intake? *Dev. Neurosci.*, 5, 13, 1982.
146. **Singh, H.H., Purohit, V., and Ahluwalia, B.S.,** Effect of methadone treatment during pregnancy on the fetal testes and hypothalamus in rats, *Biol. Reprod.*, 22. 480, 1980.
147. **Slotkin, T.A., Orband-Miller, L., and Queen, K.L.,** Development of ^{3}[H]nicotine binding sites in brain regions of rats exposed to nicotine prenatally via maternal injections or infusions, *J. Pharmacol. Exp. Ther.*, 242, 232, 1987.
148. **Sonderegger, T.B., Bromley, B., and Zimmerman, E.,** Effects of morphine pellet in plantation in neonatal rats, *Prac. Exp. Biol. Med.*, 154, 435, 1977.
149. **Sonderegger, T.B., Mesloh, B.L., Ritchie, A.J., Tharp, A., and Spencer, M.,** Reproductive behavior of female rats treated with ethanol on postnatal days 1-7 or 8-14, *Neurobehav. Toxicol. Teratol.*, 8, 415, 1986.
150. **Sparber, S.B.,** Developmental effects of narcotics, *Neurotoxicology*, 7, 335, 1986.
151. **Stahl, F., Gotz, F., Poppe, I., Amendt, P., and Dorner, G.,** Pre- and early postnatal testosterone levels in rat and human, in *Hormones and Brain Development*, Dorner, G., and Kawakami, M., Eds., Elsevier, North Holland, 1978, 99.
152. **Stewart, J., Skavarenina, A., and Pottier, J.,** Effects of neonatal androgens on open field behavior and maze learning in the prepubescent and adult rat, *Physiol. Behav.*, 14, 291, 1975.
153. **Taylor, A.N., Branch, B.J., Liu, S.H., and Kokka, N.,** Long-term effects of fetal ethanol exposure on pituitary-adrenal response to stress, *Pharmacol. Biochem. Behav.*, 16, 585, 1982.
154. **Udani, M., Parker, S., Gavaler, J., and Van Thiel, D.H.,** Effects of *in utero* exposure to alcohol upon male rats, *Alcoholism: Clin. Exp. Res.*, 9, 355, 1985.
155. **Unger, R.K., and Crawford, M.,** The troubled relationship between terms and concepts, *Psychol. Sci.*, 4, 122, 1993.
156. **Valenstein, E.S., Kakolewski, J.W., and Cox, V.C.,** Sex differences in taste preferences for glucose and saccharin solutions, *Science*, 156, 942, 1967.
157. **van Weerden, W.M., Bierings, H.G., van Steenbrugge, G.J., de Jong, F.H., and Schroder, F.H.,** Adrenal glands of mouse and rat do not synthesize androgens, *Life Sci.*, 50, 857, 1992.
158. **Vathy, I.U., Etgen, A.M., and Barfield, R.J.,** Effects of prenatal exposure to morphine on the development of sexual behavior in rats, *Pharmacol. Biochem. Behav.*, 22, 227, 1985.
159. **Vathy, I., Etgen, A.M., Rabii, J., and Barfield, R.J.,** Effects of prenatal exposure to morphine sulfate on reproductive function of female rats, *Pharmacol. Biochem. Behav.*, 19, 777, 1983.
160. **Vathy, I., and Katay, L.,** Effects of prenatal morphine on adult sexual behavior and brain catecholamines in rats, *Dev. Brain Res.*, 68, 125, 1992.
161. **Vathy, I., Katay, L., and Mini, K.N.,** Sexually dimorphic effects of prenatal cocaine on adult sexual behavior and brain catecholamines in rats, *Dev. Brain Res.*, 73, 115, 1993.
162. **Vathy, I., van der Plas, J., Vincent, P.A., and Etgen, A.M.,** Intracranial dialysis and microinfusion studies suggest that morphine may act in the ventromedial hypothalamus to inhibit female sexual behavior, *Horm. Behav.*, 25, 354, 1991.
163. **von Zigler, N.I., Schlumpf, M., and Lichtensteiger, W.,** Prenatal nicotine exposure selectively affects perinatal forebrain aromatase activity and fetal adrenal function in male rats, *Dev. Brain Res.*, 62, 23, 1991.

164. **Ward, B.O.**, Fetal drug exposure and sexual differentiation of males. Gonadal hormones and sex differences in nonreproductive behaviors, in *Handbook of Behavioral Neurobiology, Vol. 11*, Gerall, A.A., Moltz, H., and Ward, I.L., Eds., Plenum Press, New York, 1992, 181.
165. **Ward, I.L.**, Prenatal stress feminizes and demasculinizes the behavior of males, *Science*, 175, 82, 1972.
166. **Weinberg, J.**, Hyper-responsiveness to stress: Differential effects of prenatal ethanol on males and females, *Alcoholism: Clin. Exp. Res.*, 12, 647, 1988.
167. **Weinberg, J.**, Prenatal ethanol exposure: Sex differences in offspring responsiveness, *Alcohol*, 9, 219, 1992.
168. **Weisz, J., and Ward, I.L.**, Plasma testosterone and progesterone titers of pregnant rats, their male and female fetuses, and neonatal offspring, *Endocrinology*, 106, 306, 1980.
169. **Welch, A.S., and Welch, B.L.**, Isolation, reactivity and aggression: Evidence for an involvement of brain catecholamines and serotonin, in *The Physiology of Aggression and Defeat*, Eleftheriou, B.E., and Scott, J.P., Eds., Plenum Press, New York, 1971, 91.
170. **West, J.R., Hamre, K.M., and Pierce, D.R.**, Delay in brain growth induced by alcohol in artificially reared rat pups, *Alcohol*, 1, 213, 1984.
171. **Zimmerberg, B., and Reuter, J.M.**, Sexually dimorphic behavioral and brain asymmetries in neonatal rats: Effects of prenatal alcohol exposure, *Dev. Brain Res.*, 46, 281, 1989.
172. **Zimmerberg, B., and Scalzi, L.V.**, Commissural size in neonatal rats: Effects of sex and prenatal alcohol exposure, *Int. J. Dev. Neurosci.*, 7, 81, 1989.
173. **Zimmerberg, B., Sukel, H.L., and Stekler, J.D.**, Spatial learning of adult rats with fetal alcohol exposure: Deficits are sex-dependent, *Behav. Brain Res.*, 42, 49, 1991.
174. **Goy, R.W., and McEwen, B.S.**, *Sexual Differentiation of the Brain*, MIT Press, Cambridge, MA, 1980.
175. **Conover, C., Fuljis, R., Rabii, J., and Advis, J.P.**, Serial long-term assessment of *in vivo* LHRH release from a discrete area of the ewe median eminence using multiple guide type cannula assembly and removable push-pull cannulae, *Neuroendocrinology*, 57, 1119, 1993.
176. **Fueyo-Silva, A., Menendez-Patterson, A., and Marin, B.**, Effects of prenatal alcohol consumption upon fecundity, natality, growth, vaginal opening and sexual cycle in the rat, *Reproduction*, 4, 265, 1980.
177. **Fadem, B.H.**, Effects of postnatal exposure to alcohol on reproductive physiology and sexually dimorphic behavior in a marsupial, the gray, short-tailed opossum, *Alcoholism: Clin. Exp. Res.*, 17, 870, 1993.

Chapter 12

NICOTINE EFFECTS ON THE NEUROENDOCRINE REGULATION OF REPRODUCTION

D.D. Rasmussen

TABLE OF CONTENTS

0-8493-2451-3/95/$0.00+.50

ABBREVIATIONS

ACTH–adrenocorticotropic hormone
AVP–arginine vasopressin
β-END–β-endorphin
CRF–corticotropin-releasing factor
GnRH–gonadotropin-releasing hormone
HPA–hypothalamo-pituitary-adrenal
HPG–hypothalamo-pituitary-gonadal
i.p.–intraperitoneal
i.v.–intravenous
LH–luteinizing hormone
MBH–mediobasal hypothalamus
POMC–proopiomelanocortin
s.c.–subcutaneous
TH–tyrosine hydroxylase
TIDA–tuberoinfundibular dopaminergic

INTRODUCTION

It has been demonstrated that 1) women who smoke wean their babies earlier than do nonsmokers,[77] 2) there is a highly significant trend for reduced fertility with increasing numbers of cigarettes smoked per day,[9,65,171] 3) menopause occurs at an earlier age in smoking than in nonsmoking women,[171] 4) women who are habitual smokers appear to have reduced prolactin levels,[44] 5) menstrual irregularities and abnormal vaginal bleeding appear to be more common among smokers than nonsmokers,[100] and 6) testosterone levels tend to be suppressed in men who smoke compared to nonsmokers.[26] Thus, tobacco smoking probably has important effects on the neuroendocrine regulation of reproduction.

Tobacco smoke consists of a complex heterogeneous mixture of gases and particulates. Of this mixture, the physiological effects of nicotine have been most well characterized. Nicotine is the component of tobacco smoke that best correlates with the frequency of smoking[51,89,136,139] and the pleasurable sensations associated with smoking.[41,73,85] There is evidence that the dependence, tolerance and withdrawal symptoms associated with chronic cigarette smoking also result from the actions of nicotine.[24,59,68,74,82,139,153] This chapter will address the effects of nicotine on the neuroendocrine regulation of reproduction. The focus will be the effects of nicotine on brain mechanisms, even though other components of tobacco smoke (e.g., polycyclic hydrocarbons or carbon monoxide) may have roles, and nicotine also has direct (i.e., not neuroendocrine-mediated) effects on the reproductive tract.

We have previously proposed that interactions between the mediobasohypothalamic dopaminergic and β-endorphinergic neuronal systems may have key roles in the neuroendocrine regulation of reproduction.[124] Since administration of tobacco smoke or nicotine modulates activity of these neuronal systems, this chapter will focus on mechanisms by which they are likely to mediate the reproductive neuroendocrine responses to nicotine, primarily addressing regulation of pituitary gonadotropin and prolactin secretion and interactions with the hypothalamo-pituitary-adrenal (HPA) axis. We will then use these putative interactions to illustrate why interpretation of the physiological significance of much of the available data regarding nicotine effects on the neuroendocrine regulation of reproduction is especially difficult and discuss issues that must be considered when evaluating this regulation.

NICOTINE EFFECTS ON THE FOREBRAIN β-ENDORPHINERGIC NEURONAL SYSTEM

The forebrain β-endorphinergic system originates from a group of perikarya located entirely within the arcuate/periarcuate region of the mediobasal hypothalamus (MBH).[19,29,111] Although these neurons produce a variety of opiomelanocortin peptides derived from the proopiomelanocortin (POMC) precursor, the potent opioid β-endorphin (β-END) has been most thoroughly investigated.[111] Projections from these perikarya innervate many areas of the limbic system, where β-END has been proposed to function as a neurotransmitter or neuromodulator regulating or mediating a remarkable variety of brain functions, including positive reinforcement, psychomotor stimulation, thermoregulation, analgesia, eating/drinking, sexual behavior, pituitary function, attention and adaptive processes.[2,111,151,177,178,180] Although there are POMC-producing perikarya in the medullary nucleus of the solitary tract, they contribute little if any innervation to the forebrain, projecting primarily to the brainstem and spinal cord. Consequently, perikarya in the MBH are the sites of POMC (and thus β-END) synthesis for the entire forebrain.

Rewarding responses to exogenous opiates, such as morphine and heroin, are mediated by the same μ-opioid receptors that are the high affinity binding sites for endogenous β-END,[71,105,165] and administration of μ-opioid receptor antagonists, such as naloxone or naltrexone, can reduce or block self-administration of these opiates.[50,79,81,163] It has also been reported that naloxone administration reduced the number of cigarettes smoked,[52,77] that naloxone blocked the subjective pleasure response to smoking[114] and that naloxone administration to nicotine-dependent rats elicited a nicotine abstinence syndrome,[91] suggesting that some of the responses to both nicotine and exogenous opiates may be mediated by similar opioid mechanisms.

This hypothesis is consistent with a variety of other experimental evidence. For example, it has been demonstrated that nicotine administration to rats protected against subsequent brain opioid receptor inactivation by β-funaltrexamine,[38] which selectively alkylates and inactivates unoccupied μ-opioid receptors.[168] This suggests that nicotine administration induced release of endogenous opioids in the brain (probably the preferential endogenous μ-opioid receptor agonist β-END), which then occupied the opioid receptors and "protected" them from β-funaltrexamine inactivation. Furthermore, it has been reported that following s.c. administration of nicotine once per day for 30 days, mouse hypothalamic β-END content was decreased 37% after 1 week of withdrawal, but rose to almost 50% above control levels after 2 weeks of withdrawal.[131] Although it is not possible to interpret how these changes in hypothalamic β-END content reflected changes in β-END release, synthesis or both, it is clear that chronic nicotine treatment had a profound effect on the hypothalamic β-endorphinergic system, with major changes occurring during withdrawal.

We have recently demonstrated[126] that daily s.c. injections of nicotine for 4 weeks suppressed rat hypothalamic POMC mRNA content, which then returned to control levels after 1 week withdrawal from nicotine. Furthermore, it has been reported that nicotine administration acutely (within 60 minutes) decreased the concentration of β-END in the hypothalami of mice.[57] In morphine-dependent mice, an opiate withdrawal response induced by injection of naloxone could be attenuated by administration of nicotine, perhaps increasing endogenous release of β-END and decreasing the response to the constant dosage of naloxone.[23] Similarly, in a study of heroin addicts, nicotine was the most desired substitute for heroin,[20] consistent with its proposed role as a stimulator of endogenous β-END secretion. Acupuncture, which increases cerebrospinal fluid levels of β-END,[75] has been successful in helping smokers to quit tobacco smoking, and the smokers experienced an intense desire to resume smoking after administration of naloxone.[92] Finally, endogenous opioid activity has been demonstrated to be responsible for the smoking-induced respiratory suppression that occurs in some smokers.[159] Overall, these results strongly suggest that forebrain β-endorphinergic activity may have important roles in neurochemical mechanisms mediating some of the responses to nicotine.

NICOTINE EFFECTS ON THE TUBEROINFUNDIBULAR DOPAMINERGIC SYSTEM

Cigarette smoking has been demonstrated to increase activity of the tuberoinfundibular dopaminergic (TIDA) neuronal system,[6] which innervates not only the median eminence but also the arcuate region of the MBH, i.e., the site of β-endorphinergic perikarya (see Rasmussen[124] for review). Nicotine administration also increases this hypothalamic catecholaminergic activity,[44,83] whereas cigarette smoke with nicotine filtered out does not.[6] Mecamylamine, a nicotinic receptor blocker, blocks the stimulation of hypothalamic dopaminergic and noradrenergic activity by both cigarette smoking and nicotine administration;[3,44] hence, it appears that the nicotine component of cigarette smoke is responsible for the smoking-induced increases in hypothalamic catecholaminergic activity.

Nicotinic activation of TIDA activity is consistent with the presence of nicotinic cholinergic receptors as well as large numbers of cholinergic neuronal perikarya and terminals in the MBH, especially within the arcuate nucleus (see Fuxe *et al.*[44,45] for reviews). A subpopulation of the arcuate nucleus TIDA neurons even appears to synthesize acetylcholine,[45] strongly suggesting important interactions between dopaminergic and acetylcholinergic regulation in this critical focal site of neuroendocrine integration.

INTERACTIONS BETWEEN MEDIOBASAL-HYPOTHALAMIC DOPAMINERGIC AND β-ENDORPHINERGIC ACTIVITY

There is compelling evidence suggesting that MBH dopaminergic activity can stimulate activity of the forebrain β-endorphinergic system.[95,124,127,160] These MBH dopaminergic and β-endorphinergic neuronal systems both have important roles in regulating pituitary gonadotropin and prolactin secretion, and we have proposed that interaction between these systems may have a key role in integrating the neuroendocrine regulation of reproduction.[124] Since 1) nicotine stimulates MBH dopaminergic activity, 2) MBH dopaminergic stimulation can modulate forebrain β-endorphinergic activity, and 3) nicotine can also modulate forebrain β-endorphinergic activity, it is reasonable to hypothesize that nicotine may alter forebrain β-endorphinergic activity at least in part by an MBH dopaminergic mechanism. This hypothesis is consistent with the demonstration that decreased mouse hypothalamic content of β-END in response to acute nicotine administration was blocked by treatment with the dopamine receptor antagonist haloperidol.[57] Our thesis is that these MBH dopaminergic and β-endorphinergic neuronal systems likely mediate many of the nicotine-induced changes in the neuroendocrine regulation of reproduction.

NEUROENDOCRINE RESPONSES TO NICOTINE

Prolactin

A variety of studies have convincingly demonstrated that repeated administration of nicotine, either by injections or by inhalation of cigarette smoke, suppresses plasma prolactin levels in both males and females by a specific nicotinic receptor-mediated mechanism (for review, see Fuxe *et al.*[44]). In rats, this nicotine-induced inhibition of pituitary prolactin secretion is associated with increased dopaminergic activity in the median eminence.[44] Since secretion of dopamine from the median eminence into the hypophysial portal blood flow to interact with dopamine D_2 receptors at the pituitary is the principal mechanism responsible for inhibition of pituitary prolactin secretion,[86] it is thus likely that nicotine suppresses pituitary prolactin secretion by increasing TIDA activity. This is consistent with our recent demonstration that daily s.c. injections of nicotine over a 4-week period increased male rat MBH concentrations of tyrosine hydroxylase (TH) mRNA but suppressed serum prolactin levels.[126] Following withdrawal of nicotine, the MBH TH mRNA concentrations declined to control levels at between 3 and 7 days, and serum prolactin concentrations increased to control levels.[126]

Since TH is the rate-limiting enzyme in the biosynthesis of all catecholamines, and TH-immunoreactive perikarya in the MBH probably represent exclusively dopaminergic neurons,[13,156,164] these results are consistent with the hypothesis that chronic nicotine suppresses prolactin secretion by stimulating tuberoinfundibular dopamine secretion. This hypothesis is also consistent with evidence that 1) nicotine administration or exposure to tobacco smoke can block the suckling-induced rise in circulating prolactin concentrations in lactating rats,[17,158] 2) smokers who breast fed their babies had lower basal prolactin serum levels and a shortened period of lactation as compared with nonsmoking women,[110] 3) weaning has been reported to occur earlier with smokers who breast fed their babies compared to nonsmokers,[77] and 4) nicotine treatment results in reduced milk secretion and pup growth in the rat.[58]

In contrast to the consistently inhibitory prolactin response to repeated nicotine exposure, acute administration of nicotine can produce a transient increase in circulating prolactin levels.[44] There is evidence[98,145] that the acute nicotine-induced increase in prolactin secretion may be mediated by activation of adrenergic and/or noradrenergic neurons that originate in the brain stem and project to the hypothalamus where they presumably stimulate release of a prolactin-releasing factor. Other evidence suggests that this regulatory pathway may be polysynaptic and that opioid, serotonergic and dopaminergic neurons may also be involved.[43] Finally, it must be considered that the nicotine-induced increase in dopaminergic activity could also have a role because under some experimental conditions dopamine has been demonstrated to stimulate rather than inhibit prolactin release from the pituitary,[31,60,152] and activation of central D_1 dopamine receptors can temporarily increase circulating prolactin concentrations via hypothalamic prolactin-releasing or inhibiting factors.[36]

The stimulatory prolactin response to additional injections of nicotine administered 1 to 2 hours after the first is greatly diminished or absent,[69,144] indicating that the prolactin-secretory mechanism becomes rapidly desensitized or suppressed. It has been suggested that after briefly activating neuronal nicotinic cholinergic receptors, nicotine causes a prolonged desensitization or inactivation of the receptors so that its chronic effect is that of an antagonist,[69] although some investigators have reported that chronic nicotine treatment did not alter nicotinic receptor binding in any of a variety of brain regions.[116]

It has also been demonstrated that the nicotine-induced stimulation of norepinephrine secretion in the hypothalamic paraventricular nucleus is partially (approximately 50%) reduced in response to subsequent injections of nicotine,[145] so desensitization of the hypothesized[98] nicotine-induced noradrenergic stimulation of prolactin-releasing factor secretion probably has a role. However, since nicotine stimulates activity of the TIDA system,[44] and dopamine secreted into the hypophysial portal blood by this MBH neuronal system is under most conditions a potent inhibitor of pituitary prolactin secretion,[86] it is likely that this dopamine secretion has an important, perhaps primary, role in the suppression of prolactin release in response to subsequent nicotine administrations. This is consistent with the evidence that repeated intermittent administration of nicotine or cigarette smoke stimulates rat TIDA activity and inhibits prolactin secretion not only in the first set of administrations[44] but also following chronic nicotine or smoke treatment.[4] This suggests that the nicotinic stimulation of TIDA secretion does not become desensitized or suppressed with repeated administrations of nicotine, in contrast to the habituation or suppression that occurs in noradrenergic activity within the paraventricular nucleus and several other hypothalamic areas (where initial nicotine administration, but not nicotine administration after chronic exposure, increases norepinephrine turnover).[4] Thus, these results support the hypothesis that increased TIDA secretion is largely responsible for the suppression of prolactin secretion in response to repeated nicotine stimulations.

Temporary stimulation of prolactin secretion in response to an acute rapid injection of nicotine is consistent with a report that episodes of smoking to achieve relatively high plasma nicotine levels increased prolactin secretion in men.[173] Similarly, Seyler *et al.*[143] demonstrated that when subjects smoked two high-nicotine cigarettes in such a way as to reach satiation, prolactin levels were elevated. However, "satiation" was defined by the onset of nausea. At somewhat lower doses, which did not produce nausea, no observable changes in prolactin levels occurred. Plasma nicotine concentrations greater than 100 ng/ml are essentially never detected in the plasma of smokers, and concentrations as low as 60 ng/ml cause nausea,[121] so the range of nicotine concentrations that is effective but not toxic is relatively narrow. Also, the initial exposure to the unfamiliar subjective effects of even low dosages of nicotine may be stressful.

Even mild stress, such as handling or exposure to a novel environment, acutely stimulates prolactin secretion in the rat,[141] apparently, at least in part, by stimulating release of a prolactin-releasing factor.[146] It is reasonable to suggest that the stimulation of prolactin secretion in response to acute rapid initial administration of nicotine may actually reflect stress responses associated with nausea or unfamiliar subjective sensations. This would be consistent with evidence that the minimum intraperitoneal (i.p.) nicotine dosage for stimulation of prolactin release in naive (i.e., not previously treated with nicotine) rats is between 0.1 and 0.25 mg/kg body weight;[144] this is comparable to the dosage (0.1 mg/kg s.c.) that produces plasma nicotine levels (approximately 70 ng/ml)[33] associated with nausea in smokers.

Considering that 1) temporary stimulation of prolactin secretion by the single initial acute administration of nicotine may at least in part reflect a stress response or response to a novel stimulus, and 2) this stimulatory response quickly habituates or is suppressed so that it is not expressed in response to repeated nicotine administrations, the nicotinic receptor-mediated stimulation of TIDA activity and resultant suppression of pituitary prolactin secretion probably represent the most physiologically relevant responses to nicotine as voluntarily self-administered by habitual tobacco smokers.

Gonadotropins

In addition to suppressing prolactin secretion, intermittent exposure to cigarette smoke or repeated nicotine treatment also suppresses rat luteinizing hormone (LH) (and to a lesser and more variable extent, follicle-stimulating hormone) secretion by a nicotinic cholinergic receptor-mediated mechanism.[44] Because nicotine administration reduces plasma LH concentrations in ovariectomized rats in which the MBH has been deafferenated and isolated from the rest of the brain,[16] and because this response is not mediated directly at the pituitary,[14] it appears that the inhibition is mediated by a mechanism contained entirely within the MBH. This localization is consistent with evidence that the decreased LH levels are associated with increased MBH TIDA activity.[44] The inverse relationship between nicotine-induced TIDA activity and suppressed LH levels, together with evidence of an inhibitory role for this dopaminergic activity in the regulation of hypothalamic gonadotropin-releasing hormone (GnRH) secretion,[46,125,179] supports the hypothesis[44] that nicotine suppresses GnRH/LH secretion by increasing dopamine secretion in the median eminence. Because the nicotine-induced inhibition of LH secretion is blocked by the D_1 dopamine receptor antagonist SCH 23390 but not by D_2 dopamine receptor antagonists,[5] it has been proposed that the inhibition of GnRH secretion by nicotine is mediated by D_1 dopamine receptors on the GnRH neurons.[5,44] This is compatible with evidence that dopaminergic stimulation can suppress GnRH secretion *in vivo*.[179] However, it has recently been demonstrated that stimulation of D_1 dopamine receptors on clonal GnRH neurons increases rather than decreases GnRH secretion.[39] This is consistent with evidence suggesting that MBH dopaminergic mechanisms can either directly stimulate GnRH secretion or indirectly and more potently inhibit GnRH secretion by stimulating MBH β-endorphinergic activity.[124,125]

Considering the evidence discussed earlier suggesting that nicotine stimulates MBH β-endorphinergic activity, this nicotine-dopamine-βEND-GnRH sequential pathway is quite plausible, especially considering the recent evidence that naltrexone (a μ-opioid receptor antagonist) treatment attenuates the inhibitory effect of nicotine treatment on serum LH concentrations in rats.[63] Alternatively, since nicotine has been demonstrated to stimulate hypothalamic corticotropin-releasing factor (CRF) secretion,[61,161] and hypothalamic CRF has a stimulatory role in regulating MBH endogenous opioid (including β-END) release,[1,30,76,109] it would also be reasonable to hypothesize that nicotine inhibits GnRH/LH secretion by a nicotine-CRF-βEND-GnRH sequential pathway. Since hypothalamic dopaminergic activity appears to have a stimulatory role in regulating CRF secretion,[21,22,32,42] these two hypotheses are not mutually exclusive.

Similar to its effect on prolactin release, acute rapid injection of nicotine can initially induce a temporary increase in circulating LH concentrations,[44] although this stimulation has not been consistently demonstrated in all studies (e.g., see Blake *et al.*[16]). It has been suggested that an initial nicotine-induced increase in LH secretion followed by prolonged suppression of circulating LH concentrations may be, as has been proposed for the biphasic regulation of prolactin secretion, due to the existence of two forms of nicotinic cholinergic receptors within the hypothalamus.[44] However, as discussed earlier for the regulation of prolactin secretion, the initial administration of relatively high dosages of nicotine to naive rats may also be stressful due to toxicity (i.e., induction of nausea) or the novelty of unfamiliar subjective responses.

Since acute stress, including simple exposure to a novel environment, can temporarily stimulate LH secretion in rats,[27] it is likely that the acutely increased LH levels in response to the initial administration of nicotine in some experiments reflects (at least in part) a stress response, perhaps mediated by an adrenergic/noradrenergic mechanism originating from the midbrain, as seems to be the case for the initial stimulation of prolactin secretion.[98] A noradrenergic mechanism would be consistent with the evidence that noradrenergic projections from the midbrain have important stimulatory roles in GnRH/LH regulation.[123,179] Thus, as with regulation of prolactin secretion, the stimulation of LH secretion by initial exposure to relatively large dosages of nicotine in some investigations may in large part reflect a stress response or response to a novel subjective stimulus. Consequently, as for the regulation of prolactin secretion, the nicotinic receptor-mediated stimulation of TIDA activity and associated suppression of pituitary LH secretion, which are consistently demonstrated in response to repeated tobacco smoke or nicotine administration, probably represent the most physiologically relevant response to nicotine as voluntarily self-administered by habitual smokers.

A primarily inhibitory effect of nicotine on neuroendocrine regulation of the hypothalamo-pituitary-gonadal (HPG) axis is consistent with a variety of evidence. Blake *et al.*[18] provided an early demonstration of this inhibitory effect when they showed that nicotine administration can block ovulation in the rat. In humans, the number of cigarettes smoked per day has been demonstrated to be positively correlated with incidence of reduced fertility.[9,65] Several studies have shown that there is a higher infertility rate among smokers even when adjustments have been made for confounding factors such as age, education, occupation, alcohol use, frequency of marriage and husband's smoking habits, education and occupation.[171] It has also been demonstrated that menopause occurs at an earlier age in smoking than in nonsmoking women, independent of effects on body weight.[169] In addition, menstrual irregularities and abnormal vaginal bleeding appear to be more common for smokers than nonsmokers.[100]

Women who smoke had substantially suppressed levels (approximately one-third below nonsmokers) of all three major estrogens (estrone, estradiol, estriol) in the luteal but not the follicular phase of menstrual cycles in which ovulation occurred,[101] a phenomenon that may play a role in reducing risks of breast cancer and increasing risks of osteoporosis among women who smoke.[101] There have also been reports that tobacco smoking reduces levels of testosterone in men.[26,154] However, it should be noted that nicotine and other constituents of cigarette smoke have been demonstrated to have direct effects on the gonads.[169] It cannot be concluded, therefore, that apparent smoking-dependent effects on the HPG axis when comparing smokers and nonsmokers are necessarily due to neuroendocrine effects exerted directly at the hypothalamus or pituitary level.

Hypothalamo-Hypophyseal-Adrenal (HPA) Axis

Under most conditions nicotine administration acutely stimulates pituitary adrenocorticotropic hormone (ACTH) and β-END secretion, followed by an increase in circulating concentrations of adrenal glucocorticoids (primarily corticosterone in the rat).[44,122] The nicotine-induced stimulation of corticosterone secretion in the rat is consistent with reports that cigarette smoking increases circulating cortisol levels in smokers.[35,66,173,174] The stimulation appears to be mediated by a central nicotinic cholinergic mechanism that induces increased CRF and arginine vasopressin (AVP) secretion from neurons originating in the hypothalamic paraventricular nucleus.[44,122,170] In general, the sequential HPA response to acute nicotine stimulation can be summarized as follows: 1) nicotine stimulates hypothalamic secretion of CRF and AVP, 2) CRF and AVP stimulate pituitary secretion of ACTH and β-END, and 3) ACTH stimulates secretion of adrenal glucocorticoids. However, it should also be noted that nicotine can act additively with ACTH to stimulate adrenal steroidogenesis,[44] and nicotine can also stimulate release of catecholamines from the adrenal medulla, which may, in turn, exert effects on pituitary ACTH secretion.[129] Thus, peripherally mediated effects may also be important.

There is evidence that the acute stimulation of CRF secretion by nicotine is, like the stimulation of prolactin secretion, dependent at least in part on activation of brainstem catecholaminergic neurons.[96,97,99] However, it has also been demonstrated that acetylcholine stimulates release of CRF and AVP from the rat isolated hypothalamus *in vitro*.[53,61,149,161] Since nicotine has been demonstrated to act presynaptically at cholinergic synapses in both the brain and the autonomic nervous system to stimulate acetylcholine release,[7,25,34,133] it is clear that nicotine could also act at a site(s) directly within the hypothalamus to stimulate CRF and AVP secretion, mediated either by presynaptic stimulation of endogenous acetylcholine release or directly through postsynaptic nicotinic receptors. Finally, there is evidence that dopaminergic neurons innervate hypothalamic CRF neurons[64] and that release of dopamine by hypothalamic neurons stimulates CRF release,[22] as does administration of dopamine receptor agonists.[21,32,42] This means the nicotine-induced increases in hypothalamic dopaminergic activity may also play an intrahypothalamic role in mediating some nicotine-induced increases in CRF neurosecretion.

Since CRF has a role as an intrahypothalamic modulator that increases activity of the forebrain β-endorphinergic system,[1,30,76,109] nicotine-induced CRF neurosecretion is consistent with and supportive of the proposed β-endorphinergic response to nicotine. There is also evidence that CRF-induced increases in hypothalamic β-endorphinergic activity are mediated by hypothalamic AVP neurosecretion,[30] consistent with the stimulation of AVP neurosecretion by nicotine.[44,122]

Like LH and prolactin, the amplitude of the increase in adrenal glucocorticoid secretion in response to acute nicotine administration decreases with repeated treatments.[55,144] However, complete tolerance apparently does not develop following normal daily smoking patterns since cigarette smoking can still increase cortisol levels in men who are chronic smokers.[173,174] In rats, although adrenalectomy did not enhance the stimulatory prolactin response to a second administration of nicotine (i.e., did not block the suppression of the prolactin responsiveness that occurs following the stimulatory first administration), it did partially restore the ACTH response.[144] Also, pretreatment with corticosterone did not modify the prolactin stimulatory response to a single dose of nicotine but it partially suppressed the ACTH response.[144] These results suggest that the partial desensitization of the HPA axis (but not that of prolactin regulation) to the stimulatory effect of nicotine is at least partly dependent on glucocorticoid negative feedback. This is consistent with evidence that the desensitization in the responses of a variety of other behavioral and physiological functions following repeated nicotine administrations is also mediated by glucocorticoid feedback.[55,117] Similarly, we have recently demonstrated that month-long daily administration of nicotine to rats produced significantly increased circulating corticosterone levels, which were associated with decreased MBH POMC mRNA concentrations. Following cessation of nicotine treatment, the POMC mRNA concentrations increased at the same time (i.e., between 3 and 7 days after treatment) that the corticosterone levels declined.[126]

Since administration of exogenous glucocorticoids has been demonstrated to suppress MBH POMC mRNA concentrations,[12] it is likely that the nicotine-induced increase in circulating corticosterone levels suppressed activity of the MBH POMC-producing neurons. Glucocorticoids can suppress LH/GnRH secretion under some conditions,[28] so the nicotine-induced adrenal glucocorticoid hypersecretion (due to either direct adrenal stimulation or mediated by CRF/ACTH) may also have a role in mediating the inhibitory effect on LH secretion in response to repetitive nicotine administration.

The pronounced activation of the HPA axis in response to initial acute administration of nicotine may partly reflect a stress response or a change associated with subjective responses to the novel nicotine stimulus, as we have proposed to probably have a role in the nicotine-induced initial stimulation of prolactin and LH secretion. In early studies, Hökfelt[67] found that smoking appeared to increase urinary 17-hydroxysteroids in nonsmokers but not in chronic smokers. Because smoking also produces pallor, sweating and nausea in nonsmokers,[174] however, these results were probably confounded by a stress response. Kershbaum *et al.*[78] later reported an increase of plasma cortisol in response to smoking by habitual smokers, but a heavy exposure was used (five cigarettes in 30 minutes). With light to moderate smokers, Tucci and Sode[162] failed to find significant changes in either plasma cortisol or urinary steroids in smokers as compared with nonsmokers. However, Winternitz and Quillen[174] demonstrated that smoking high-nicotine cigarettes produced a pronounced rise in plasma cortisol levels in habitual smokers, apparently without producing nausea.[174]

More recent studies have confirmed that smoking high-nicotine cigarettes increases circulating cortisol and β-END levels, even in habitual smokers, without manifesting overt nausea,[121,122,173] although significant plasma ACTH and AVP increases occurred only with the nausea associated with very high plasma nicotine levels.[121,132] This is consistent with evidence that the nicotine threshold for acutely stimulating ACTH (as well as prolactin) release in naive rats appears to be between 0.1 and 0.25 mg/kg b.w., administered i.p.,[144] a dose that should produce plasma nicotine levels greater than 70 ng/ml (the level produced by 0.1 mg/kg[33]), which produces nausea in smokers.[121] It appears likely that nicotine administered by the normal smoking pattern of habitual smokers moderately stimulates the HPA axis, but that the pronounced acute stimulation produced by rapid administration of relatively large dosages of nicotine to naive subjects in many experimental paradigms may be largely due to stress associated with nausea or the subjective response to a novel stimulus.

Implications for Smokers

With the nicotine self-administration pattern and dosages characteristic of chronic habitual smokers, the prevailing effects on these neuroendocrine systems that control reproduction would appear to be 1) inhibition of HPG activity, 2) inhibition of prolactin secretion, and 3) moderate stimulation of HPA activity. On occasions of heavy smoking and resultant high nicotine intake, especially after a period of abstinence (e.g., overnight), prolactin and LH secretion may be acutely but only temporarily stimulated, and HPA activation accentuated. Moreover, interaction with additional factors, such as stress, drugs of abuse or other medications that alter CNS catecholaminergic activity, along with individual variability, could clearly augment or diminish these responses.[147]

Hypothesized Integrating Mechanisms

A plausible set of mechanisms can be reasonably proposed to integrate these effects of nicotine administration on hypothalamic dopaminergic and β-endorphinergic activity, the HPG and HPA axes, and prolactin secretion. These putative mechanisms are summarized here in an attempt to reveal the most parsimonious hypothesis that still comprehends all of the experimental evidence and characterizations that we have discussed. As such, it provides a provocative heuristic basis for productive further investigations of the roles of hypothalamic mechanisms in mediating the effects of nicotine on the neuroendocrine regulation of reproduction.

We hypothesize that the initial temporary stimulation of prolactin, LH and ACTH secretion in response to rapid administration of relatively large dosages of nicotine is due to releasing factor stimulation, probably mediated by an adrenergic/noradrenergic mechanism originating in the midbrain. However, the principal response to the repetitive low-dose nicotine administration characteristic of habitual smokers is probably largely mediated by increased MBH dopaminergic activity. The nicotine-induced dopamine secretion probably suppresses pituitary prolactin secretion by an action at the pituitary, and also stimulates forebrain β-endorphinergic activity, likely through a CRF-mediated mechanism. The increased release of β-END would then be expected to suppress GnRH/LH secretion. Nicotine-induced hypothalamic CRF secretion into the hypothalamo-hypophyseal portal flow stimulates ACTH and β-END secretion from the pituitary, with the ACTH in turn stimulating glucocorticoid secretion from the adrenal glands. Elevation of circulating glucocorticoid levels likely feeds back to attenuate CRF/ACTH secretion and forebrain β-endorphinergic activity (see Figure 1).

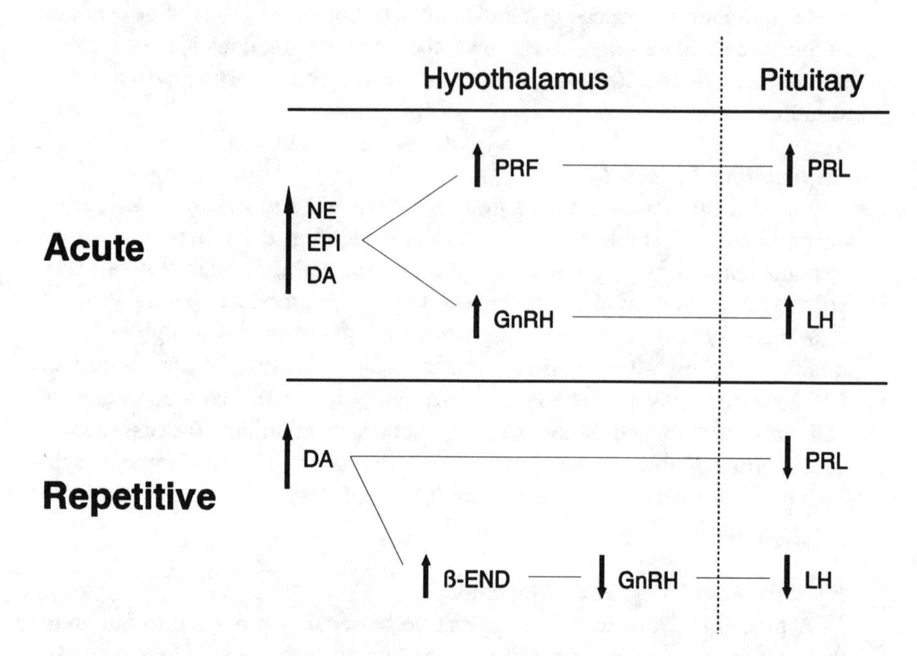

FIGURE 1. Hypothesized mechanisms mediating the opposite effects of acute vs. repetitive nicotine administration on prolactin and luteinizing hormone secretion, as discussed in the text. ***Abbreviations:*** NE, norepinephrine; EPI, epinephrine; DA, dopamine; PRF, prolactin-releasing factor; GnRH, gonadotropin-releasing factor; PRL, prolactin; LH, luteinizing hormone; β-END, β-endorphin.

LONG-TERM EFFECTS OF FETAL NICOTINE EXPOSURE ON THE NEUROENDOCRINE REGULATION OF REPRODUCTION

Nicotine readily crosses the placental barrier; levels in fetal animals[93,155] and in human umbilical cord serum[90] have been demonstrated to be comparable to or higher than those of the maternal blood. Maternal cigarette smoking during pregnancy, and pre- and perinatal exposure of experimental animals to nicotine or cigarette smoke, even at low doses, can induce a wide variety of neurobiological and behavioral changes in children and adult rat offspring (for reviews see Martin,[94] Lichtensteiger and Schlumpf,[88] and Navarro *et al.*[106]). For example, it has been demonstrated that pre- or perinatal administration of nicotine exerts complex effects on forebrain dopaminergic and noradrenergic activity in both fetal and adult rats.[87,88,107,130,142]

With respect to the TIDA activity that appears to play such a key role in the neuroendocrine regulation of reproduction,[124] it has been demonstrated that administration of nicotine to pregnant rats, and while the mother is still suckling pups, leads to increased TIDA activity in the pups that is maintained 6 months later in adulthood.[45] This suggests that these neurons may have been permanently altered.[45]

Since our studies suggest that this TIDA activity is an important regulator of forebrain β-endorphinergic activity and that changes in TIDA activity appear to be positively correlated with changes in forebrain opiomelanocortinergic activity,[124,125,127,128] it is reasonable to hypothesize that this altered tuberoinfundibular activity in the adult also induces corresponding changes in forebrain β-endorphinergic activity and the diverse functions that this system regulates. Consistent with this hypothesis, perinatal administration of nicotine to rats has been demonstrated to alter subsequent prepubertal LH regulation and pubertal development[103] as well as adult sexual behavior,[140] plasma gonadal steroid levels,[140] serum prolactin levels[72] and LH regulation[62,103] (i.e., reproductive functions regulated by activity of the hypothalamic TIDA and β-endorphinergic systems).[124,148] Thus, even though it is clearly simplistic to suggest, based on the minimal available experimental evidence, that fetal nicotine-induced developmental changes in neuroendocrine mechanisms that regulate pubertal and adult reproductive function are mediated by changes in TIDA/β-endorphinergic interactions, such a hypothesis does provide a reasonable basis for further investigations. It is not currently known whether the effects of experimental fetal/neonatal nicotine administration on pubertal and adult reproductive function in rats can be reliably extrapolated to adolescent and adult human long-term responses to maternal smoking, but this remains a provocative, and likely, possibility.

CAVEATS

Although it seems clear that nicotine can have important effects on the neuroendocrine regulation of reproduction, and that these effects may be largely mediated by hypothalamic dopaminergic and β-endorphinergic interactions, many of the studies on which these hypotheses are based have significant inherent limitations that may compromise their physiological relevance. In addition, some of the experimental models may be physiologically meaningful only for specific conditions or subsets of the smoking population. Finally, nicotine exerts effects on a wide variety of functions that may directly or indirectly influence neuroendocrine regulation, or that may be incorrectly assumed to reflect neuroendocrine mediation. Some of the most significant of these limitations due to experimental design, restricted inference space and confounding variables will be discussed here, using the putative hypothalamic dopaminergic/β-endorphinergic interactions as examples.

Pattern of Nicotine Administration

Receptor-mediated functions often respond differently, even in opposite directions, to pulsatile vs. constant administration of agonists. Each plasma nicotine "peak" in response to smoking a cigarette is due to serial bolus nicotine administrations corresponding to each puff (25% of inhaled nicotine reaches a smoker's brain in only 7–10 seconds).[134] Compelling evidence suggests that this abruptly pulsatile pattern is an important factor in providing a rewarding response to smoking, and that the drug-like subjective effect is dependent on the acute rate of increase in brain nicotine rather than the steady-state level.[134] This evidence suggests that the effects of nicotine in the rat brain, including the development of dependence, are most reliably expressed if the nicotine is administered in repeated bolus form,[10,134] which is in accord with human studies.[135]

We have hypothesized that the rewarding response to nicotine is mediated, at least in part, by activation of the same forebrain β-endorphinergic system that appears to play an important role in the neuroendocrine regulation of reproduction. This suggests that the most appropriate model (i.e., the model that most accurately reproduces the responses characteristic of the self-administration of nicotine by smokers) for investigating the neuroendocrine effects of nicotine is to administer the nicotine in repeated small pulses instead of maintaining constant levels (such as by implanted osmotic pumps) or utilizing routes of administration (such as including nicotine in the drinking water), which produce very gradual changes in plasma nicotine levels.[10]

Smokers administer nicotine only while awake, and the nicotine self-injection rate in rats (which are nocturnal) is much higher in the dark phase of the light cycle (see Dougherty *et al.*[40]). Consequently, in addition to the acutely pulsatile nature of "normal" nicotine exposure, self-administration of nicotine is also characterized by a daily diurnal pattern. This pattern of alternating daily stimulation and withdrawal may be critical for maintaining normal response mechanisms. There are also pronounced diurnal rhythms in rat hypothalamic catecholaminergic activity,[15] MBH POMC mRNA and β-END contents[48,175] and HPA activity,[84] so disruption of normal rhythms by administering nicotine during the sleep period (i.e., during the daytime for rats) could produce changes in critical hypothalamic mechanisms not directly related to those normally expressed in smokers. Disruption of rats' sleep by administering nicotine during the light phase (sleep period) may also be stressful. Finally, the diurnal nature of rat forebrain catecholaminergic and β-endorphinergic activity, as well as HPA regulation, reveals that regulatory mechanisms are different between the day and night.

The pattern of nicotine administration is probably also important for studies investigating the long-term effects of fetal nicotine exposure. It has been demonstrated that prenatal administrations of nicotine by daily s.c. bolus injections vs. constant administration by subcutaneously implanted osmotic minipump differ in their delayed effects on catecholamine systems.[107,150]

Interactions with Stress

Anecdotal evidence suggests that drugs, including nicotine, are often taken to avoid or relieve stress, suggesting probable interaction of mechanisms. Since stress activates hypothalamic CRF and opiomelanocortinergic systems,[8,112,115] but increased glucocorticoid levels (characteristic of the stress response) can decrease brain opiomelanocortinergic activity,[12] it is clear that stress could modulate nicotine-induced changes in a β-endorphinergic mechanism. Thus, it is significant that chronic stress produces subsensitivity to nicotine.[118]

Injection even of saline alone is stressful and can alter brain β-END content,[172] as can exposure to a novel environment for even a few minutes.[37] Thus, if experimental subjects are not first habituated to the procedure of nicotine administration or to the subjective response to nicotine exposure, initial administration may induce a stress or arousal response that can clearly influence these hypothalamic mechanisms.

As noted earlier, plasma nicotine concentrations as low as 60 ng/ml cause nausea in smokers (presumably eliciting a stress response), and concentrations of greater than 100 ng/ml are essentially never detected in smokers' plasma, thus the range of appropriately nonstressful nicotine concentrations is narrow. Although 0.1 mg nicotine/kg b.w. administered s.c. produces plasma concentrations of approximately 70 ng/ml,[33] most studies have used much larger dosages (e.g., up to 2 mg/kg). Forcing rats to inhale tobacco smoke is probably at least initially stressful, and also commonly produces plasma nicotine levels well over 100 ng/ml (up to 200 ng/ml in some cases).[3,4,6]

Studies investigating the long-term consequences of perinatal nicotine administration on the neuroendocrine regulation of reproduction may also be confounded by stress associated with the route of nicotine administration, novelty of the subjective response to nicotine, toxicity (nausea) associated with the dose of nicotine, or disruption of sleep associated with the timing of nicotine administration. For example, it has been demonstrated that prenatal maternal stress can alter the regulation of the HPA axis,[138] sex behavior,[166] brain catecholaminergic activity,[157] brain opioid receptors,[70] puberty and fertility,[167] and LH regulation[80,137] during postnatal development and in adult rats and mice.

Determination of Appropriate Dosage

The dosages of nicotine used in some studies appear to be too high to be physiologically relevant when extrapolated from human plasma nicotine concentrations, as well as when based on behavioral indices (in our studies, dosages as low as 0.6 mg/kg b.w. administered as a s.c. bolus often induce temporary immobility and gasping in rats). However, the problem of determining the physiologically appropriate dosage is a difficult one for some studies. The ratio of brain:blood nicotine concentrations was approximately two- to threefold higher after acute i.v. bolus than after i.p. injection even though blood levels were comparable,[135] suggesting that proportionately more is taken up into the brain following discrete bolus administration than with the more gradual absorption associated with i.p. administration. Consequently, the plasma nicotine level may be misleading as an index of brain stimulation in response to i.p. and s.c. administration, as compared to stimulation provided by nicotine that is inhaled or administered by rapid i.v. bolus.

Biphasic Response Pattern

Nicotine's biphasic behavioral response pattern, in which enhanced arousal and alertness is followed by calming and tension reduction, can apparently be modified by changes in self-dosing to produce selective emphasis of one response component vs. the other, under the apparent control of demands of the setting and/or personality characteristics.[122] Likewise, it has been demonstrated that lower dosages of nicotine (200–400 g/kg) stimulated spontaneous motor activity in rats, whereas higher dosages (800 g/kg) depressed this activity.[122] Thus, the direction (stimulation vs. inhibition) of responses to nicotine can be both time and dose dependent and may vary greatly between individuals as well as within individuals under differing circumstances. This variability in response may be due to activation of multiple mechanisms (e.g., it appears that the stimulation of prolactin and LH secretion may be mediated by initial activation of a noradrenergic mechanism originating in the midbrain whereas the subsequent inhibition appears to be mediated by stimulation of hypothalamic TIDA activity, as we have discussed) and/or time and dose-dependent changes in nicotinic receptor mechanisms.[122]

Sex Differences

Male rats have lower MBH β-END content, MBH POMC mRNA content, TIDA turnover and MBH TH (rate limiting enzyme in catecholamine synthesis) mRNA content than do females, and both hypothalamic β-endorphinergic and TIDA activity are suppressed by elevated gonadal steroid levels (reviewed by Rasmussen[124]). This is consistent with a hypothesis that drug use and "sensation seeking," which are higher in men and are positively correlated with gonadal steroid levels in both sexes,[176] may be relatively greater due to the lower forebrain β-endorphinergic activity that makes it more desirable to "self-medicate" with nicotine, other drugs of abuse, or intense sensations (e.g., those associated with the intense arousal, stress, aggression, social conflict, etc., which appear to activate the forebrain β-endorphinergic system[111,112]). Thus, the sex-dependent differences in MBH dopaminergic and β-endorphinergic regulation may provide bases for the sex-dependent differences in responses to smoking[56,102] and must be considered when investigating neuroendocrine responses that appear to also be mediated by these same hypothalamic mechanisms. These differences are also important for studies investigating the long-term effects of fetal nicotine exposure, because prenatal nicotine exposure disturbs male sexual brain differentiation, apparently by altering fetal testosterone secretion and brain aromatase activity.[88]

Confounding Responses

Nicotine clearly has effects on various functions that may indirectly alter neuroendocrine regulation, such as alteration of blood flow and direct effects on the gonads and adrenal glands.[121,122,169] A confounding variable, which is particularly important for studies of the effects of long-term nicotine administration, is the effect of nicotine on eating/drinking, metabolism and weight gain.[54,119] Since the forebrain β-endorphinergic system appears to have an important role in regulating appetite,[104] effects of nicotine on these parameters may be associated with alterations in the regulation of the various other systems modulated by this endogenous opioid system, including the neuroendocrine regulation of reproduction.

CONCLUSION

It appears that nicotine self-administration may affect the neuroendocrine regulation of reproduction throughout life, with effects on the regulation of prolactin secretion and on the HPG and HPA axes, which can affect fertility and sexual function, menstrual regulation, lactation and even, probably, the subsequent neuroendocrine regulation of reproduction in offspring. Although the effects of nicotine are clearly more complex than can be explained solely by the hypothalamic mechanisms proposed here, and more rigorous investigations are required to further characterize even these, it nonetheless does appear that hypothalamic dopaminergic and β-endorphinergic neuronal systems have key central integrating roles in this regulation.

REFERENCES

1. **Almeida, O.F.X., Nikolarakis, K.E., and Herz, A.,** Evidence for the involvement of endogenous opioids in the inhibition of luteinizing hormone by corticotropin-releasing factor, *Endocrinology*, 122, 1034, 1988.
2. **Amalric, M., Cline, E.J., Martinez Jr., J.L., Bloom, F.E., and Koob, G.F.,** Rewarding properties of β-endorphin as measured by conditioned place preference, *Psychopharmacology*, 91, 14, 1987.
3. **Andersson, K.,** Mecamylamine pretreatment counteracts cigarette smoke induced changes in hypothalamic catecholamine neuron systems and in anterior pituitary function, *Acta Physiol. Scand.*, 125, 445, 1985.
4. **Andersson, K., Eneroth, P., Fuxe, K., Mascagni, F., and Agnati, L.F.,** Effects of chronic exposure to cigarette smoke on amine levels and turnover in various hypothalamic catecholamine nerve terminal systems and on the secretion of pituitary hormones in the male rat, *Neuroendocrinology*, 41, 462, 1985.
5. **Andersson, K., Fuxe, K., Eneroth, P., Harfstrand, A., and Agnati, L.F.,** Involvement of D_1 dopamine receptors in the nicotine-induced neuro-endocrine effects and depletion of diencephalic catecholamine stores in the male rat, *Neuroendocrinology*, 48, 188, 1988.
6. **Andersson, K., Fuxe, K., Eneroth, P., Mascagni, F., and Agnati, L.F.,** Effects of acute intermittent exposure to cigarette smoke on catecholamine levels and turnover in various types of hypothalamic DA and NA nerve terminal systems as well as on the secretion of adenohypophyseal hormones and cortisone, *Acta Physiol. Scand.*, 124, 277, 1985.
7. **Armitage, A.K., Hall, G.H., and Morrison, C.F.,** Pharmacological basis for the smoking habit, *Nature*, 125, 331, 1968.
8. **Axelrod, J., and Reisine, T.D.,** Stress hormones: Their interaction and regulation, *Science*, 224, 452, 1984.
9. **Baird, D.D., and Wilcox, A.J.,** Cigarette smoking associated with delayed conception, *J. Am. Med. Assoc.*, 253, 2979, 1985.
10. **Balfour, D.J.K.,** Studies on the biochemical and behavioural effects of oral nicotine, *Arch. Int. Pharmacodyn.*, 245, 95, 1980.

11. **Barbarino, A., De Marinis, L., Tofani, A., Della Casa, S., D'Amico, C., Mancini, A., Corsello, S.M., Sciuto, R., and Barini, A.,** Corticotropin-releasing hormone inhibition of gonadotropin release and the effect of opioid blockade, *J. Clin. Endocrinol. Metab.*, 68, 523, 1989.
12. **Beaulieu, S., Gagne, B., and Barden, N.,** Glucocorticoid regulation of proopiomelanocortin messenger ribonucleic acid content of rat hypothalamus, *Mol. Endocrinol.*, 2, 727, 1988.
13. **Björklund, A., Lindvall, O., and Nobin, A.,** Evidence of an incerto-hypothalamic dopamine neuron system in the rat, *Brain Res.*, 89, 29, 1975.
14. **Blake, C.A.,** Parallelism and divergence in luteinizing hormone and follicle stimulating hormone release in nicotine-treated rats, *Proc. Soc. Exp. Biol. Med.*, 145, 706, 1974.
15. **Blake, C.A., Campbell, G.T., Mascagni, F., Culler, M.D., and Negro-Vilar, A.,** Luteinizing hormone-releasing hormone (LHRH) and gonadotropin-releasing hormone associated peptide (GAP) and the appearance of luteinizing hormone (LH) and follicle-stimulating hormone (FSH) immunoreactivity in the rat anterior pituitary gland (APG), *Proc. Soc. Neurosci. Mtg. Abstr.*, 16, 397, 1990.
16. **Blake, C.A., Norman, R.L., and Sawyer, C.H.,** Localization of the inhibitory actions of estrogen and nicotine on release of luteinizing hormone in rats, *Neuroendocrinology*, 16, 22, 1974.
17. **Blake, C.A., and Sawyer, C.H.,** Nicotine blocks the suckling-induced rise in circulating prolactin in lactating rats, *Science*, 177, 619, 1972.
18. **Blake, C.A., Scaramuzzi, R.J., Norman, R.L., Kanematsu, S., and Sawyer, C.H.,** Effect of nicotine on the proestrous ovulatory surge of LH in the rat, *Endocrinology*, 91, 1253, 1972.
19. **Bloch, B., Bugnon, C., Fellmann, D., and Lenys, D.,** Neuro-endocrinologie. Presence de determinants antigeniques de la β-LPH, de la β-MSH, de l'α-endorphine, de l'ACTH et de l'αMSH dans les neurones reveles par l'anti-β-endorphine au niveau du noyau infundibulaire de l'homme, *C.R. Acad. Sci. (Paris)*, 287, 1019, 1978,
20. **Blumberg, H.H., Cohen, S.D., Dronfield, B.E., Mordecai, E.A., Roberts, C.J., and Hawks, D.,** British opiate users: I. People approaching London drug treatment centres, *Int. J. Addict.*, 1, 1, 1974.
21. **Borowsky, B., and Kuhn, C.M.,** D_1 and D_2 dopamine receptors stimulate hypothalamo-pituitary-adrenal activity in rats, *Neuropharmacology*, 31, 671, 1992.
22. **Borowsky, B., and Kuhn, C.M.,** GBR12909 stimulates hypothalamo-pituitary-adrenal activity by inhibition of uptake at hypothalamic dopamine neurons, *Brain Res.*, 613, 251, 1993.
23. **Brase, D.A., Tseng, L.F., Loh, H.H., and Way, E.L.,** Cholinergic modification of naloxone-induced jumping in morphine dependent mice, *Eur. J. Pharmacol.*, 26, 1, 1974.
24. **Brentmark, B., Ohlin, P., and Westling, H.,** Nicotine-containing chewing gum as an anti-smoking aid, *Psychopharmacologia*, 31, 191, 1973.
25. **Briggs, C.A., and Cooper, J.R.,** Cholinergic modulation of the release of ^{3}H-acetylcholine from synaptosomes of the myenteric plexus, *J. Neurochem.*, 38, 501, 1982.
26. **Briggs, W.J.,** Cigarette smoking and infertility in men, *Med. J. Aust.*, 1, 616, 1973.
27. **Briski, K.P., and Sylvester, P.W.,** Effect of specific acute stressors on luteinizing hormone release in ovariectomized and ovariectomized estrogen-treated female rats, *Neuroendocrinology*, 47, 194, 1988.
28. **Briski, K.P., and Sylvester, P.W.,** Acute inhibition of pituitary LH release in the male rat by the glucocorticoid agonist decadron phosphate, *Neuroendocrinology*, 54, 313, 1991.
29. **Bugnon, C., Bloch, B., Lenys, D., Gouget, A., and Fellmann, D.,** Comparative study of the neuronal populations containing beta-endorphin, corticotropin and dopamine in the arcuate nucleus of the rat hypothalamus, *Neurosci. Lett.*, 14, 43, 1979.
30. **Burns, G., Almeida, O.F.X., Passarelli, F., and Herz, A.,** A two-step mechanism by which corticotropin-releasing hormone releases hypothalamic β-endorphin: The role of vasopressin and G-proteins, *Endocrinology*, 125, 1365, 1989.

31. **Burris, T.P., Stringer, L.C., and Freeman, M.E.**, Pharmacologic evidence that a D_2 receptor subtype mediates dopaminergic stimulation of prolactin secretion from the anterior pituitary gland, *Neuroendocrinology*, 54, 175, 1991.
32. **Calogero, A.E., Galluci, W.T., Chrousos, G.P., and Gold, P.W.**, Catecholamine effects upon rat hypothalamic corticotropin-releasing hormone secretion *in vitro*, *J. Clin. Invest.*, 82, 839, 1988.
33. **Carr, L.A., Rowell, P.P., and Pierce, W.M.**, Effects of subchronic nicotine administration on central dopaminergic mechanisms in the rat, *Neurochem. Res.*, 14, 511, 1989.
34. **Chiou, C.Y., Long, J.P., Potrepka, R., and Spratt, J.L.**, The ability of various nicotinic agents to release acetylcholine from synaptic vesicles, *Arch. Int. Pharmacodyn. Ther.*, 187, 88, 1970.
35. **Cryer, P.E., Haymond, M.W., Santiago, J.V., and Shah, S.D.**, Norepinephrine and epinephrine release and the adrenergic mediation of smoking-associated hemodynamics and metabolic events, *New Engl. J. Med.*, 295, 573, 1976.
36. **Curlewis, J.D., Clarke, I.J., and McNeilly, A.S.**, Dopamine D_1 receptor analogues act centrally to stimulate prolactin secretion in ewes, *J. Endocrinol.*, 137, 457, 1993.
37. **Dalmaz, C., Netto, C.A., Oliveira-Netto, C.B., Fin, C.A., and Izuierdo, E.**, Chronic ethanol treatment: Effects on the activation of brain beta-endorphin system by novelty, *Psychobiology*, 19, 70, 1991.
38. **Davenport, K.E., Houdi, A.A., and Van Loon, G.R.**, Nicotine protects against μ-opioid receptor antagonism by β-funaltrexamine: Evidence for nicotine-induced release of endogenous opioids in brain, *Neurosci. Lett.*, 113, 40, 1990.
39. **de la Escalera, G.M., Gallo, F., Choi, A.L.H., and Weiner, R.I.**, Dopaminergic regulation of the GT_1 gonadotropin-releasing hormone (GnRH) neuronal cell lines: Stimulation of GnRH release via D_1-receptors positively coupled to adenylate cyclase, *Endocrinology*, 131, 2965, 1992.
40. **Dougherty, J., Miller, D., Todd, G., and Kostenbauder, H.B.**, Reinforcing and other behavioral effects of nicotine, *Neurosci. Biobehav. Rev.*, 5, 487, 1981.
41. **Ejrup, B.**, The role of nicotine in smoking pleasure, nicotinism, and treatment, in *Tobacco Alkaloids and Related Compounds*, Von Euler, V.S., Ed., Pergamon Press, Oxford, 1965, 333.
42. **Farah, J.M., and Mueller, G.P.**, A D-2 dopaminergic agonist stimulates secretion of anterior pituitary immunoreactive β-endorphin in rats, *Neuroendocrinology*, 50, 26, 1989.
43. **Flores, C.M., Hulihan-Giblin, B.A., Hornby, P.J., Lumpkin, M.D., and Kellar, K.J.**, Partial characterization of a neurotransmitter pathway regulating the *in vivo* release of prolactin, *Neuroendocrinology*, 55, 519, 1992.
44. **Fuxe, K., Andersson, K., Eneroth, P., Harfstrand, A., and Agnati, L.F.**, Neuroendocrine actions of nicotine and of exposure to cigarette smoke: Medical implications, *Psychoneuroendocrinology*, 14, 19, 1989.
45. **Fuxe, K., Andersson, K., Eneroth, P., Jansson, A., von Euler, G., Tinner, B., Bjelke, B., and Agnati, L.F.**, Neurochemical mechanisms underlying the neuroendocrine actions of nicotine: Focus on the plasticity of central cholinergic nicotinic receptors, *Prog. Brain Res.*, 79, 197, 1989.
46. **Fuxe, K., Andersson, K., Lofstrom, A., Hökfelt, T., Ferland, L., Agnati, L.F., Perez de la Mora, M., and Schwarcz, R.**, Neurotransmitter mechanisms in the control of the secretion of hormones from the anterior pituitary, in *Central Regulation of the Endocrine System*, Fuxe, K., Hökfelt, T., and Luft, R., Eds., Plenum Press, New York, 1979, 349.
47. **Gambacciani, M., Yen, S.S.C., and Rasmussen, D.D.**, GnRH release from the mediobasal hypothalamus: *In vitro* inhibition by corticotropin-releasing factor, *Neuroendocrinology*, 43, 533, 1986.
48. **Genazzani, A.R., Trentini, G.P., Petraglia, F., De Gaetani, C.F., Criscuolo, M., Ficarra, G., De Ramundo, B.M., and Cleva, M.**, Estrogens modulate the circadian rhythm of hypothalamic beta-endorphin contents in female rats, *Neuroendocrinology*, 52, 221, 1990.

49. **Gindoff, P.R., and Ferin, M.,** Endogenous opioid peptides modulate the effect of corticotropin-releasing factor on gonadotropin release in the primate, *Endocrinology*, 121, 837, 1987.
50. **Goldberg, S.R., Woods, J.H., and Schuster, C.R.,** Nalorphine-induced changes in morphine self-administration in rhesus monkeys, *J. Pharmacol. Exp. Ther.*, 176, 464, 1971.
51. **Goldfarb, T., Gritz, E.R., Jarvik, M.E., and Stolerman, I.P.,** Reactions to cigarettes as a function of nicotine and "tar", *Clin. Pharmacol.*, 19, 767, 1976.
52. **Gorelick, D.A., Rose, J., and Jarvik, M.E.,** Effect of naloxone on cigarette smoking, *J. Subst. Abuse*, 1, 153, 1989.
53. **Gregg, C.M.,** The compartmentalized hypothalamo-neurohypophyseal system: Evidence for a neurohypophyseal action of acetylcholine on vasopressin release, *Neuroendocrinology*, 40, 423, 1985.
54. **Grunberg, N.E., Winders, S.E., and Popp, K.A.,** Sex differences in nicotine's effects on consummatory behavior and body weight in rats, *Psychopharmacology*, 91, 221, 1987.
55. **Grun, E.A., Pauly, J.R., and Collins, A.C.,** Adrenalectomy reverses chronic injection-induced tolerance to nicotine, *Psychopharmacology*, 109, 299, 1992.
56. **Grunberg, N.E., Winders, S.E., and Wewers, M.E.,** Gender differences in tobacco use, *Health Psychol.*, 10, 143, 1991.
57. **Gudehithlu, K.P., Hubble, J.P., Hadjiconstantinou, M., and Tejwani, G.A.,** Nicotine alters opioid peptide levels in striatum and hypothalamus of mouse brain, *Proc. Soc. Neurosci. Mtg. Abstr.*, 15, 146, 1989.
58. **Hamosh, M., Simon, M.R., and Hamosh, P.,** Effect of nicotine on the development of fetal and suckling rats, *Biol. Neonate*, 35, 290, 1979.
59. **Hanson, H.M., Ivester, C.A., and Morton, B.R.,** Nicotine self-administration in rats, in *NIDA Research Monograph 23: Cigarette Smoking as a Dependent Process*, Kranegor, N.A., Ed., U.S. Department of Health, Education, and Welfare, 1979, 70.
60. **Hill, J.B., Nagy, G.M., and Frawley, L.S.,** Suckling unmasks the stimulatory effect of dopamine on prolactin release: Possible role for α-melanocyte-stimulating hormone as a mammotrope responsiveness factor, *Endocrinology*, 129, 843, 1991.
61. **Hillhouse, E.W., Burden, J., and Jones, M.T.,** The effects of various putative neurotransmitters on the release of corticotrophin-releasing hormone from the hypothalamus of the rat *in vitro*. I. The effect of acetylcholine and noradrenaline, *Neuroendocrinology*, 17, 1, 1975.
62. **Hodson, C.A., Davenport, A., Price, G., and Burden, H.W.,** Administration of nicotine to neonatal rats alters adult luteinizing hormone regulation, *Med. Sci. Res.*, 20, 491, 1992.
63. **Hodson, C.A., Davenport, A., Price, G., and Burden, H.W.,** Naltrexone treatment attenuates the inhibitory effect of nicotine treatment on serum LH in rats, *Life Sci.*, 53, 839, 1993.
64. **Hornby, P.J., and Piekut, D.T.,** Opiocortin and catecholamine input to CRF-immunoreactive neurons in rat forebrain, *Peptides*, 10, 1139, 1989.
65. **Howe, G., Westhoff, C., Vessey, M., and Yeates, D.,** Effects of age, cigarette smoking, and other factors on fertility: Findings in a large prospective study, *Br. Med. J.*, 290, 1697, 1985.
66. **Hökfelt, B.,** The effect of smoking on the production of adrenocorticoid hormones, *Acta Med. Scand. Suppl.*, 369, 123, 1961.
67. **Hökfelt, B.,** The effect of smoking on the production of adrenocortical hormones, *Acta Med. Scand.*, 170, 123, 1961.
68. **Hubbard, J.E., and Gohd, R.S.,** Tolerance to the arousal effects of nicotine, *Pharmacol. Biochem. Behav.*, 3, 471, 1975.
69. **Hulihan-Giblin, B.A., Lumpkin, M.D., and Kellar, K.J.,** Acute effects of nicotine on prolactin release in the rat: Agonist and antagonist effects of a single injection of nicotine, *J. Pharmacol. Exp. Ther.*, 252, 15, 1990.
70. **Insel, T.R., Kinsley, C.H., Mann, P.E., and Bridges, R.S.,** Prenatal stress has long-term effects on brain opiate receptors, *Brain Res.*, 511, 93, 1990.
71. **Inturrisi, C.E., Schultz, M., Shin, S., Umans, J.G., Angel, L., and Simon, E.J.,** Evidence from opiate binding studies that heroin acts through its metabolites, *Life Sci.*, 33, 773, 1983.

72. **Jansson, A., Andersson, K., Bjelke, B., Eneroth, P., and Fuxe, K.**, Effects of a postnatal exposure to cigarette smoke on hypothalamic catecholamine nerve terminal systems and on neuroendocrine function in the postnatal and adult male rat. Evidence for long-term modulation of anterior pituitary function, *Acta Physiol. Scand.*, 144, 453, 1992.
73. **Jarvik, M.E.**, Further observations on nicotine as the reinforcing agent in smoking, in *Smoking Behavior: Motives and Incentives*, Dunn, W.L., Ed., Winston and Sons, Washington, D.C., 1973, 33.
74. **Jarvik, M.E., Glick, S.D., and Nakamura, R.K.**, Inhibition of cigarette smoking by orally administered nicotine, *Clin. Pharmacol. Ther.*, 11, 574, 1970.
75. **Jones, V.C., Tomlin, S., Rees, L.H., McLoughlin, L., Besser, G.M., and Wen, H.L.**, Increased β-endorphin but not met-enkephalin levels in human cerebrospinal fluid after acupuncture for recurrent pain, *Lancet*, 946, 1980.
76. **Kapcala, L.P., and Weng, C.-F.**, *In vitro* regulation of immunoreactive β-endorphin secretion from adult and fetal hypothalamus by sequential stimulation with corticotropin-releasing hormone, *Brain Res.*, 588, 13, 1992.
77. **Karras, A., and Kane, J.M.**, Naloxone reduces cigarette smoking, *Life Sci.*, 27, 1541, 1980.
78. **Kershbaum, A., Pappajohn, D.J., Bellet, S., Hirabayashi, M., and Shafiha, H.**, Effect of smoking and nicotine on adrenocortical secretion, *J. Am. Med. Assoc.*, 203, 275, 1968.
79. **Killian, A., Bonese, K., Rothberg, R.M., Wainer, B.H., and Schuster, C.R.**, Effects of passive immunization against morphine on heroin self-administration, *Biochem. Behav.*, 9, 347, 1987.
80. **Kinsley, C.H., Mann, P.E., and Bridges, R.S.**, Luteinizing hormone (LH) release is attenuated in prenatally-stressed (P-S) male rats exposed to sexually-receptive females, *Proc. Soc. Neurosci. Mtg. Abstr.*, 16, 744, 1990.
81. **Koob, G.F., and Bloom, F.E.**, Cellular and molecular mechanisms of drug dependence, *Science*, 242, 715, 1988.
82. **Kozlowski, L.T., Jarvik, M.E., and Gritz, E.R.**, Nicotine regulation and cigarette smoking, *Clin. Pharmacol. Ther.*, 17, 93, 1975.
83. **Kubo, T., Amano, H., Kurahashi, K., and Misu, Y.**, Nicotine-induced regional changes in brain noradrenaline and dopamine turnover in rats, *J. Pharmacobiodyn.*, 12, 107, 1989.
84. **Kwak, S.P., Young, E.A., Morano, I., Watson, S.J., and Akil, H.**, Diurnal corticotropin-releasing hormone mRNA variation in the hypothalamus exhibits a rhythm distinct from that of plasma corticosterone, *Neuroendocrinology*, 55, 74, 1992.
85. **Larson, P.S., and Silvette, H.**, *Tobacco: Experimental and Clinical Studies (Suppl. 3)*, Williams and Wilkins, Baltimore, 1975.
86. **Leong, D.A., Frawley, L.S., and Neill, J.D.**, Neuroendocrine control of prolactin secretion, *Annu. Rev. Physiol.*, 45, 109, 1983.
87. **Lichtensteiger, W., Ribary, U., Schlumpf, M., Odermatt, B., and Widmer, H.R.**, Prenatal adverse effects of nicotine on the developing brain, *Prog. Brain Res.*, 73, 137, 1988.
88. **Lichtensteiger, W., and Schlumpf, M.**, Prenatal nicotine exposure: Biochemical and neuroendocrine bases of behavioral dysfunction, *Dev. Brain Dysfunction*, 6, 279, 1993.
89. **Lucchesi, B.R., Schuster, C.R., and Emley, G.S.**, The role of nicotine as a determinant of cigarette smoking frequency in man with observations of certain cardiovascular effects associated with the tobacco alkaloid, *Clin. Pharmacol. Ther.*, 8, 789, 1967.
90. **Luck, W., Nau, H., Hansen, R., and Steldinger, R.**, Extent of nicotine and cotinine transfer to the human fetus, placenta and amniotic fluid of smoking mothers, *Dev. Pharmacol. Ther.*, 8, 384, 1985.
91. **Malin, D.H., Carter, V.A., Lake, J.R., Cunningham, J.S., and Wilson, O.B.**, Naloxone precipitates nicotine abstinence syndrome in rat, *Soc. Neurosci. Abstr.*, 18, 545, 1992.
92. **Malizia, E., Andreucci, G., Cerbo, R., and Colombo, G.**, Effect of naloxone on acupuncture-elicited analgesia in addicts, *Adv. Biochem. Psychopharmacol.*, 18, 361, 1978.
93. **Manning, F., Walker, D., and Feyerabend, C.**, The effect of nicotine on fetal breathing movements in conscious pregnant ewes, *Obstet. Gynecol.*, 52, 563, 1978.

94. **Martin, J.C.**, Irreversible changes in mature and aging animals following intrauterine drug exposure, *Neurobehav. Toxicol. Teratol.*, 8, 335, 1986.
95. **Matera, C., and Wardlaw, S.L.**, Dopamine regulation of POMC gene expression in the hypothalamus, *Proc. Endocr. Soc. Mtg. Abstr.*, 74, 149, 1992.
96. **Matta, S.G., Foster, C.A., and Sharp, B.M.**, Nicotine stimulates the expression of cFos protein in the parvocellular paraventricular nucleus and brainstem catecholaminergic regions, *Endocrinology*, 132, 2149, 1993.
97. **Matta, S.G., McAllen, K.M., and Sharp, B.M.**, Role of the fourth cerebroventricle in mediating rat plasma ACTH responses to intravenous nicotine, *J. Pharmacol. Exp. Ther.*, 252, 623, 1990.
98. **Matta, S.G., and Sharp, B.M.**, The role of the fourth cerebroventricle in nicotine-stimulated prolactin release in the rat: Involvement of catecholamines, *J. Pharmacol. Exp. Ther.*, 260, 1285, 1992.
99. **Matta, S.G., Singh, J., and Sharp, B.M.**, Catecholamines mediate nicotine-induced adrenocorticotropin secretion via α-adrenergic receptors, *Endocrinology*, 127, 1646, 1990.
100. **Mattison, D.R.**, The effects of smoking on fertility from gametogenesis to implantation, *Environ. Res.*, 28, 410, 1982.
101. **McMahon, B., Trichopoulos, D., Cole, P., and Brown, J.**, Cigarette smoking and urinary estrogens, *N. Engl. J. Med.*, 307, 1062, 1993.
102. **Meliska, C.J., and Gilbert, D.G.**, Hormonal and subjective effects of smoking the first five cigarettes of the day: A comparison in males and females, *Pharmacol. Biochem. Behav.*, 40, 229, 1991.
103. **Meyer, D.C., and Carr, L.A.**, The effects of perinatal exposure to nicotine on plasma LH levels in prepubertal rats, *Neurotoxicol. Teratol.*, 9, 95, 1987.
104. **Morley, J.E.**, Neuropeptide regulation of appetite and weight, *Endocr. Rev.*, 8, 256, 1987.
105. **Mucha, R.F., and Herz, A.**, Motivational properties of kappa and mu opioid receptor agonists studied with place and taste preference conditioning, *Psychopharmacology*, 86, 274, 1985.
106. **Navarro, H.A., Seidler, F.J., Schwartz, R.D., Baker, F.E., Dobbins, S.S., and Slotkin, T.A.**, Prenatal exposure to nicotine impairs nervous system development at a dose which does not affect viability or growth, *Brain Res. Bull.*, 23, 187, 1989.
107. **Navarro, H.A., Siedler, F.J., Whitmore, W.L., and Slotkin, T.A.**, Prenatal exposure to nicotine via maternal infusions: Effects on development of catecholamine systems, *J. Pharmacol. Exp. Ther.*, 244, 940, 1988.
108. **Nikolarakis, K.E., Almeida, O.F.X., and Herz, A.**, Hypothalamic opioid receptors mediate the inhibitory actions of corticotropin-releasing hormone on luteinizing hormone release: Further evidence from a morphine-tolerant animal model, *Brain Res.*, 450, 360, 1988.
109. **Nikolarakis, K.E., Almeida, O.F.X., Sirinathsinghji, D.J.S., and Herz, A.**, Concomitant changes in the *in vitro* and *in vivo* release of opioid peptides and luteinizing hormone-releasing hormone from the hypothalamus following blockade of receptors for corticotropin-releasing factor, *Neuroendocrinology*, 47, 545, 1988.
110. **Nyboe-Andersen, A., Lund-Andersen, C., Falck-Larsen, J., Juel Christensen, N., Legros, J.J., Louis, F., Angelo, H., and Molin, J.**, Suppressed prolactin but normal neurophysin levels in cigarette smoking breast-feeding women, *Clin. Endocrinol.*, 17, 363, 1982.
111. **O'Donohue, T.L., and Dorsa, D.M.**, The opiomelanotropinergic neuronal and endocrine systems, *Peptides*, 3, 353, 1982.
112. **Olson, G.A., Olson, R.D., and Kastin, A.J.**, Endogenous opiates: 1989, *Peptides*, 11, 1277, 1990.
113. **Ono, N., Lumpkin, M.D., Samson, W.K., McDonald, J.K., and McCann, S.M.**, Intrahypothalamic action of corticotropin-releasing factor (CRF) to inhibit growth hormone and LH release in the rat, *Life Sci.*, 35, 1117, 1984.
114. **Palmer, R.F., and Berens, A.**, Double blind study of the effects of naloxone on the pleasure of cigarette smoking, *Fed. Proc. Abstr.*, 42, 654, 1983.

115. **Patel, V.A., and Pohorecky, L.A.**, Interaction of stress and ethanol: Effect on β-endorphin and catecholamines, *Alcoholism: Clin. Exp. Res.*, 12, 785, 1988.
116. **Pauly, J.R., Grun, E.U., and Collins, A.C.**, Tolerance to nicotine following chronic treatment by injections: A potential role for corticosterone, *Psychopharmacology*, 108, 33, 1992.
117. **Pauly, J.R., Grun, E.U., and Collins, A.C.**, Glucocorticoid regulation of sensitivity to nicotine, in *The Biology of Nicotine*, Lippiello, P.M., Collins, A.C., Gray, J.A., and Robinson, J.H., Eds., Raven Press, New York, 1992, 121.
118. **Peck, J.A., Dilsaver, S.C., and McGee, M.**, Chronic forced swim stress produces subsensitivity to nicotine, *Pharmacol. Biochem. Behav.*, 38, 501, 1991.
119. **Perkins, K.A., Epstein, L.H., Sexton, J.E., Solberg-Kassel, R., Stiller, R.L., and Jacob, R.G.**, Effects of nicotine on hunger and eating in male and female smokers, *Psychopharmacology (Berlin)*, 106, 53, 1992.
120. **Petraglia, F., Sutton, S., Vale, W., and Plotsky, P.**, Corticotropin-releasing factor decreases plasma luteinizing hormone levels in female rats by inhibiting gonadotropin-releasing hormone release into hypophysial-portal circulation, *Endocrinology*, 120, 1083, 1987.
121. **Pomerleau, O.F.**, Nicotine and the central nervous system: Biobehavioral effects of cigarette smoking, *Am. J. Med.*, 93 (Suppl. 1A), 2S, 1992.
122. **Pomerleau, O.F., and Rosecrans, J.**, Neuroregulatory effects of nicotine, *Psychoneuroendocrinology*, 14, 407, 1989.
123. **Ramirez, V.D., Feder, H.H., and Sawyer, C.H.**, The role of brain catecholamines in the regulation of LH secretion: A critical inquiry, in *Frontiers in Neuroendocrinology*, Martini, L., and Ganong, W.F., Eds., Raven Press, New York, 1984, 27.
124. **Rasmussen, D.D.**, The interaction between mediobasohypothalamic dopaminergic and endorphinergic neuronal systems as a key regulator of reproduction: An hypothesis, *J. Endocrinol. Invest.*, 14, 323, 1991.
125. **Rasmussen, D.D.**, Dopamine-opioid interaction in the regulation of hypothalamic gonadotropin-releasing hormone (GnRH) secretion, *Neuroendocrinol. Lett.*, 13, 419, 1991.
126. **Rasmussen, D.D.**, Effects of chronic nicotine treatment and withdrawal on neuroendocrine regulation and mediobasohypothalamic proopiomelanocortin and tyrosine hydroxylase mRNA content, *Endocr. Soc. Abstr.*, 75, 321, 1993.
127. **Rasmussen, D.D., Jakubowski, M., Allen, D.L., and Roberts, J.L.**, Positive correlation between proopiomelanocortin and tyrosine hydroxylase mRNA levels in the mediobasohypothalamus of ovariectomized rats: Response to estradiol replacement and withdrawal, *Neuroendocrinology*, 56, 285, 1992.
128. **Rasmussen, D.D., Liu, J.H., Wolf, P.L., and Yen, S.S.C.**, Neurosecretion of human hypothalamic immunoreactive beta-endorphin: *In vitro* regulation by dopamine, *Neuroendocrinology*, 45, 197, 1987.
129. **Reisine, T.D., Mezey, E., Palkovits, M., Heisler, S., and Axelrod, J.**, Beta adrenergic control of adrenocorticotropic hormone release from the anterior pituitary, in *Catecholamines: Neuropharmacology and Central Nervous System Theoretical Aspects*, Usdin, E., Carlsson, A., Dahlstrom, A., and Engel, J., Eds., Alan R. Liss, New York, 1984, 419.
130. **Ribary, U., and Lichtensteiger, W.**, Effects of acute and chronic prenatal nicotine treatment on central catecholamine systems of male and female rat fetuses and offspring, *J. Pharmacol. Exp. Ther.*, 248, 786, 1989.
131. **Rosecrans, J.A., Hendry, J.S., and Hong, J.-S.**, Biphasic effects of chronic nicotine treatment on hypothalamic immunoreactive β-endorphin in the mouse, *Pharmacol. Biochem. Behav.*, 23, 141, 1985.
132. **Rowe, J.W., Kolgore, A., and Robertson, G.**, Evidence in man that cigarette smoke induces vasopressin release via an airway-specific mechanism, *J. Clin. Endocrinol. Metab.*, 51, 170, 1980.
133. **Rowell, P., and Winkler, D.L.**, Nicotine stimulation of ^{3}H-acetylcholine release from mouse cerebral cortical synaptosomes, *J. Neurochem.*, 43, 1593, 1984.

134. **Russell, M.A.H.,** Nicotine replacement: The role of blood nicotine levels, their rate of change, and nicotine tolerance, in *Nicotine Replacement: A Critical Evaluation*, Pomerleau, O.F., and Pomerleau, C.F., Eds., Alan R. Liss, New York, 1988, 63.
135. **Russell, M.A.H., and Feyerabend, C.,** Cigarette smoking: Dependence on high nicotine boli, *Drug Metab. Rev.*, 8, 29, 1978.
136. **Russell, M.A.H., Wilson, C., Feyerabend, C., and Cole, P.V.,** Effect of nicotine chewing-gum on smoking behavior and as an aid to cigarette withdrawal, *Br. Med. J.*, 2, 391, 1976.
137. **Salisbury, R., Reed, J., Ward, I.L., and Weisz, J.,** Plasma luteinizing hormone levels in normal and prenatally stressed male and female rat fetuses and their mothers, *Biol. Reprod.*, 40, 111, 1989.
138. **Sánchez, M.D., Milanés, M.V., Fuente, T., and Laorden, M.L.,** The β-endorphin response to prenatal stress during postnatal development in the rat, *Dev. Brain Res.*, 74, 142, 1993.
139. **Schachter, S.,** Regulation, withdrawal and nicotine addiction, in *NIDA Research Monograph 23: Cigarette Smoking as a Dependent Process*, Kranegor, N.A., Ed., U.S. Department of Health, Education, and Welfare, 1979, 123.
140. **Segarra, A.C., and Strand, F.L.,** Perinatal administration of nicotine alters subsequent sexual behavior and testosterone levels of male rats, *Brain Res.*, 480, 151, 1989.
141. **Seggie, J.A., and Brown, G.M.,** Stress response patterns of plasma corticosterone, prolactin, and growth hormone in the rat following handling or exposure to novel environment, *Can. J. Physiol. Pharmacol.*, 53, 629, 1975.
142. **Seidler, F.J., Levin, E.D., Lappi, S.E., and Slotkin, T.A.,** Fetal nicotine exposure ablates the ability of postnatal nicotine challenge to release norepinephrine from rat brain regions, *Dev. Brain Res.*, 69, 288, 1992.
143. **Seyler, L.E., Pomerleau, O.F., Fertig, J.B., Hunt, D., and Parker, K.,** Pituitary hormone response to cigarette smoking, *Pharmacol. Biochem. Behav.*, 24, 159, 1986.
144. **Sharp, B.M., and Beyer, H.S.,** Rapid desensitization of the acute stimulatory effects of nicotine on rat plasma adrenocorticotropin and prolactin, *J. Pharmacol. Exp. Ther.*, 238, 486, 1986.
145. **Sharp, B.M., and Matta, S.G.,** Detection by *in vivo* microdialysis of nicotine-induced norepinephrine secretion from the hypothalamic paraventricular nucleus of freely moving rats: Dose-dependency and desensitization, *Endocrinology*, 133, 11, 1993.
146. **Shin, S.H.,** Prolactin secretion in acute stress is controlled by prolactin releasing factor, *Life Sci.*, 25, 1829, 1979.
147. **Siegel, R.A., Andersson, K., Fuxe, K., Eneroth, P., Lindbom, L.-O., and Agnati, L.F.,** Rapid and discrete changes in hypothalamic catecholamine nerve terminal systems induced by audiogenic stress, and their modulation by nicotine-relationship to neuroendocrine function, *Eur. J. Pharmacol.*, 91, 49, 1983.
148. **Sirinathsinghji, D.J.S.,** Modulation of lordosis behavior of female rats by naloxone, beta-endorphin and its antiserum in the mesencephalic central gray: Possible mediation via GnRH, *Neuroendocrinology*, 39, 222, 1984.
149. **Sladek, C.D., and Joynt, R.J.,** Characterization of cholinergic control of vasopressin release by the organ-cultured rat hypothalamo-neurohypophyseal system, *Endocrinology*, 104, 659, 1979.
150. **Slotkin, T.A., Cho, H., and Whitmore, W.L.,** Effects of prenatal nicotine exposure on neuronal development: Selective actions on central and peripheral catecholaminergic pathways, *Brain Res. Bull.*, 18, 601, 1987.
151. **Spanagel, R., Herz, A., Bals-Kubik, R., and Shippenberg, T.S.,** β-Endorphin-induced locomotor stimulation and reinforcement are associated with an increase in dopamine release in the nucleus accumbens, *Psychopharmacology (Berlin)*, 104, 51, 1991.
152. **Stirling, R.G., and Shin, S.H.,** A high concentration of dopamine preferentially permitted release of newly synthesized prolactin, *Mol. Cell. Endocrinol.*, 70, 65, 1990.
153. **Stolerman, I.P., Fink, R., and Jarvik, M.E.,** Acute and chronic tolerance to nicotine measured by activity in rats, *Psychopharmacology*, 30, 329, 1973.

154. **Surgeon General of the United States,** *The Health Consequences of Smoking for Women: A Report of the Surgeon General*, Office on Smoking and Health, Public Health Service, U.S. Department of Health, Rockville, MD, 1980, 235.
155. **Suzuki, K., Horiguchi, T., Comas-Urrutia, A.C., Mueller-Heubach, E., Morishima, H., and Adamsons, K.,** Placental transfer and distribution of nicotine in the pregnant rhesus monkey, *Am. J. Obstet. Gynecol.*, 119, 253, 1974.
156. **Swanson, L., Sawchenko, P., Berod, A., Hartman, B., Helle, K., and Vanorden, D.,** An immunohistochemical study of the organization of catecholaminergic cells and terminal fields in the paraventricular and supraoptic nuclei of the hypothalamus, *J. Comp. Neurol.*, 196, 271, 1981.
157. **Takahashi, L.K., Turner, J.G., and Kalin, N.H.,** Prenatal stress alters brain catecholaminergic activity and potentiates stress-induced behavior in adult rats, *Brain Res.*, 574, 131, 1992.
158. **Terry, J.D., McLean, B.K., and Nikitovich, M.B.,** Tobacco-smoke inhalation delays suckling-induced prolactin release in the rat, *Proc. Soc. Exp. Biol. Med.*, 147, 110, 1974.
159. **Tobin, M.J., Jenouri, G., and Sackner, M.A.,** Effect of naloxone on change in breathing pattern with smoking. A hypothesis on the addictive nature of cigarette smoking, *Chest*, 5, 530, 1982.
160. **Tong, Y., and Pelletier, G.,** Role of dopamine in the regulation of proopiomelanocortin (POMC) mRNA levels in the arcuate nucleus and pituitary gland of the female rat as studied by *in situ* hybridization, *Mol. Brain Res.*, 15, 27, 1992.
161. **Tsagarakis, S., Holly, J.M.P., Rees, L.H., Besser, G.M., and Grossman, A.,** Acetylcholine and norepinephrine stimulate the release of corticotropin-releasing factor-41 from the rat hypothalamus *in vitro*, *Endocrinology*, 123, 1962, 1988.
162. **Tucci, J.R., and Sode, J.,** Chronic cigarette smoking. Effect on adrenocortical and sympathoadrenomedullary activity in man, *J. Am. Med. Assoc.*, 221, 282, 1972.
163. **Vaccarino, F.J., Bloom, F.E., and Koob, G.F.,** Blockade of nucleus accumbens opiate receptors attenuates intravenous heroin reward in the rat, *Psychopharmacology*, 86, 37, 1985.
164. **van den Pol, A., Herbst, R., and Powel, J.,** Tyrosine hydroxylase immunoreactive neurons of the hypothalamus. A light and electron microscopic study, *Neuroscience*, 13, 1117, 1984.
165. **Van Ree, J.M.,** Reward and abuse: Opiates and neuropeptides, in *Brain Reward Systems and Abuse*, Engel, J., Oreland, L., Ingvar, D., Pernow, B., Rossner, S., and Pellborn, L.A., Eds., Raven Press, New York, 1987, 75.
166. **Velazquez-Moctezuma, J., Dominguez Salazar, E., and Cruz Rueda, M.L.,** The effect of prenatal stress on adult sexual behavior in rats depends on the nature of the stressor, *Physiol. Behav.*, 53, 443, 1993.
167. **Vom Saal, F.S., Even, M.D., and Quadagno, D.M.,** Effects of maternal stress on puberty, fertility and aggressive behavior of female mice from different intrauterine positions, *Physiol. Behav.*, 49, 1073, 1991.
168. **Ward, S.J., Portoghese, P.S., and Takemori, A.E.,** Pharmacological characterization *in vivo* of the novel opiate, β-funaltrexamine, *J. Pharmacol. Exp. Ther.*, 220, 494, 1982.
169. **Weathersbee, P.S.,** Nicotine and its influence on the female reproductive system, *J. Reprod. Med.*, 25, 243, 1980.
170. **Weidenfeld, J., Bodoff, M., Saphier, D., and Brenner, T.,** Further studies on the stimulatory action of nicotine on adrenocortical function in the rat, *Neuroendocrinology*, 50, 132, 1989.
171. **Weisberg, E.,** Smoking and reproductive health, *Clin. Reprod. Fertil.*, 3, 175, 1985.
172. **Wiegant, V.M., Sweep, C.G.J., and Nir, I.,** Effect of acute administration of delta 1-tetrahydrocannabinol on β-endorphin levels in plasma and brain tissue of the rat, *Experientia*, 43, 413, 1987.
173. **Wilkins, J.N., Carlson, H.E., Van Vunakis, H., Hill, M.A., Gritz, E., and Jarvik, M.E.,** Nicotine from cigarette smoke increases circulating levels of cortisol, growth hormone, and prolactin in male chronic smokers, *Psychopharmacology*, 78, 305, 1982.

174. **Winternitz, W.W., and Quillen, D.,** Acute hormonal response to cigarette smoking, *J. Clin. Pharmacol.*, 17, 389, 1977.
175. **Wise, P.M., Scarbrough, K., Weiland, N.G., and Larson, G.H.,** Diurnal pattern of proopiomelanocortin gene expression in the arcuate nucleus of proestrous, ovariectomized, and steroid-treated rats: A possible role in cyclic luteinizing hormone secretion, *Mol. Endocrinol.*, 4, 886, 1990.
176. **Zuckerman, M.,** Sensation seeking and the endogenous deficit theory of drug abuse, *NIDA Res. Monogr.*, 74, 59, 1986.
177. **Conover, C.D., Kuljis, R.O., Rabii, J., and Advis, J.P.,** Beta-endorphin regulation of luteinizing hormone-releasing hormone release at the median eminence in ewes: Immunocytochemical and physiological evidence, *Neuroendocrinology*, 57, 1182, 1993.
178. **Kalra, S.P., and Kalra, P.S.,** Opioid-adrenergic-steroid connection in regulation of luteinizing hormone secretion in the rat, *Neuroendocrinology*, 38, 418, 1984.
179. **Sarkar, D.K., and Fink, G.,** Gonadotropin-releasing hormone surge: Possible modulation through postsynaptic α-adrenoreceptors and two pharmacologically distinct dopamine receptors, *Endocrinology*, 108, 862, 1981.
180. **Sarkar, D.K., and Yen, S.S.C.,** Changes in β-endorphin-like immunoreactivity in pituitary portal blood during the estrous cycle and after ovariectomy in rats, *Endocrinology*, 166, 2075, 1985.

Chapter 13

ALCOHOL AND COCAINE ABUSE EFFECTS ON THE MALE REPRODUCTIVE NEUROENDOCRINE AXIS

M.A. Emanuele, M.M. Halloran, S. Uddin, J.J. Tentler, A.M. Lawrence, N.V. Emanuele and M.R. Kelley

TABLE OF CONTENTS

0-8493-2451-3/95/$0.00+.50

ABBREVIATIONS

cAMP–cyclic adenosine monophosphate
ETOH–ethanol
FSH–follicle-stimulating hormone
GHF–growth hormone-releasing factor
hCG–human chorionic gonadotropin
LH–luteinizing hormone
LHRH–luteinizing hormone-releasing hormone
NAD–nicotinamide adenine dinucleotide (oxidized form)
NADH–nicotinamide adenine dinucleotide (reduced form)
NADPH–nicotinamide adenine dinucleotide phosphate
PCR–polymerase chain reaction
PKC–protein kinase C
POU–transcriptional regulators, including PIT-1, OCT-1, OCT-2 and VNC-86
PRL–prolactin
RT–reverse transcription
TIF– testicular interstriatal fluid
TSH– thyroid-stimulating hormone

INTRODUCTION

The effects of ethanol (ETOH) on the reproductive axis are multiple and complex. Male gonadal dysfunction after ETOH intake has been well documented, resulting in the clinical picture of decreased libido, impotence and infertility. Although direct gonadal toxicity after ETOH exposure has been well demonstrated, the failure of serum luteinizing hormone (LH) and serum follicle-stimulating hormone (FSH) levels to rise in the presence of blunted testosterone values supports a central neuroendocrine defect as well. It appears that the neuroendocrine target of ETOH's action is mediated at the pituitary as well as the suprapituitary level. This chapter, which will be limited solely to male reproductive dysfunction, will attempt to summarize the available data on this topic.

DIRECT GONADAL EFFECTS OF ETHANOL (ETOH)

The clinical picture of male hypogonadism with ETOH first appeared more than 40 years ago.[100] These early studies dealt with chronic ETOH-abusing individuals, but the contributing roles of malnutrition, liver disease, medications, other substance abuse problems, and concurrent medical problems were not addressed. Thirty years later, other investigators[75] attempted to further characterize the gonadal defect in a healthy population of normal males given ETOH over a 4-week period. A drop in testosterone was noted as early as 5 days after ETOH exposure, with progressive decline over the study period. An increase in the metabolic clearance rate of testosterone was documented, similar to what had been previously reported in alcoholics with cirrhosis.[143]

Since these pioneering investigations, the suppressive effect of ETOH on serum testosterone in both humans and animals has been confirmed many times.[2,4,6,10,16,24,30,35,63,67,69,86,108,109,114,129,155,157,160] A decrease in testosterone production rate, increased testosterone clearance, and increased peripheral conversion to estradiol have all been reported. Time course analysis revealed that the fall in serum testosterone occurred rapidly after ETOH exposure without an antecedent decrement in LH levels, suggesting that ETOH had a primary effect at the testicular level in decreasing testosterone production.[108] In addition, in the hands of most investigators the ETOH-induced suppression of serum testosterone could not be reversed with exogenous human chorionic gonadotropin (hCG)[2,4,27,63,114] (Figure 1).

This lack of reversal of the ETOH-induced fall in serum testosterone with hCG has strengthened the argument that ETOH is directly toxic to the gonad, as has the discovery that ETOH decreased gonadotropin receptors.[67,129] It has also been noted that hCG failed to reverse other effects of ETOH, including reduced testicular interstriatal fluid (TIF) volume and TIF testosterone concentration,[3] again implying direct ETOH inhibition of testosterone biosynthesis and secretion.[2,113] Direct gonadal ETOH effects are supported by other evidence. For example, in a series of *in vitro* studies using perfused testes and

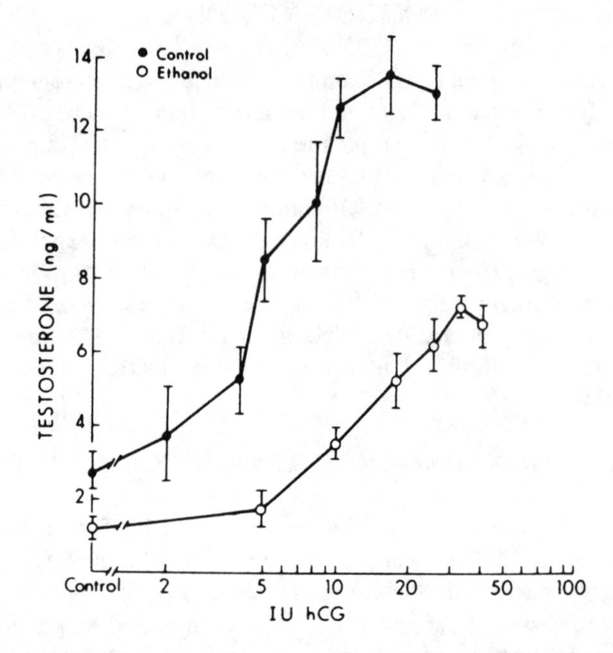

FIGURE 1. The effects of various doses of hCG (international units on the mean ± SEM) serum testosterone levels in male rats with water (designated control, ●) or 2.5 g/kg of ETOH (o). $N = 6$ at each time point. (Taken from Cicero *et al.*[27] See page 299 for citation credit.)

isolated Leydig cells in culture,[28,34,51,88] a sharp fall in secreted testosterone levels was seen with a dose of ETOH as low as 100 mg% with a return to normal after ETOH was discontinued.

Most investigators agree that ETOH must be metabolized to acetaldehyde to exert its testosterone suppressive effect.[28,30,34,88,114] This is particularly well shown in *in vitro* studies of Leydig cells where the addition of 4-methylpyrazole, which blocks the metabolism of ETOH to acetaldehyde but has in itself no effect on testosterone secretion, results in a reversal of ETOH-induced blunting of testosterone levels[69] (Figure 2).

The metabolism of ETOH results in a diminished level of NAD cofactor, since the conversion of ETOH to acetaldehyde and subsequently to acetate are both oxidation reactions during which NAD is reduced. This leads to a resultant alteration in the NAD/NADH redox ratio of the testes. This may be a partial explanation for the reduced steroidogenic enzyme activity which leads to the consistent fall in testosterone noted. In fact, inhibition of three enzyme activities have been described, including 3β-hydroxysteroid dehydrogenase, C17, 20-lyase and 17β-hydroxysteroid dehydrogenase.[6] A lack of citrate, glutamate and pyruvate has also been invoked as playing a role,[115] as the ETOH-induced inhibition of testosterone could be abolished by the addition of these metabolites. It has been proposed that the decreased availability of these key metabolites, important in maintaining the NAD/NADPH redox

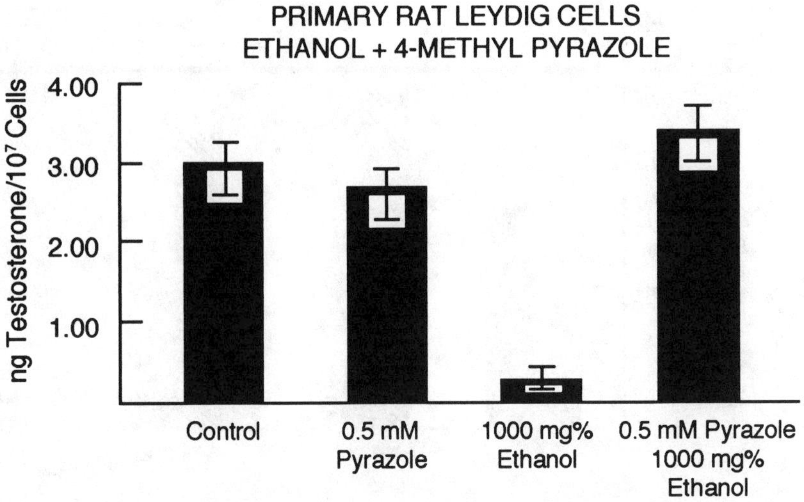

PRIMARY RAT LEYDIG CELLS
ETHANOL + 4-METHYL PYRAZOLE

ng Testosterone/10^7 Cells

4.00
3.00
2.00
1.00

Control
0.5 mM Pyrazole
1000 mg% Ethanol
0.5 mM Pyrazole 1000 mg% Ethanol

FIGURE 2. Testosterone production by cultured rat Leydig cells in the presence and absence of ETOH with and without the addition of 4-methylpyrazole 0.5 mM to the medium. Bars represent mean values for 6 separate flasks. Brackets represent SEM. (Redrawn from Gavaler, J.S. *et al.* See page 299 for citation credit.)

states between the mitochondrial and smooth endoplasmic reticulum, may account for the effect of ETOH on steroidogenesis.[115] Finally, a reduction in testicular phosphodiesters and adenosine triphosphate has also been found after short-term ETOH feeding.[64]

It is interesting that after orchidectomy an increase in the rate of ETOH metabolism has been observed. This appears to be related to removal of dihydrotestosterone, a compound which causes inhibition of hepatic alcohol dehydrogenase. Dihydrotestosterone, given simultaneously with ETOH to normal volunteers, resulted in decreased ETOH metabolism.[157] Thus, *in vivo*, a vicious circle could occur in which ETOH is metabolized to acetaldehyde, causing a "chemical castration," disinhibiting alcohol dehydrogenase and accelerating production of gonadally toxic acetaldehyde, or other potentially toxic agents, such as salsolinol, a condensation product of acetaldehyde and dopamine.

Other interesting mechanisms by which ETOH might decrease testicular steroidogenesis include increased corticosterone and increased testicular β-endorphin. ETOH is known to increase circulating corticosterone via activation of the hypothalamic-pituitary-adrenal axis, and corticosterone, acting at the Leydig cell level, may inhibit testosterone synthesis and/or release.[12,160] It has also been demonstrated that ETOH causes sharp increases in β-endorphin in testicular interstitial fluid, and this testicular opiate may act locally to inhibit testosterone release and/or synthesis as well (Figure 3).[2]

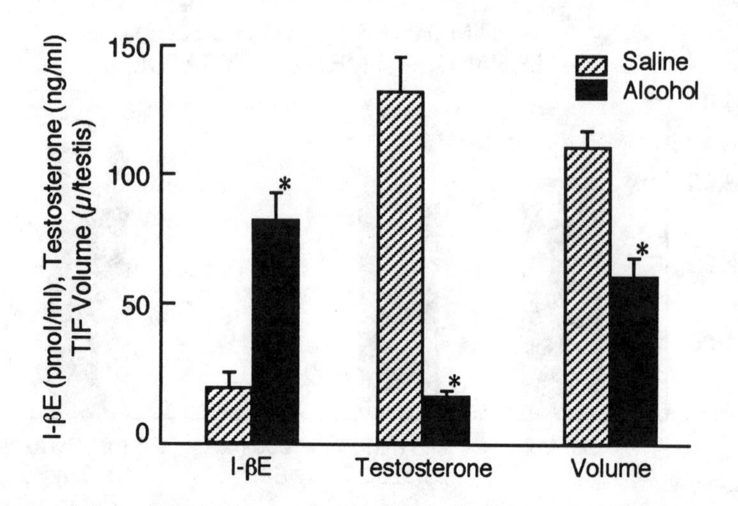

FIGURE 3. Testicular interstitial fluid (TIF) I-βE in picomoles (pmol)/ml of TIF (mean ± SEM), testosterone in nanograms (ng)/nl, of TIF, and collectable TIF volume in microliters (μl)/testis 4 hours after intraperitoneal saline or alcohol (3 g/kg). *$p < 0.05$. (Redrawn from Adams and Cicero.[2] See page 299 for citation credit.)

ETOH AND PITUITARY GONADOTROPIN FUNCTION

ETOH and Serum LH

Despite the consistent fall in serum testosterone levels after ETOH, the expected increase in LH was not seen, suggesting that, in addition to the peripheral gonadal effect, ETOH had an additional effect centrally on the hypothalamic-pituitary unit.

Numerous investigations have demonstrated that acute[26,29,123,160] and chronic[62,63,129,148,155] ETOH exposure significantly depresses serum LH levels in gonadally intact male rats and that ETOH also markedly suppresses the expected increase in serum LH following castration.[3,18,26,46,47,152,156] Our own studies have confirmed this serum LH fall, seen after ETOH injection in castrated male rats (Figure 4). We have found that 1.5 and 3 hours after a single i.p. injection of ETOH (3 g/kg), LH fell by 45–50% compared with saline injected controls. By 24 and 72 hours after injection, with clearance of ETOH, LH values had returned to control levels.[57,60] One laboratory, however, has reported that the time interval after castration may be important,[33] because serum LH levels actually rose if ETOH was given 2 weeks after castration but fell if given 3 weeks after the surgery. Additionally, the amount of the ETOH dose also appears to be important, with higher doses actually causing an unexpected and unexplained rise in serum LH.[33] Others[54] have noted a fall in serum LH only if the animals were stressed at the time they were given ETOH. Overall, the effect of ETOH at doses usually achieved by drinkers on serum LH appears to be one of attenuation.

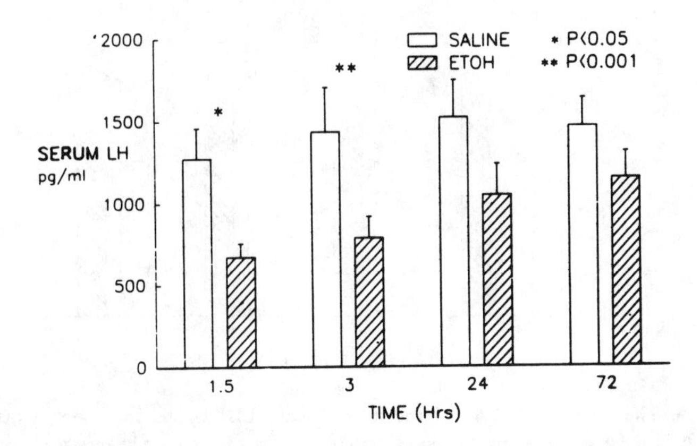

FIGURE 4. ETOH's effect on serum LH levels in castrated male rats. Serum LH levels were quantitated using RIA. Ten rats were used at each time point, five for saline (control) and five for ETOH-injected rats. (Taken with permission from Emanuele *et al.*[60] See page 299 for citation credit.)

ETOH and Pituitary LH Content

Despite fairly consistent falls in serum LH after ETOH, pituitary LH content has been increased[60,129] —or unchanged[33]— but, in any case, nondepleted. Our own data are shown in Figure 5. In light of the fall in serum LH levels, this lack of pituitary LH depletion suggests impaired LH release. This is not entirely surprising since ETOH has been shown to alter the transport of other proteins in other cells such as hepatocytes.[79,151] A variety of *in vitro* and *in vivo* studies have demonstrated that acute or chronic ETOH exposure impairs hepatic protein secretion,[94,126,142,150,151] resulting in intracellular glycoprotein accumulation. Subcellular fractionation studies revealed that about 50% of these accumulated secretory proteins in ETOH-exposed livers were in Golgi, 25% in endoplasmic reticulum, and 25% in cytosol.[151] Thus, by analogy, ETOH could be disruptive to normal posttranslational trafficking of LH, resulting in accumulation of this glycoprotein in the pituitary.

Several mechanisms may be involved in the apparent ETOH inhibition of pituitary LH secretion. LH, as well as FSH, thyroid-stimulating hormone (TSH), and chorionic gonadotropin are glycoprotein heterodimers composed of α and β subunits, which are products of different genes.[112] A number of studies have shown that LH and the other glycoproteins cannot be secreted unless the α and β subunits are assembled, and efficient assembly cannot take place unless the subunits are correctly glycosylated.[40,85,90-92] It may be pertinent, then, that ETOH results in striking morphologic as well as cytochemical changes in the Golgi (of the hepatocyte, at least), the intracellular site where

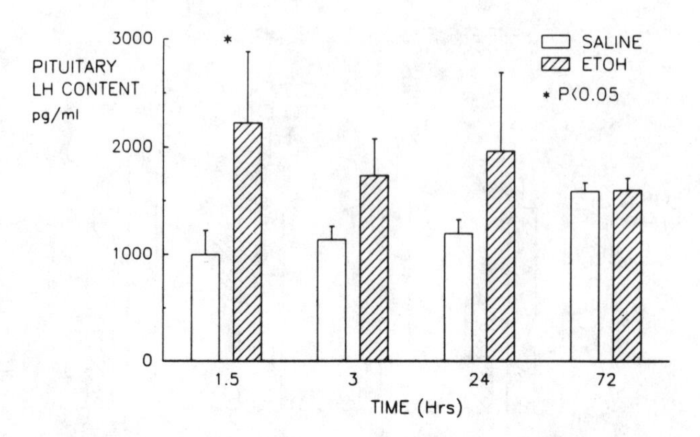

FIGURE 5. The effect of ETOH on the intrapituitary LH content in castrated male rats. Intrapituitary LH levels were quantitated using RIA. Ten rats were used at each time point, five for the saline (control) and five for the ETOH i.p.-injected rats. (Used with permission from Emanuele *et al.*[60] See page 299 for citation credit.)

glycosylation is completed.[124] The possible functional importance of this apparent damage in subcellular machinery is supported by the report of ETOH-induced attenuation of glycosylation of apoprotein E in hepatocytes.[73] There is, therefore, some rationale to speculate that ETOH may alter LH glycosylation, impairing subunit assembly and resulting in the impaired LH secretion noted after ETOH exposure.

Another possible mechanism by which ETOH could impair protein secretion is by its metabolism to acetaldehyde, which would, in turn, bind to tubulin. Tubulin is very reactive toward acetaldehyde and such a modification might functionally impair microtubule-dependent events such as intracellular trafficking and secretion.[151]

α and β-LH Gene Transcription and ETOH

It has been well demonstrated that castration results in a rise not only in serum LH and FSH, but also in the β-LH and β-FSH mRNA concentrations in male rats.[1,116,117] The timing and magnitude of the increase in serum gonadotropins and subunit mRNA concentrations are different, however, and suggest differential regulation of gonadotropin subunit gene expression.[1,39,72]

At 1.5 and 3 hours after an i.p. injection of 3 g/kg ETOH, there was a marked suppression in the mRNA for LH,[60] which returned to normal in 24 hours (Figures 6 and 7). This paralleled the changes seen in serum LH. The mRNA for the common subunit was not affected by ETOH at any time point, thus demonstrating that ETOH's effect on the gonadotrope is selective for β-LH (data not shown). The effect of ETOH on pituitary steady-state β-LH mRNA levels appears to differ depending on whether the exposure is acute or chronic, because the same suppression was not seen in β-LH mRNA after 5 weeks of chronic ETOH feeding.[130] The β-LH mRNA actually rose in the

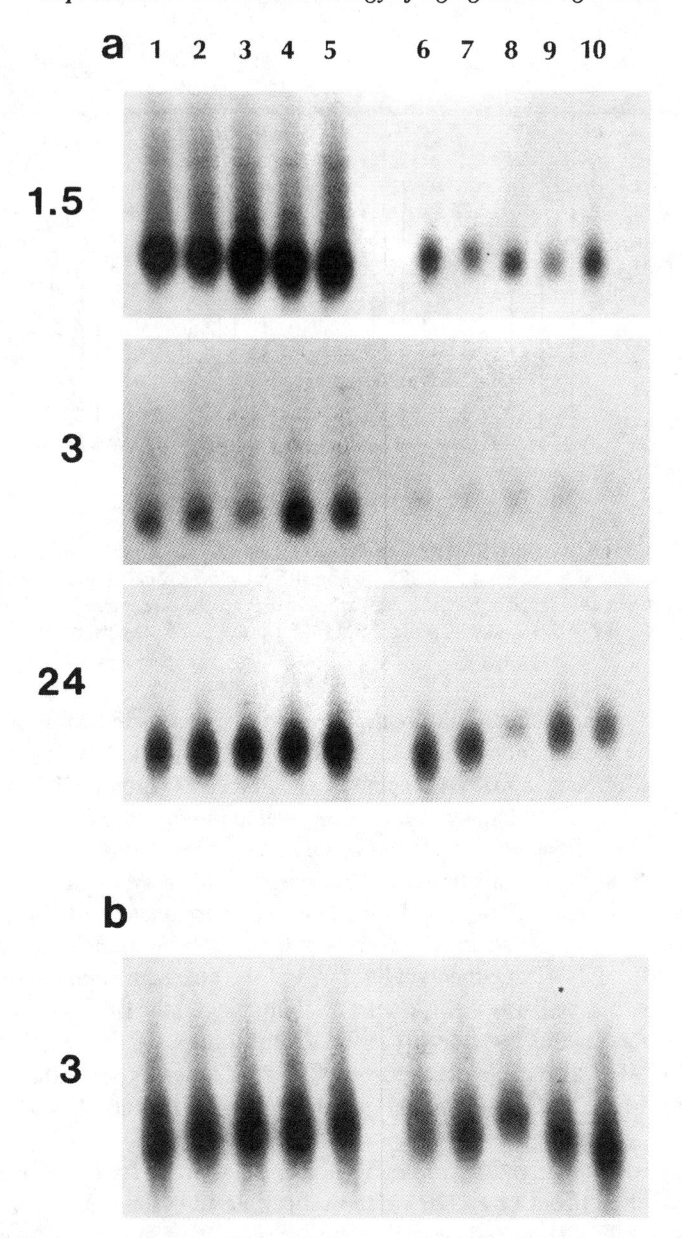

FIGURE 6. Northern blot analysis of α-subunit and β-LH mRNA. Lanes 1–5 are i.p. saline-injected rats and lanes 6–10 are i.p. ETOH-injected rats. Times after i.p. injection, in hours, are given on the left side (1.5, 3 and 24 hours). (***a***) Northern blot probed with β-LH cDNA, (***b***) Northern blot in (a) was stripped and reprobed with common α-subunit cDNA clone. Only the blot at 3 hours is shown. Loading differences were normalized using a 28S ribosomal clone, which has been shown not to change with ETOH treatment (unpublished data). The blot at 3 hours appears to have less hybridization, but this is only due to a shorter exposure to the X-ray film. Quantitative comparisons cannot be made between the time points, only between the controls and ETOH samples at each time point. (Used with permission from Emanuele *et al.*[60] See page 299 for citation credit.)

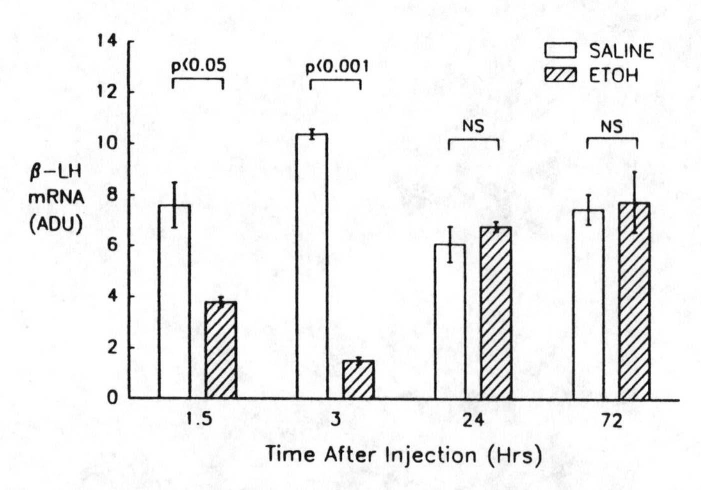

FIGURE 7. The effect of ETOH (i.p.) on gonadectomy-induced rise in mRNA for β-subunit of LH. Data corrected for loading with 28S ribosomal RNA clone. Animals were sacrificed at designated times after saline or ETOH injection. These data are the densitometer readings of the blots shown in Figure 6. (Used with permission from Emanuele *et al.*[60] See page 299 for citation credit.)

ETOH-fed group compared to both the pair-fed controls and chow-fed animals, while the common subunit did not change (Figure 8). It seems possible that this difference in mRNA response to acute versus chronic ETOH represents an example of tolerance at the tissue level to prolonged ETOH exposure.

The suppression seen after acute ETOH could be explained by decreased mRNA synthesis, by enhanced mRNA degradation, or by a combination of both. Since the half-life of the β-LH message is approximately 24 hours[17] and it has been shown that steady-state levels of β-LH mRNA plummet by 1.5 and 3 hours after ETOH treatment, at least part of the mechanism underlying the fall in steady-state β-LH mRNA must be shortening in the β-LH half-life. Of great importance in this context is the demonstration of the length of the poly(A) tail as a possible determinant of hormone message stability.[21,93,105] Conceivably, ETOH could be altering the length of the poly(A) tail of LH, making it more susceptible to degradation.

In addition to alterations in steady-state mRNA levels for β-LH, we have explored the impact of ETOH on the subcellular distribution of the β-LH message. In ETOH-treated animals, a greater percentage of message was seen in fractions not containing polysomes than was the case in control animals. This suggests that in addition to diminishing levels of LH mRNA, ETOH also decreases translational efficiency, since mRNA must be polysome associated for translation to occur (unpublished results).

ETOH and Follicle-Stimulating Hormone (FSH)

Whereas a great deal of information is available on the effect of ETOH on LH, there is very little information on FSH. The studies dealing with FSH have

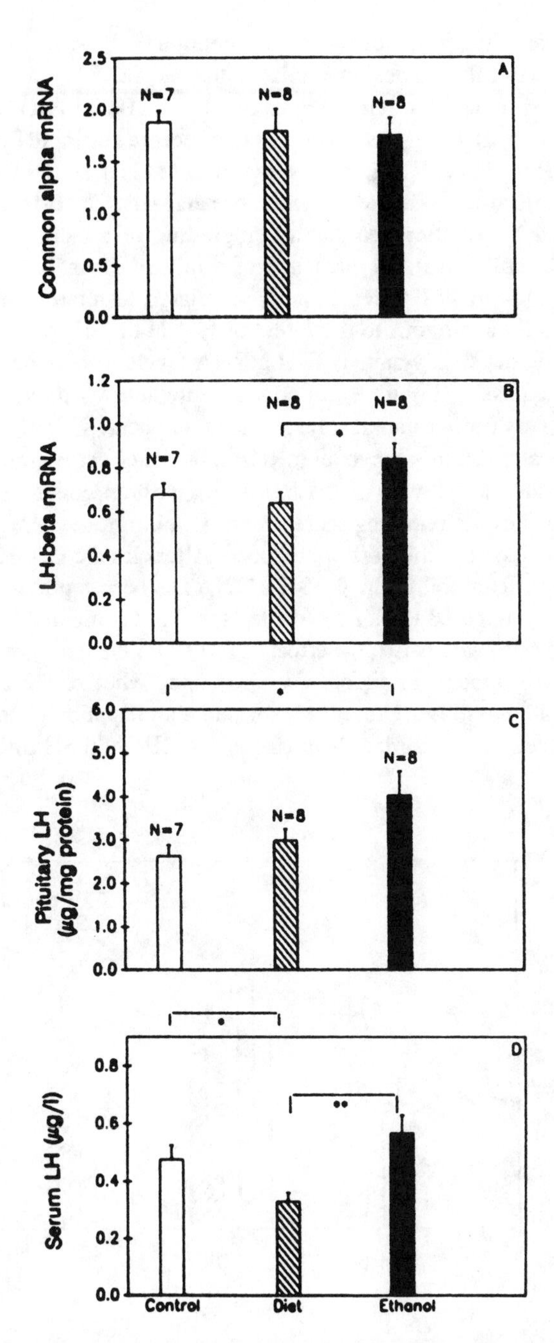

8. Pituitary common α-subunit mRNA (*A*; $F = 0.13$, ANOVA), LH-β (*B*; $F = 3.67$) and = 3.58), and serum LH (*D*; $F = 6.07$) ***turn*** controls (blank), diet controls (hatched), and ETOH, weeks (black). The mRNA levels are in arbitrary densitometric or pituitary. *$p < .01$. (Used with permission from Salonen ***et al.***[130] See page 299 for citation credit.)

noted a decrease in serum levels of this peptide[61,129] with no change in pituitary FSH content. In a design similar to that used in LH studies (above), we examined serum and pituitary immunoreactive FSH and FSH mRNA 1.5, 3, 6, and 24 hours after castrated male rats had received a single i.p. injection of either ETOH or saline. ETOH caused significant falls in serum FSH 1.5 and 3 hours after injection, compared to saline-treated rats. The effect had dissipated 6 and 24 hours after treatment. Our finding of a lack of concurrent pituitary FSH depletion in the presence of a fall in serum FSH levels has suggested a block in FSH release, perhaps related to altered intracellular trafficking, totally analogous to the effect of ETOH on LH release.

Interestingly, the steady-state β-FSH mRNA levels were unchanged after acute ETOH exposure (Figure 9), in contrast to what was previously noted with β-LH.[60] This further underscores the idea that acute ETOH administration is not globally suppressive to all mRNAs, but its effect is quite selective. It reduces steady-state levels of β-LH mRNA without changing levels of β-FSH mRNA, despite reducing secretion of both hormones. Also, the message for the common α-subunit is not reduced. After chronic ETOH exposure, however, a significant fall in the β-FSH mRNA has been reported in ETOH-treated animals compared to pair-fed controls, but not compared to chow-fed animals.[130] Thus, as with β-LH, the effect of ETOH on the steady-state mRNA levels for β-FSH appears to differ depending upon whether the exposure is acute or chronic, possibly representing a tissue manifestation of tolerance to ETOH. However, the discordant behavior of the LH and FSH mRNAs after

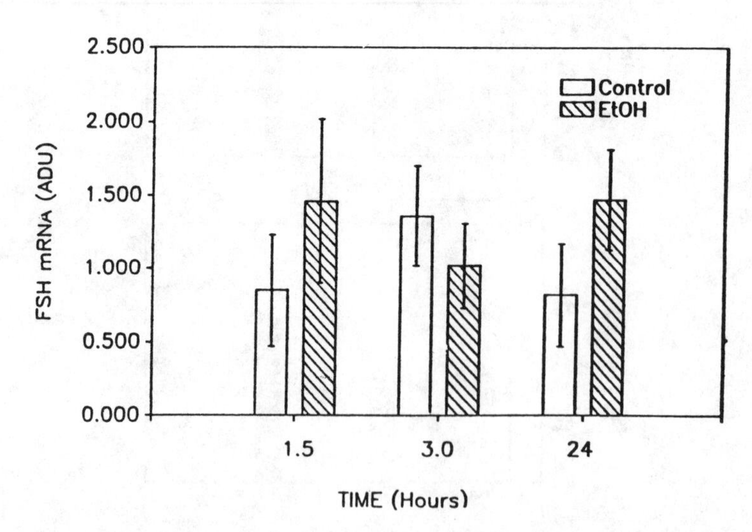

FIGURE 9. The effect of ETOH (i.p.) on gonadectomy-induced rise in mRNA for β-subunit of FSH. Data were corrected for loading with 28S ribosomal RNA. Animals were killed at designated times after saline (open bars) or ETOH (cross-hatched bars) injection. Data expressed as arbitrary densitometer units (ADU). (Used with permission from Emanuele *et al.*[61] See page 300 for citation credit.)

ETOH exposure is not the first example of such differences in response as there are other situations described where this occurs.[22,140,141,158] After castration, for example, the β-FSH mRNA rises rapidly for up to 5 days, and then actually begins to fall, in contrast to β-LH where the mRNA levels rise at 2 weeks and plateau at this elevated level.[141] Closer investigation of the reasons for the differences in LH and FSH message responses to ETOH may shed light on important differences in the ways that these genes are regulated.

ETOH and Pituitary Transmembrane Signal Processing

Alterations in events prior to β-LH gene transcription, LH secretion or FSH secretion could also explain the generally deleterious effects on gonadotropin physiology noted. Preliminary work has been initiated to analyze ETOH's effects on the transmembrane signaling process. These events include G-protein activation, increased phosphoinositol turnover, calcium-calmodulin interaction, and protein kinase C (PKC) activation.[37] Also recently described as important in LH synthesis and secretion are caldesmon, spectrin and calcineurin.[38]

The possibility that ETOH may interfere with one or more of these processes in the pituitary is supported by data showing that ETOH disrupts some of these events in other tissues, particularly in the hepatocyte[65,79,80,127] and in brain tissue.[48,81,99] However, there is a relative paucity of data on ETOH's effects on these systems in the pituitary, and thus the possible and likely role of ETOH in perturbing intracellular signal transduction in pituitary would seem to be a fertile area for future research.

Some of these studies have demonstrated a selective effect of ETOH on the synthesis of G_s, a stimulatory G-protein[82] and on the function of G_o[20,159] in a brain region and exposure-specific manner.[19] There are no such published studies utilizing anterior pituitary cells.

In the anterior pituitary gonadotropes, the phosphoinositol second messenger system may also transduce the effects of LHRH on LH secretion.[37,89] Although ETOH effects on this pathway have been worked out in hepatocytes,[79] platelets,[128] and different brain regions,[43,74,99,101,146] no one has ever evaluated the effects of ETOH on phosphoinositol turnover in the pituitary or, more selectively, in the gonadotrope.

It also appears likely that LHRH-induced changes in intracellular calcium levels and calmodulin translocation are important in gonadotropin physiology.[37,38,89] To date there are no studies available examining the effect of ETOH on pituitary intracellular calcium levels or calmodulin translocation. This certainly would be a fertile area for investigative research in the future.

An important role for PKC in LHRH stimulation of β-LH mRNA levels (but perhaps not secretion) has also been proposed.[8,9,37,50,87,89] There is evidence suggesting that ETOH does interact with PKC in other cell systems. ETOH in concentrations as low as 20 mM caused a significant and rapid activation of cytosolic PKC in hepatocytes,[79] mouse brain,[48,82] human platelets[128] and lymphocytes.[49] This stimulation in hepatocytes, evident within

20–30 seconds, followed closely by a rise in cytosolic calcium that rapidly decays within minutes,[78] is similar to findings reported in the mouse brain.[48,82] Preliminary data in our laboratory have demonstrated no alteration in PKC activity levels in total pituitary cells after ETOH exposure. Such data should be viewed as very preliminary because it would seem important to study the impact of ETOH not only on total PKC activity in the gonadotrope, but also the different isoforms and translocation of PKC. Recently it has been noted that PKC-β is the predominant subspecies of PKC in the anterior pituitary.[68] Conceivably, ETOH could be affecting the transmembrane signaling of only this particular PKC isoform, and this effect would be missed if total PKC activity were assessed. Furthermore, gonadotropes comprise only 7–10% of the total pituitary cell population and it is quite conceivable that ETOH-induced changes in gonadotropes could be obscured by counterbalancing changes in other pituitary cell types. Purification of gonadotropes from the entire cell population is time consuming and expensive. The imminent availability of a β-LH producing cell line will make such studies much more feasible in the near future.[84]

ETOH and Pituitary LH and FSH *In Vitro*

The direct effect of ETOH on dispersed pituitary cells in culture has also been addressed.[58,120,133] These studies showed that LH release caused by LHRH was significantly attenuated by prior and co-incubation with ETOH at concentrations of 200 mg% or greater. There was no effect of any ETOH concentration, even as high as 800 mg%, on basal LH release. Thus, at least some of the LH suppressive effects of ETOH are mediated at the pituitary level.[58,120,133] More recent studies in our laboratory, again using primary pituitary cell culture, have also indicated that ETOH can directly suppress FSH release.[139]

Hypothalamic LHRH and ETOH

The effect of ETOH on LHRH is another area of active investigation. The synthesis and secretion of LH and FSH is regulated by hypothalamic LHRH, which is secreted into the portal blood in a pulsatile fashion. This pulsatile secretion of LHRH is required for maintaining gonadotropin secretion. Both the amplitude and the frequency of LHRH secretion vary in different physiologic circumstances, and the manner of LHRH stimulation appears to be important in regulating gonadotropin synthesis and secretion.[103] A continuous infusion of LHRH desensitizes the gonadotrope, and secretion, especially of LH, is diminished.[11,95] Several lines of largely indirect evidence suggest that part of ETOH's neuroendocrine impact on reproduction is via inhibition of release of LHRH.

First, peripheral *in vivo* administration of LHRH overcomes the LH suppressive effects of ETOH.[26,123] This has been interpreted as indicating that ETOH did not reduce the ability of the pituitary to respond to LHRH during ETOH exposure and, thus, the lowered LH levels seen with ETOH were due to diminished LHRH secretion. A second line of evidence supporting a CNS locus for ETOH action is the report that ETOH reduces pulsatile LH release, as it is believed that the pulsatile nature of LH release is dependent upon episodic release of LHRH from the hypothalamus.[123] Some (although not all) authors have reported that multiple ETOH injections, which cause diminished serum LH levels, are associated with elevated hypothalamic content of LHRH.[46,47] This elevated LHRH suggested impaired release. The ability of ETOH to reduce the LH stimulatory effects of the opiate receptor antagonist naloxone may also support a CNS site of action.[31] The concept here is that CNS opiates, notably hypothalamic β-endorphin, inhibit LHRH release. Naloxone, by blocking the opiate, stimulates release of LHRH and leads to a measured increase in LH. Thus, ETOH's partial block of the LH response to naloxone was thought due to ETOH-induced inhibition of LHRH.

The notion that ETOH blocked LHRH release, supported by all of the above indirect data, was buttressed by the report that animals given ETOH had significantly less LHRH in hypothalamic-pituitary portal blood than did control animals receiving injections of water (Figure 10).[23]

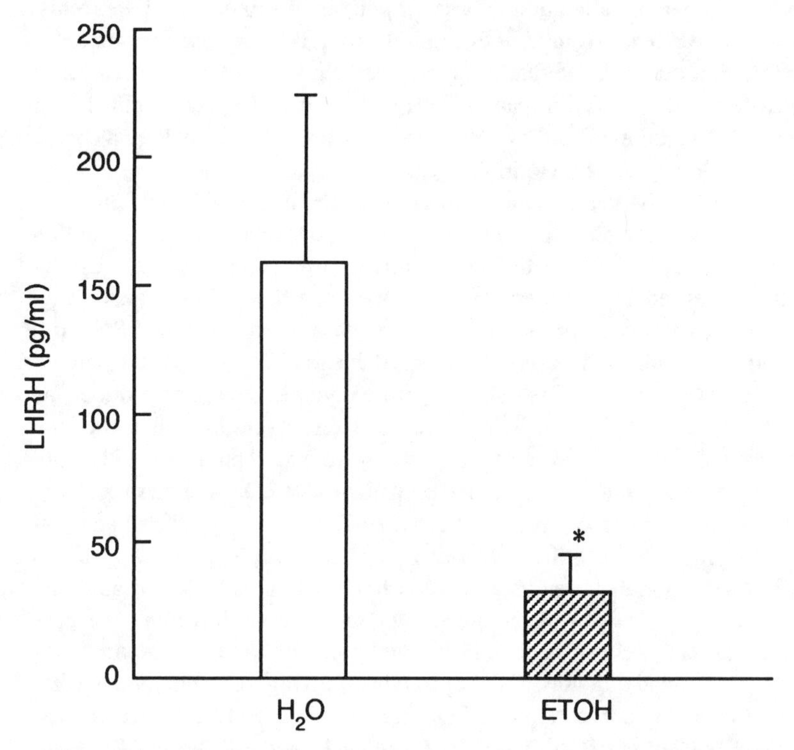

FIGURE 10. ETOH treatment lowers LHRH levels in the hypophyseal portal plasma of orchidectomized rats ($N = 6.7$). The asterisk (*) indicates this level of LHRH was significantly different ($p < 0.05$) from those of H_2O-injected control rats. (Redrawn with permission from Ching *et al.*[23] See page 300 for citation credit.)

By measuring release of immunoreactive LHRH from hypothalamic slices in culture, our laboratory has examined the question of whether ETOH's effect on LHRH was exerted *directly* at the hypothalamic level.[56,59]

First, utilizing hypothalami from ETOH-naive animals cultured in a static incubation system, we found that coincubation of 50 mg%, 100 mg%, 200 mg%, or 400 mg% ETOH was unable to reduce LHRH release stimulated by depolarizing concentrations of potassium. Furthermore, basal release of LHRH was not affected by 400 mg% ETOH.[56] Next, again utilizing hypothalami from ETOH-naive animals, but in a perfusion system rather than in static incubation, we were unable to demonstrate any LHRH suppressive effect of 300 mg% ETOH compared to control. This was true for basal LHRH release as well as for the release provoked by a variety of physiologically important secretogogues including dopamine, norepinephrine, naloxone, prostaglandin E_2 and potassium.[59] Finally, we studied the impact of *in vivo* ETOH treatment on subsequent release of LHRH, *in vitro*. We evaluated castrated male rats after both acute (a single i.p. injection) or chronic (3 days of ETOH administration via a permanent gastric cannula) ETOH compared to controls. Despite the fact that both ETOH treatment protocols significantly reduced serum LH levels, there was no alteration in basal or potassium stimulated LHRH release compared to release from hypothalami of non-ETOH exposed animals.[56]

Based on our data and that of others, we believe that the most reasonable conclusion is that ETOH impairs LHRH release, but its effect is neither extra-hypothalamic nor mediated by a metabolite such as acetaldehyde or a condensation product such as salsolinol.

We have embarked on studies to examine the effect of ETOH on LHRH synthesis, as reflected by measuring steady-state levels of hypothalamic pre-pro-LHRH mRNA. Since this is a relatively low abundance message, not easily measured by Northern analysis, we used the reverse transcription/polymerase chain reaction (RT/PCR) technique. We extracted total RNA from individual hypothalami, reverse transcribed this to cDNA, amplified manyfold a portion of pre-pro-LHRH cDNA by using synthetic oligonucleotides, and separated the PCR products by agarose gel electrophoresis. Further confirmation that the PCR products of expected size were indeed pre-pro-LHRH cDNA was obtained by Southern hybridization with labeled cDNA clones for LHRH. Thus, we measured pre-pro-LHRH cDNA reverse transcribed from and, thus, reflecting mRNA. With the inclusion of internal controls, we have verified in several ways that this is a quantitative technique. For example, the amount of PCR product increases with an increasing number of amplification cycles, and there is a very highly correlated linear relationship between the amount of starting RNA and amounts of PCR product. Utilizing this technique, we have found that a single injection of ETOH significantly reduced LHRH message in intact but not castrated male rats (Figures 11 and 12). Serum LH levels, however, were reduced by this treatment in both gonadally intact and castrated animals. Thus, we must tentatively conclude that while acute ETOH (and/or its metabolites) may diminish LHRH release, it does not affect synthesis in

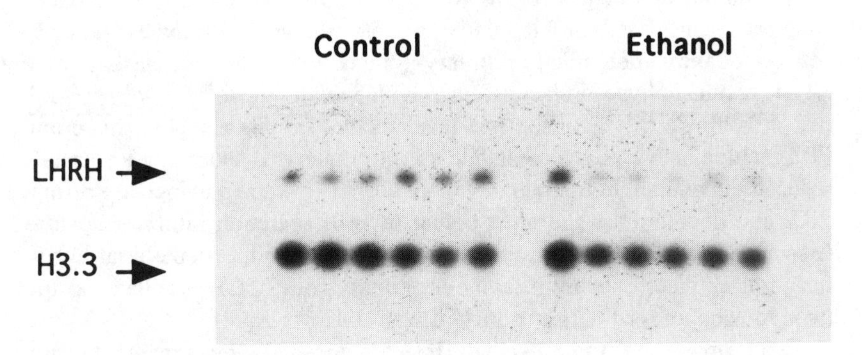

FIGURE 11. The effect of acute ETOH exposure on hypothalamic LHRH mRNA levels as assessed by RT/PCR.

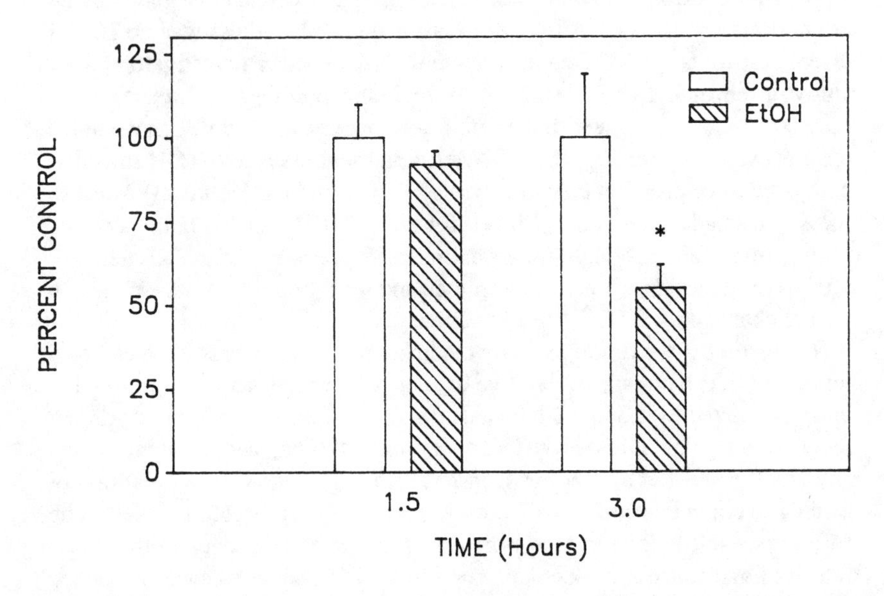

FIGURE 12. Densitometric analysis of the effect of acute ETOH exposure on hypothalamic LHRH mRNA levels, as assessed by RT/PCR. Autoradiogram of 3-hour data is shown in Figure 11.

castrated rats, but may in intact animals. Studies on the effect of chronic ETOH exposure on LHRH message are in progress.

ETOH and Prolactin (PRL)

An important consideration that must be addressed in any discussion of ETOH and the neuroendocrine reproductive axis is that of PRL-ETOH interactions, because hyperprolactinemia in males (as well as females) reduces activity of the hypothalamic-pituitary-gonadal axis. The consensus is that ETOH results in hyperprolactinemia in both humans[12,52,55,86,118,141,161,162] and rats.[18,45,55,63,104,129,136,138] Proposed mechanisms for this elevation in serum PRL include increased hypothalamic and/or pituitary[2] β-endorphin levels and reduced dopaminergic inhibition.[138] β-Endorphin is known to increase serum PRL and dopamine is the most potent of PRL secretion inhibitors. Also, substance P, which has been found to be decreased in the mediobasal hypothalamus and increased in the anterior pituitary with ETOH, has been postulated to account for the rise in PRL after ETOH exposure.[138]

In our own acute studies we have injected gonadally intact male rats with i.p. ETOH or saline and studied pituitary and serum at various time points up to 24 hours later. Serum PRL values were significantly higher in the ETOH animals, reaching values two times control at 1.5 hours and approximately three times control at 3 hours after injection. The effect had dissipated by 24 hours (data not shown). Although in our acute studies there was no ETOH effect on pituitary PRL content (data not shown), other investigators, after studying chronic ETOH exposure, have found pituitary PRL content decreased.[129] Concomitant with the significant increase in serum PRL, we noted a significant 30% fall in PRL mRNA 1.5 and 3 hours after ETOH injection compared to control, but the effect was gone by 24 hours (Figure 13). We also measured steady-state message levels for Pit 1 (GHF1), a POU/homeodomain transcription factor thought important in the expression of the PRL gene, as well as in expression of growth hormone and thyrotropin.[102] Pit 1 expression was unchanged by ETOH (data not shown).

Interestingly, we found pituitary cyclic amp (cAMP) levels were elevated tenfold by ETOH treatment relative to control 30 minutes after injection, but had normalized by 1.5 and 3 hours. The sharp tenfold rise in pituitary cAMP levels is intriguing. Our data reflect total pituitary cAMP and, thus, we cannot say whether ETOH's effect is seen in all the different cell populations or pituitary (e.g., lactotropes, gonadotropes, etc.) or is restricted to a few. The former possibility is more likely the case since data from several other laboratories have consistently demonstrated that ETOH acutely causes enhanced cAMP accumulation in a diverse variety of cells including hepatocytes,[78] NIE-115 neuroblastoma cells,[71] PC12 cells [121] and platelets.[77] Thus, it seems reasonable to suppose that the cAMP response to ETOH in pituitary cells is a generalized cellular phenomenon, and our findings of increased total pituitary cAMP reflect rises in each of the specific hormone-secreting cellular subtypes of the anterior pituitary. It is relevant that cAMP has been shown to

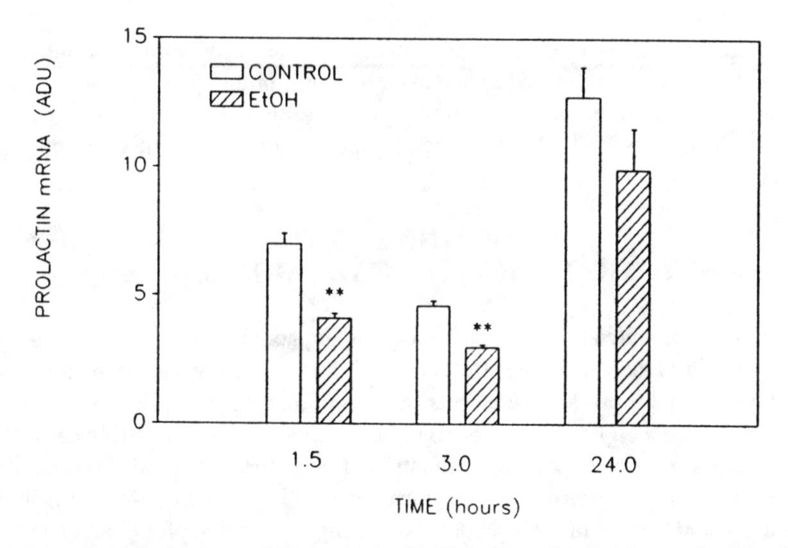

FIGURE 13. Prolactin (PRL) mRNA levels after ETOH exposure. The levels were significantly decreased at 1.5 and 3 hours, with a return to the control level by 24 hours. Data were corrected for loading and expressed in arbitrary densitometer units. The results are expressed as a percentage of the control value. $N = 5–8$ per group. $^{**}p < 0.01$. (Used with permission from Emanuele *et al.*[55] See page 300 for citation credit.)

be important in growth hormone synthesis[66] and to stimulate transcription of the Pit-1 and PRL genes, as well as secretion of PRL. The elevated cAMP levels seen in our ETOH-treated rats, compared to control rats, might then be expected to result in substantial increases in steady-state levels of Pit-1 and PRL mRNAs. Thus, the 30% decrement in PRL message in ETOH-treated rats and the lack of apparent change in Pit-1 (GHF-1) message might be inappropriate in light of the sharp cAMP rise. It may be that ETOH caused a *marked* reduction in cAMP-mediated Pit-1 and PRL gene transcription, and/or diminished stability of these messages.

In vitro ETOH exposure of dispersed normal male rat pituitary cells has consistently effected a rise in PRL secretion.[55,137] Thus, at least part of ETOH's effect on PRL release (just as on LH and FSH secretion) is a direct pituitary effect, but this does not rule out a hypothalamic locus of action as well. This increase in PRL release with ETOH exposure may be secondary to cell swelling with subsequent hormone secretion, an effect found to be dependent on calcium mechanisms.[132]

While elevated PRL is associated with attenuation of LHRH and thus LH, this is probably *not* the explanation or at least not the total explanation for the lowering of LH reported after ETOH exposure. In a series of elegant studies addressing this possibility, utilizing apomorphine to maintain low PRL levels, Chapin *et al.*[18] concluded that neither dopaminergic neuronal activation nor hyperprolactemia were the explanation for the LH depression seen after ETOH.

Summary

It appears that ETOH is disruptive to the male neuroendocrine reproductive axis at several points, including the hypothalamus, pituitary and gonad. Future studies combining molecular biology techniques with those of classic endocrine physiology will allow a more in-depth exploration of this fascinating area.

COCAINE AND THE HYPOTHALAMIC-PITUITARY-GONADAL AXIS

In general, much less is known about the impact of cocaine on neuroendocrine function than about the effects of ETOH. Remarkably, there are only two studies on the reproductive effects of cocaine in male rats. Gordon *et al.*[76] demonstrated a biphasic effect of 30 mg/kg cocaine, given intraperitoneally on plasma testosterone, showing stimulation of testosterone 90 and 120 minutes and inhibition 180 minutes after cocaine injection (Figure 14). Berul and Harclerode[13] tested the effects of 10 or 40 mg/kg cocaine on LH and testosterone at 1, 2, and 3 hours after injection. Plasma testosterone first increased, then decreased, qualitatively similar to the findings of Gordon *et al.*, while both doses caused decreases in LH at all time points. Analysis of this paper is limited because the authors' statistical presentation is unclear and it does not appear that there were vehicle-injected controls at all time points. In the second experiment in the same paper, the effect of chronic cocaine administration was examined in rats that had received 15 daily injections of 10 or 40 mg/kg cocaine and were sacrificed 2 hours after the last injection. The lower dose caused no hormonal change, but the higher dose of cocaine depressed both LH and testosterone. The authors noted that the rats chronically treated with the higher dose of cocaine gained significantly less weight than controls and that cocaine was a strong appetite suppressant. Because fasting may reduce gonadotropins, the effect seen in the chronic experiments may be secondary to nutritional deficiency, as well as to cocaine itself. Indeed, the authors pointed out that several animals died during prolonged cocaine treatment.

Three studies in monkeys, one that included males as well as females,[106,107] seemed to consistently demonstrate that single intravenous injections of cocaine acutely raised basal and LHRH-stimulated LH levels, as well as raised LHRH-stimulated (but not basal) FSH levels.

In humans hospitalized for cocaine abuse, there were no major differences in LH levels, LH pulsatility or testosterone levels during the cocaine-free interval of study.[14,41] Only limited conclusions can be drawn from data in humans for several reasons. First, we are unaware of data on neuroendocrine studies *during* cocaine ingestion. There are only studies on chronic cocaine abuses during periods of abstinence. Second, it seems likely that these individuals abused multiple drugs, not just cocaine, and that they may be malnourished or have other diseases, further impairing interpretation.

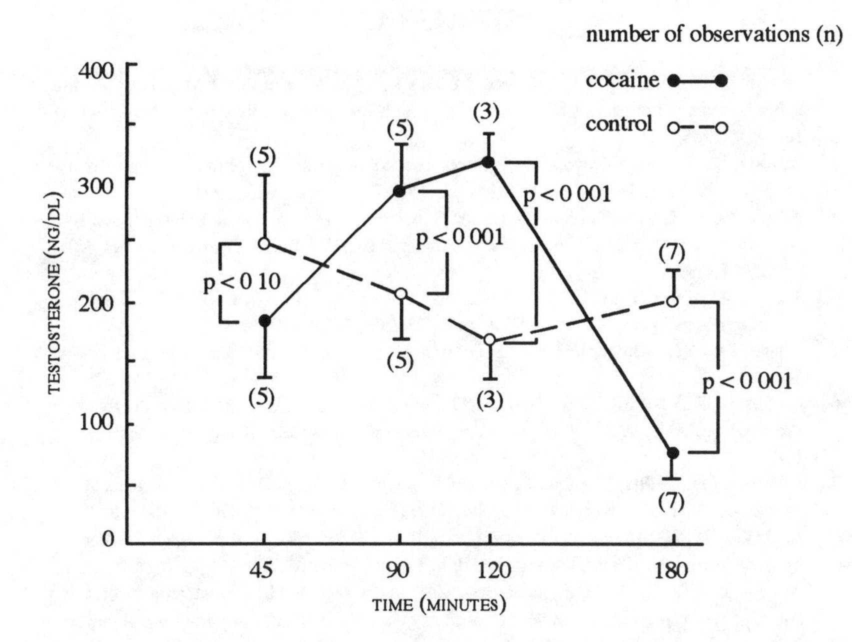

FIGURE 14. Effect of cocaine HCl on plasma testosterone in male albino rats. (Redrawn with permission from Gordon *et al.*[76] See page 300 for citation credit.)

Cocaine and PRL

Studies in male and female rats generally showed a suppressive effect of cocaine on PRL whether this be after a single or multiple administration.[14,15,97,98,119,122,145] In only two studies, one in female rats and one in male rats,[125] was cocaine found without effect on PRL. Suppression of plasma PRL was similarly seen after cocaine injection in male and female monkeys.[106,107]

There are no human data on PRL levels after cocaine infusions, but studies done in drug-free hospitalized cocaine abusers do not present any consistent picture. PRL levels lower,[70] higher[36,42,110,111,149] or no different from[83,96,111,131,147,149] controls have been reported.

ACKNOWLEDGMENTS

This work was supported by NIAAA grants, R01 AA08661 and R01 AA06755 to Mary Ann Emanuele, M.D., Nicholas V. Emanuele, M.D. and Mark R. Kelley, Ph.D.; V.A. Medical Research Service Grant to Nicholas V. Emanuele, M.D., Predoctoral Fellowship from NIH awarded to Margaret Halloran and John Tentler, and a Schweppe Career Development Award to Mark R. Kelley, Ph.D.

REFERENCES

1. **Abbot, S.D., Docherty, K., Roberts, J.L., Tepper, M.A., Chinn, W.W., and Clayton, R.N.**, Castration increases LH subunit mRNA levels in male rat pituitaries, *Endocrinology,* 107, R1, 1985.
2. **Adams, M.L., and Cicero, T.J.**, Effects of alcohol on β-endorphin and reproductive hormones in the male rat, *Alcoholism: Clin. Exp. Res.*, 15(4), 685, 1991.
3. **Adams, M.L., Little, P.J., and Cicero, T.J.**, Alcohol effects rat testicular interstriatal fluid volume and testicular secretion of testosterone and β-endorphin, *J. Pharmacol. Exp. Ther.*, 258, 1008, 1991.
4. **Akane, A., Fakushina S., Shiomi, H., and Fukui, Y.**, Effects of ethanol on testicular steroidogenesis in the rat, *Alcohol Alcohol.*, 23, 203, 1988.
5. **Aliev, N.A.**, Neuroendocrine regulation of immunogenesis in alcoholism, *Prob. Endocrinol.*, 34, 29, 1988.
6. **Anderson, R.A., Willis, B.R., Oswald, C., and Zineweld, L.J.D.**, Male reproductive tract sensitivity to ethanol: A critical overview, *Pharmacol. Biochem. Behav.*, 18 (Suppl.), 305, 1983.
7. **Anderson, R., Willis, B., Oswald, C., and Zaneveld, L.**, The reproductive tract sensitivity to ethanol: A critical overview, *Pharmacol. Biochem. Behav.*, 18 Suppl., 305, 1983.
8. **Andrews, W.V., Maurer, R.A., and Conn, P.M.**, Stimulation of Rat LHβ mRNA levels by gonadotropin releasing hormone, *J. Biol. Chem.*, 263, 13755, 1988.
9. **Andrews, W.V., Hansen, J., Janovick, J.A., and Conn, P.M.**, Gonadotropin-releasing hormone modulation of protein kinase-C activity in perifused anterior pituitary cell cultures, *Endocrinology,* 127, 2393, 1990.
10. **Bannister, P., and Lowosky, M.S.**, Ethanol and hypogonadism, *Alcohol Alcohol.*, 22, 213, 1987.
11. **Belchatz, P.E., Plant, T.M., Nakai, Y., Keogh, E.G., and Knobil, E.**, Hypophysial responses to continuous and intermittent delivery of hypothalamic gonadotropic-releasing hormones, *Science,* 202, 631, 1978.
12. **Bertello, P., Guioli, L., Faffiuolo, R., Veglio, F., Terragnone, C., and Angeli, A.**, Effect of ethanol infusion on the pituitary-testicular responsiveness to gonadotropin releasing hormone and thyrotropin releasing hormone in normal males and in chronic alcoholism with hypogonadism, *J. Endocrinol. Invest.*, 6, 413, 1983.
13. **Berul, C.I., and Harclerode, J.E.**, Effects of cocaine hydrochloride on the male reproductive system, *Life Sci.*, 45, 91, 1989.
14. **Borowsky, B., and Kuhn, C.M.**, Chronic cocaine administration sensitizes behavioral but not neuroendocrine responses, *Brain Res.*, 543, 301, 1991.
15. **Borowsky, B., and Kuhn, C.M.**, Monoamine mediation of cocaine-induced hypothalamo-pituitary-adrenal activation, *J. Pharmacol. Exp. Ther.*, 256, 204, 1991.
16. **Boyden, T.W., and Pamenter, R.W.**, Effects of ethanol on the male hypothalamic-pituitary-gonadal axis, *Endocr. Rev.*, 4, 389, 1983.
17. **Carroll, R., Corrigan, A., Vale, M., and Chin, W.**, Activin stabilizes FBS mRNA levels, *Endocrinology,* 129, 1721, 1991.
18. **Chapin, R.E., Breese, C.R., and Mueller, R.D.**, Possible mechanism of reduction of plasma luteinizing hormone by ethanol, *J. Pharmacol. Exp. Ther.*, 212, 6, 1980.
19. **Charness, M.E., Querimit, L.A., and Henteleff, M.**, Ethanol differentially regulates G proteins in neural cells, *Biochem. Biophys. Res. Commun.*, 155, 138, 1988.
20. **Charness, M., Roa, S., Hu, G., Hilton-Clarke, I., and Henteleff, M.**, Ethanol up regulates $G_{o\alpha}$ in NIE-115 cells, *Alcoholism: Clin. Exp. Res.*, 15, 334 (#136), 1991.
21. **Chin, W.W.**, Hormonal regulation of thyrotropin and gonadotrophin gene expression, *Clin. Res.*, 36(5), 484, 1988.
22. **Chin, W.W., Gharib, S.D., Kowash, P., Schwall, R.A., Corrigan, A.Z., Vale, W., and Carroll, R.S.**, Hormonal regulation of FSH subunit gene expression, *Symp. Reg. Act. FSH,* Evanston, IL, 1990.

23. **Ching, M., Valenca, M., and Negro-Vilar, A.**, Acute ethanol treatment lowers hypophyseal portal plasma LHRH and systemic plasma LH levels in orchidectomized rats, *Brain Res.*, 443, 325, 1988.
24. **Chung, K.W.**, Effect of ethanol on androgen receptors in the anterior pituitary, hypothalamus and brain cortex in rats, *Life Sci.*, 44, 273, 1989.
25. **Cicero, T.J., and Badger, J.R.**, Effects of alcohol on the hypothalamic-pituitary-gonadal axis in the male rat, *J. Pharmacol. Exp. Ther.*, 201, 427, 1977.
26. **Cicero, T.J., Bernstein, D., and Badger, T.**, Effects of acute alcohol administration on reproductive endocrinology in the male rat, *Alcoholism: Clin. Exp. Res.*, 2, 249, 1978.
27. **Cicero, T.J., Meyer, E.R., and Bell, R.D.**, Effects of ethanol on the hypothalamic-pituitary-luteinizing hormone axis and testicular steroidogenesis, *J. Pharmacol. Exp. Ther.*, 208, 210, 1979.
28. **Cicero, T.J., Bell, R.D. Meyer, E.R., and Badger, T.M.**, Ethanol and acetaldehyde directly inhibit testicular steroidogenesis, *J. Pharmacol. Exp. Ther.*, 213, 228, 1980.
29. **Cicero, T.J.**, Neuroendocrinological effects of alcohol, *Ann. Rev. Med.*, 32, 123, 1981.
30. **Cicero, T.J., Newman, K.S., and Meyer, E.R.**, Ethanol-induced inhibitions of testicular steroideogenesis in the male rat. Mechanisms of action, *Life Sci.*, 28, 871, 1981.
31. **Cicero, T., Newman, K.S., Gerity, M., Schmaeker, P.F., and Bell, R.D.**, Ethanol inhibits the naloxone-induced release of LHRH from the hypothalamus of the male rat, *Life Sci.*, 31, 1587, 1982.
32. **Cicero, T.J.**, Alcohol-induced deficits in the hypothalamic-pituitary-luteinizing hormone axis in the male, *Alcoholism: Clin. Exp. Res.*, 6, 207, 1982.
33. **Cicero, T.J., Greenwald, J., Nock, B., and O'Connor, C.**, Castration-induced changes in the response of the hypothalamic-pituitary axis to alcohol in the male rats, *J. Pharmacol. Exp. Ther.*, 252, 456, 1990.
34. **Cobb, C.F., Ennis, M.F., Van Thiel, D.H., Gavaler, J.S., and Lester, R.**, Acetaldehyde and ethanol are testicular toxins, *Gastroenterology*, 75, 958, 1978.
35. **Cobb, C.F., Gavaler, J.S., and Van Thiel, D.H.**, Is ethanol a testicular toxin? *Clin. Toxicol.*, 18, 149, 1981.
36. **Cocores, J.A., Dackis, C.A., and Gold, M.S.**, Sexual dysfunction secondary to cocaine abuse in two patients, *J. Clin. Psychiatry*, 47, 384, 1986.
37. **Conn, P.M., McArdle, C., Andrews, W., and Hucke, W.**, The molecular basis of GnRH action in the pituitary gonadotrope, *Biol. Reprod.*, 36, 17, 1987.
38. **Conn, P.M.**, Molecular mechanism of action of GnRH, in *Symposium on Modes of Action of GnRH and GnRH Analogs*, sponsored by Serono, USA. Scottsdale, AZ, 1991, 4.
39. **Corbani, M., Counis, R., Starzec, A., and Jutisz, M.**, Effect of gonadectomy on pituitary levels of mRNA encoding gonadotropin subunits and secretion of luteinizing hormone, *Mol. Cell. Endocrinol.*, 35, 83, 1984.
40. **Corless, C.L., Matzuk, M.M., Triprayar, V., Ramabhadran, T.V., Krichevsky, A., and Boime, I.**, Gonadotropin β-subunits determine the rate of assembly and the oligosaccharide processing of hormone dimer in transfected cells, *J. Cell Biol.*, 104, 1173, 1987.
41. **Cregler, L.L., and Mark, H.**, Medical complications of cocaine abuse, *N. Engl. J. Med.*, 315, 1495, 1986.
42. **Dackis, C.A., and Gold, M.S.**, New concepts in cocaine addiction: The dopamine depletion hypothesis, *Neurosci. Biobehav. Rev.*, 9, 469, 1985.
43. **Daniell, L.C., and Harris, R.A.**, Ethanol and inositol 1,4,5-triphosphate release calcium from separate stores of brain microsomes, *J. Pharmacol. Exp. Ther.*, 250, 875, 1989.
44. **Dees, W.L., and Kozlowski, G.P.**, Quantitative and qualitative effects of ethanol on the LHRH-LH axis of the rat, *Peptides*, 5 (Suppl. 1), 209, 1984.
45. **Dees, W.L., Skelly, C.W., and Kozlowski, G.P.**, Intragastric cannulation as a method of ethanol administration for neuroendocrine studies, *Alcohol*, 1, 177, 1984.
46. **Dees, W.L., McArthur, N.H., and Harns, P.G.**, Effects of ethanol on LHRH in the male rat: An immunocytochemical study, *Exp. Brain Res.*, 54, 197, 1984.

47. **Dees, W.L., McArthur, N.H, Farr, K.L., Culler, M.D., and Harns, P.G.,** Effects of ethanol on rat LHRH: A study utilizing radioimmunoassay, *Biol. Reprod.,* 28, 1066, 1983.
48. **Deitrich, R.A., and Menez, J.P.,** Investigations of the role of protein kinase C in the acute sedative effects of ETOH, *Alcoholism: Clin. Exp. Res.,* 13, 737, 1989.
49. **DePetrillo, P.B., and Liou, C.S.,** Ethanol exposure increases total protein kinase C activity in human lymphocytes, *Alcoholism: Clin. Exp. Res.,* 17, 351, 1993.
50. **Drouva, S.V., Laplante, G.E., Rerat, E., Enjalbert, A., and Cordon, C.,** Estradiol modulates protein kinase-C activity in the rat pituitary *in vivo* and *in vitro, Endocrinology,* 126, 536, 1990.
51. **Ellingboe, J., and Varanelli, C.G.,** Ethanol inhibits testosterone biosynthesis by direct action on Leydig cells, *Res. Commun. Chem. Pathol. Pharmacol.,* 24, 87, 1979.
52. **Ellingboe, J., Mendelson, J.H., Kuehnle, J.C., Skupny, A.S., and Miller, K.D.,** Effect of acute ethanol ingestion in integrated plasma prolactin levels in normal men, *Pharmacol. Biochem. Behav.,* 12, 297, 1980.
53. **Ellingboe, J., Mendelson, J.H, Kuehnle, J.D., Skupny, A.S.T., and Miller, K.D.,** Effect of acute ethanol ingestion on integrated plasma prolactin in normal men, *Pharmacol. Biochem. Behav.,* 12, 297, 1980.
54. **Ellingboe, J.,** Acute effects of ethanol on sex hormones in non-alcoholic men and women, *Alcohol Alcohol.,* 1 (Suppl.), 109, 1987.
55. **Emanuele, M.A., Tentler, J.J., Kirsteins, L., Emanuele, N.V., Lawrence, A., and Kelley, M.R.,** Effect of "binge" ethanol exposure on growth hormone and prolactin gene expression and secretion, *Endocrinology,* 131, 2079, 1992.
56. **Emanuele, M.A., Tentler, J.J., Reda, D., Kirsteins, L., Emanuele, N., and Lawrence, A.M.,** Failure of *in vitro* ethanol to inhibit LHRH release from the hypothalamus, *Alcohol,* 6, 263, 1989.
57. **Emanuele, M.A., Tentler, J., Reda, D., Kirsteins, L., Emanuele, N.V., and Lawrence, A.M.,** *In vivo* effect of ethanol in release of LHRH and LH in rats, *J. Endocrinol.,* 121, 37, 1989.
58. **Emanuele, M.A., Tentler, J., Emanuele, N.V., Kirsteins, L., and Lawrence, A.M.,** *In vitro* effect of ethanol exposure on basal and GnRH-stimulated LH secretion from pituitary cells, *Endocrinol. Res.,* 15, 383, 1989.
59. **Emanuele, M.A., Tentler, J., Reda, D., Kirsteins, L., Emanuele, N., and Lawrence, A.M.,** The effect of *in vitro* ethanol exposure on LHRH release from perifused rat hypothalami, *Endocrinol. Res.,* 16, 313, 1990.
60. **Emanuele, M.A., Tentler, J., Halloran, M., Emanuele, N.V., and Kelley, M.R.,** *In vivo* effects of acute ETOH on rat α and β luteinizing hormone gene expression, *Alcohol,* 8, 345, 1991.
61. **Emanuele, M.A., Tentler, J.J., Halloran, M.M., Emanuele, N.V., Wallock, L., and Kelley, M.R.,** The effect of acute *in vivo* ethanol exposure on follicle-stimulating hormone transcription and translation, *Alcoholism: Clin. Exp. Res.,* 16(4), 776, 1992.
62. **Eskay, R., Majchrowicz, E., Goldman, M., and Ryback, R.,** Effect of chronic ethanol administration on the hypothalamic-pituitary-gonadal axis, *Fed. Proc.,* 37, 478, 1978.
63. **Esquifino, A.I., Mateos, A., Agrosal, C., Martin, I., Canovas, J.M., and Feimoso, J.,** Time-dependent effects of alcohol on hypothalamic-hypophyseal-testicular function in the rat, *Alcoholism: Clin. Exp. Res.,* 13, 219, 1989.
64. **Farghali, H., Williams, D.S., Gavaler, J., and Van Thiel, D.H.,** Effect of short-term ETOH feeding on rat testes as assessed by 31P NMR spectroscopy, 1H NMR imaging and biochemical methods, *Alcoholism,* 15, 1018, 1991.
65. **Franks, N.P., and Lieb, W.R.,** Are the biological effects of ethanol due to primary interactions with lipids or with proteins? *Alcohol Alcohol.,* 1 (Suppl.), 139, 1987.
66. **Frohman, L.A., and Jamsson, J.O.,** Growth hormone-releasing hormone, *Endocrinol. Res.,* 7, 223, 1986.
67. **Gantt, P.A., Tho, P.T., Bhalla, V.K, McDonough, P.G., Costoff, A., and Mahesh, V.B.,** Effect of ethanol-containing diet upon gonadotrophin receptor depletion in rat testes, *J. Pharmacol. Exp. Ther.,* 223, 848, 1982.

68. **Garcia-Navarro, S., Kalina, M., and Naor, Z.**, Immunocytochemical localization of PKC subtypes in anterior pituitary cells: Colocalization in hormone-containing cells reveals heterogeneity, *Endocrinology*, 129, 2780, 1991.
69. **Gavaler, J.S., Van Thiel, D.H., and Lester, R.**, Ethanol: A gonadal toxin in the mature rat of both sexes, *Alcoholism: Clin. Exp. Res.*, 4, 271, 1980.
70. **Gawin, F.H., and Kleber, H.D.**, Neuroendocrine findings in chronic cocaine abusers: A preliminary report, *Br. J. Psychiatry*, 147, 569, 1985.
71. **Gayer, G., Gordon, A., and Miles, M.**, Ethanol increases tyrosine hydroxylase gene expression in NIE-115. Neuroblastoma cells, *J. Biol. Chem.*, 266, 22279, 1991.
72. **Gharib, S.D., Bower, S.M., Need, L.R., and Chin, W.W.**, Regulation of rat luteinizing hormone subunit messenger ribonucleic acids by gonadal steroid hormones, *J. Clin. Invest.*, 77, 582, 1986.
73. **Ghosh, P., Chirtel, S.J., and Lakshman, M.R.**, Effect of chronic ethanol on apolipoprotin E synthesis and glycosylation in rats, *Alcoholism: Clin. Exp. Res.*, 15, 725, 1991.
74. **Gonzales, R.A., and Crews, F.T.**, Effects of ethanol *in vivo* and *in vitro* on stimulated phosphoinositide hydrolysis in rat cortex and cerebellum, *Alcoholism: Clin. Exp. Res.*, 12, 94, 1988.
75. **Gordon, G.G., Altman, K., Southern, A.L., Rubin, E., and Lieber, C.S.**, The effects of alcohol (ethanol) administration on sex hormone metabolism in normal men, *N. Engl. J. Med.*, 295, 793, 1976.
76. **Gordon, L.A., Mostofsky, D.I., and Gordon, G.G.**, Changes in testosterone levels in the rat following intraperitoneal cocaine HCl, *Int. J. Neurosci.*, 11, 139, 1980.
77. **Gordon, A.S., Collier, K., and Diamond, I.**, Ethanol regulation of adenosine receptor stimulated cAMP levels in a clonal neural cell line: An *in vitro* model of cellular tolerance to ETOH, *Proc. Natl. Acad. Sci. USA*, 83, 2105, 1986.
78. **Hoek, J.B., Thomas, A.P., Rubin, R., and Rubin, E.** Ethanol-induced mobilization of calcium by activation of phosphoinisitide-specific phospholipase C in intact hepatocytes, *J. Biol. Chem.*, 262, 682, 1987.
79. **Hoek, J.B., and Rubin, E.**, Alcohol and membrane-associated signal transduction, *Alcohol Alcohol.*, 25, 143, 1990.
80. **Hoek, J.B., and Higashi, K.**, Effects of alcohol on polyphosphoinositide-mediated intracellular signaling, *Ann. NY Acad. Sci.*, 625, 375, 1992.
81. **Hoffman, P.L., Moses, F., Luthin, G.R., and Tabakoff, B.**, Acute and chronic effects of ethanol on receptor-mediated phosphatidylinositol 4,5 biphosphate breakdown in mouse brain, *J. Pharmacol. Exp. Ther.*, 30, 13, 1986.
82. **Hoffman, P.L., and Tabakoff, B.**, Ethanol and guanine nucleotide binding proteins: A selective interaction, *FASEB J.*, 4, 2612, 1990.
83. **Hollander, E., Nunes, E., DeCaria, C.M., Quitkin, F.M., Cooper, T., Wager, S., and Klein, D.F.**, Dopaminergic sensitivity and cocaine abuse: response to apomorphine, *Psychiatry Res.*, 33, 161, 1990.
84. **Horn, F., Bilezikijian, L., Perrin, M., Bosman, M., Windle, J., Huber, K., Blount, A., Hille, B., Vale, W., and Mellon, P.**, Intracellular responses to gonadotropin-releasing hormone in a clonal cell line of the gonadotrope lineage, *Mol. Endocrinol.*, 5, 347, 1991.
85. **Hoshina, H., and Boime, I.**, Combination of rat lutropin subunits occurs early in the secretory pathway, *Proc. Natl. Acad. Sci. USA*, 79, 7649, 1982.
86. **Ida, Y., Tsujimaru, S., Nakamura, K., Shirao, I., Mukasa, H., Egami, H., Nakazawa, Y.**, Effects of acute and repeated alcohol ingestion on hypothalamic-pituitary gonadal and adrenal functioning in normal males, *Drug Alcohol Depend.*, 31, 57, 1992.
87. **Johnson, M.S., Mitchell, R., and Fink, G.**, The role of protein kinase C in LHRH-induced LH and FSH release and LHRH self-priming in anterior pituitary glands *in vitro*, *J. Endocrinol.*, 116, 231, 1987.
88. **Johnston, D.E., Chao, Y.B., Gavaler, J.S., and Van Thiel, D.H.**, Inhibition of testosterone synthesis by ethanol and acetaldehyde, *Biochem. Pharmacol.*, 30, 1827, 1981.

89. **Huckle, W.R., and Conn, P.M.**, Molecular mechanisms of GnRH action: The effector system, *Endocrinol. Rev.*, 9, 387, 1988.
90. **Kaetzel, D.M., and Nilson, J.H.**, Methotrexate-induced amplification of the bovine lutropin genes in Chinese hamster ovary cells, *J. Biol. Chem.*, 263, 6344, 1988.
91. **Kaetzel, D.M., Virgin, J., Clay, C., and Nilson, J.**, Disruption of N-linked glycosylation by site directed mutagenesis dramatically increases its intracellular stability but does not affect biological activity of the secreted heterodimer, *Mol. Endocrinol.*, 3, 1765, 1989.
92. **Keene, J.L., Matzuk, M.M., Otani, T., Fauser, B.C.J.M., Galway, A.B., Hsueh, A.J.W., and Boime, I.**, Expression of biologically active human follitropin in Chinese hamster ovary cells, *J. Biol. Chem.*, 264, 4769, 1989.
93. **Krane I., Spindel, E., and Chin, W.W.**, Thyroid hormone decreases the stability and the poly A tract length of rat-thyrotropin β subunit mRNA, *Mol. Endocrinol.*, 5, 469, 1991.
94. **Lakshman, M.R., Ghosh, P., Chirtel, S.J., and Sapp, R.**, Effect of chronic ethanol in hepatic synthesis and glycosylation of transferin, *Alcoholism: Clin. Exp. Res.*, 15, (#195), 1991.
95. **Lalloz, M.R., Detta, A., and Clayton, R.N.**, Gonadotropin-releasing hormone desensitization preferentially inhibits expression of the LH-β gene *in vivo, Endocrinology*, 122, 1689, 1988.
96. **Lee, M.A., Bowers, M.M., Nash, J.F., and Meltzer, H.Y.**, Neuroendocrine measures of dopaminergic function in chronic cocaine users, *Psychiatry Res.*, 33, 151, 1990.
97. **Levy, A.D., Li, Q., Alvarez Sanz, M.D., Rittenhouse, P.A., Kerr, J.E., and Van de Kar, L.D.**, Neuroendocrine responses to cocaine do not exhibit sensitization following repeated cocaine exposure, *Life Sci.*, 51, 887, 1992.
98. **Levy, A.D., Rittenhouse, P.A., Li, Q., Bonadonna, A.M., Alvarez Sanz, M.C., Kerr, J.E., Bethea, C.L., and Van de Kar, L.D.**, Repeated injections of cocaine inhibit the serotonergic regulation of prolactin and renin secretion in rats, *Brain Res.*, 580, 6, 1992.
99. **Lin, T.A., Navidi, M., James, W., Lin, T.-N., and Sun, G.**, Effects of acute ethanol administration on polyphosphoinositide turnover and levels of inositol 1,4,5- triphosphate in mouse cerebrum and cerebellum, *Alcoholism: Clin. Exp. Res.*, 17, 401, 1993.
100. **Lloyd, C.W., and Williams, R.H.**, Endocrine changes associated with Laennec's cirrhosis of the liver, *Am. J. Med.*, 4, 315, 1948.
101. **Lucchi, L., Govoni, S., Battaini, F., Pasinetti, G., and Trabucchi, M.**, Ethanol administration *in vivo* alters calcium ions control in rat striatum, *Brain Res.*, 332, 376, 1985.
102. **Mangalam, H., Albert, V., Ingraham, H., Kapiloff, M., Wilson, L., Nelson, C., Elsholtz, H., and Rosenfeld, M.G.**, A pituitary POU domain, Pit-1, activates both GH and prolactin promoter transcription, *Genes Dev.*, 3, 946, 1989.
103. **Marshall, J., Haisenleder, D.L., Dalkin, A., Paul, S., and Ortolano, G.**, Regulation of gonadotropin subunit gene expression, in *Neuroendocrine Regulation of Reproduction*, Samuel, S., Ed., Yen and Wiley Vale, Sorono Symposium SYMPOSIA, 1990.
104. **Mateos, A., Fernoso, J., Agrosal, C., Martin, J., Pez-Bouza, J., Tresqueven, J.A., and Esquifino, A.I.**, Effect of chronic consumption of alcohol on the hypothalamo-pituitary-testicular axis in the rat, *Rev. Esp. Fisiol.*, 43, 33, 1987.
105. **Matzuk, M.M., Kornmeier, C.M., Whitfield, G.K., Kourides, I.A., and Boime, I.**, The glycoprotein β-subunit is critical for secretion and stability of the human thyrotropin β-subunit, *Mol. Endocrinol.*, 2, 95, 1988.
106. **Mello, N.K., Mendelson, J.H., Drieze, J., and Kelley, M.**, Cocaine effects on luteinizing hormone-releasing hormone-stimulated anterior pituitary hormones in female Rhesus monkey, *J. Clin. Endocrinol. Metab.*, 71, 1434, 1990.
107. **Mello, N.K., Mendelson, J.H., Drieze, J., and Kelly, M.**, Acute effects of cocaine on prolactin and gonadotropins in female Rhesus monkey during the follicular phase of the menstrual cycle, *J. Pharmacol. Exp. Ther.*, 254, 815, 1990.
108. **Mendelson, J.H., Mello, N.K., and Ellingboe, J.**, Effects of acute alcohol intake on the pituitary-gonadal hormones in normal human males, *J. Pharmacol. Exp. Ther.*, 202, 676, 1977.

109. **Mendelson, J.H., Ellingboe, J., Mello, N.K., and Kuehnle, J.,** Effects of alcohol on plasma testosterone and luteinizing hormone levels, *Alcoholism: Clin. Exp. Res.*, 2, 255, 1978.
110. **Mendelson, J.H., Siew, M.D., Teoh, S.K., Lange, U., Mellow, N.K., Weiss, R., Skupny, A., and Ellingboe, J.,** Anterior pituitary, adrenal, and gonadal hormones during cocaine withdrawal, *Am. J. Psychiatry*, 145, 1094, 1988.
111. **Mendelson, J.H., Mello, N.K., Koon, Teoh, S., Ellingboe, J., and Cochin, J.,** Cocaine effects on pulsatile secretion of anterior pituitary, gonadal, and adrenal hormones, *J. Clin. Endocrinol. Metab.*, 69, 1256, 1989.
112. **Nilson, J.H., Nejedlik, M.T., Virgin, J.B., Crowder, M.E., and Nett, T.M.,** Expression of subunit and LHβ genes in the bovine anterior pituitary, *J. Biol. Chem.*, 258, 12087, 1983.
113. **Orpana, A.K., H:Ark:Onen, M., and Eriksson, C.J.,** Ethanol-induced inhibition of testosterone biosynthesis in rat Leydig cells: Role of mitochondrial substrate and citrate, *Alcohol Alcohol.*, 25, 499, 1990.
114. **Orpana, A.K., Vrava, M.M., Vihko, R.K., H:Ark:Onen, M., and Eriksson, C.J.,** Role of ethanol metabolism in the inhibition of testosterone biosynthesis in rats *in vivo*: Importance of gonadotrophin stimulation, *J. Steroid Biochem. Mol. Biol.*, 37, 273, 1990.
115. **Orpana A.K., Orava, M.M., Vihkork, H:Ark:Onen, M., and Eriksson, C.J.,** Ethanol-induced inhibition of testosterone biosynthesis in rat Leydig cells: Role of L-glutamate and pyruvate, *J. Steroid Biochem. Mol. Biol.*, 36, 473, 1990.
116. **Papavasiliou, S.S., Zmeili, S., Herbon, L., Duncan-Weldon, J., Marshall, J.C., and Landefeld, T.D.,** α- and β-LH mRNA of male and female rats after castration: Quantitation using an optimized RNA dot blot hydribization assay, *Endocrinology*, 119, 691, 1986.
117. **Peters, B.P., Krzesicki, R.F., Hartle, R.J., Perini, F., and Ruddon, W.R.,** A kinetic comparison of the processing and secretion of the α β dimer and the uncombined α and β subunits of chorionic gonadotropin synthesized by human choriocarcinoma cells, *J. Biol. Chem.*, 259, 15123, 1984.
118. **Phipps, W.E., Lukas, S.E., Mendelson, J.H., Ellingboe, J., Palmieri, S.L., and Schiff, J.,** Acute ethanol administration enhances plasma testosterone following gonadotropin stimulation in men, *Psychoneuroendocrinology*, 12, 459, 1987.
119. **Pilotte, N.S., Sharper, L.G., and Dax, E.M.,** Multiple, but not acute, infusions of cocaine alter the release of prolactin in male rats, *Brain Res.*, 512, 107, 1990.
120. **Pohl, C.R., Guilinger, R.A., and Van Thiel, D.H.,** Inhibitory action of ethanol on LH secretion of rat anterior pituitary cells in culture, *Endocrinology*, 120, 849, 1987.
121. **Rabe, C., and Weight, F.,** Effects of ethanol on neurotransmitter release and intracellular free calcium in PC12 cells, *J. Pharmacol. Exp. Ther.*, 244, 417, 1988.
122. **Ravitz, A.J., and Moore, K.E.,** Effects of amphetamine, methylphenidate and cocaine on serum prolactin concentrations in the male rats, *Life Sci.*, 21, 267, 1977.
123. **Redmond, G.P.,** Effect of ethanol on pulsatile gonadotropin secretion in the male rat, *Alcoholism: Clin. Exp. Res.*, 4, 226, 1980.
124. **Renau-Piqueres, J., Sancho-Tello, M., Cerevellera, R., and Guerri, C.,** Prenatal exposure to ethanol alters the synthesis and glycosylation of proteins in fetal hepatocytes, *Alcoholism: Clin. Exp. Res.*, 13, 817, 1989.
125. **Rivier, C., Imaki, T., and Vale, W.,** Prolonged exposure to ethanol: Effect on CRF mRNA levels and CRF and stress-induced ACTH secretion in the rat, *Brain Res.*, 520, 1, 1990.
126. **Rothschild, M.A., Oratz, M., and Schreiber, S.S.,** Effects of ethanol on protein synthesis, *Ann. N.Y. Acad. Sci.*, 492, 233, 1987.
127. **Rubin, R., and Hoek, J.,** Ethanol-induced stimulation of PI turnover and calcium influx in isolated hepatocytes, *Biochem. Pharmacol.*, 37, 2461, 1988.
128. **Rubin, R., and Hoek, J.,** Alcohol-induced stimulation of phospholipase C in human platelets requires G protein activation, *J. Biochem.*, 254, 147, 1988.
129. **Salonen, I., and Huhtaniemi, I.,** Effect of chronic ethanol diet on pituitary-testicular function of the rat, *Biol. Res.*, 42, 55, 1990.

130. **Salonen, I., Pakarinen, P., and Huhtaniemi, I.**, Effect of chronic ethanol diet on expression of gonadotropin genes in the male rat, *J. Pharmacol. Exp. Ther.*, 260(2), 463, 1992.
131. **Satel, S.L., Price, L.H., Palumbo, J.M., McDougle, C.J., Krystal, J.H., Gawin, F., Charney, D.S., Heninger, G.R., and Kleber, H.D.**, Clinical phenomenology and neurobiology of cocaine abstinence: A prospective inpatient study, *Am. J. Psychiatry*, 148, 1712, 1991.
132. **Sato, N., Wang, X.B., and Green, M.A.**, Hormone secretion stimulated by ethanol-induced cell swelling in normal rat adenohypophyseal cells, *Am. J. Physiol.*, 260, 946, 1991.
133. **Schade, R.R., Bonner, G., Goy, V.L., and Van Thiel, D.H.**, Evidence for a direct inhibitory effect of ethanol upon gonadotrophin secretion at the pituitary level, *Alcoholism: Clin. Exp. Res.*, 7, 150, 1983.
134. **Scher, P.M., Almirez, R.G., Steger, R.W., and Smith, C.G.**, The effects of cocaine on reproductive hormones in the primate, *Pharmacology*, 24, 185A, 1981.
135. **Schulz, R., Wuster, M., Duka, T., and Herz, A.**, Acute and chronic ethanol treatment changes endorphin levels in brain and pituitary, *Psychopharmacology*, 68, 221, 1980.
136. **Seilicovich, A., Rettori, V., Koch, O.R., Duvilinski, B., Diaz, M.C., and Debeljuk, L.**, The effect of acute and chronic ethanol administration on prolactin secretion in male rats, *J. Androl.*, 3, 344, 1982.
137. **Seilicovich, A., Duvilonski, B.H., Debeljuk, L., Diaz, M.C., Munoz-Maines, V., and Rettori, V.**, The effect of ethanol on prolactin secretion *in vitro*, *Life Sci.*, 35, 1931, 1984.
138. **Seilicovich, A., Rubio, M., Duvilonski, B., Munoz-Maines, V., and Rettori, V.**, Inhibition of naloxone of the rise in hypothalamic dopamine and serum prolactin produced by ethanol, *Psychopharmacology*, 87, 461, 1985.
139. **Uddin, S., Emanuele, M.A., Emanuele, N., and Kelley, M.R.**, The effect of *in vitro* ethanol exposure on pituitary LH gonadotropin synthesis and secretion, *Endocrinol. Res.*, 20, 2, 1994.
140. **Shupnik, M., Gharib, S., and Chin, W.W.**, Estrogen suppresses rat gonadotropin gene transcription *in vivo*, *Endocrinology*, 122, 1842, 1988.
141. **Shupnik, M.**, Effects of gonadotropin-releasing hormone on rat gonadotropin gene transcription *in vitro*: Requirement for pulsatile administration for luteinizing hormone β-gene stimulation, *Mol. Endocrinol.*, 4, 1444, 1990.
142. **Slomiany, A., Grzelinska, E., Tamura, S., and Slomiany, B.L.**, Effect of ethanol on the release and transport of apoprotein from ER to Golgi, *Alcoholism: Clin. Exp. Res.*, 15, 329 (Abstr. 108), 1991.
143. **Southern, A.L., Gordon, G.G., and Olivo, J.**, Androgen metabolism in cirrhosis of the liver, *Metabolism*, 22, 695, 1973.
144. **Soyka, M., Gorig, M., and Naber, D.**, Serum prolactin increase induced by ethanol—a dose dependent effect not related to stress, *Psychoneuroendocrinology*, 16, 441, 1991.
145. **Steger, R.W., Silverman, A.Y., Johns, A., and Asch, R.H.**, Interactions of cocaine and D^9-tetrahydrocannabinol with the hypothalamic-hypophysial axis of the female rat, *Fertil. Steril.*, 36, 567, 1981.
146. **Sun, G.Y., Huang, H.M., Chandrasekhar, R., Lee, D., and Sun, A.**, Effects of chronic ethanol administration on rat brain phospholipid metabolism, *J. Neurochem.*, 48, 974, 1987.
147. **Swartz, C.M., Breen, K., and Leone, F.**, Serum prolactin levels during extended cocaine abstinence, *Am. J. Psychiatry*, 147, 777, 1990.
148. **Symonds, A.M., and Marks, V.**, The effects of alcohol on weight gain and the hypothalamic-pituitary-gonadotropin axis in the maturing male rat, *Biochem. Pharmacol.*, 24, 955, 1975.
149. **Teoh, S.K., Mendelson, J.H., Mello, N.K., Weiss, R., McElroy, S., and McAfee, B.**, Hyperprolactinemia and risk for relapse of cocaine abuse, *Biol. Psychiatry*, 28, 824, 1990.
150. **Tuma, D.L., Mailliard, M.E., Casey, C.A., Volentine, G.D., and Sorell, M.F.**, Ethanol-induced alterations of plasma membrane assembly in the liver, *Biochim. Biophys. Acta*, 856, 571, 1986.
151. **Tuma, D.L., and Sorell, M.F.**, Effect of ethanol in protein trafficking in the liver, *Semin. Liver Dis.*, 8, 69, 1988.

152. **Van Thiel, D.H., Lester, R., and Sherins, R.J.,** Hypogonadism in alcoholic liver disease: Evidence for a dual defect, *Gastroenterology,* 67, 1188, 1974.
153. **Van Thiel, D.H., Lester, R., and Vaitukaitis, J.,** Evidence for a defect in pituitary secretion of LH in chronic alcoholic men, *J. Clin. Endocrinol. Metab.,* 47, 499, 1978.
154. **Van Thiel, D.H., and Lester, R.** Further evidence for hypothalamic-pituitary dysfunction in alcoholic men, *Alcoholism: Clin. Exp. Res.,* 2, 265, 1978.
155. **Van Thiel, D.H., Gavaler, J.S., Cobb, C.H., Sherins, R.J., and Lester, R.,** Alcohol-induced testicular atrophy in the adult male rat, *Endocrinology,* 105, 888, 1979.
156. **Van Thiel, D.H.,** Ethanol: Its adverse effect upon the hypothalamic-pituitary-gonadal axis, *J. Lab. Clin. Med.,* 101, 21, 1983.
157. **Vaubourdolle, M., Guechot, J., Chazouilleres, O., and Poupon Giboudeau, J.,** Effect of dihydrotestosterone on the rate of ethanol elimination in healthy men, *Alcoholism: Clin. Exp. Res.,* 15, 238, 1991.
158. **Weiss, J., Jameson, J.C., Burrin, J.M., and Crowley, W.F.,** Divergent responses of gonadotropin subunit messenger RNAs to continuous versus pulsatile gonadotropin-releasing hormone *in vitro, Mol. Endocrinol.,* 4, 557, 1990.
159. **Whelen, J.P., Hoffman, P., and Tabakoff, B.,** Effects of chronic ethanol exposure on G-proteins in mouse brain membranes, *Alcoholism: Clin. Exp. Res.,* 15, 333 (#129), 1991.
160. **Widenius, T.V., Eriksson, C.J., Ylikahri, R.H., and H:Ark:Onen, M.,** Inhibition of testosterone synthesis by ethanol: Role of luteinizing hormone, *Alcohol,* 6, 241, 1989.
161. **Ylikahri, R.H., Huttunen, M.O., H:Ark:Onen, M., Leino, T., Helenius, T., Liewendahl, K., and Karonen, S.L.,** Acute effects of alcohol on anterior pituitary secretion of the tropic hormones, *J. Clin. Endocrinol. Metab.,* 46, 715, 1978.
162. **Ylikahri, R.H., Huttinen, M.O., and H:Ark:Onen, M.,** Hormonal changes during alcohol intoxication and withdrawal, *Pharm. Biochem. Behav.,* 13 (Suppl 1), 131, 1980.

FIGURE CREDITS

Figure 1 taken from Cicero, T.J., Meyer, E.R., and Bell, R.D., Effects of ethanol on the hypothalamic-pituitary-luteinizing hormone axis and testicular steroidogenesis, *J. Pharmacol. Exp. Ther.,* 208, 210, 1979, with kind permission from the American Society for Pharm. and Exp. Therapeutics, Williams & Wilkins.

Figure 2 redrawn from Gavaler, J.S., Urso, T., and VanThiel, D.H., Ethanol: Its adverse effects upon the hypothalamic-pituitary-gonadal axis, *Alcohol,* 4, 97, 1983, with kind permission from Pergamon Press Ltd., Headington Hill Hall, Oxford 0X3 0BW, UK.

Figure 3 redrawn from Adams, M.L., and Cicero, T.J., Effects of alcohol on β-endorphin and reproductive hormones in the male rat, *Alcoholism: Clin. Exp. Res.,* 15(4), 685, 1991, with kind permission from the American Society for Pharm. and Exp. Therapeutics, Williams & Wilkins.

Figures 4, 5, 6, and **7** taken from Emanuele, M.A., Tentler, J., Halloran, M., Emanuele, N.V., and Kelley, M.R., *In vivo* effects of acute ETOH on rat α and β luteinizing hormone gene expression, *Alcohol,* 8, 345, 1991, with kind permission from Pergamon Press Ltd., Headington Hill Hall, Oxford 0X3 0BW, UK.

Figure 8 taken from Salonen, I., Pakarinen, P., and Huhtaniemi, I., Effect of chronic ethanol diet on expression of gonadotropin genes in the male rat, *J. Pharmacol. Exp. Ther.,* 260(2), 463, 1992, with kind permission from the American Society for Pharm. and Exp. Therapeutics.

Figure 9 taken from Emanuele, M.A., Tentler, J.J., Halloran, M.M., Emanuele, N.V., Wallock, L., and Kelley, M.R., The effect of acute *in vivo* ethanol exposure on follicle-stimulating hormone transcription and translation, *Alcoholism: Clin. Exp. Res.*, 16(4), 776, 1992, with kind permission from the Research Society on Alcoholism.

Figure 10 taken from Ching, M., Valenca, M., and Negro-Vilar, A., Acute ethanol treatment lowers hypophyseal portal plasma LHRH and systemic plasma LH levels in orchidectomized rats, *Brain Res.*, 443, 325, 1988, with kind permission from Elsevier Science Publishers B.V.

Figure 13 taken from Emanuele, M.A., Tentler, J.J., Kirsteins, L., Emanuele, N.V., Lawrence, A., and Kelley, M.R., Effect of "binge" ethanol exposure on growth hormone and prolactin gene expression and secretion, *Endocrinology*, 131, 2079, 1992, with kind permission from the Endocrine Society.

Figure 14 taken from Gordon, L.A., Mostofsky, D.I., and Gordon, G.G., Changes in testosterone levels in the rat following intraperitoneal cocaine HCl, *Int. J. Neurosci.*, 11, 139, 1980, with kind permission from Gordon and Breach Science Publishers.

Chapter 14

EFFECTS OF ETHANOL ON THE REPRODUCTIVE NEUROENDOCRINE AXIS OF PREPUBERTAL AND ADULT FEMALE RATS

W.L. Dees, J.K. Hiney and C.L. Nyberg

TABLE OF CONTENTS

0-8493-2451-3/95/$0.00+.50

ABBREVIATIONS

cAMP–cyclic adenosine monophosphate
CGRP–calcitonin gene-related peptide
CNS–central nervous system
D_1–first diestrus
E_2–estradiol
ETOH–ethanol
FSH–follicle-stimulating hormone
GH–growth hormone
GRF or GHRH–growth hormone-releasing hormone
LH–luteinizing hormone
LHRH–luteinizing hormone-releasing hormone
ME–median eminence
NE–norepinephrine
NMA–N-methyl-DL-aspartic acid
NMDA–N-methyl-D-aspartic acid
NPY–neuropeptide Y
P–progesterone
PGE_2–prostaglandin E_2
PMSG–pregnant mare serum gonadotropin
PRL–prolactin
SP–substance P
SRIF–somatostatin
T–testosterone
VIP–vasoactive intestinal peptide
VO–vaginal opening

INTRODUCTION

The central nervous system (CNS) plays a crucial role in bringing together events that lead to the onset of puberty by controlling both the anterior pituitary function via the secretion of hypothalamic hormones and the ovarian function via pituitary hormone secretions and direct neural inputs. A drug that is capable of affecting any one of these portions of the reproductive axis during the late juvenile and peripubertal periods of development could cause serious manifestations, which may affect the otherwise normal series of events leading to the onset of puberty. Based upon what we know about the effects of ethanol (ETOH) in adult humans and laboratory animals, this is one such drug that has that potential, and many surveys indicate an alarming increase in its abuse, especially by adolescents. Although the potential effects of ETOH on human adolescent development is acknowledged, virtually nothing is known at the present time about the magnitude of any specific physiological effects of the drug on the sexual maturation process. Conversely, researchers have known since the early 1980s that exposure to ETOH can cause delayed puberty in laboratory animals. Until recently, however, little was known of the extent of these effects or of ETOH's mechanism(s) of action.

During the last few years, we have used both immature and mature female rats to further investigate how serious the effects of ETOH are and what the physiological mechanism(s) of its action might be. In order to better understand the basis of ETOH's effects on female maturation and reproductive processes, we have utilized experiments designed around both *in vivo* and *in vitro* approaches. Additionally, we have sought to assess the effects of ETOH on hormonal secretions at all three levels of the hypothalamic-pituitary-ovarian axis. Thus, this chapter demonstrates an account of our efforts, as well as others, to understand the detrimental effects and mechanisms of action of ETOH exposure on the onset of female puberty, as well as on its detrimental effects following attainment of sexual maturity in the rat.

ONSET OF FEMALE PUBERTY IN THE RAT

Before discussing the effects of ETOH on the pubertal process, it is necessary, in the context of this report, to review briefly some of the most crucial events leading to the onset of female puberty. For more in-depth discussions regarding factors controlling the onset of puberty, see reviews by Ojeda *et al.*,[94] Odell,[87] and chapter 2 of this volume.

The interval between birth and puberty has been categorized into four phases: a neonatal period consisting of the first week of life; an infantile period from days 7 through 21; a juvenile period from days 21 to approximately 32; and a peripubertal period signified as the days encompassing the first ovulation.[88] The capacity of the rat hypothalamus to release luteinizing hormone-releasing hormone (LHRH) has been shown *in vitro* to increase during the second week of life,[65] and again during the juvenile period.[11] At the end of the juvenile period and the beginning of the peripubertal period, the pulsatile release of luteinizing hormone (LH) becomes more prominent in the afternoon than during the morning hours.[10,119] This period of activation, evidently resulting from the synchronization of pulsatile discharges of LHRH, has been determined to be a centrally driven, gonadal-independent event[121] and reflects the final stages of hypothalamic maturation of the LHRH releasing system to a mode of release that is essential for normal pubertal development.

The exact signal that initiates this centrally derived event is not presently known; however, it has been hypothesized that this change in the pattern of LHRH release may be related to the removal of a hypothalamic inhibitory tone responsible for suppressing LHRH secretion and/or to the completion of the development, or to the activation of neurotransmitter and excitatory inputs that contribute to enhanced secretion of the peptide during the peripubertal period. In this regard, at the beginning of the peripubertal period, a less inhibitory opioid effect has been shown,[14] as well as an increase in hypothalamic norepinephrine (NE) and dopamine turnover.[4,106,128] In addition to the well-known effects of NE on cyclic LH release in the adult, this neurotransmitter appears to participate in the first preovulatory surge of LH.[110] Furthermore, it is important to note that the activation of a specific class of excitatory amino acid receptors in the brain, namely N-methyl-D-aspartic acid (NMDA) receptors, plays a role in the onset of female puberty in both the rat[120,122] and monkey.[56] Additionally, recent evidence has been presented demonstrating a possible involvement of neuropeptide Y (NPY) in the pubertal process.[80]

Along with the above-mentioned events, there are also increased secretions of both prolactin (PRL) and growth hormone (GH) during pubertal development. PRL has been shown to advance puberty by acting at the level of the CNS,[126] as well as at the level of the ovary.[2,3] GH secretion is also necessary for the normal progression of events leading to the onset of puberty, as shown by the fact that suppression of GH release causes delayed ovarian maturity, resulting in delayed puberty.[5]

At this point it is important to stress the importance of the timely maturation of the ovary. The ovary not only grows and matures under the influence of the gonadotropins, PRL and GH, but there is now evidence to show that the maturing ovary is also influenced by direct neural inputs. In recent years, we have described the immunocytochemical localization of adrenergic- and peptidergic-containing nerve fibers in the prepubertal rat ovary. We have also discussed the putative functions of these nerves with regard to the onset of female puberty.[7,37,40,70,76]

As the pubertal process progresses and the pulsatile mode of LH release develops into a well-established AM-low amplitude, PM-high amplitude pulse pattern and PRL as well as GH levels rise, an increased amount of estradiol (E_2) is produced by the ovary. As the serum levels of E_2 become elevated, the central component of E_2-positive feedback is activated. During this phase of development there is an increase in hypothalamic capacity to synthesize prostaglandin E_2 (PGE_2),[89] a compound known to mediate the NE-induced release of LHRH in peripubertal rats.[93] E_2 appears to not only increase PGE_2 production prior to the first preovulatory surge of gonadotropins[31] but also aid in neuronal growth and maturation of synaptic connections.[74,82] After E_2 reaches a sufficient level,[12] it is then able to act at the level of the hypothalamus to produce the preovulatory LHRH and gonadotropin surges and subsequently, first ovulation; an event designating attainment of reproductive maturity.

CHRONIC EFFECTS OF ETOH ON THE FEMALE PUBERTAL PROCESS

Animal Model and ETOH Administration

In recent years we have used a novel approach of administering ETOH for studying its effects on the onset of female puberty.[44] Because of its usefulness, we felt it important to include here a brief description, as well as point out the advantages of this methodology. Briefly, for this procedure 23-day-old rats were first implanted with a permanent gastric cannula by a procedure described previously.[46] When 28 days old, the rats were divided into three groups, all matched closely in weight. Group I consisted of animals that received a 5% ETOH liquid diet (Bio-Serve, Frenchtown, NJ), in which ETOH provided 36% of the total caloric intake. Group II consisted of animals that received the companion control liquid diet in which dextramaltose was isocalorically substituted for the ETOH. Group III consisted of additional control animals that had cannula implants but received lab chow and water, *ad libitum*. The following day, animals in groups I and II began receiving their respective liquid-diet regimen. These diets were administered such that a portion of the respective diet was injected via the intragastric cannula dispersed equally over the lights-on period, with the remainder being provided *ad libitum* (bottle) during the lights-off period. Further details of this diet regimen and of its use have been described previously.[44] This method of ETOH administration has several advantages. As shown in Figure 1a, animals receiving the liquid diet grow at the same rate as the animals receiving the lab chow diet. Additionally, this method not only ensures that all liquid diet-fed animals receive the same amount of diet and, hence, the same number of calories per 24-hour period but also ensures that all of the ETOH-treated animals receive the same amount of ETOH. Thus, we feel this regimen is particularly suited for puberty-related studies.

Several additional measures were also useful to more precisely determine ETOH's effects on the pubertal process. In this regard, body weights were recorded daily and, beginning on day 32, the rats were inspected for vaginal opening (VO). After VO, changes in vaginal cytology were determined daily. Using these and other criteria reported previously,[1] the animals killed throughout the peripubertal period were then classified into the different phases of pubertal development. Importantly, by using this overall experimental design we have been able to circumvent variables associated with diet consumption

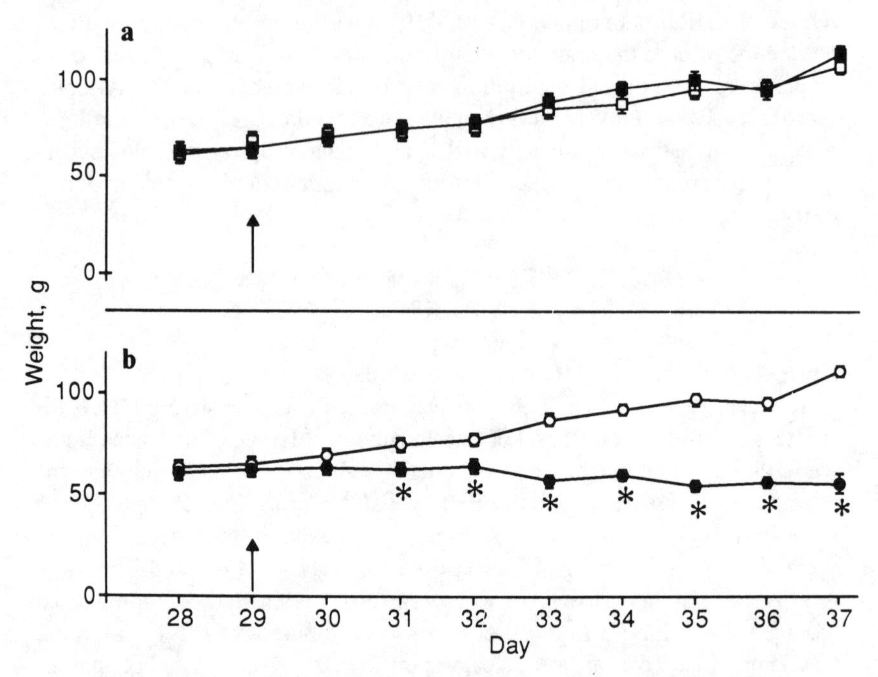

FIGURE 1. Effects of the respective diet regimens on mean (± SEM) body growth. (*a*) Note that there are no differences between the animals fed the control liquid diet (□) and those fed laboratory chow (■) with regard to daily weight gain. (*b*) The combined control animals (○) showed a steady increase in weight throughout the study, whereas the ETOH-treated animals (●) only maintained the weight at which they started their diet regimen. Beginning on day 31, the body weights were lower in the ETOH-treated animals. $^*p < 0.001$, ETOH vs. same-day control; arrow represents beginning day of respective diet. (Reprinted from Dees *et al.*[44] with permission from S. Karger AG, Basel, Switzerland.)

and nutrition, as well as with animal growth rates and the possibility of analyzing indices of puberty from animals of the same age but actually in different phases of the maturation process. Utilization of the above experimental protocol has resulted in our description of several specific changes in puberty-related events, some of which are described in the following section.

ETOH Affects Growth, Pituitary Hormone Secretion and the Onset of Puberty

About 10 years ago it was shown that the administration of a 5% ETOH diet to prepubertal rats caused a significant delay in VO.[16] Although that report demonstrated ETOH's ability to alter the onset of puberty, nothing was known at that time about the site(s) of its action, which hormones were involved, and what stage of the pubertal process was most vulnerable to the drug's effects. Recently, we have attempted to address these issues. In this regard, animals from each of the three diet groups described above were killed by decapitation throughout the peripubertal period. Because no significant differences were

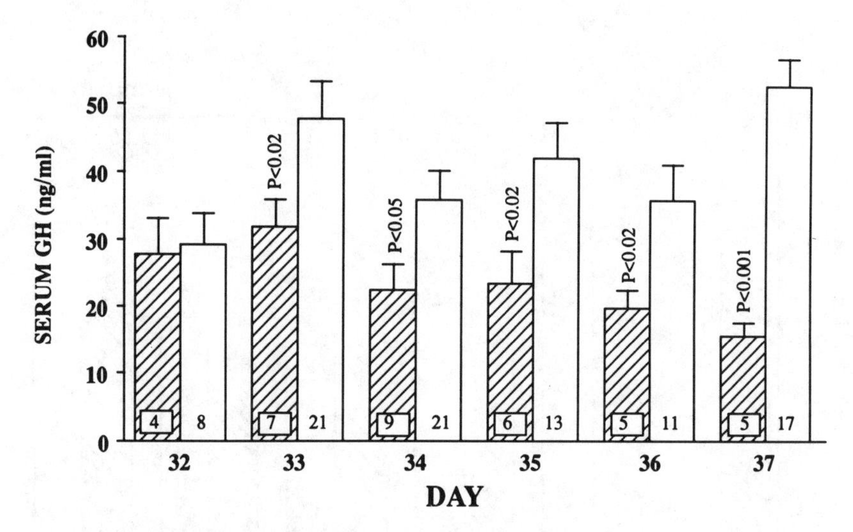

FIGURE 2. Effect of ETOH on mean (± SEM) serum GH levels. Note the significantly lower GH levels in the ETOH-treated animals, as compared to control animals, each day from day 33 through day 37. Numbers within bars represent the number of animals per group. (Reprinted from Dees *et al.*[44] with permission from S. Karger AG, Basel, Switzerland.)

detected in growth weights or any other parameters measured between the rats fed lab chow or control liquid diet, their data were combined in the graphs in Figures 2 and 3 to simplify comparative descriptions.

We have shown[44] that despite receiving an adequate food supply, the ETOH-treated rats did not grow (Figure 1b). In addition, serum GH levels were also affected by the ETOH. Figure 2 demonstrates that serum GH levels were similar between control and ETOH-treated rats on day 32, but the levels in the control rats increased at the beginning of the peripubertal period (day 33). By contrast, this peripubertal rise in GH did not occur on day 33 in the ETOH-treated animals. Beginning on day 34, GH levels began to decline in the ETOH-treated animals, remaining lower than in the controls each day through day 37.

In these same animals, ETOH did not affect follicle-stimulating hormone (FSH) levels during the peripubertal period (not shown), but did alter LH levels as well as the onset of puberty (Figure 3). Importantly, on day 32, three days after beginning the respective diet regimen, all of the control and ETOH-treated rats were anestrous; however, from day 33 through 35 there was a marked increase in the number of control vs. ETOH-treated animals showing

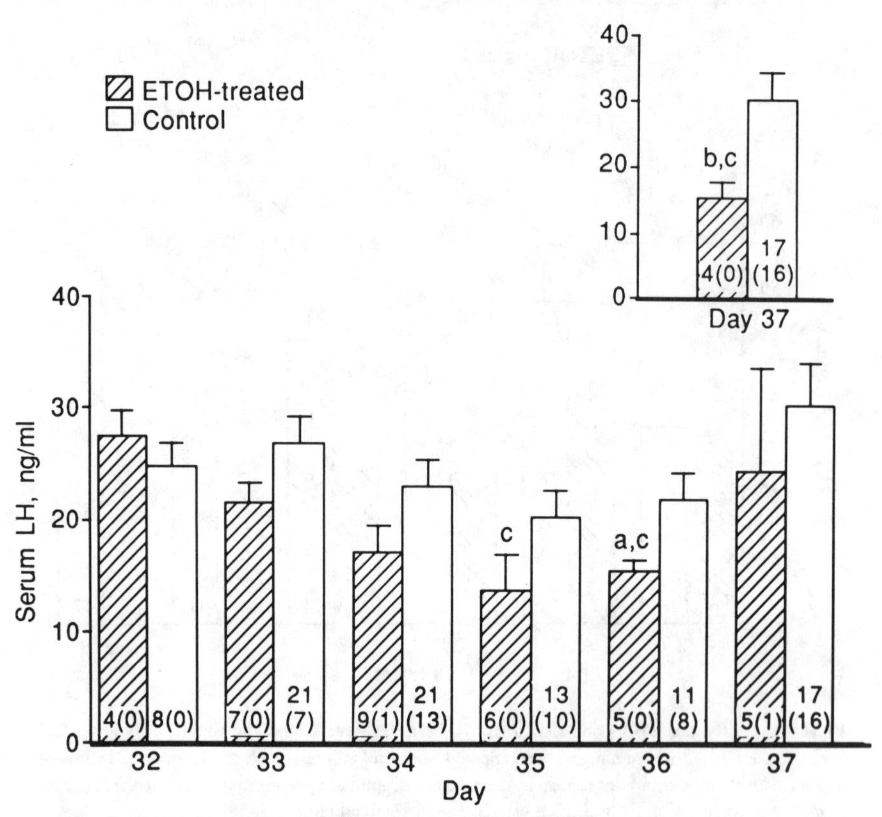

FIGURE 3. Effect of ETOH on mean (± SEM) serum LH levels. Numbers within the bars indicate the number of animals per group (without parentheses) and the number of those animals showing signs of having entered the peripubertal period (in parentheses). Beginning on day 33, note the increasing numbers of animals per day showing signs of pubertal onset in control vs. ETOH-treated animals. Also, LH levels in the ETOH-treated animals declined slightly each day and reached significantly lower levels than the controls on day 36. Because the day 37 group contained the only ETOH-treated, late-proestrous animal in this study, and because serum LH from this animal was not very representative of the other 31 ETOH-treated anestrous animals, we have, alternatively, plotted these day 37 data after deleting this animal (insert). $^{a}p < 0.05$, $^{b}p < 0.01$: ETOH-treated vs. same-day controls, $^{c}p < 0.05$: ETOH-treated vs. day 32 ETOH-treated. (Reprinted from Dees *et al.*[44] with permission from S. Karger AG, Basel, Switzerland.)

signs of maturing (Figure 3). As depicted, only 1 of the 22 ETOH-treated rats (days 33–35) had entered the peripubertal period, as compared to 30 out of the 55 control animals. This figure also demonstrates that only 1 out of the 10 ETOH-treated animals killed on days 36 and 37 had entered the peripubertal period, as compared with 24 of the 28 control animals, further demonstrating the ETOH effects on female puberty. Furthermore, additional animals which received the ETOH diet through day 41 confirmed the ETOH-induced pubertal delay. In an experiment in which the ETOH diet was removed and replaced by the control diet, VO did not occur in every animal, but when VO was recorded, first diestrus (D_1) was delayed. In this regard, the mean (± SEM)

VO-D_1 interval was longer in the ETOH-treated animals (ETOH: 7.9 ± 1.4 days vs. control: 1.4 ± 0.2 days; $p < 0.005$). Interestingly, we also found that when ETOH-treated animals were reintroduced to the control liquid diet on day 41, they began to grow rapidly. These animals attained puberty, as did the controls, at a body weight between 90 and 100 g; although they were approximately 8 days older than the controls at this time due to the ETOH-induced delay. Thus, these results demonstrate (a) that ETOH administration during the peripubertal period delays the onset of female puberty in the rat; (b) that this effect is associated with deficiencies in both GH and LH secretion; and (c) that, although delayed, growth and sexual maturity can occur after removal of the drug from the diet.

ETOH Affects Entry into the Peripubertal Period: Hypothalamic Actions

Based on the above hormone data and because 30 of the 32 ETOH-treated animals that were killed between 33 and 37 days of age were still in the anestrous phase, as opposed to only 29 of the 83 control animals, we surmised that ETOH may be blocking entry into the peripubertal period. To investigate this further, we assessed body and reproductive organ weights, as well as specific hormones from only control and ETOH-treated immature (anestrous) animals that had not entered the peripubertal period.[45] Body, uterine, and ovarian weights were reduced significantly in the ETOH-treated rats (Figure 4). Additionally, we found that ETOH did not alter somatostatin (SRIF) but did cause an increase in growth hormone-releasing hormone (GHRH) content

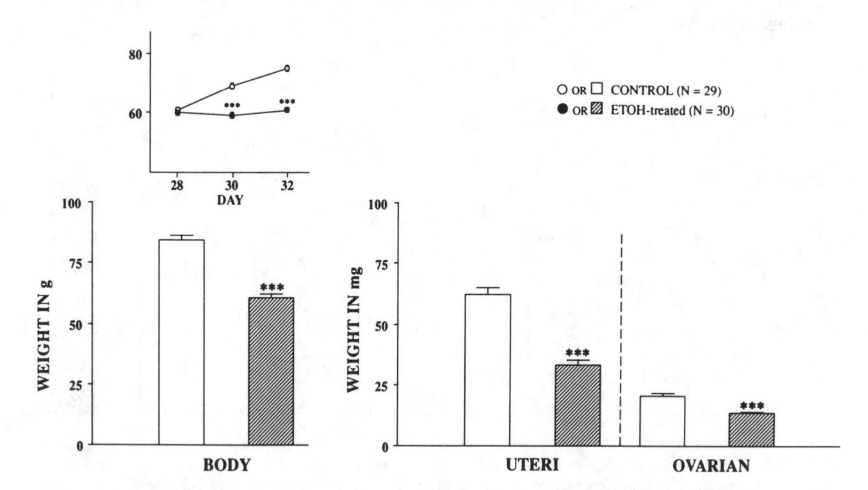

FIGURE 4. Effects of ETOH on mean (± SEM) body, uterine and ovarian weights at the time the animals were killed. Body weights on preceding days are depicted in the insert.***$p < 0.001$, ETOH vs. respective control. (Reprinted from Dees, W.L., Skelley, C.W., Hiney, J.K., and Johnston, C.A., Actions of ethanol on hypothalamic and pituitary hormones in prepubertal female rats, *Alcohol*, 7, 21, 1990, with kind permission from Pergamon Press Ltd., Headington Hill Hall, Oxford 0X3 0BW, U.K.)

with a concomitant decrease in serum GH (Figure 5). Furthermore, the hypothalamic content of LHRH was increased in the ETOH-treated animals with concomitant decreases in serum LH (but not FSH; Figure 6) and E_2 levels. These results suggest that depressed GH and LH levels during the anestrous phase may be altering the animals' timely entry into the peripubertal phase of development and also imply that this is due, at least in part, to alterations in specific hypothalamic peptides controlling GH and LH release.

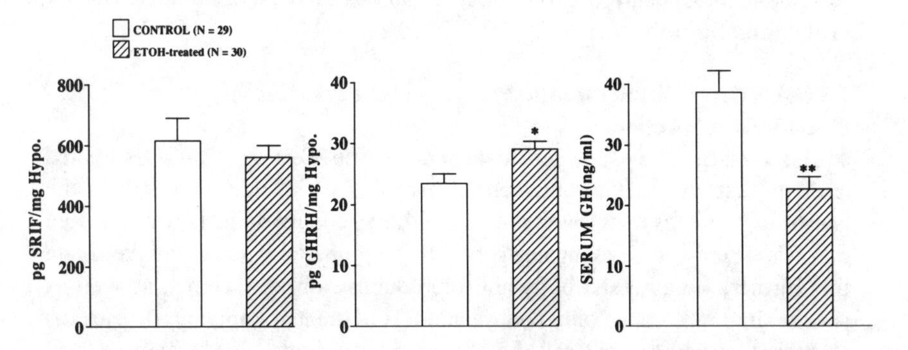

FIGURE 5. Effects of ETOH on mean (±SEM) concentrations of somatostatin (SRIF) and growth hormone-releasing hormone (GHRH), as well as serum GH. Note that ETOH did not significantly alter hypothalamic SRIF content but caused an increase (*$p < 0.02$) in hypothalamic GHRH with a concomitant decrease (**$p < 0.01$) in serum GH. (Reprinted from Dees, W.L., Skelley, C.W., Hiney, J.K., and Johnston, C.A., Actions of ethanol on hypothalamic and pituitary hormones in prepubertal female rats, *Alcohol*, 7, 21, 1990, with kind permission from Pergamon Press Ltd., Headington Hill Hall, Oxford 0X3 0BW, U.K.)

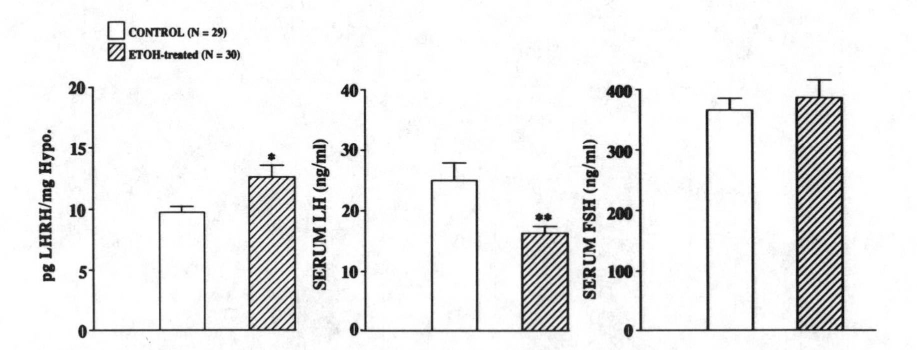

FIGURE 6. Effect of ETOH on mean (±SEM) concentrations of LHRH, as well as LH and FSH. The hypothalamic content of LHRH was increased (*$p < 0.02$) in the ETOH-treated animals with a concomitant decrease (**$p < 0.01$) in serum LH, but not FSH. (Reprinted from Dees, W.L., Skelley, C.W., Hiney, J.K., and Johnston, C.A., Actions of ethanol on hypothalamic and pituitary hormones in prepubertal female rats, *Alcohol*, 7, 21, 1990, with kind permission from Pergamon Press Ltd., Headington Hill Hall, Oxford 0X3 0BW, U.K.)

Overview

The above examples depict the detrimental effects of prepubertal ETOH exposure on hormones that are involved in the onset of female puberty. The delay in puberty we have described is in agreement with previous studies using female[16] and male[103] rats and male mice.[9] Our results represent an important addition, since they focused on the effects of ETOH during the different phases of female pubertal development and correlated the effects with hormone secretions of hypothalamic, pituitary and ovarian origin. In a similar study, specific indices of puberty were assessed from prepubertal male rats.[26] Although the timing of male puberty was not mentioned, the authors did report depressed testosterone (T) levels, testicular weights and hypothalamic β-endorphin levels. They also reported that LH levels were suppressed significantly after the first week of ETOH, but not thereafter. This is in agreement with the depressed LH levels we observed at the end of our 1-week treatment period in prepubertal females. Taken together, these results offer convincing evidence that ETOH can alter specific puberty-related events in both female and male rats. Although we are only beginning to understand the effects of ETOH on the onset of puberty, some important considerations with regard to ETOH's effects on prepubertal GH and LH secretion, as well as its potential effects on prepubertal ovarian maturation, warrant further discussion.

GH and LH Secretions

The notion that a portion of ETOH's effects on the onset of puberty may be due to depressed GH secretion is supported by several lines of evidence that suggest a physiological role for GH in the timing of puberty. First, depressed serum GH induced by the somatomedin-producing worm, *Spirometra mansonoides*, has been shown to cause a delay in the onset of female puberty.[104] Similarly, GH implants within the ME have also been shown to depress GH levels and delay the onset of female puberty, an event associated with blunted ovarian responsiveness to gonadotropins.[5] Additionally, it is known that GH levels increase as the animal approaches maturity, attaining adult levels on the peripubertal days preceding the first gonadotropin surge.[90] Our results also lend support to the hypothesis that there is a physiological involvement for GH on the timing of puberty and, additionally, indicate that ETOH administration may negate this involvement by causing a GH deficiency. The site of this ETOH-induced deficiency in GH secretion following short-term or chronic ETOH administration is not well understood; however, acute studies, which circumvent problems associated with long-term exposure to the drug, have led to some success in determining whether the principal effect is at the hypothalamic or pituitary level. These studies, which analyze the pulsatile release of GH, will be discussed in a later section.

With regard to serum LH, previous reports have indicated that chronic ETOH exposure lowers LH in adult female monkeys[78] and rats.[109] We now know that prepubertal administration of ETOH, in addition to causing depressed reproductive organ weights, causes a significant increase in the hypothalamic content of LHRH and a significant decrease in LH, but not FSH, levels in immature female rats. This differential effect of ETOH on LH and FSH secretion is in agreement with that shown previously in adult female rats.[47,109] We suggest that ETOH's action to suppress prepubertal LH is at the hypothalamic level, since the increased content of LHRH likely reflects diminished secretion of this neuropeptide, a hypothesis also supported by other lines of evidence. In other studies, chronic ETOH exposure was found to reduce the number of ova shed following treatment of 30-day-old rats with pregnant mare serum gonadotropin (PMSG)[15] and to block the PMSG-induced surge of LH in immature female rats.[116] Other evidence for a hypothalamic effect comes from studies in which rats were prenatally exposed to ETOH. In one study[36] it was concluded that delay in puberty in these animals was not directly due to pituitary pathology but to a hypothalamic action to impair gonadotropin secretion. In another study it was suggested that the pubertal delay may be a result of subtle alterations in LHRH neuronal morphology.[77] Additional support for this hypothesis comes from studies which have analyzed the pulsatile release of LH following acute ETOH administration. This topic will be discussed in a later section.

Ovarian Maturity and Steroid Secretion

The observation that chronic ETOH administration suppresses prepubertal serum E_2 levels indicates that the drug may alter hypothalamic/pituitary secretions and/or have a direct ovarian action. Even though GH and LH are important for ovarian maturity, it is now clear that in addition to its hormonal control, the ovary is regulated by direct neural inputs. Since ETOH, in addition to causing hypothalamic-pituitary impairments as described above, is considered to be a direct gonadal toxin in men,[79,115,125] as well as both adult male and female rats,[13,27,32,49,55,59,60,67] it is possible that this drug may be having direct detrimental effects on the prepubertal ovary as well.

Using the rat, we have observed the presence of ovarian nerve fibers containing specific peptides or catecholamines.[7,37,40,70,76] We were unable to demonstrate any significant effect of substance P (SP) or NPY on ovarian steroidogenesis, but because both peptides innervate predominately ovarian blood vessels, it was suggested that they may participate in regulating ovarian blood flow. In contrast, vasoactive intestinal peptide (VIP) and catecholamines are potent stimuli for T, progesterone (P) and E_2 secretion, as assessed by the *in vitro* incubation of immature ovaries with the peptide.[6,7] Based on these facts, it is possible that ETOH could affect ovarian maturation and the pubertal process, at least in part, by altering specific sympathetic or sensory innervations to the prepubertal ovary and/or by altering ovarian responsiveness to specific innervating substances, specifically VIP. With regard to ovarian innervation, it is well known that the ovary is a target field for sympathetic and sensory neurons.[23] The sympathetic neurons are represented by catecholaminergic and NPY fibers, and the sensory neurons by SP and calcitonin gene-related peptide (CGRP) fibers. VIP-ergic fibers of both sympathetic and sensory nature also innervate the ovary. Of potential importance is our preliminary finding that chronic ETOH administration during the late juvenile and peripubertal periods, which causes delayed puberty and depressed serum E_2 levels, is associated with a decreased number of ovarian immunoreactive VIP-ergic fibers versus controls of the same age and pubertal stage (Dees, W.L., and Hiney, J.K., unpublished observation). Therefore, it is possible that ETOH could be causing decreased synthesis of the peptide within the neurons providing the extrinsic innervation, or, it could possibly be altering the synthesis, or action, of nerve growth factor. This target-derived neurotropin is involved in promoting growth and survival of peripheral sympathetic and sensory neurons and is likely involved in maintaining ovarian innervation and its density.

Regarding prepubertal ovarian responsiveness, the fact that the VIP facilitatory influence on ovarian steroidogenesis varies in relation to the phase of female puberty[7] implies that VIP may play a physiological role during the last stages of ovarian development. Because of the potential importance of VIP on the ovarian maturation process, we tested whether or not ETOH could interfere with VIP-induced steroid secretion *in vitro*.[38] Those results demonstrated that ETOH did not affect either basal or VIP-induced steroid release from ovaries collected from anestrous (late juvenile period) animals. Likewise, the ETOH did not alter basal steroid secretion from peripubertal animals in their "first" proestrus; however, the drug reduced the VIP-stimulated release of both T and E_2 from the proestrous ovaries (Figure 7). Thus, these data demonstrate for the first time that ETOH is capable of altering prepubertal ovarian responsiveness to VIP, a peptide known to be involved in the developmental regulation of ovarian function.

CHRONIC EFFECTS OF ETOH ON THE ADULT FEMALE RAT ESTROUS CYCLE

The detrimental effects of chronic ETOH exposure on male reproductive function have been studied extensively in men, as well as in several laboratory animal species; however, very few studies have been conducted to determine the effects of ETOH abuse on female reproductive function. Chronic ETOH exposure has been associated with amenorrhea and depressed LH levels in female macaque monkeys.[78] Also in women, signs of persistent amenorrhea, infertility and spontaneous abortions have been described.[18,61,66,81,107] In the rat, short-term (2–3 days) ETOH exposure was shown to reduce LH and elevate PRL levels in ovariectomized rats.[47] A more chronic exposure to ETOH is known to cause ovarian failure[124] and disruption of the estrous cycle[54] in intact rats.

Although it was known for some years that the cumulative effects of prolonged exposure to ETOH led to impaired reproductive function in females of all species studied, conclusions regarding the mechanism(s) and sites of ETOH's actions on the hypothalamic-pituitary-ovarian axis were limited by

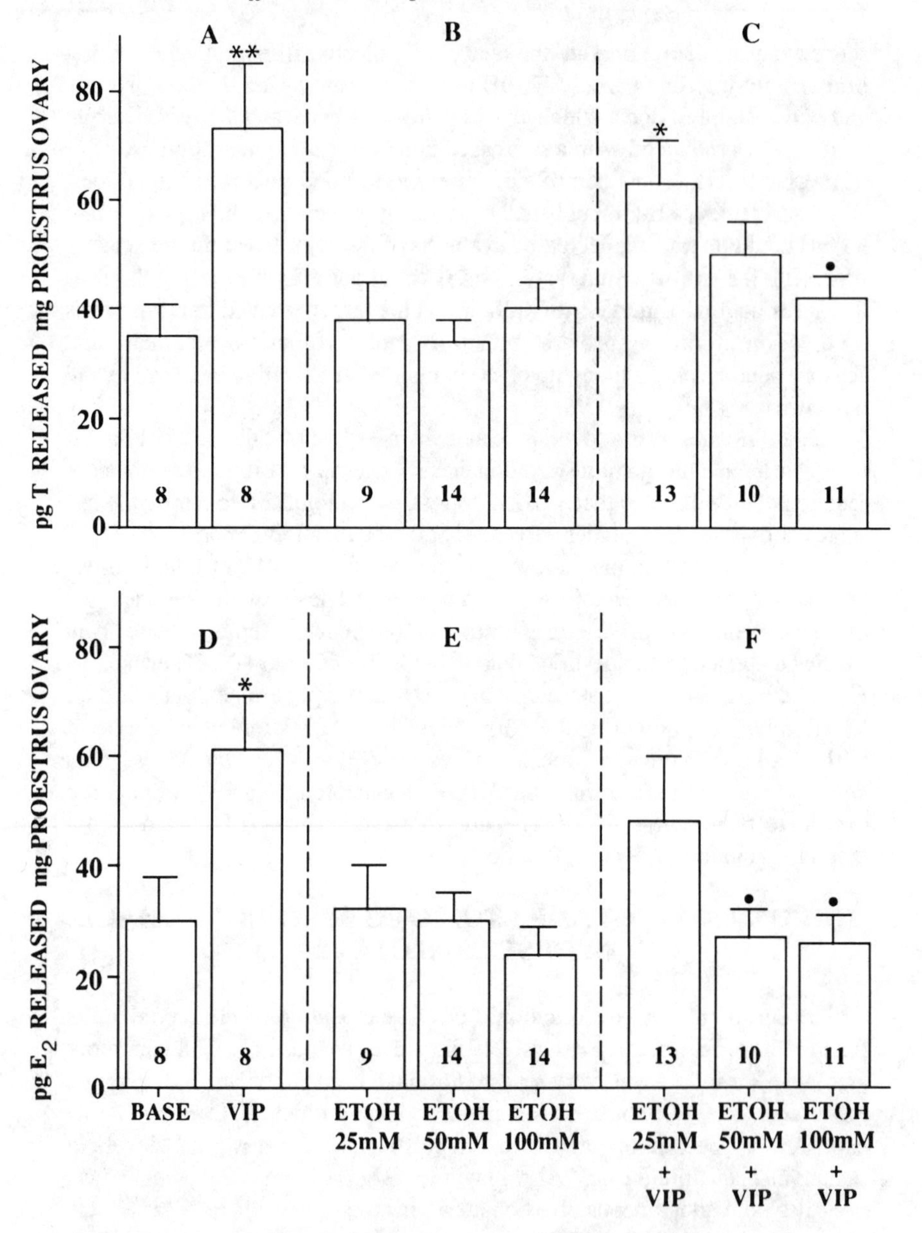

FIGURE 7. Effects of VIP and ETOH on T and E_2 released from early proestrus ovaries. Panels *A* and *D* depict significant VIP-induced release of both T and E_2, respectively. As depicted by panels *B* and *E*, none of the ETOH doses altered basal steroid secretion. Panels *C* and *F* demonstrate the ETOH dose response decrease in VIP-induced steroid release. $^{**}p < 0.02$; $^{*}p < 0.05$ denote a significant stimulation of steroid released into the medium when compared with base values. ● = $p < 0.05$ denote values significantly different than those of the VIP-only group. The number of ovaries per group is indicated by the number within each bar. (Reprinted from Dees, W.L., Hiney, J.K., Fuentes, F., and Forrest, D.W., Ethanol alters vasoactive intestinal peptide-induced steroid release from immature rat ovaries *in vitro, Life Sci.*, 46, 165, 1989, with kind permission from Pergamon Press Ltd., Headington Hill Hall, Oxford 0X3 0BW, U.K.)

the fact that only pituitary and/or gonadal hormones were measured. In a more recent study,[109] we attempted to assess more completely the overall activity of this system following short-term ETOH exposure by monitoring the estrous cycle and, for the first time, measuring the hypothalamic content of LHRH, as well as serum levels of LH, FSH, PRL, and P. Importantly, rats that received a specific ETOH-diet regimen exhibited signs of persistent diestrus.[109] Furthermore, as described above for immature female rats, and as we have also observed previously in adult male rats,[41,42] a significant increase in hypothalamic LHRH content was observed. Additionally, along with this hypothalamic effect, we observed a significant decrease in serum LH, but not FSH, levels (Figure 8). Elevated serum PRL levels were also observed with P levels in the normal range for diestrous rats.

Interestingly, the short-term exposure to ETOH did not modify the content of pituitary LHRH receptors. This lack of change in receptor numbers at a time when LH was reduced suggests that although LHRH release may have been impaired, enough must have been secreted for the maintenance of adequate numbers of its receptors in the pituitary. It is known that LHRH up regulates its receptors and, also, that the number of receptors declines in the absence of the peptide.[99] Based on this, we suggest that ETOH does not inhibit but rather diminishes LHRH release. Importantly, it is also known that pituitary respon-

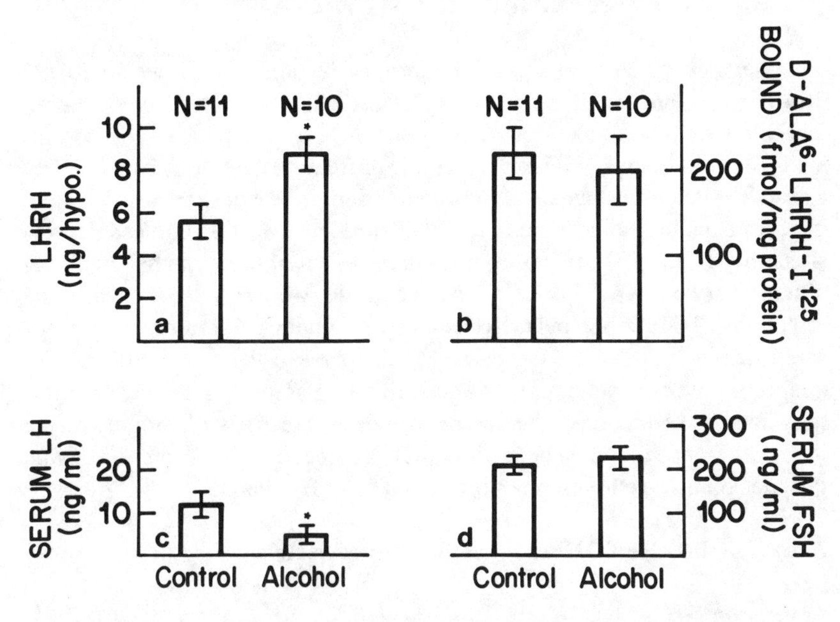

FIGURE 8. Short-term effects of ETOH administration on mean (± SEM) concentrations of (*a*) hypothalamic LHRH, (*b*) pituitary LHRH receptors, (*c*) serum LH and (*d*) serum FSH. Note the increase (*$p < 0.05$) in LHRH, along with a concomitant decrease (*$p < 0.01$) in serum LH levels in the ETOH-treated animals. (Reprinted from Rettori *et al.*[109] with permission from the Society for the Study of Reproduction, Madison, WI.)

siveness to exogenous LHRH is not altered following the acute administration of ETOH. This, as well as other considerations supporting a principal hypothalamic action to alter LH secretion, will be discussed in a subsequent section.

We also determined that persistent diestrus due to ETOH is not a result of pseudopregnancy and can be reversed following removal of the drug from the diet, suggesting that spontaneous recovery from short-term ETOH exposure can occur in adult female rats. A spontaneous recovery from ETOH-induced infertility has been reported for male mice.[8] Those authors suggested that male alcoholic patients with reproductive disorders could have a prognosis of partial recovery following abstinence from the drug. Our results using female rats support this hypothesis, at least with regard to short-term alcohol exposure. Whether or not a recovery would be observed in the female following a longer exposure to ETOH is not known.

In summary, it is clear that chronic exposure can alter the female reproductive cycle resulting in depressed reproductive function. Some of the effects are likely due, as stated earlier, to ETOH's ability to act as a direct gonad toxin.[13,27,32,49,55,59,60,67,79,115,125] Additionally, it now appears that at least some of its effects are exerted within the hypothalamic-pituitary-LH axis.

MECHANISMS OF ACTION

Studies designed to analyze the chronic effects of ETOH are obviously important and are useful in identifying ETOH's ability to alter the circulating levels of specific hormones; however, because of confounding factors associated with long-term ETOH exposure, it is difficult to determine conclusively a specific site or mechanism of hormone action. In recent years investigators have used acute studies to more closely determine if ETOH-induced depressions in GH and LH are due to a primary deficit at the hypothalamus, the pituitary, or both. Specifically, *in vivo* techniques were first used in which the effects of ETOH on the pulsatile secretion of pituitary hormones were measured, then *in vitro* methods were used to assess ETOH's effects on the release of specific hypothalamic releasing and inhibiting hormones that govern pituitary hormone secretions. This section will deal primarily with progress made in identifying sites and mechanisms of ETOH's action with regard to GH and gonadotropin secretion in prepubertal and adult female rats.

Acute Effects of ETOH on Pulsatile Hormone Release

Actions on GH

Although the site of its action was not investigated, acute ETOH administration was first shown to alter pulsatile GH secretion in adult male rats.[108] More recently, we showed that ETOH altered the pulsatile secretion of GH in prepubertal as well as adult female rats.[39] In this regard, it was shown that ETOH completely blocked the pulsatile secretion of GH and that responsiveness to a challenge dose of GHRH was similar in both the ETOH-treated and control animals from both age groups. Figure 9 depicts representative examples of these effects from prepubertal female rats. Because pituitary responsiveness was not altered, these results suggest that ETOH's acute inhibitory effect on GH secretion is due to an action at the hypothalamic level. Further support of this proposed site of action has recently been demonstrated using adult male[75] and female[34] rats. In the latter study, these authors suggest there may be a sexually dimorphic effect in that ETOH appears to alter the SRIF system, which is evident when GHRH concentrations are low. They also pointed out that ETOH likely interferes with the GHRH system as well, but this influence does not appear to involve an α_2-GHRH component.

Taken together, these considerations support a hypothalamic site of action for ETOH to alter GH secretion. There also appears to be some evidence that a pituitary site of action exists as well, at least in adult male rats. In this regard, ETOH has been shown to suppress GH release from primary cultures of anterior pituitary cells.[50] Additionally, the same authors also reported[51] that depressed GH levels occurred following an acute i.p. injection of ETOH and that this effect was correlated with a fall in GH mRNA levels but no change in pituitary GH content.

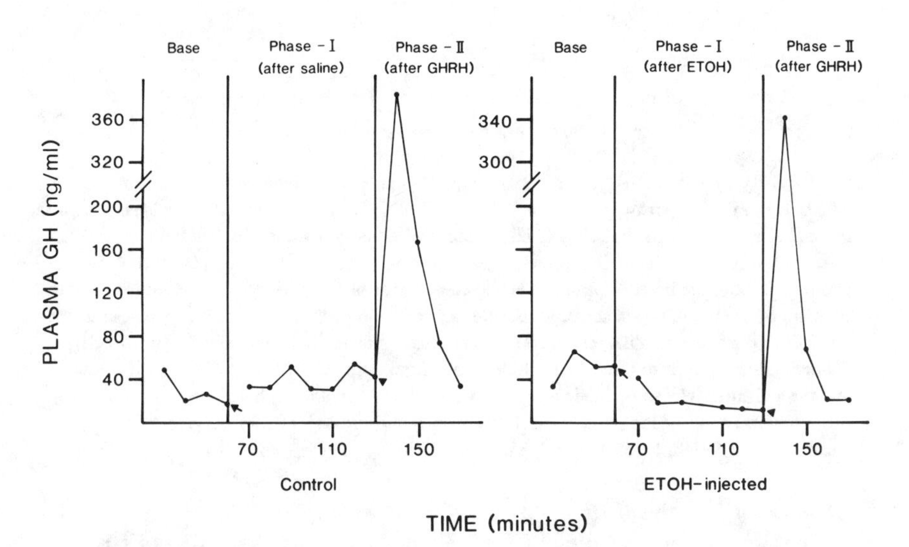

FIGURE 9. Representative GH secretory profiles from individual saline- and ETOH-injected immature female rats before and after receiving a challenge dose of exogenous GHRH. Four 10-minute base samples were taken, followed by a 70-minute absorption interval prior to the phase I sampling period. The arrows represent the time of saline or ETOH injections, and the arrowheads represent the time at which the GRF (GHRH) challenge was administered. (Reprinted from Dees *et al.*[39] with permission from S. Karger AG, Basel, Switzerland.)

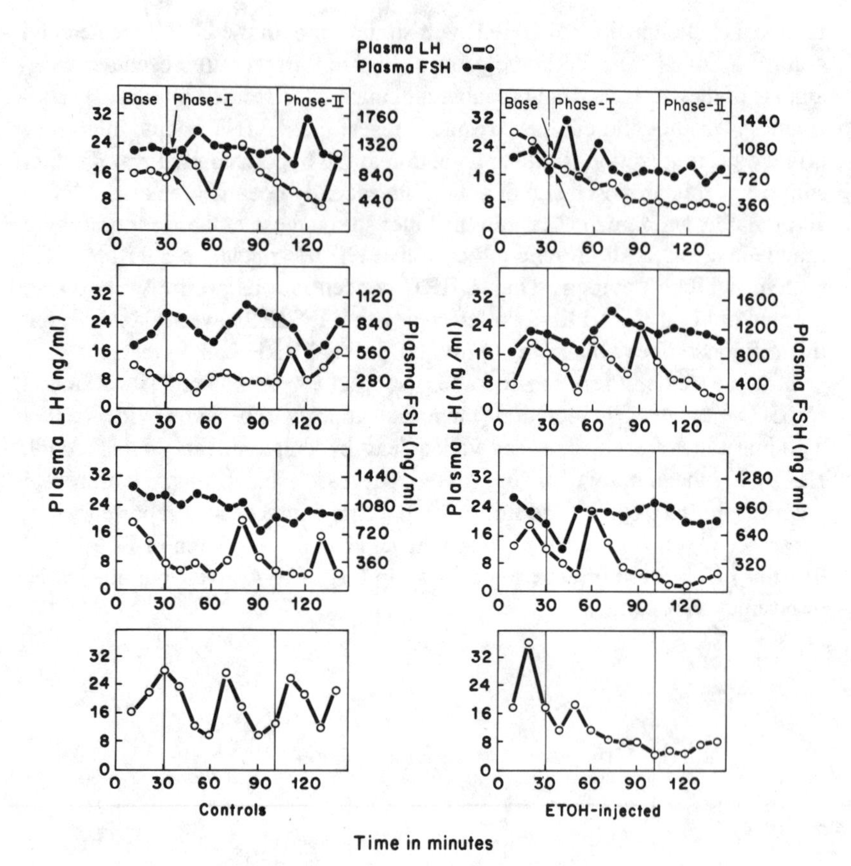

FIGURE 10. Representative LH and FSH secretory profiles from individual ovariectomized rats which were injected (arrow) with either saline or ETOH. The saline-injected animals showed no marked alterations in the pulsatile release pattern of either gonadotropin. By contrast, ETOH-injected animals showed a disruption in the secretion pattern of LH, but not FSH. Note the decrease in both the level of plasma LH and the number of LH pulses, especially during phase II. Also, by examining each secretory profile, several examples can be observed which indicate that LH and FSH have different secretory patterns. Animals 1–4, left column and 5–8, right column. (Reprinted from Dees, W.L., Rettori, V., Kozlowski, G.P., and McCann, S.M., Ethanol and the pulsatile release of luteinizing hormone follicle stimulating hormone and prolactin in ovariectomized rats, *Alcohol*, 2, 641, 1985, with kind permission from Pergamon Press Ltd., Headington Hill Hall, Oxford OX3 OBW, UK.)

Actions on LH and FSH

We first demonstrated that ETOH alters the pulsatile secretion of LH, but not FSH, using adult ovariectomized rats.[43] Representative examples of LH and FSH secretory profiles from individual adult rats are shown in Figure 10. These results clearly demonstrate that ETOH alters the pulsatile pattern of LH secretion without affecting the pattern of FSH secretion and, hence, may explain the dissociation in gonadotropin secretion as described above for both immature and mature female rats following their chronic exposure to ETOH.

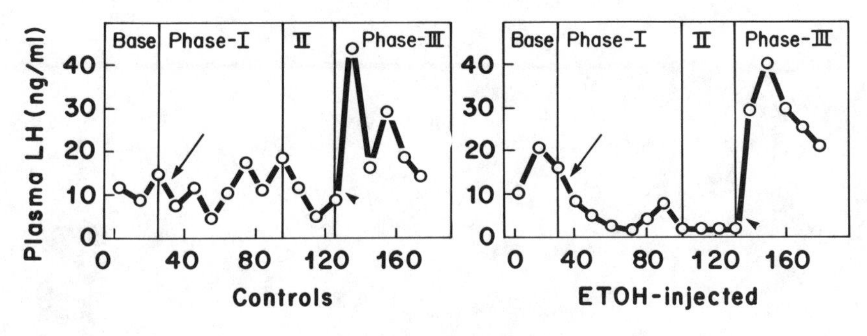

FIGURE 11. Representative LH secretory profiles from individual saline and ETOH-injected adult rats before and after receiving a challenge dose (arrowheads) of exogenous LHRH. A similar reaction to the LHRH injections can be observed in both animals, indicating that pituitary responsiveness was not modified by ETOH. The arrows represent the time of saline or ETOH injections. (Reprinted from Dees, W.L., Rettori, V., Kozlowski, G.P., and McCann, S.M., Ethanol and the pulsatile release of luteinizing hormone follicle stimulating hormone and prolactin in ovariectomized rats, *Alcohol*, 2, 641, 1985, with kind permission from Pergamon Press Ltd., Headington Hill Hall, Oxford 0X3 0BW, U.K.)

Several possibilities for this ETOH-induced differential effect, including the hypothesis for a separate hypothalamic control of LH and FSH release, have been discussed previously.[43,47] It has also been determined that saline and ETOH-treated animals show comparable increases in plasma LH levels after an LHRH challenge to test pituitary responsiveness (Figure 11). These results confirm and extend a previous report that indicated pituitary responsiveness in adult male rats was also unchanged by ETOH.[28] Others, however, have noted a somewhat reduced LH response to LHRH in male rats.[123]

More recently, we demonstrated that ETOH also blocks pulsatile LH release in prepubertal female rats and that, as shown for adult females, the ETOH did not alter pituitary responsiveness to LHRH. In addition, it is important to note that acute ETOH administration can block the afternoon (PM) increase in LH levels in intact, immature female as well as in immature ovariectomized (Figure 12) rats. The normal PM increase in LH is necessary for the onset of puberty and has been shown to occur just before puberty in the rat[119] and monkey.[118] As covered earlier, it has been shown that the ovary matures under the influence of GH, in part, and that the LH pulsatile secreting system matures under the influence of ovarian steroids; but, the activation of the low AM/high PM mode of LH secretion, which initiates pubertal onset, is due to the increased pulsatile release of LHRH and that this activation during the peripubertal period is ovarian independent.[121]

The above-mentioned results strongly imply that the depressed LH levels observed following acute ETOH administration in both adult and prepubertal female rats are due to a disruption in the pulsatile secretory pattern of LH. Furthermore, since all LH pulses are due to previous LHRH pulses,[84] the

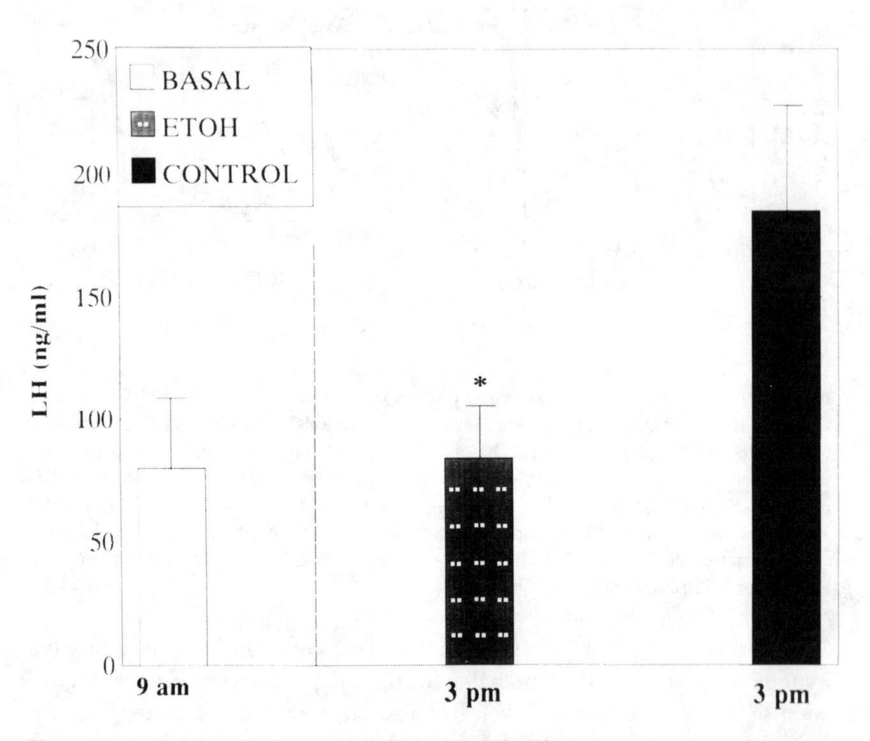

FIGURE 12. Effect of ETOH on daytime LH secretion in immature ovariectomized rats. Note the blockade of the gonadal-independent afternoon increase in LH secretion in the ETOH-treated rats.

combined results presented here suggest that ETOH must be primarily causing a disruption in the pulsatile release of hypothalamic LHRH release. This does not, however, rule out the possibility of some direct pituitary effects as well.[101,111]

ETOH and LHRH Secretion: Prepubertal Effects

There are numerous important neuromodulators associated with the control of hypothalamic LHRH secretion. However, only a few of these substances and their pathways of action have been investigated with regard to determining ETOH's ability to alter their function. In this section we will discuss some of these recent and relevant findings and point out their potential importance, especially as they relate to our attempt to understand the effects of ETOH on events leading to the onset of female puberty.

Norepinephrine (NE)-Prostaglandin E_2 (PGE_2) Pathway

Studies have shown that NE can elicit the release of both LHRH and PGE_2 from hypothalamic-ME fragments.[83,93] Additionally, PGE_2 not only is synthesized in large quantities in the hypothalamus and ME[57,91] but also elicits LHRH release[69,92,96] and mediates the NE-induced release of LHRH from the

ME.[93] Therefore, we investigated the effects of ETOH on NE- and PGE_2-induced LHRH release from the immature female rat ME using a modification[62] of an *in vitro* protocol that has been used for almost 15 years.[83] An advantage of this protocol, if used properly, is that it enables the investigator to use a very small volume (250–400 μl) of medium, which will then allow for the accurate measurement of picogram amounts of peptide released.

Using this methodology, we have demonstrated specific effects of ETOH on the NE-PGE_2-LHRH releasing system.[62] As depicted in Figure 13, ETOH did not effect the basal release of LHRH nor did it affect the NE-induced increase in LHRH secretion from those MEs exposed to the lowest (15 mM) dose of ETOH. Conversely, the 30-mM dose of ETOH blunted the NE-induced release of LHRH, whereas both the 50- and 70-mM doses completely blocked this response. Therefore, our results demonstrate convincingly that ETOH alters the NE-induced release of LHRH.

Additionally, we investigated whether ETOH would affect the PGE_2-induced LHRH release. Figure 14 demonstrates that ETOH had no effect in this regard, since both control and ETOH-exposed MEs released similar amounts of LHRH in response to their respective PGE_2 stimulation. These data indicate that the presence of PGE_2 in the medium can override this inhibitory effect of ETOH. This is important, since it has been shown that PGE_2 mediates the NE-induced release of LHRH, as evidenced by the fact that indomethacin blocks the NE-induced release, but not the PGE_2-induced release, of this peptide.[93]

NE stimulation of α-adrenergic receptors is known to cause sequential increases in PGE_2 and cyclic adenosine monophosphate (cAMP) formation.[93,95,98,105] Unlike other neurotransmitters in the CNS, PGE_2 is synthesized from arachidonic acid and released continuously into the interstitial space.[33] Our results suggested that ETOH may act to block PGE_2 synthesis within the ME, and in a subsequent experiment this was tested. As depicted in Figure 15, we observed that the NE stimulation of PGE_2 synthesis/release was indeed blocked by ETOH. Several possible mechanisms exist by which ETOH could be exerting this effect. One of ETOH's initial actions is to fluidize membranes,[24,58,73] which can alter protein-lipid interactions and could produce alterations in mobilization and metabolism of membrane phospholipids. In addition, ETOH may be altering cholesterol esterase, prostaglandin synthetase, or phospholipase-A_2, which are rate limiting enzymes for PGE_2 synthesis. Importantly, phospholipase-A_2, the enzyme that catalyzes the hydrolysis of arachidonic acid from membrane phospholipids, is activated by Ca^{2+} in a calmodulin-dependent manner;[96] thus, the possibility exists that ETOH could be altering the activation of phospholipase-A_2 and thereby affecting the liberation of arachidonic acid from membrane phospholipids. Another possibility is that the formation of one or more arachidonyl metabolites could be altered. Furthermore, it may alter Ca^{2+} mobilization, a limiting step in PGE_2 formation and subsequent LHRH release.[96] Additionally, we cannot rule out the possibility that ETOH may first be affecting the adrenergic receptor, although it has

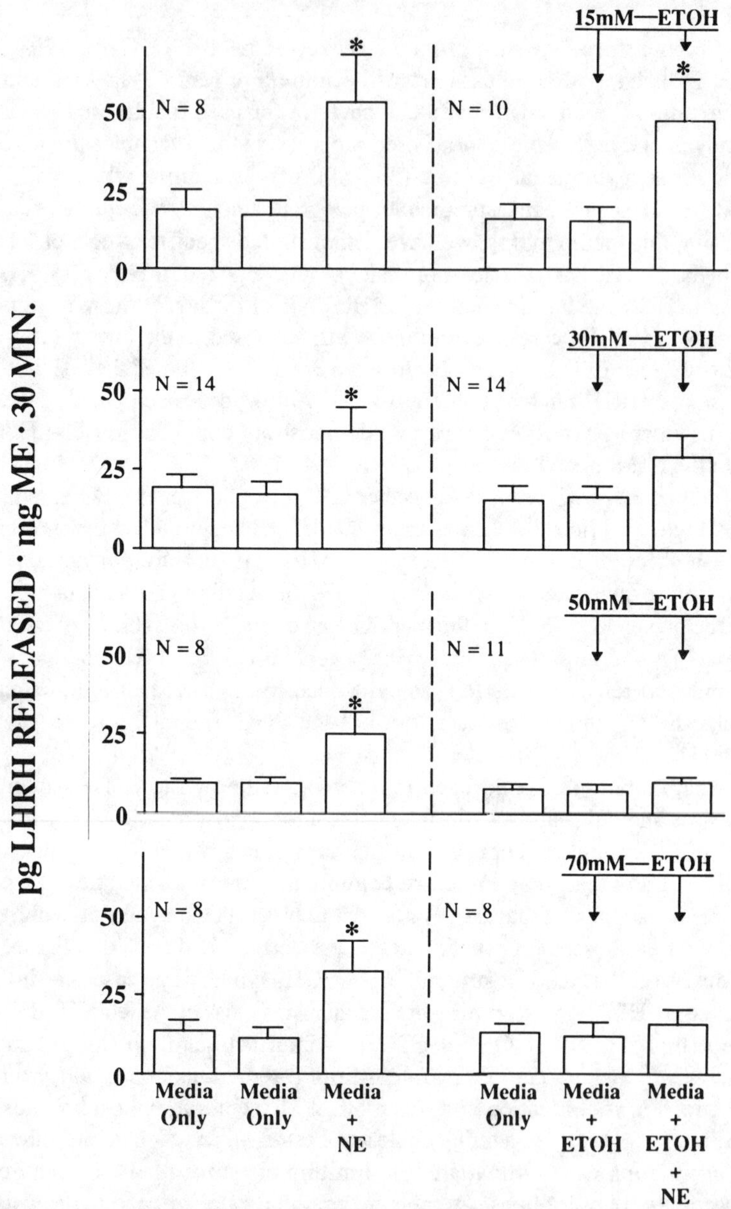

FIGURE 13. Effects of ETOH on basal and NE-induced LHRH release from the ME of 30-day-old female rats. Note the consistent basal (medium only) secretion of LHRH achieved in both control (*left panel*) as well as test (*right panel*) MEs. Also note that the medium containing ETOH had no effect on basal LHRH secretion. However, during the challenge period, the addition of NE (60 M) significantly (*$p < 0.05$) elicited LHRH release from MEs in the control vials (*left panel*) and those MEs exposed to the lowest (15 mM) concentration of ETOH (*right panel*). Importantly, this NE-induced LHRH release from the nerve terminals was blunted ($p > 0.05$) by the 30-mM dose of ETOH and completely blocked by the 50-mM dose. Values are the mean ± SEM. (Reprinted, with permission, from Hiney, J.K., and Dees, W.L., Ethanol inhibits luteinizing hormone-releasing hormone from the median eminence of prepubertal female rats *in vitro*: Investigation of its actions on norepinephrine and prostaglandin-E_2, *Endocrinology*, 128, 1404, 1991. © The Endocrine Society.)

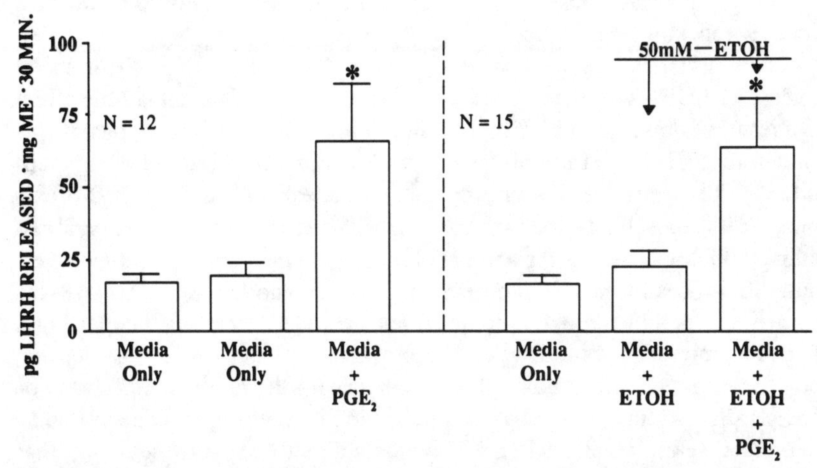

FIGURE 14. Effects of ETOH on basal and PGE_2-induced LHRH release from the ME of 30-day-old female rats. Note again that the basal (medium only) release of LHRH was not affected by the medium containing ETOH. The addition of PGE_2 (2.8 M) during the challenge period resulted in a significant (*$p < 0.05$) increase in LHRH secretion from both the MEs in control vials (*left panel*) and those exposed to the 50-mM dose of ETOH (*right panel*). Values are the mean ± SEM. (Reprinted, with permission, from Hiney, J.K., and Dees, W.L., Ethanol inhibits luteinizing hormone-releasing hormone from the median eminence of prepubertal female rats *in vitro*: Investigation of its actions on norepinephrine and prostaglandin-E_2, *Endocrinology*, 128, 1404, 1991. © The Endocrine Society.)

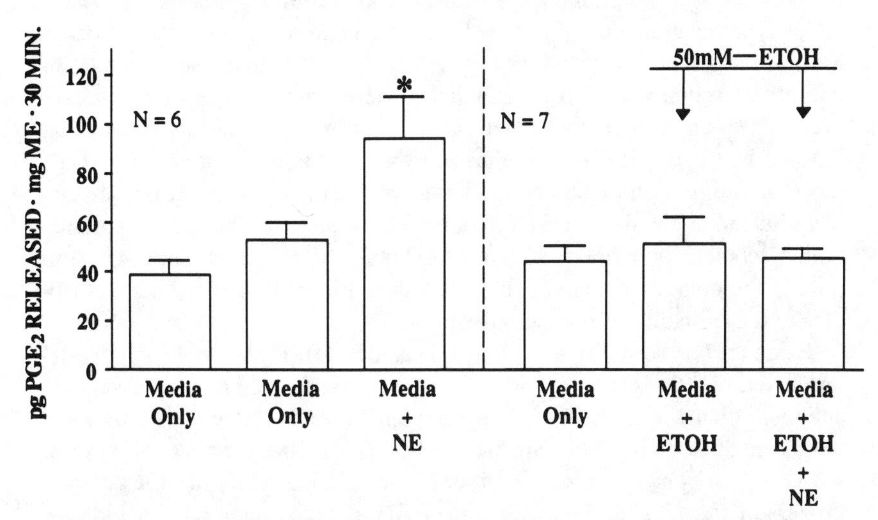

FIGURE 15. Effects of ETOH on basal and NE-induced PGE_2 release from the MEs of 30-day-old female rats. Note that basal (medium only) PGE_2 release was not affected by the medium containing ETOH. However, during the challenge period, addition of NE (60 M) elicited a significant (*$p < 0.05$) release of PGE_2 from the MEs in control vials (*left panel*), whereas this response was blocked in the MEs exposed to the 50-mM ETOH dose (*right panel*). Values are the mean ± SEM. (Reprinted, with permission, from Hiney, J.K., and Dees, W.L., Ethanol inhibits luteinizing hormone-releasing hormone from the median eminence of prepubertal female rats *in vitro*: Investigation of its actions on norepinephrine and prostaglandin-E_2, *Endocrinology*, 128, 1404, 1991. © The Endocrine Society.)

been suggested previously that ETOH's initial effects are not likely to be a specific receptor-ligand effect.[73,85]

The above data clearly demonstrate that ETOH blocks the NE-induced release of LHRH from the ME of prepubertal female rats and that this effect is a result of diminished PGE_2 synthesis/release. Using adult male rats, it was shown that ETOH did not affect the *in vitro* stimulated release of the peptide.[52,53] The cause of this discrepancy is unclear, but as described previously,[62] it is most likely due to numerous differences in the *in vitro* experimental methodology, such as the amount of incubation medium, incubation times, and differences in the age and sex of the rats. In another study, ETOH was shown to block the naloxone-induced release in LH secretion, supporting a hypothalamic site of action.[30] Furthermore, in a more recent study, again using adult male rats, it was shown that ETOH causes the diminished release of LHRH directly into hypophyseal portal blood, resulting in concomitantly depressed serum levels of LH.[25] These data directly support the notion that ETOH alters hypothalamic LHRH release in adult males, as we have shown in prepubertal females. Further support for an ETOH-induced alteration in LHRH secretion is described below.

Excitatory Amino Acid Pathway

NMDA and N-methyl-DL-aspartic acid (NMA) are excitatory amino acid analogs of aspartate that have been shown to stimulate the release of LH in both rats[97,102,120] and primates.[56,127] This action to stimulate LH release occurs at the hypothalamic level via an NMA-induced release of LHRH, and not via a direct action at the pituitary.[19,31,56,97,102,117,120,127] Because the onset of puberty is accompanied by a change in the diurnal pattern of LH secretion,[21,118,119] an event that, as stated earlier, is centrally originated and gonadal independent,[118,121] it was hypothesized that activation of specific NMDA receptors might initiate this peripubertal change in pulsatile LH release and contribute to pubertal onset. Indeed, it has been shown that NMA is associated with precocious pubertal development in both rats[120,122] and monkeys,[56] supporting the notion that activation of hypothalamic NMDA receptors contributes to the initiation of female puberty.

Evidence has been discussed that indicates ETOH is able to delay pubertal onset, and that this event is associated with depressed GH and LH levels via actions, at least in part, at the hypothalamic level. More recently, we have documented that ETOH is capable of altering the NMA-induced release of LHRH *in vitro* and interfering with the ability of NMA to initiate the pubertal process *in vivo*.[86] The *in vitro* effects of ETOH on basal and NMA-induced release of LHRH were assessed from arcuate nucleus-ME fragments obtained from immature female rats.[86] It was shown that ETOH did not alter basal LHRH release, but dose-dependently blocked the NMA-induced release of the peptide during anestrus (Figure 16). The effects of ETOH on LHRH released from fragments during either first proestrus or estrus were very similar to those shown for the anestrous animals.

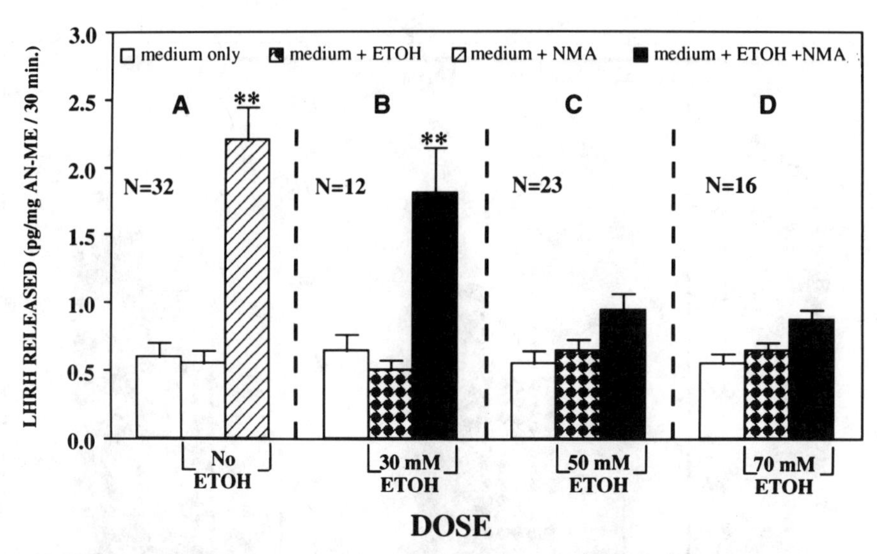

FIGURE 16. Effects of ETOH on basal and NMA-induced LHRH release from AN-ME fragments obtained from 30-day-old (anestrous) rats. Note that basal (medium only) secretion of LHRH was similar in both control (*A*) as well as test (*B–D*) vials. Medium containing ETOH had no effect on basal secretion. However, during the challenge period, the addition of NMA (20 mM) significantly (*$p < 0.01$) elicited LHRH release from AN-MEs in the control vials (*A*) and those AN-MEs exposed to the lowest (30 mM) dose of ETOH (*B*). This NMA-induced LHRH release was blocked ($p < 0.05$) by the 50- and 70-mM doses of ETOH. Values are the mean ± SEM. (Reprinted from Nyberg *et al.*[86] with permission from S. Karger AG, Basel, Switzerland.)

To address the question of whether ETOH's effect on LHRH secretion would alter the precocious puberty induced by NMA administration, female sibling rats were divided among 3 treatment groups. Beginning at 26 days of age, rats were weighed and assigned to one of the following groups: (a) rats receiving a 3-g/kg dose of ETOH in saline at 12:30 pm via gastric gavage, followed by subcutaneous injections of NMA (40 mg/kg) at 2:00 pm and again at 4:00 pm; (b) rats receiving an intragastric injection of an equal volume of saline only, followed by the afternoon subcutaneous injections of NMA, as described above; and (c) rats receiving saline, both gastrically and subcutaneously. This protocol enabled us to discern a distinct effect of ETOH exposure on NMA-induced puberty.[86] As shown (Figure 17), by assessing both VO and D_1, the timing of puberty was advanced significantly in the NMA-treated rats when compared to the control rats receiving saline only. Importantly, this NMA-induced advancement in puberty was attenuated significantly by ETOH, therefore suggesting that the drug had interfered with NMA's ability to elicit afternoon discharges of LHRH. Additionally, we have very recently shown that ETOH can diminish the ability of NMA to induce an increase in pulsatile LH release *in vivo* (Nyberg, C., Hiney, J., and Dees, L., unpublished observation). These results, therefore, suggest that ETOH alters hypothalamic excitatory amino acid neurotransmission at the time of puberty. Importantly, studies have

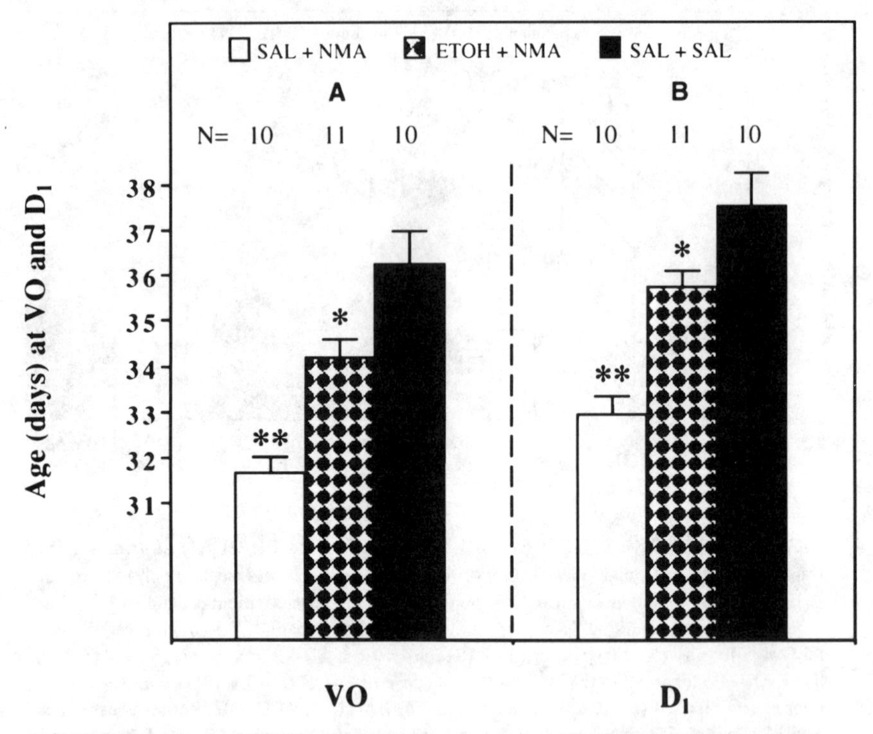

FIGURE 17. Effects of ETOH on NMA-induced puberty. (*A*) Mean (± SEM) age at VO. (*B*) Mean (± SEM) age at D_1. Note that both indices of pubertal development indicate that ETOH was able to attenuate significantly NMA's ability to advance puberty. Number of animals in parentheses, ***p* < 0.01, vs. control; **p* < 0.01, vs. NMA or control. (Reprinted from Nyberg *et al.*[86] with permission from S. Karger AG, Basel, Switzerland.)

demonstrated that NMDA receptors involved in controlling LHRH secretion are markedly and transiently activated around the time of the onset of puberty.[20] Furthermore, the preovulatory gonadotropin surge in the female rat is dependent on NMDA neurotransmission for its expression.[22] Taken together, these results support the hypothesis for the activation of excitatory inputs within or associated with the reproductive hypothalamus at the time of puberty.

The hypothalamus is a very complex region containing numerous neurotransmitter and neuromodulator substances that influence LHRH secretion. Therefore, ETOH may alter the relationship between NMA-sensitive neuronal pathways and other neuronal systems involved with LHRH release. For example, ETOH may alter LHRH release by interfering with the proposed interaction between NMA-sensitive systems and endogenous opioid-containing systems.[29] In this regard, ETOH has been shown to alter prepubertal β-endorphin levels in the hypothalamus of male rats.[26] Also, it is feasible that ETOH may alter the ability of NMA to stimulate NE release from the hypothalamic-ME region, because this effect has been demonstrated recently using hippocampal synaptosomes.[100] It is also conceivable that ETOH may be exerting a more specific effect. The action of NMA depends not only on activation of its receptor, but also on the number of its receptors.[35] Thus, it is tempting to speculate that ETOH may block NMDA receptors in the hypothalamus, as shown previously in other brain regions such as the hippocampus,[72] cerebral cortex,[71] and the medial septum.[113] The possibility also exists that ETOH may be altering the synthesis of new receptors. These possibilities are important since, as stated above, NMDA receptors are markedly activated during pubertal onset. Another conceivable mechanism of action may be that ETOH causes a sufficient alteration in membrane permeability,[24,58,73] which could possibly disrupt Ca^{2+} mobilization. Importantly, in other brain systems[48,64] ETOH has been shown to be capable of blocking the NMDA-induced Ca^{2+} influx. In summary, the activation of the NMDA receptor appears to play a role in the onset of female puberty, and results at this time suggest that ETOH can affect the pubertal process via an action at the hypothalamic level to diminish excitatory amino acid-activated secretion of LHRH.

Opioid Pathway

Endogenous opioids have been shown to cause diminished LH secretion via an action to diminish LHRH release.[68] These peptides have also been implicated in playing a role in the timing of sexual maturation.[14,114] One theory is that there may be a removal of the opioid inhibition of LHRH release at puberty. It has been suggested recently, at least for adult male rats, that opioid-induced LHRH inhibition may be exerted either directly at the level of the LHRH neuron and/or by causing a diminished excitatory amino acid efflux, resulting in decreased LH release.[17] The effects of acute and chronic ETOH administration on hypothalamic β-endorphin content from adult male rats was shown a number of years ago.[112] More recently, ETOH has been shown to cause an increase in hypothalamic β-endorphin levels following its administration to prepubertal male rats.[26] Furthermore, similar results have been shown recently in prepubertal female rats which were prenatally exposed to ETOH.[36] These results show that ETOH does have the potential to affect the hypothalamic opioid system and suggest that further identification of interactions between ETOH, β-endorphin and the NMDA receptor may prove useful in understanding the drug's effects on the pubertal process.

SUMMARY

For many years we have been aware of the effects of ETOH exposure regarding male reproductive function. However, it is only in recent years that we have become aware of the fact that ETOH exposure can affect the female sexual maturation process as well as adult female reproductive function. We now know that ETOH has the capability to alter hormonal secretions at all three levels of the hypothalamic-pituitary-ovarian axis, but we are only now beginning to understand more about ETOH's sites and mechanisms of action. As research on the mechanisms of these ETOH-induced neuroendocrine problems continues, a better understanding of how this drug produces these effects will continue to emerge. Even now it appears that several diverse mechanisms may be involved, suggesting a very complex profile of events that may be even more complicated around the time of puberty. Additional *in vivo* and *in vitro* approaches, as well as technological advances in molecular biology, should provide us with the tools to make even more progress in the coming years with regard to understanding the biomedical effects of ETOH on the neuroendocrinology of female puberty and subsequent reproductive function.

ACKNOWLEDGMENTS

This work was supported by NIAAA grants AA07216 and AA06014. Also, Dr. Dees is the recipient of an NIAAA Research Scientist Development Award, AA00104. The authors wish to acknowledge Drs. G.P. Kozlowski, Valeria Rettori and S.M. McCann (all from the Division of Neuropeptides, Southwestern Medical Center, Dallas, TX) for their important contributions over the years. We also appreciate the long-term collaboration with Dr. Sergio R. Ojeda (Division of Neuroscience, Oregon Regional Primate Center, Beaverton, OR) regarding our interest in the neural control of ovarian function. We thank Carl Skelley for technical assistance and Jaide Hazelwood for editorial and typing assistance.

REFERENCES

1. **Advis, J.P., Andrews, W.W., and Ojeda, S.R.,** Changes in ovarian steroidal and prostaglandin E responsiveness to gonadotropins during the onset of puberty in the female rat, *Endocrinology*, 104, 653, 1979.
2. **Advis, J.P., and Ojeda, S.R.,** Hyperprolactinemia-induced precocious puberty in the female rat: Ovarian site of action, *Endocrinology*, 103, 924, 1978.
3. **Advis, J.P., Richards, J.S., and Ojeda, S.R.,** Hyperprolactinemia-induced precocious puberty: Studies on the mechanism(s) by which prolactin enhances ovarian progesterone responsiveness to gonadotropins in prepubertal rats, *Endocrinology*, 108, 1333, 1981.
4. **Advis, J.P., Simpkins, J.W., Chen, H.T., and Meites, J.,** Relation of biogenic amines to onset of puberty in the female rat, *Endocrinology*, 103, 11, 1978.
5. **Advis, J.P., White, S.S., and Ojeda S.R.,** Activation of growth hormone short-loop negative feedback delays puberty in the female rat, *Endocrinology*, 108, 1343, 1981.
6. **Aguado, L.I., Petrovic, S.L., and Ojeda, S.R.,** Ovarian-adrenergic receptors during the onset of puberty: Characterization, distribution and coupling to steroidogenic response, *Endocrinology*, 110, 1124, 1982.
7. **Ahmed, C.E., Dees, W.L., and Ojeda, S.R.,** The immature rat ovary is innervated by vasoactive intestinal peptide (VIP)-containing fibers and responds to VIP with steroid secretion, *Endocrinology*, 118, 1682, 1986.
8. **Anderson, R.A., Willis, B.R., and Oswald, C.,** Spontaneous recovery from ethanol-induced male infertility, *Alcohol*, 2, 479, 1985.
9. **Anderson, R.A., Willis, B.R., Oswald, C., Gupta, A., and Zaneveld, L.,** Delayed male sexual maturation induced by chronic ethanol ingestion, *Fed. Proc.*, 40, 825, 1981.

10. **Andrews, W.W., and Ojeda, S.R.,** A detailed analysis of the serum LH secretory profile in conscious, free-moving female rats during the time of puberty, *Endocrinology*, 109, 2032, 1981.
11. **Andrews, W.W., Heiman, M., Porter, J.R., and Ojeda, S.R.,** The infantile female rat: *In vivo* ovarian adrenal steroidogenic responses to exogenous administration or to endogenously induced elevations of gonadotropins and ACTH, *Biol. Reprod.*, 24, 597, 1981.
12. **Andrews, W.W., Mizejewski, G.J., and Ojeda, S.R.,** Development of estradiol-positive feedback on luteinizing hormone release in the female rat: A quantitative study, *Endocrinology*, 109, 1404, 1981.
13. **Badr, F.M., Bartke, A., Dalterio, S., and Bulger, W.,** Suppression of testosterone production by ethyl alcohol. Possible mode of action, *Steroid*, 30, 647, 1977.
14. **Blank, M.S., Panerai, E.A., and Friesen, F.G.,** Opioid peptides modulate luteinizing hormone secretion during sexual maturation, *Science*, 203, 1129, 1979.
15. **Bo, W.J., Krueger, W.A., and Rudeen, P.K.,** Effects of ethanol on superovulation in the immature rat following pregnant mare's serum gonadotropin (PMSG) or PMSG and human chorionic gonadotropin, *Biol. Reprod.*, 28, 956, 1983.
16. **Bo, W.J., Krueger, W.A., Rudeen, P.K., and Symmes, S.K.,** Ethanol-induced alterations in the morphology and function of the rat ovary, *Anat. Rec.*, 202, 255, 1982.
17. **Bonavera, J.J., Kalra, S.P., and Kalra, P.S.,** Evidence that luteinizing hormone suppression in response to inhibitory neuropeptides, β-endorphin, Interleukin-1β, and Neuropeptide-K, may involve excitatory amino acids, *Endocrinology*, 133, 178, 1993.
18. **Borhanmanesh, R., and Haghighi, P.,** Pregnancy in patients with cirrhosis of the liver, *Obstet. Gynecol.*, 36, 315, 1970.
19. **Bourguignon, J.P., Gerard, A., and Franchimont, P.,** Direct activation of gonadotropin-releasing hormone secretion through different receptors to neuroexcitatory amino acids, *Neuroendocrinology*, 49, 402, 1989.
20. **Bourguignon, J.P., Gerard, A., Mathieu, J., Mathieu, A., and Franchimont, P.,** Maturation of the hypothalamic control of pulsatile gonadotropin-releasing hormone secretion at onset of puberty. I. Increased activation of N-methyl-D-aspartate receptors, *Endocrinology*, 127, 873, 1990.
21. **Boyar, R.M., Finkelstein, J., Roffwarg, H., Kapen, S., Weitzman, E., and Hellman, L.,** Synchronization of augmented luteinizing hormone secretion with sleep during puberty, *N. Engl. J. Med.*, 287, 582, 1972.
22. **Brann, D.W., and Mahesh, V.B.,** Endogenous excitatory amino acid involvement in the preovulatory and steroid-induced surge of gonadotropins in the female rat, *Endocrinology*, 128, 1541, 1991.
23. **Burden, H.W.,** The adrenergic innervation of mammalian ovaries, in *Catecholamines as Hormone Regulators, Vol. 18*, Ben-Jonathan, N., Ed., Raven Press, New York, 1985, 261.
24. **Chin, J.H., and Goldstein, D.B.,** Effects of low concentrations of ethanol on the fluidity of spin-labeled erythrocyte and brain membranes by a series of short-chain alcohols, *Mol. Pharmacol.*, 13, 435, 1977.
25. **Ching, M., Valencia, M., and Negro-Vilar, A.,** Acute ethanol treatment lowers hypophyseal portal plasma luteinizing hormone-releasing hormone (LHRH) and systemic plasma, LH levels in orchidectomized rats, *Brain Res.*, 443, 325, 1988.
26. **Cicero, T.J., Adams, M.L., O'Connor, L., Nock, B., Meyer, E.R., and Wozniak, D.,** Influence of chronic alcohol administration on representative indices of puberty and sexual maturation in male rats and the development of their progeny, *J. Pharmacol. Exp. Ther.*, 255, 707, 1990.
27. **Cicero, T.J., and Bell, R.D.,** Effects of ethanol and acetaldehyde on the biosynthesis of testosterone in the rodent testes, *Biochem. Biophys. Res. Commun.*, 94, 814, 1980.
28. **Cicero, T.J., Bernstein, D., and Badger, T.M.,** Effects of acute alcohol administration on reproductive endocrinology in the male rat, *Alcoholism: Clin. Exp. Res.*, 2, 249, 1978.
29. **Cicero, T.J., Meyer, E.R., and Bell, R.D.,** Characterization and possible opioid modulation of N-methyl-D-aspartic acid induced increases in serum luteinizing hormone levels in the developing male rat, *Life Sci.*, 42, 1725, 1988.

30. **Cicero, T.J., Newman, K.S., Gerrity, M., Schumoeker, P.F., and Bell, R.D.,** Ethanol inhibits the naloxone-induced release of luteinizing hormone-releasing hormone from the hypothalamus of the male rat, *Life Sci.*, 31, 1587, 1982.
31. **Claypool, L.E., and Terasawa, E.,** N-methyl-DL-aspartate (NMDA) induces LHRH release as measured by *in vivo* push-pull perfusion in the stalk-median eminence of pre- and peripubertal female rhesus monkeys, *Biol. Reprod.*, 40 (Suppl.), 83, 1989.
32. **Cobb, C.F., Ennis, M.F., Van Thiel, D.H., Galvaler, J.S., and Lester, R.,** Acetaldehyde and ethanol are testicular toxins, *Gastroenterology*, 75, 958, 1978.
33. **Coceani, F.,** Prostaglandins and the central nervous system, *Arch. Intern. Med.*, 133, 119, 1974.
34. **Conway, S., and Mauceri, H.,** The influence of acute ethanol exposure on growth hormone release in female rats, *Alcohol*, 8, 159, 1991.
35. **Cotman, C.W., Bridges, R.J., Taube, J.S., Clark, A.S., Geddes, J.W., and Monaghan, D.T.,** The role of the NMDA receptor in central nervous system plasticity and pathology, *J. NIH Res.*, 1, 65, 1989.
36. **Creighton-Taylor, J.A., and Rudeen, P.K.,** Prenatal ethanol exposure and opiatergic influence on puberty in the female rat, *Alcohol*, 8, 187, 1991.
37. **Dees, W.L., Ahmed, C.E., and Ojeda, S.R.,** Substance P- and vasoactive intestinal peptide-containing fibers reach the ovary independent routes, *Endocrinology*, 119, 638, 1986.
38. **Dees, W.L., Hiney, J.K., Fuentes, F., and Forrest, D.W.,** Ethanol alters vasoactive intestinal peptide-induced steroid release from immature rat ovaries *in vitro*, *Life Sci.*, 46, 165, 1989.
39. **Dees, W.L., Skelley, C.W., Rettori, V., Kentroti, M.S., and McCann, S.M.,** Influence of ethanol on growth hormone secretion in adult and prepubertal female rats, *Neuroendocrinology*, 48, 495, 1988.
40. **Dees, W.L., Kozlowski, G.P., Dey, R., and Ojeda, S.R.,** Evidence for the existence of substance P in the prepubertal rat ovary. II. Immunocytochemical localization, *Biol. Reprod.*, 33, 471, 1985.
41. **Dees, W.L., McArthur, N.H., Farr, K.L., Culler, M.D., and Harms, P.G.,** Effects of ethanol on rat hypothalamic luteinizing hormone-releasing hormone. A study utilizing radioimmunoassay, *Biol. Reprod.*, 28, 1066, 1983.
42. **Dees, W.L., McArthur, N.H., and Harms, P.G.,** Effects of ethanol on hypothalamic luteinizing hormone-releasing hormone (LHRH) in the male rat. An immunocytochemical study, *Exp. Brain Res.*, 54, 1972, 1984.
43. **Dees, W.L., Rettori, V., Kozlowski, G.P., and McCann, S.M.,** Ethanol and the pulsatile release of luteinizing hormone, follicle stimulating hormone and prolactin in ovariectomized rats, *Alcohol*, 2, 641, 1985.
44. **Dees, W.L., and Skelley, C.W.,** Effects of ethanol during the onset of female puberty, *Neuroendocrinology*, 51, 64, 1990.
45. **Dees, W.L., Skelley, C.W., Hiney, J.K., and Johnston, C.A.,** Actions of ethanol on hypothalamic and pituitary hormones in prepubertal female rats, *Alcohol*, 7, 21, 1990.
46. **Dees, W.L., Skelley, C.W., and Kozlowski, G.P.,** Intragastric cannulation as a method of ethanol administration for neuroendocrine studies, *Alcohol*, 1, 177, 1984.
47. **Dees, W.L., and Kozlowski, G.P.,** Differential effects of ethanol on luteinizing hormone, follicle stimulating hormone and prolactin secretion in the female rat, *Alcohol*, 1, 429, 1984.
48. **Dildy, J.E., and Leslie, S.W.,** Ethanol inhibits NMDA-induced increases in free intracellular CA^{2+} in dissociated brain cells, *Brain Res.*, 499, 383, 1989.
49. **Ellingboe, J., and Varanelli, C.C.,** Ethanol inhibits testosterone biosynthesis by direct action on Leydig cells, *Res. Commun. Chem. Pathol. Pharmacol.*, 24, 87, 1979.
50. **Emanuele, M.A., Kirsteins, L., Reda, D., Emanuele, N.V., and Lawrence, A.M.,** The effect of *in vitro* ethanol exposure on basal growth hormone secretion, *Endocrinol. Res.*, 14, 283, 1989.

51. **Emanuele, M.A., Tentler, J.J., Kirsteins, L., Emanuele, N.V., Lawrence, L.A., and Kelley, M.R.**, The effect of "binge" ethanol exposure on growth hormone and prolactin gene expression and secretion, *Endocrinology*, 131, 2077, 1992.
52. **Emanuele, M.A., Tentler, J., Reda, D., Kirsteins, L., Emanuele, N.V., and Lawrence, A.M.**, The effect of *in vitro* ethanol exposure on LHRH release from perifused rat hypothalami, *Endocr. Res.*, 16, 313, 1990.
53. **Emanuele, M.A., Tentler, J., Reda, D., Kirsteins, L., Emanuele, N.V., and Lawrence A.M.**, Failure of *in vitro* ethanol to inhibit LHRH release from the hypothalamus, *Alcohol*, 6, 263, 1989.
54. **Eskay, R.L., Rayback, R.S., Goldman, M., and Majchrowica, E.**, Effect of chronic ethanol administration on plasma levels of LH and the estrous cycle in the female rat, *Alcoholism: Clin. Exp. Res.*, 5, 204, 1981.
55. **Gavaler, J., Van Thiel, D.H., and Lester, R.**, Ethanol: A gonadal toxin in the mature rat of both of sexes, *Alcoholism: Clin. Exp. Res.*, 4, 217, 1980.
56. **Gay, V.L., and Plant, T.M.**, N-methyl-DL-aspartate elicits hypothalamic gonadotropin-releasing hormone release in prepubertal male rhesus monkey (*Macaca mulatta*), *Endocrinology*, 120, 2289, 1987.
57. **Gerozissis, K., Savedra, J.M., and Drayu, F.**, Prostanoid profile in specific brain areas, pituitary and pineal gland of the male rat. Influence of experimental conditions, *Brain Res.*, 279, 133, 1983.
58. **Goldstein, D.B., Chin, J.H., and Lyon, R.C.**, Ethanol disordering of spin-labeled mouse brain membranes: Correlation with genetically determined ethanol sensitivity of mice, *Proc. Natl. Acad. Sci. USA*, 79, 4231, 1982.
59. **Gordon, G.G, Vittek, J., Southern, A.L., Munnangi, P., and Lieber, C.S.**, Effect of chronic alcohol ingestion on the biosynthesis of steroids in rat testicular homogenate *in vitro*, *Endocrinology*, 106, 1880, 1980.
60. **Gordon, G.G., Southern, A.L., Vittek, J., and Lieber, C.S.**, The effect of alcohol ingestion on hepatic aromatase activity and plasma steroid hormones on the rat, *Metabolism*, 28, 20, 1979.
61. **Harlap, S., and Shiono, P.H.**, Alcohol, smoking and incidence of spontaneous abortion in the first and second trimester, *Lancet*, 2, 173, 1980.
62. **Hiney, J.K., and Dees, W.L.**, Ethanol inhibits luteinizing hormone-releasing hormone from the median eminence of prepubertal female rats *in vitro*: Investigation of its actions on norepinephrine and prostaglandin-E_2, *Endocrinology*, 128, 1404, 1991.
63. **Hiney, J.K., Ojeda, S.R., and Dees, W.L.**, Insulin-like growth factor I: A possible metabolic signal involved in the regulation of female puberty, *Neuroendocrinology*, 54, 420, 1991.
64. **Hoffman, P.L., Rabe, C.S., Grant, K.A., Valverius, P.I., Hudspith, M., and Tabakoff, B.**, Ethanol and the NMDA receptor, *Alcohol*, 7, 229, 1990.
65. **Hompes, P., Vermes, I., Tilders, F., and Schoemaker, J.**, The *in vitro* release of LHRH from the hypothalamus of female rats during prepubertal development, *Neuroendocrinology*, 35, 8, 1982.
66. **Hugues, J.N., Coste, T., Perret, G., Jayle, M.T., Sebaoun, J., and Modigliani, E.**, Hypothalamo-pituitary ovarian function in thirty-one women with chronic alcoholism, *Clin. Endocrinol.*, 12, 543, 1980.
67. **Johnston, D.E., Chiao, Y.B., Gavaler, J.S., and Van Thiel, D.H.**, Inhibition of testosterone synthesis by ethanol and acetaldehyde, *Biochem. Pharmacol.*, 30, 1827, 1981.
68. **Kalra, S.P., Allen, L.G., and Kalra, P.S.**, Opioids in the steroid-adrenergic circuit regulating LH secretion: Dynamics and diversities, in *Brain Opioid Systems in Reproduction*, Dyer, R.G., and Bicknell, R.J., Ed., Oxford University Press, Oxford, 1989, 95.
69. **Kim K., and Ramirez, V.D.**, Effects of prostaglandin E_2, forskolin and cholera toxin on cAMP production and *in vitro* LH-RH release from the rat hypothalamus, *Brain Res.*, 386, 258, 1986.

70. **Lara, H.E., Dees, W.L., Hiney, J.K., Rivier, C., and Ojeda, S.R.,** Functional recovery of the developing rat ovary after transplantation: Contribution of the extrinsic innervation, *Endocrinology,* 129, 1849, 1991.
71. **Leslie, W., Brown, L.M., Dildy, J.E., and Sims, J.S.,** Ethanol and neuronal calcium channels, *Alcohol,* 7, 233, 1990.
72. **Lovinger, D.M., White, G., and Weight, F.F.,** Ethanol inhibits NMDA-activated ion current in hippocampal neurons, *Science,* 243, 1721, 1989.
73. **Lyon, R.C., Mc Comb, J.A., Schreurs, J., and Goldstein, D.B.,** A relationship between alcohol intoxication and the disordering of brain membranes by a series of short-chain alcohols, *J. Pharmacol. Exp. Ther.,* 218, 669, 1981.
74. **Matsumaoto, A., and Arai, Y.,** Precocious puberty and synaptogenesis in the hypothalamic arcuate nucleus in pregnant mare serum gonadotropin (PMSG)-treated immature female rats, *Brain Res.,* 129, 375, 1977.
75. **Mauceri, H., and Conway, S.,** The effects of acute ethanol exposure on clonidine-induced growth hormone release in male rats, *Alcohol,* 8, 7, 1991.
76. **McDonald, J.K., Dees, W.L., Ahmed, C.E., Noe, B.D., and Ojeda, S.R.,** Biochemical and immunocytochemical characterization of neuropeptide Y in the immature rat ovary, *Endocrinology,* 120, 1703, 1987.
77. **McGivern, R.F., and Yellon, S.M.,** Delayed onset of puberty and subtle alterations in GnRH neuronal morphology in female rats exposed to ethanol, *Alcohol,* 9, 335, 1992.
78. **Mello, N.K., Bree, M.P., Mendelson, J.H., Ellingboe, J., King, N.W., and Sehgal, P.,** Alcohol self-administration disrupts female reproductive function in primates, *Science,* 221, 677, 1983.
79. **Mendelson, J.H., Mello, N., and Ellingboe, J.,** Effects of acute alcohol intake on pituitary-gonadal hormones in normal human males, *J. Pharmacol. Exp. Ther.,* 202, 676, 1977.
80. **Minami, S., and Sarkar, D.K.,** Central administration of neuropeptide Y induces precocious puberty in female rats, *Neuroendocrinology,* 59, 930, 1992.
81. **Moskovic, S.,** Effect of chronic alcohol intoxication on ovarian dysfunction, *Srp. Arkh. Tselok Lek.,* 103, 751, 1975.
82. **Naftolin, F., and Brawer, J.R.,** Sex hormones as growth promoting factors for the endocrine hypothalamus, *J. Steroid Biochem.,* 8, 339, 177.
83. **Negro-Vilar, A., Ojeda, S.R., and McCann, S.M.,** Catecholaminergic modulation of luteinizing hormone-releasing hormone release by median eminence terminals *in vitro, Endocrinology,* 104, 1749, 1979.
84. **Neil, J.D., Patton, J.M., Dailey, R.A., Tson, R.C., and Tindall, G.T.,** Luteinizing hormone-releasing hormone (LHRH) in pituitary stalk blood of rhesus monkeys: Relationship to level of LH release, *Endocrinology,* 101, 430, 1977.
85. **Noe, M., Oliva, D., Corsini, A., Soma, M., Fumogalli, R., and Nicosia, S.,** Differential effects of *in vitro* ethanol on prostaglandin E_1-sensitive adenylate cyclase from smooth muscle cells and platelets, *J. Cyclic Nucleotide Protein Phosphor. Res.,* 10, 293, 1985.
86. **Nyberg, C.L., Hiney, J.K., Minks, J.B., and Dees, W.L.,** Ethanol alters N-methyl-DL-aspartic acid-induced secretion of luteinizing hormone and the onset of puberty in the female rat, *Neuroendocrinology,* 57, 863, 1993.
87. **Odell, W.D.,** Sexual maturation in the rat, in *Control of the Onset of Puberty,* Grumbach, M.M., Sizomenko, P.C., and Aubert, M.L., Eds., Williams & Wilkins, Baltimore, MD, 1990, 183.
88. **Ojeda, S.R., Andrews, W.W., Advis, J.P., and Smith-White, S.,** Recent advances in the endocrinology of puberty, *Endocr. Rev.,* 1, 228, 1980.
89. **Ojeda, S.R., and Campbell, W.B.,** An increase in hypothalamic capacity to synthesize prostaglandin E_2 proceeds the first preovulatory surge of gonadotropins, *Endocrinology,* 111, 1031, 1982.
90. **Ojeda, S.R., and Jameson, H.E.,** Developmental patterns of plasma and pituitary growth hormone in the female rat, *Endocrinology,* 100, 881, 1977.
91. **Ojeda, S.R., Naor, Z., and McCann, S.M.,** Prostaglandin E levels in hypothalamus, median eminence and anterior pituitary of rats of both sexes, *Brain Res.,* 149, 274, 1978.

92. **Ojeda, S.R., and Negro-Vilar, A.,** Prostaglandin E_2-induced luteinizing hormone-releasing hormone release involves mobilization of intracellular Ca^{+2}, *Endocrinology,* 116, 1763, 1985.
93. **Ojeda, S.R., Negro-Vilar, A., and McCann, S.M.,** Release of prostaglandin Es by hypothalamic tissue: Evidence of their involvement in catecholamine-induced luteinizing hormone-releasing hormone release, *Endocrinology,* 104, 617, 1979.
94. **Ojeda, S.R., Urbanski, H.F., and Ahmed, C.E.,** The onset of female puberty: Studies in the rat, *Recent Prog. Horm. Res.,* 42, 385, 1986.
95. **Ojeda, S.R., Urbanski, H.F., Katz, K.H., and Costa, M.E.,** Stimulation of cyclic adenosine 3', 5'-monophosphate production enhances hypothalamic luteinizing hormone-releasing hormone release without increasing prostaglandin E_2 synthesis: Studies in prepubertal female rats, *Endocrinology,* 117, 1175, 1985.
96. **Ojeda, S.R., Urbanski, H.F., Katz, K.H., and Costa, M.E.,** Prostaglandin E_2 releases luteinizing hormone-releasing hormone from the juvenile hypothalamus through a Ca^{2+}-dependent, calmodulin-independent mechanism, *Brain Res.,* 441, 339, 1988.
97. **Ondo, J.G., Wheeler, D.D., and Dom, R.M.,** Hypothalamic site of action for N-methyl-D-aspartate (NMDA) on LH secretion, *Life Sci.,* 43, 2283, 1988.
98. **Partington, C.R., Edwards, M.W., and Daly, J.W.,** Regulation of cyclic AMP formation in brain tissue by α-adrenergic receptors: Requisite intermediacy of prostaglandins of the E series, *Proc. Natl. Acad. Sci. USA,* 77, 3024, 1980.
99. **Pieper, D.R., Gala, R.R., Rigiani, S.R., and Marshal, J.C.,** Dependence of pituitary gonadotropin-releasing hormone (GHRH) receptors on GHRH secretion from the hypothalamus, *Endocrinology,* 110, 749, 1982.
100. **Pittaluga, A., and Raiteri, M.,** N-methyl-D-aspartic acid (NMDA) and non-NMDA receptors regulating hippocampal norepinephrine release. I. Location on axon terminals and pharmacological characterization, *J. Pharmacol. Exp. Ther.,* 260, 232, 1992.
101. **Pohl, C.R., Guilinger, R.A., and Van Thiel, D.H.,** Inhibitory action of ethanol on luteinizing hormone secretion by rat anterior pituitary cells in culture, *Endocrinology,* 120, 849, 1987.
102. **Price, M.T., Olney, J.W., and Cicero, T.J.,** Acute elevations of serum luteinizing hormone induced by kainic acid, N-methyl-aspartic acid, or homocysteic acid, *Neuroendocrinology,* 26, 352, 1978.
103. **Ramaley, J.A.,** The regulation of gonadotropin secretion in immature ethanol-treated rats, *J. Androl.,* 3, 248, 1982.
104. **Ramaley, J.A., and Phares, C.K.,** Delay of puberty onset in females due to suppression of growth hormone, *Endocrinology,* 106, 1989, 1980.
105. **Ramirez, V.D., Kim, K., and Dluzen, D.,** Progesterone action on the LHRH and the nigrostriatal dopamine neuronal systems: *In vitro* and *in vivo* studies, *Recent Prog. Horm. Res.,* 41, 421, 1985.
106. **Raum, W.J., Glass, A.R., and Swerdloff, R.S.,** Changes in hypothalamic catecholamine neurotransmitter and pituitary gonadotropins in the immature female rat: Relationships to the gonadostat theory of puberty onset, *Endocrinology,* 106, 1253, 1980.
107. **Rayback, R.S.,** Chronic alcohol consumption and menstruation, *J. Am. Med. Assoc.,* 238, 2143, 1977.
108. **Redmond, G.P.,** Effect of ethanol on endogenous rhythms of growth hormone secretion, *Alcoholism: Clin. Exp. Res.,* 4, 50, 1980.
109. **Rettori, V., Skelley, C.W., McCann, S.M., and Dees, W.L.,** Detrimental effects of short-term ethanol exposure on reproductive function in the female rat, *Biol. Reprod.,* 37, 1089, 1987.
110. **Sarkar, D.K., Smith, G.C., and Fink, G.,** The effect of manipulating central catecholamines on puberty and the surge of luteinizing hormone and gonadotropin-releasing hormone induced by pregnant mare serum gonadotropin in female rats, *Brain. Res.,* 213, 335, 1981.
111. **Schade, R.R., Bonner, G., Gay, V.C., and Van Thiel, D.H.,** Evidence for a direct inhibitory effect of ethanol upon gonadotropin secretion at the pituitary level, *Alcoholism: Clin. Exp. Res.,* 7, 150, 1983.

112. **Schultz, R., Wester, M., Duka, T., and Herz, A.,** Acute and chronic ethanol treatment changes endorphin levels in brain and pituitary, *Psychopharmacology*, 68, 221, 1980.
113. **Simson, P.E., Criswell, H.E., Johnson, K.B., Hicks, R.E., and Breese, G.R.,** Ethanol inhibits NMDA-evoked electrophysiological activity *in vivo*, *J. Pharmacol. Exp. Ther.*, 257, 225, 1991.
114. **Sirinathsinghji, D.J., Motta, M., and Martini, L.,** Induction of precocious puberty in the female rat after chronic naloxone administration during the neonatal period: The opiate "brake" on pubertal gonadotropin secretion, *J. Endocrinol.*, 104, 299, 1985.
115. **Southern, A.L., Gordon, G.G., Olivo, J., Rafii, F., and Rosenthal, W.S.,** Androgen metabolism in cirrhosis of the liver, *Metabolism*, 22, 695, 1973.
116. **Sundberg, D.K., Bo, W.J., and Reilly, J.,** Effect of chronic alcohol consumption on the pregnant mare serum gonadotropin-induced luteinizing hormone surge, *Neuroendocrinology*, 46, 283, 1987.
117. **Tal, J., Price, M.T., and Olney, J.W.,** Neuroactive amino acids influence gonadotropin output by a suprapituitary mechanism in either rodents or primates, *Brain Res.*, 273, 179, 1983.
118. **Terasawa, E., Bridson, W.E., Naso, T.E., Noonan, J.J., and Dierschke, D.J.,** Developmental changes in the luteinizing hormone secretory pattern in peripubertal female monkeys: Comparisons between gonadally intact and ovariectomized animals, *Endocrinology*, 115, 2233, 1984.
119. **Urbanski, H.F., and Ojeda, S.R.,** The juvenile peripubertal transition period in the female rat: Establishment of a diurnal pattern of pulsatile LH secretion, *Endocrinology*, 117, 644, 1985.
120. **Urbanski, H.F., and Ojeda, S.R.,** Activation of luteinizing hormone-releasing hormone release advances the onset of female puberty, *Neuroendocrinology*, 46, 273, 1987.
121. U**rbanski, H.F., and Ojeda, S.R.,** Gonadal-independent activation of enhanced afternoon luteinizing hormone release during pubertal development in the female rat, *Endocrinology*, 129, 907, 1987.
122. **Urbanski, H.F., and Ojeda, S.R.,** A role for N-methyl-D-aspartate (NMDA) receptors in the control of LH secretion and initiation of female puberty, *Endocrinology*, 126, 1774, 1990.
123. **Van Thiel, D.H., Gavaler, J.S., Cobb, C.F., Sherins, R.J., and Lester, R.,** Alcohol-induced testicular atrophy in the adult male rat, *Endocrinology*, 105, 888, 1979.
124. **Van Thiel, D.H., Gavalier, J., Lester, R., and Sherins, R.J.,** Alcohol-induced ovarian failure in the rat, *J. Clin. Invest.*, 61, 624, 1978.
125. **Van Thiel, D.H., Lester, R., and Sherins, R.J.,** Hypogonadism in alcoholic liver disease: Evidence for a double defect, *Gastroenterology*, 67, 1188, 1974.
126. **Voogt, J.L., Clemens, J.A., and Meites, J.,** Stimulation of pituitary FSH release in immature female rats by prolactin implant in median eminence, *Neuroendocrinology*, 4, 157, 1969.
127. **Wilson, R.C., and Knobil, E.,** Acute effects of N-methyl-D-aspartate on the release of pituitary gonadotropins and prolactin in the adult female rhesus monkey, *Brain Res.*, 248, 177, 1982.
128. **Wuttke, W., Honma, K., Lamberts, R., and Hohn, K.G.,** The role of monoamines in female puberty, *Fed. Proc.*, 39, 2378, 1980.

Chapter 15

THE EFFECT OF ALCOHOL ABUSE ON THE MENSTRUAL CYCLE

S.K. Teoh, N.K. Mello and J.H. Mendelson

TABLE OF CONTENTS

ABBREVIATIONS

CRF–corticotropin-releasing factor
E_2–estradiol
FSH–follicle-stimulating hormone
hCG–human chorionic gonadotropin
LH–luteinizing hormone
LHRH–luteinizing hormone-releasing hormone
NAD–nicotinamide adenine dinucleotide
NADH–nicotinamide adenine dinucleotide (reduced form)

INTRODUCTION

Alcohol abuse and alcoholism are associated with a broad spectrum of disorders of reproductive function in women. Amenorrhea, anovulation, luteal phase dysfunction, ovarian pathology and hyperprolactinemia may occur in alcohol-dependent women and alcohol abusers.[41,86,105,111-113,133] Persistent *hyperprolactinemia* is sometimes observed in alcoholic women studied during alcohol abstinence.[126,132] Luteal phase dysfunction, anovulation and persistent hyperprolactinemia have also been observed in social drinkers studied under clinical research ward conditions and in animal models of alcohol dependence. A higher frequency of *irregular menstrual cycles* (abnormal cycle duration and/or menstrual flow) has been reported by alcoholic women than by age-matched control women.[4,47] In some instances, alcohol abuse is associated with *early menopause*.[32-34]

The alcohol dose, frequency and duration of drinking necessary to disrupt menstrual cycle regularity is unknown. The extent to which tolerance may occur to alcohol's effects on reproductive function is unclear, but tolerance may be inferred from the fact that alcohol-dependent women do become pregnant. The reproductive consequences of alcohol abuse and alcoholism range from infertility and increased risk for spontaneous abortion to impairments of fetal growth and development.[74]

Examination of the endocrine consequences of alcohol abuse is important because it is increasingly recognized that many women of reproductive age have alcohol-related problems. For example, alcoholism and alcohol abuse were the fourth most frequent of all psychiatric disorders among young women aged 18 to 24 according to a survey of major metropolitan areas sponsored by the National Institute of Mental Health.[90] Women accounted for 24 percent of 10,000 first admissions to a proprietary hospital for alcoholism treatment.[76] In a 1992 National Household Survey,[1] more than half of all women ages 18 to 25 and 26 to 34 years reported that they had used alcohol during the previous month. In the general population, survey evidence suggests that heavy alcohol consumption is associated with menstrual cycle disorders as well as high rates of gynecological and obstetrical surgery.[140] This chapter focuses on recent studies of alcohol's effects on neuroendocrine function in social drinkers and in women who are alcohol dependent or alcohol abusers.

Information about alcohol's effects on reproductive function has been obtained from endocrine evaluations made during treatment for alcohol-related problems and from clinical histories of menstrual cycle abnormalities. However, alcoholic women often have a number of medical disorders such as liver disease or pancreatitis, sometimes complicated by malnutrition or infectious disease.[76] Since these medical disorders can also contribute to reproductive dysfunction, it is not possible to attribute abnormal menstrual cycles in alcoholic women to alcohol alone. However, recent replications of these reproductive disorders in healthy social drinkers[79] and in animal models of alcoholism under controlled conditions[26,35,55,67,72,106,137] indicate the generality of observations on alcoholic women with other medical complications.[66,73,74]

ALCOHOLISM AND REPRODUCTIVE SYSTEM DYSFUNCTION IN WOMEN

Amenorrhea

Amenorrhea, or the complete cessation of menses, may persist for months or years.[41,86,105,111-113,133] To the best of our knowledge, there have been no longitudinal studies of amenorrheic alcohol-dependent women to determine if tolerance to alcohol's effects develops over time. Case reports on two women suggest that amenorrhea may remit during alcohol abstinence, but menstrual cycles were not evaluated for normalcy and only one woman continued to menstruate.[105] Seven of eight amenorrheic women reported spontaneous resumption of menses after treatment for alcoholism but no endocrine measures were reported.[111,112]

The clinical literature on the endocrine concomitants of alcohol-related amenorrhea is quite limited. Twenty-two alcoholic women admitted for the treatment of liver disease or pancreatitis were studied in Paris (N = 13) and in Finland (N = 9).[41,133] Twenty-three women admitted for treatment of alcoholism were studied in Japan.[111,112] The Japanese women were not cirrhotic but did have hepatitis or fatty liver. Normal estrogen levels were measured in eight of the twenty-two European alcoholic amenorrheic women, and there was a positive estradiol response to stimulation with clomiphene or human chorionic gonadotropin (hCG).[41] But the other amenorrheic women had endocrine profiles similar to menopausal women. Fourteen European women had lower levels of estrogens and higher levels of luteinizing hormone (LH) and follicle-stimulating hormone (FSH) than normal controls. The amenorrheic Japanese women also had low estrogen levels and high FSH levels.[111,112] Estrogen levels were lower in women with the most severe amenorrhea than in women who responded to progesterone treatment with withdrawal bleeding (22.3 ± 9.4 pg/ml vs. 29.7 ± 12 pg/ml).[111,112] Ovarian pathology has been reported in postmortem studies of alcoholic women, alcohol-dependent rhesus monkeys and rats.[48,67,137] But abnormally low estrogen levels could reflect either impairment of ovarian function or disruption of gonadotropin secretory activity or both.

Pituitary function in abstinent alcoholic women was evaluated with synthetic luteinizing hormone-releasing hormone (LHRH). The LH and FSH response to LHRH stimulation (100 mcg) in European amenorrheic alcoholic women did not differ significantly from normal controls.[41,133] LHRH (100 mcg) also stimulated a rapid increase in LH and FSH in the amenorrheic Japanese women but the magnitude of the gonadotropin increase was significantly higher in women with less severe amenorrhea.[111,112] A normal gonadotropin response to synthetic LHRH stimulation suggested that the anterior pituitary may not be the primary site of alcohol's toxic effects in amenorrheic alcoholic women.

Possible Mechanisms Underlying Amenorrhea

The endocrine pathology underlying amenorrhea is poorly understood. A number of conditions other than alcohol abuse can contribute to the development of amenorrhea. Clinical disorders involving severe weight loss, such as anorexia nervosa, are often associated with persistent amenorrhea,[115] but obesity may also lead to amenorrhea.[30] In otherwise healthy, normal women, amenorrhea may be associated with weight reduction, jogging, or professional athletic pursuits.[9,30,31,61,115] It is unlikely that a single endocrine disruption accounts for amenorrhea of such disparate origins. A number of medical disorders (e.g., liver, kidney, and/or thyroid disease; polycystic ovaries and pituitary adenoma) also can result in persistent amenorrhea.[114] A discussion of all of these pathological conditions is beyond the scope of this review. We have focused on the possible contribution of alcohol-induced abnormalities in gonadotropin secretory patterns and the possible contribution of elevated levels of prolactin and corticotropin-releasing factor (CRF).

Amenorrhea and Gonadotropin Secretory Activity

It is possible that alcohol may suppress hypothalamic release of endogenous LHRH with concomitant suppression of gonadotropin secretory activity. Contemporary understanding of the neuroendocrine regulation of the menstrual cycle is based upon the fundamental discovery of the importance of pulsatile gonadotropin secretion for normal reproductive function (see Knobil[51,52] and Knobil and Hotchkiss[53] for review). When hypothalamic release of endogenous LHRH was disrupted in ovariectomized rhesus monkeys by lesions of the hypothalamic arcuate nucleus and the median eminence, LH and FSH secretory activity was abolished. Pulsatile administration of synthetic LHRH restored LH and FSH secretory patterns whereas continuous administration of LHRH did not (see Knobil[51,52] and Knobil and Hotchkiss[53] for review).

These preclinical data have been confirmed and extended by clinical studies that suggest that primary amenorrhea and secondary hypothalamic amenorrhea are associated with suppression of gonadotropin secretory activity.[5,16,19,98,108] A low *frequency* of LH pulses was most commonly associated with secondary hypothalamic amenorrhea, but low *amplitude* LH pulses were sometimes observed.[5,16,19,98,108] The most severe abnormalities were associated with a complete absence of LH pulses and low LH levels.[19,108] Normal ovulatory function can be restored in amenorrheic patients by pulsatile infusion of synthetic LHRH.[16,19,42,58,107,108]

Amenorrhea also developed in the female macaque monkey alcohol self-administration model.[67,72] Average LH levels were significantly lower during amenorrheic cycles (16.9 [±1.2] to 24 [±1.4] ng/ml) than during nonalcoholic control cycles (28 [±1.2] to 30 [±2.2] ng/ml).[72] These data in primates are consistent with the hypothesis that amenorrhea may be related to suppression of gonadotropin levels. However, there have been no systematic studies to confirm or refute the hypothesis that alcohol-induced amenorrhea reflects abnormal gonadotropin secretory patterns. At present, it is not known if alcohol suppresses gonadotropin secretory activity by suppressing hypothalamic LHRH release or by stimulation of prolactin or CRF.

Hyperprolactinemia and Alcohol-Related Amenorrhea

Hyperprolactinemia associated with normal postpartum lactation or with pituitary adenomas may cause amenorrhea and other disruptions of the menstrual cycle.[8,60,109,130] However, hyperprolactinemia is not invariably associated with amenorrhea.[8,60,109,130] Amenorrhea with normal prolactin levels was observed in alcoholic women with liver disease.[133] These alcoholic women (ages 23–40) reported amenorrhea of 3 to 12 months' duration, and their basal prolactin levels averaged 10.6 [±1.1] ng/ml.[133] Hyperprolactinemia without amenorrhea has also been reported in alcoholic women (ages 18–46) during abstinence[132] and in healthy social drinkers during daily consumption of between 4.24 and 8.24 drinks per day.[79] Hyperprolactinemia also was observed in 19 of 23 Japanese alcoholic women with amenorrhea.[111] Prolactin values averaged 92.9 [±9.9] ng/ml upon admission for alcoholism treatment. These women (ages 20–40) reported amenorrhea of 7–38 months' duration.[111]

In one amenorrheic alcohol-dependent macaque monkey, prolactin levels increased from 16.5 to 63 ng/ml during chronic high-dose alcohol self-administration (3.4 g/kg/day), and immunocytochemical examination of the anterior pituitary showed apparent hyperplasia of the lactotrophs.[67] Another monkey developed *galactorrhea* during a 97-day amenorrheic cycle when alcohol self-administration averaged 3.35 g/kg/day. These data suggested that hyperprolactinemia might contribute to alcohol-induced amenorrhea in this model, but this hypothesis was not confirmed in subsequent studies.[72] Examination of four other amenorrheic cycles (85–194 days) indicated that although prolactin levels were intermittently elevated above 20 ng/ml, average prolactin levels during the amenorrheic cycles (14.7 [±1.8] to 19.6 [±1.5] ng/ml) did not differ significantly from prolactin levels during normal ovulatory menstrual cycles when no alcohol was available (19.7 [±0.63] ng/ml).[72] These data suggest that hyperprolactinemia probably is not the primary mechanism underlying alcohol-induced amenorrhea in the female macaque monkey model.[72]

Acute Effects of Alcohol on Prolactin

Studies of the effects of *acute* alcohol intoxication on basal levels of prolactin in normal human subjects have yielded conflicting findings. In Finland, acute alcohol administration to normal women during the mid-luteal phase of the menstrual cycle significantly decreased basal prolactin levels over the first four hours of observation.[131] In Japan, 1.2 g/kg of alcohol given to normal women during the luteal phase of the menstrual cycle was followed by an increase in prolactin of 150 percent above baseline, and prolactin levels remained elevated throughout the 180-minute sampling period.[113] Blood alcohol levels remained above 100 mg/dl for 40 to 180 minutes after alcohol ingestion.[113] A comparable alcohol dose given during mid-follicular phase had no effect on basal prolactin when blood alcohol levels averaged 88 mg/dl, except in women who complained of nausea and vomiting.[82]

When prolactin levels are stimulated by a provocative test, alcohol tends to augment increases in prolactin levels. Alcohol increased prolactin stimulation by naloxone in mid-luteal phase women.[80] However, in early follicular phase women, higher peak blood alcohol levels (123 [±4.3] mg/dl) did not enhance naltrexone-stimulated prolactin levels.[128] When alcohol was given simultaneously with hCG, prolactin levels increased within 30 minutes in mid-luteal phase women, whereas hCG and placebo-alcohol administration was not followed by an increase in prolactin.[129] The basis for alcohol's stimulation of prolactin after naloxone and hCG stimulation (but not after naltrexone stimulation) is unknown, but an antecedent alcohol-related increase in estradiol may have affected the prolactin response. Prolactin increased 20 minutes after estradiol increased following concurrent hCG and alcohol administration.[129] Estradiol is known to decrease the sensitivity of the pituitary lactotrophs to dopamine suppression.[96] Consequently, an alcohol-related elevation in estradiol may modulate lactotroph sensitivity to the inhibitory effects of dopamine, which results in prolactin elevations.

Corticotropin-Releasing Factor and Amenorrhea

Another possibility is that alcohol may stimulate CRF, ACTH and adrenal hormones, which in turn suppress gonadotropin secretory activity and lead to amenorrhea. Alcohol, as well as stress, can stimulate CRF, ACTH and cortisol.[100-102] Administration of synthetic CRF inhibited pulsatile release of LH and FSH in ovariectomized rhesus females,[92] but administration of ACTH and cortisol did not.[142] Synthetic CRF administration also suppressed endogenous LHRH measured in rat portal blood.[94] These data suggest that CRF-induced suppression of LH and FSH is a central effect mediated through the hypothalamic/pituitary axis rather than through adrenal activation.[142] The role of alcohol-related increases in CRF to amenorrhea in alcohol-dependent women remains to be determined.

Anovulation and Luteal Phase Dysfunction

Alcoholic women who continue to menstruate may have anovulatory cycles or luteal phase dysfunction[41,86,132] (see Mello[66] and Mello *et al.*[73,74] for review). Alcoholic women also report more menstrual cycle abnormalities than age-matched controls.[4,47] The clinical significance of these disorders is that fertility is impaired either by prevention of pregnancy or by an increased risk for spontaneous abortion.

Anovulation, or failure to ovulate, is inferred from the absence of a mid-cycle gonadotropin surge and a subsequent elevation in progesterone during the luteal phase. Luteal phase dysfunction is defined either as a *short luteal phase defect* (eight days or less from ovulation to menses) or an *inadequate luteal phase* (when progesterone levels are abnormally low but the interval from ovulation to menstruation is of normal length).[23,36,115,123] In the normal cycle, the luteal phase lasts for 14 to 16 days after ovulation during which the corpus luteum develops from the post-ovulatory follicle, then regresses if pregnancy does not occur. In a fertile cycle, the corpus luteum is essential for maintaining the endocrine milieu of early pregnancy.

Spontaneous abortion is often associated with luteal phase defects. Some investigators argue that low progesterone levels may be more detrimental to pregnancy than a short luteal phase since implantation may occur within 8 days.[3] The prevalence of luteal phase defects in the general population is debated in part because accurate diagnosis is difficult.[3,63] One problem is that conclusions drawn from single samples of progesterone may be misleading because progesterone is secreted in a pulsatile pattern during the mid- and late-luteal phase of the menstrual cycle (see Filicori *et al.*,[27] McNeely and Soules,[63] and Soules *et al.*[120] for review). Moreover, it has been difficult to define the minimum parameters of luteal function that are necessary for initiation and maintenance of pregnancy.[123]

Only one endocrine evaluation of alcoholic women with anovulatory cycles or luteal phase defects has been reported.[41] Four women with *anovulatory* cycles had severe oligomenorrhea (scanty menses) and intermittent amenorrhea. Clomiphene administration induced a significant increase in LH and estradiol. Three of these alcoholic women had pancreatitis and one had cirrhosis.[41] Six women with *luteal phase inadequacy* had mild oligomenorrhea and low plasma progesterone levels during the luteal phase. Menstrual cycles were of normal length and gonadotropin and estradiol levels during the late follicular phase were normal. The hCG stimulation during the luteal phase increased progesterone levels to above 10 ng/ml in three of six women. Two of these women had cirrhosis and four had pancreatitis.[41]

Possible Mechanisms Underlying Anovulation and Luteal Phase Dysfunction

The Role of Impaired Folliculogenesis

The factors that account for alcohol-related anovulation and luteal phase dysfunction are poorly understood. As with amenorrhea, a number of factors including systemic diseases and exercise may contribute to these disorders.[9,63,123] Analysis of these disorders is complicated since each results from abnormalities that occur earlier in the sequence of hormonal events preceding ovulation. For example, abnormalities in follicle development during the follicular phase may cause luteal phase defects as well as anovulation.[36,63,123] Moreover, the relative contribution of hypothalamic-pituitary and ovarian factors to anovulation and luteal phase defects is unclear.

Although FSH is not the sole determinant of folliculogenesis, adequate FSH levels are necessary for normal follicular development and maturation during the follicular phase.[36,103] It is well established that suppression of FSH during the follicular phase may delay follicle maturation and ovulation or result in luteal phase dysfunction after timely ovulation.[23,24,36,139] Systematic studies of folliculogenesis in the primate ovarian cycle suggest that recruitment of the dominant follicle occurs during menstrual cycle days 1–4; a single follicle is selected during days 5–7, and that follicle achieves dominance during cycle days 8–12.[23,36,39] Although the determinants of the selection and dominance of a single ovulatory follicle are unclear, it has been postulated that non-steroidal ovarian peptides, the inhibins, critically affect this process through titration of FSH levels during folliculogenesis.[36] Low levels of inhibin during the early and mid-follicular phase were measured in women with luteal phase deficiency.[121] If alcohol intoxication suppresses FSH directly or modulates inhibin to down-regulate FSH secretory activity, this could produce aberrations in folliculogenesis, which culminate in anovulation or luteal phase dysfunction (see Mello[66] and Mello *et al.*[73,74] for review).

Alcohol's Effects on Anterior Pituitary Gonadotropins During the Follicular Phase

Clarification of alcohol's effects during the follicular phase is important because normal follicle growth and development are essential for a normal menstrual cycle. There is evidence that alcohol prevents stimulation of FSH by synthetic LHRH during the follicular phase of the menstrual cycle in normal female rhesus monkeys.[71] In contrast to FSH, LH increased significantly within 15 minutes after synthetic LHRH stimulation when blood alcohol levels averaged 184 and 276 mg/dl. When an isocaloric sucrose solution was substituted for alcohol, both FSH and LH increased significantly after LHRH stimulation.[71] If alcohol also inhibits FSH responsivity to *endogenous* LHRH stimulation during the follicular phase, this could result in menstrual cycle irregularities commonly seen in alcohol-dependent females.

Inferential evidence for the importance of *ovarian factors* in modulating alcohol's effects on pituitary FSH is based on the finding that alcohol did not suppress LHRH-stimulated FSH in *ovariectomized* rhesus females.[70] After synthetic LHRH stimulation, LH and FSH increased significantly in ovariectomized females when blood alcohol levels averaged 242 and 296 mg/dl.[70] The non-steroidal ovarian peptide, inhibin, suppressed FSH without affecting LH.[11] In the normal human menstrual cycle, inhibin is inversely related to FSH during the mid- to late-follicular phase.[62] It is possible that alcohol may suppress LHRH-stimulated FSH by stimulating ovarian inhibin, but data on alcohol's effects on inhibin are not yet available.

It is important to emphasize that clinical studies have not consistently implicated decrements in FSH to account for luteal phase deficiencies. Administration of rapid pulses of synthetic LHRH during the follicular phase induced a luteal phase deficiency in normal women.[118] High LH pulse frequency during the early follicular phase has also been observed in luteal phase deficiency patients, and this may also compromise follicular development (see McNeely and Soules[63] for review). For example, 20 women with diagnosed luteal phase deficiency had FSH levels that were equivalent to FSH levels in 21 control women during the early- and late-follicular phase.[119] A significantly higher LH pulse frequency during the early follicular phase distinguished patients from controls. This LH pulse frequency (12.8 [±1.4] pulses/12 hours) persisted throughout the follicular phase in the patients, whereas LH pulse frequency in controls increased from 8.2 ± 0.7 pulses per 12 hours to approximately 15 pulses per 12 hours during the late follicular phase. LH pulse amplitude was also lower in some patients.[119] These findings suggest that stable high frequency LH pulses during the early follicular phase may also contribute to luteal phase defects that are secondary to impaired follicular development. There remain many controversies and unresolved questions concerning the pathogenesis, differential diagnosis and prevalence of luteal phase dysfunction. These issues have been critically examined in recent reviews by Stouffer,[123] McNeely and Soules[63] and Brodie and Wentz.[7]

Alcohol's Effects on Ovarian Hormones During the Follicular Phase

An alternative possibility is that alcohol may increase *estradiol* levels, which in turn suppress FSH during the follicular phase and impair or delay follicle maturation and ovulation (see Hutz *et al.*[45] for review). There is considerable evidence that an increase in estradiol levels during the early follicular phase suppresses FSH and inhibits pre-ovulatory follicular growth and prolongs the follicular phase.[21,22,145] Luteal phase defects were consistently observed after 6, 12 and 24 hours of exposure to estradiol during the follicular phase (day 6 or 7 of the menstrual cycle).[21] An increase in estradiol levels of about 30 pg/ml significantly reduced FSH concentrations and prolonged the follicular phase.[145]

Clinical studies in normal women have reported significant increases in estradiol levels during alcohol intoxication. Acute alcohol administration (0.695 g/kg) induced a significant increase of 19.5 [±4.1] pg/ml in estradiol levels under basal (non-stimulated) conditions.[77] Plasma estradiol reached peak levels within 25 minutes after initiation of drinking when blood alcohol levels were relatively low and averaged 34 mg/dl. These data are shown in Figure 1. Collection of plasma samples every 5 minutes permitted detection of an alcohol-related increase in plasma estradiol levels during the ascending phase of the blood alcohol curve. Previous studies of acute alcohol effects on estradiol levels used 20-minute integrated sample collection procedures,[82] and no significant changes in plasma estradiol levels were detected during the ascending, peak or descending phase of the blood alcohol curve. Alcohol (2.5 g/kg) also significantly increased estradiol within 150 to 210 minutes under basal (non-stimulated) conditions in mid-luteal phase rhesus females.[68]

Increases in estradiol levels during alcohol intoxication were even higher following gonadotropin stimulation by naloxone and naltrexone.[80,128] The alcohol-related augmentation of opioid antagonist-stimulated estradiol was 45 to 50 pg/ml. These estradiol levels are equivalent to those shown to selectively suppress FSH secretion in clinical studies (40 to 50 pg/ml).[59]

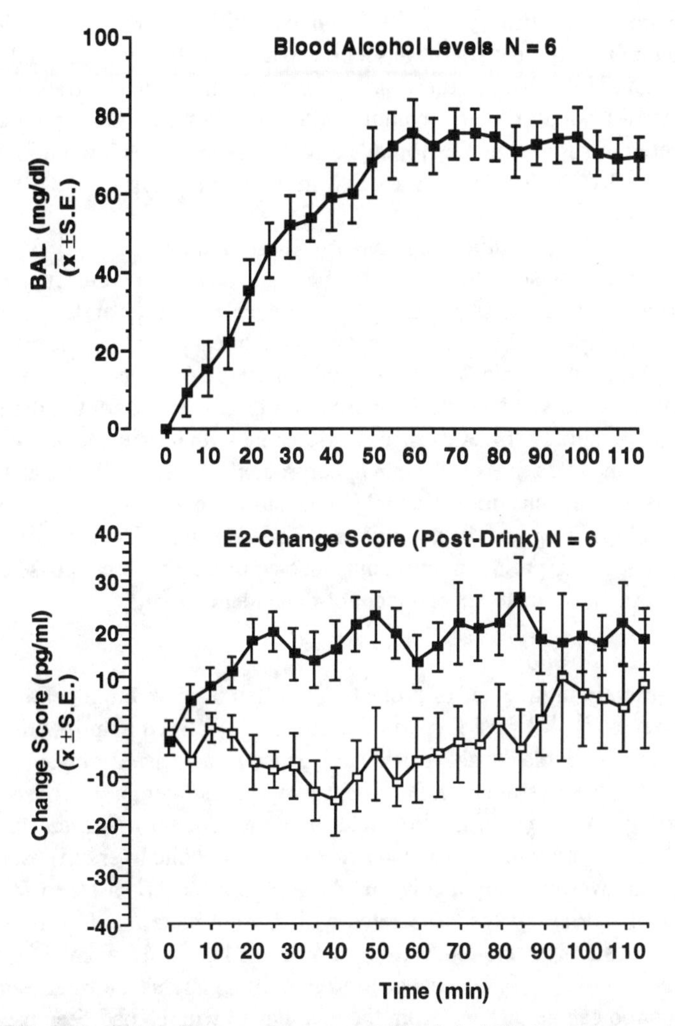

FIGURE 1. Acute alcohol effects on plasma estradiol levels in follicular phase women. Blood alcohol levels (mg/dl) following acute administration of 0.695 g/kg alcohol (top panel) and estradiol levels following alcohol (black square) or placebo (white square) administration (lower panel). Each data point represents six subjects (mean ± SEM). Data are presented as delta change scores from baseline. Integrated plasma samples were collected at 5-minute intervals. (Reprinted from Mendelson, J.H. *et al., Psychopharmacology* 94, 464, 1988, with permission of Springer-Verlag.)

Alcohol's Effects During the Luteal Phase

Alcohol's Effects on Ovarian Hormones During the Luteal Phase

Alternatively, alcohol-related increases in estradiol during the luteal phase could also produce the functional equivalent of a luteal phase defect. Administration of estrogen and progesterone capsules to rhesus females on luteal phase days 2–6 resulted in low progesterone, and menses occurred 5 to 6 days earlier than in control cycles.[44] In monkeys with hypothalamic lesions or hypothalamic-pituitary stalk transection, where ovulatory menstrual cycles were restored by pulsatile administration of synthetic LHRH, estradiol administration did not result in premature luteal regression.[44] These data were interpreted to suggest that estrogen's effects on the corpus luteum are mediated by the hypothalamic-pituitary axis in intact monkeys.

Luteal Phase Dysfunction and Prolactin Abnormalities

Luteal phase dysfunction may also be associated with either increases or decreases in prolactin from normal levels (see McNeely and Soules[63] for review). Low levels of progesterone during the luteal phase may be associated with hyperprolactinemia (20 to 40 ng/ml) in 10 to 20 percent of luteal phase deficiency patients.[63] Decreases in prolactin levels secondary to administration of a dopamine agonist were also associated with low levels of progesterone during the luteal phase.[110] Both hyperprolactinemia and decreased prolactin levels may occur during alcohol intoxication, depending on the dose and duration of drinking and the conditions of gonadotropin stimulation. Alcohol-related changes in prolactin levels may also contribute to luteal phase defects observed in social drinkers and alcohol-dependent women.[41,79]

Hyperprolactinemia

Abnormally high levels of prolactin are often observed in alcohol-dependent women.[111-113,126,132] Hyperprolactinemia is defined as prolactin levels above 20 ng/ml. Prolactin abnormalities also may be associated with amenorrhea and luteal phase dysfunction. Persistent hyperprolactinemia was observed in 16 alcoholic women during 6 weeks of treatment on a clinical research ward.[132] None of these women had evidence of alcoholic liver cirrhosis. They reported an average daily alcohol intake of 170 g for the past 2 to 16 years. Regular menstrual cycles, associated with normal patterns of gonadotropin and ovarian steroid secretion, were observed in 14 of the 16 women (ages 18–46 years). Anovulatory cycles occurred in only 2 patients. Normal reproductive function can be inferred from the fact that 13 women had been pregnant. Of 30 reported pregnancies, 16 were completed successfully; 2 were terminated by spontaneous abortions and 12 by legal abortions.[132]

In Japan, 22 of 23 women admitted for alcoholism treatment had prolactin levels above 25 ng/ml upon admission.[111,112] Six women had prolactin levels above 100 ng/ml (115–184 ng/ml) and 10 women had prolactin levels above 50 ng/ml (59–97 ng/ml). Of the remaining 7 women, 6 had elevated prolactin levels ranging between 27 and 38 ng/ml.[111,113] However, in contrast to the Finnish sample, prolactin levels returned to normal after up to 3 months of treatment.[111] None of the Japanese women had cirrhosis but 10 had hepatitis and the rest, fatty liver. These women (ages 20–40) met DSM III-R criteria for alcoholism and reported drinking an average of 84.1 g of alcohol each day for at least 7 years. All women had oligomenorrhea ($N = 2$) or amenorrhea ($N = 21$) of 7 to 38 months' duration. Their reproductive history was not described.[111-113]

Six of twelve alcohol-dependent women admitted to a Massachusetts hospital for treatment under civil commitment had hyperprolactinemia ranging from 22.3 to 87.5 ng/ml.[126] Prolactin determinations were based on a single sample collected 7 to 10 days after admission to the treatment facility. The hyperprolactinemic women reported a 7- to 33-year history of regular drinking of 75.74 to 247.2 g of alcohol per day. Four women were post-menopausal and two women were of reproductive age. Each of the latter reported regular menses and one reported three successful pregnancies. The clinical significance of hyperprolactinemia in this subset of alcohol-dependent women is unclear. Of the women who had normal prolactin levels, one reported irregular menses and one reported amenorrhea for the past 12 months and her LH and estradiol levels were low. However, this report of amenorrhea cannot be attributed to alcoholism per se since the patient reported a history of anorexia nervosa. In another subgroup of six polysubstance abusers, five reported regular menses and four reported live births despite a history of abuse of alcohol, cocaine, opiates, marijuana and amphetamines.[126] This small sample of socially and economically disadvantaged women[57] illustrates the relative resilience of the reproductive system despite chronic alcoholism and/or polysubstance abuse and intercurrent medical problems.

ALCOHOL'S EFFECTS ON REPRODUCTIVE FUNCTION IN SOCIAL DRINKERS

Alcohol-related *luteal phase dysfunction, anovulation,* and *hyperprolactinemia* have also been observed in healthy, well-nourished women during residence on a clinical research ward for 35 days.[79] After a 7-day alcohol-free baseline period, these social drinkers could self-administer alcohol for 21 consecutive days. Women could earn alcohol (beer, wine, or distilled spirits) or money ($0.50) for 30 minutes of performance on a simple operant task, a second-order fixed-ratio 300, fixed-interval 1 second schedule of reinforcement (FR 300 FI 1 sec:S). Points earned for alcohol and for money were not interchangeable.[65,79] Following 3 weeks of alcohol availability, women remained on the clinical research ward for an additional 7 days.

Women were classified as heavy, social or occasional alcohol users on the basis of the actual number of drinks consumed during three consecutive weeks of alcohol availability. Five women who consumed an average of 7.8 [±0.69] drinks per day were classified as heavy drinkers. Twelve women who consumed an average of 3.84 [±0.19] drinks per day were classified as moderate drinkers and nine women who consumed an average of 1.22 [±0.21] drinks per day were classified as occasional drinkers. These drinking patterns were consistent with each subject's self report of alcohol use before admission to the clinical research ward. The heavy, social and occasional alcohol users reported an average drinking history of 7.5, 6.6, and 6.9 years respectively.

During the 21 days of alcohol availability, average peak blood alcohol levels measured in the moderate and heavy drinkers ranged from 109 [±16] to

199 [±13] mg/dl. Peak blood alcohol levels measured in the occasional drinkers averaged between 48 [±10] and 87 [±22] mg/dl. Individual women sometimes achieved higher blood alcohol levels than the group average. Peak blood alcohol levels for individual heavy alcohol users ranged between 69 and 196 mg/dl. Peak blood alcohol levels for individual moderate alcohol users ranged between 27 and 233 mg/dl. Individual peak blood alcohol levels for the occasional alcohol users ranged between 5 and 159 mg/dl.[79]

Sixty percent of the heavy drinkers and fifty percent of the social drinkers who consumed more than three drinks per day had significant derangements of the menstrual cycle.[79] Three heavy drinkers and one moderate social drinker who consumed between 4.24 and 8.24 drinks per day, had persistent *hyperprolactinemia* (defined as elevations in plasma prolactin levels above 20 ng/ml) during at least 7 of the 21 days of alcohol consumption. Hyperprolactinemia in a heavy drinker who consumed an average of 8.24 drinks per day is shown in Figure 2. Plasma prolactin levels were significantly elevated within 5 days after initiation of drinking and reached peak levels at the same time as a normal LH surge on study day 16. This woman's prolactin levels remained elevated throughout the luteal phase and after cessation of drinking.

Alcohol consumption was also associated with disruption of folliculogenesis. An example of a prolonged follicular phase with *delayed ovulation* in a woman who drank an average of 4.10 (±0.77 drinks per day) is shown in Figure 3. This subject did not ovulate until the 28th day of her menstrual cycle. Three moderate social drinkers who consumed between 3.48 and 4.05 drinks per day had *anovulatory* cycles.

These alcohol-related menstrual cycle disorders appear to be alcohol dose dependent. There was no evidence of menstrual cycle dysfunction or abnormal hormone levels in the occasional drinkers or in two of the moderate social drinkers who consumed less than an average of three drinks per day. However, five of ten social drinkers who drank more than three drinks per day and three of the five heavy drinkers had significant derangements of the menstrual cycle and reproductive hormone function. The contrast between the occasional and heavy social drinkers suggests that these abnormalities can be attributed to alcohol and not to living conditions on the research ward per se. Since these women were otherwise healthy and well nourished, it appears that alcohol and not extraneous factors accounts for the menstrual cycle derangements observed.[79]

It is important to emphasize that alcohol did not invariably cause menstrual cycle abnormalities. The woman who consumed the most alcohol (10 [±0.69] drinks per day) did not have abnormal menstrual cycles or hyperprolactinemia. It is possible that this heavy drinker had developed tolerance for alcohol.[79] The existence of reproductive system tolerance for alcohol can be inferred from the fact that many alcohol abusers and alcohol-dependent women have normal pregnancies.[37,40,126]

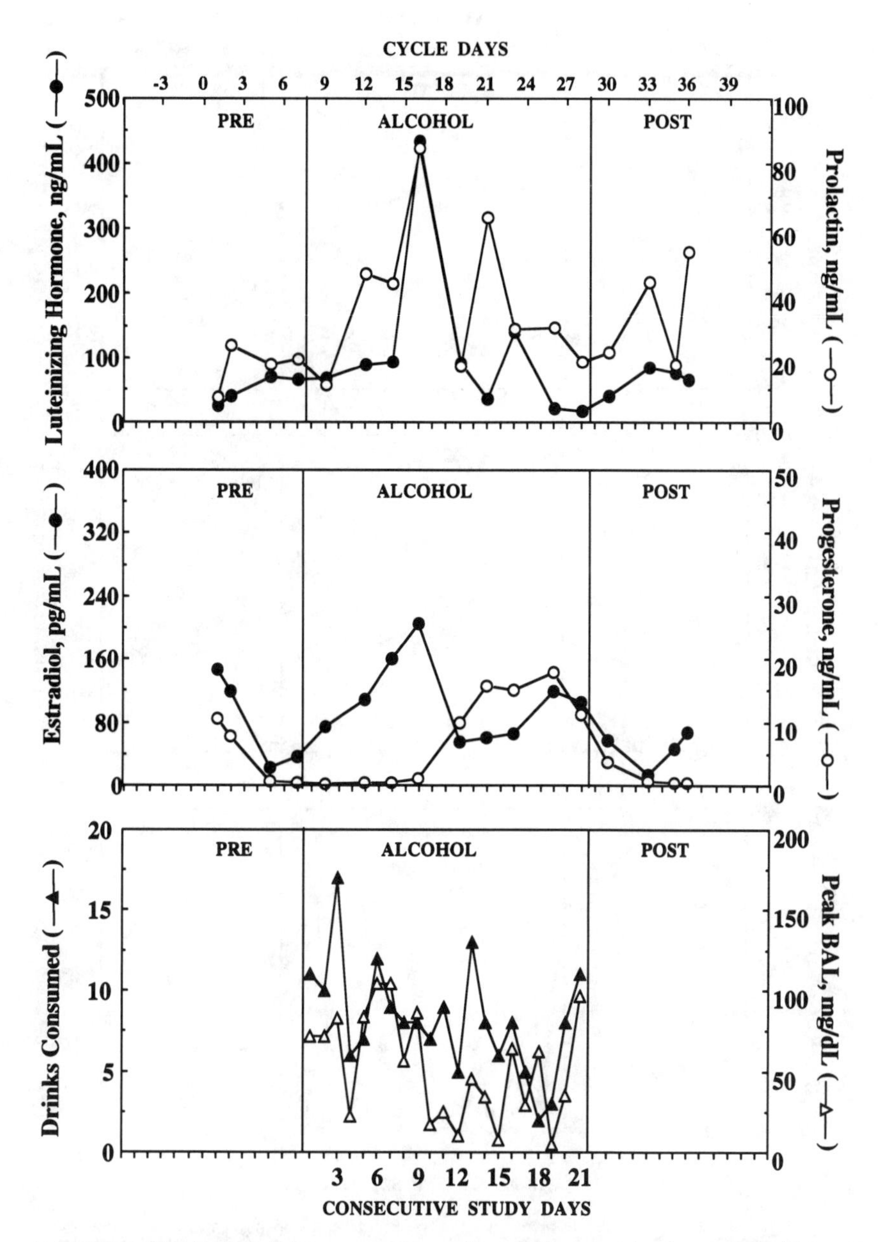

FIGURE 2. Chronic alcohol effects on pituitary and gonadal hormones in a heavy social drinker. LH (ng/ml), prolactin (ng/ml), estradiol (pg/ml) and progesterone (ng/ml) levels measured before, during and after 21 days of alcohol self-administration (top and middle panel). The number of drinks consumed (mean±8.24 [±0.75] drinks per day) and peak blood alcohol levels (mg/dl) during operant response-contingent alcohol self-administration on a clinical research ward are shown in the lower panel. This woman had recurrent hyperprolactinemia during daily alcohol consumption. (Reprinted with permission from Mendelson, J.H., and Mello, N.K., *J. Pharmacol. Exp. Ther.*, 245, 407, 1988.)

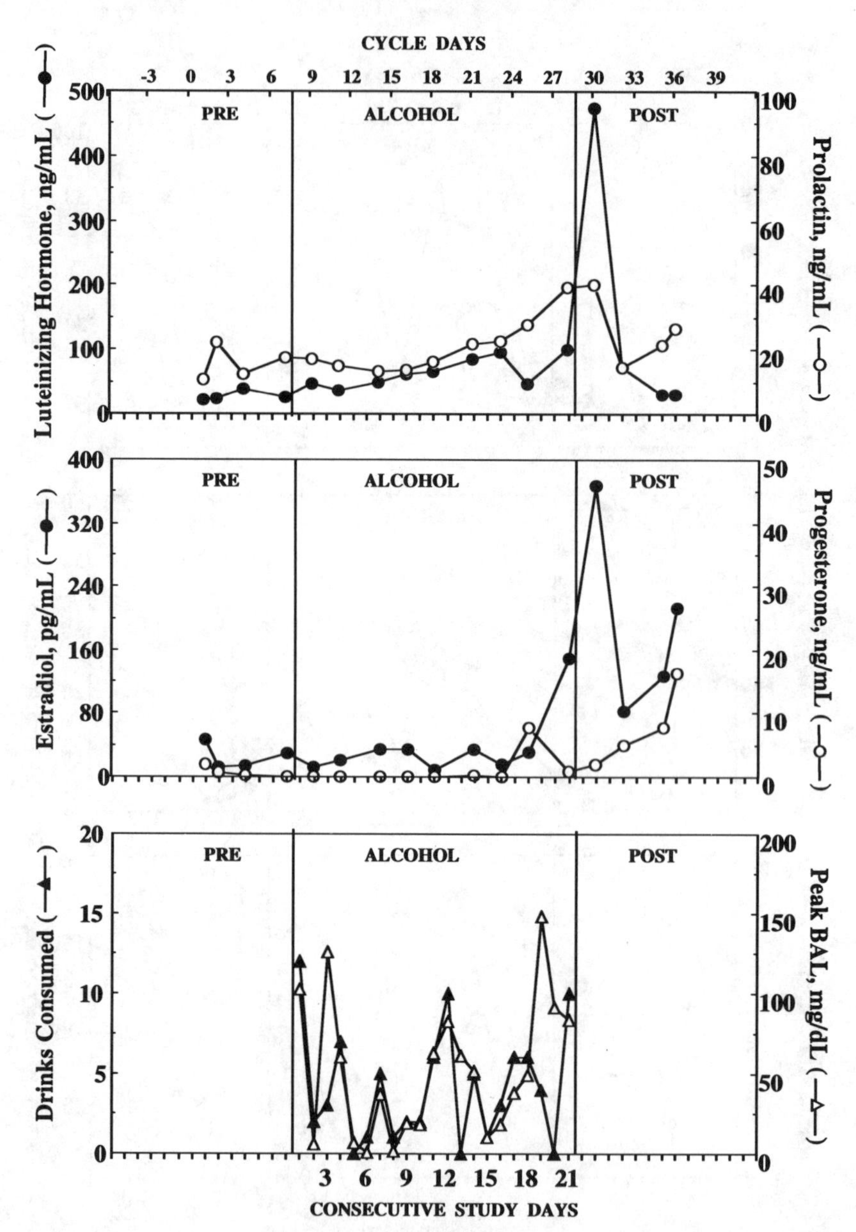

FIGURE 3. Chronic alcohol effects on pituitary and gonadal hormones in a moderate social drinker. LH (ng/ml), prolactin (ng/ml), estradiol (pg/ml) and progesterone (ng/ml) levels were measured before, during and after 21 days of alcohol self-administration (top and middle panel). The number of drinks consumed (mean±4.10 [±0.77] drinks per day) and peak blood alcohol levels (mg/dl) during operant response-contingent alcohol self-administration on a clinical research ward is shown in the lower panel. This woman did not ovulate until day 28 of this menstrual cycle. (Reprinted with permission from Mendelson, J.H., and Mello, N.K., *J. Pharmacol. Exp. Ther.*, 245, 407, 1988.)

One implication of these data on the adverse effects of chronic alcohol intake on anterior pituitary and ovarian hormones in healthy women is that alcohol-related menstrual cycle abnormalities and reproductive hormone dysfunctions may be more prevalent among social and heavy drinkers than is usually assumed. This conclusion is concordant with survey evidence that alcohol drinking and reproductive dysfunction are related in the general, non-clinical population.[140] A stratified household sample of 917 women showed a strong association between alcohol consumption and several menstrual disorders, including dysmenorrhea, heavy menstrual flow, and premenstrual discomfort. The incidence of these disorders increased as a concomitant of reported drinking levels. Women who consumed six or more drinks each day at least five times a week had elevated rates of gynecological surgery (other than hysterectomy) and obstetrical disorders.[140]

ANALYSIS OF THE PATHOGENESIS OF ALCOHOL-RELATED REPRODUCTIVE SYSTEM DYSFUNCTIONS

The Contribution of Animal Models

Although there is increasing evidence that a variety of disorders of reproductive function are associated with alcohol dependence and alcohol abuse, as well as moderate social drinking, very little is known about the mechanisms of alcohol's effects on the neuroendocrine regulation of the menstrual cycle (see Mello[66] and Mello *et al.*[73,74] for review). Since data from human alcohol abusers are often complicated by liver disease and other medical problems, as well as polydrug abuse, animal models of alcoholism are essential for a systematic analysis of alcohol's effects on the hypothalamic-pituitary-gonadal-adrenal axis. Alcohol's effects can be studied in animal models under controlled conditions where other substance abuse, malnutrition and/or intercurrent illness cannot contribute to results obtained. Female rhesus monkeys offer the additional advantage that neuroendocrine regulation of their menstrual cycle is very similar to that of human females.[51,52] Rhesus monkeys have long been the model of choice in reproductive biology.[36,51,52]

Chronic alcohol self-administration resulted in amenorrhea, atrophy of the uterus and decreased ovarian mass in otherwise healthy female macaque monkeys.[67] When monkeys were taught to self-administer alcohol intravenously on a simple operant task, daily self-administration of high doses of alcohol (2.9 to 4.4 g/kg/day) was accompanied by amenorrhea that persisted for 84 to over 200 days.[67,72] Amenorrheic monkeys developed blood alcohol levels ranging from 266 to 438 mg/dl immediately following an alcohol self-administration session.[67] These blood alcohol levels are comparable to those observed in alcoholic men during intoxication.[69] In contrast, monkeys that self-administered relatively low doses of alcohol (1.3 and 1.6 g/kg/day) for 119 and 173 days, respectively, continued to have stable ovulatory menstrual cycles.[67] Our ongoing studies with the primate alcohol self-administration model have shown that chronic alcohol administration also suppresses ovulation and results in luteal phase dysfunction in otherwise healthy animals. Replication of endocrine pathology observed in alcoholic women in animal models of alcoholism, under controlled experimental conditions, attests to the generality of the clinical findings and to the role of alcohol in the development of these disorders.

Provocative Tests of Hormonal Function

Examination of alcohol's effects on pituitary, hypothalamic, ovarian and adrenal function in animal models and in clinical studies has been greatly facilitated by the availability of provocative tests commonly used in clinical endocrinology.[99,143,144] Each component of the hypothalamic-pituitary-gonadal-adrenal axis can be selectively stimulated to evaluate the primary site or sites of alcohol's toxic effects. Provocative tests of endocrine release provide a tool for analysis of alcohol's effects on the various facets of this complex, interrelated system. For example, *synthetic LHRH* can be used to directly stimulate pituitary release of LH and FSH (see Filicori *et al.*[28] and Yen[143] for review), and the effects of alcohol on LHRH-stimulated gonadotropins can be examined.[66,70,71,73,74,83,95]

Opioid antagonist drugs also stimulate release of pituitary gonadotropins, presumably by antagonism of endogenous opioid peptides which modulate the inhibitory regulation of endogenous LHRH in the hypothalamus.[78,81,84,85,144] Two opioid antagonists, naloxone and naltrexone, are used to stimulate hypothalamic release of endogenous LHRH followed by pituitary release of LH, FSH and prolactin.[144] One disadvantage of the short-acting narcotic antagonist, naloxone, is that it is only effective during the late follicular and luteal phase of the menstrual cycle.[144] In contrast, naltrexone, a long-acting opioid antagonist, significantly stimulates FSH and LH during the early follicular phase in women.[81] Naltrexone also stimulates release of ACTH and cortisol in women.[81]

Human chorionic gonadotropin (hCG) stimulates release of ovarian steroid hormones and can be used to evaluate the effects of alcohol on ovarian function.[129] hCG also can be used to simulate the endocrine milieu of early pregnancy in the primate model.[93,138] In early pregnancy, endogenous chorionic gonadotropin, secreted by the conceptus, prolongs the life span of the corpus luteum until the placenta begins to secrete progesterone and estrogens.[50,87]

ALCOHOL-INDUCED STIMULATION OF PITUITARY AND GONADAL HORMONES

Traditionally, disorders of reproductive function associated with alcoholism have been attributed to alcohol's *suppressive* effects on pituitary and gonadal hormones in both women and men. For example, alcohol-induced amenorrhea is usually attributed to a decreased frequency and/or amplitude of gonadotropin pulsatile secretory activity and/or suppression of ovarian hormones (estradiol, progesterone) essential for normal menstrual cycle function (see Mello[66] and Mello *et al.*[73,74] for review). In alcoholic men, impotence, testicular atrophy, and gynecomastia are associated with low testosterone levels reflecting inhibition of testosterone biosynthesis in the testes.[6,12-14,25,91,134-136,141] However, with the availability of provocative tests for evaluation of pituitary, hypothalamic and ovarian function and improved techniques for rapid blood sample collection, a more complex picture of alcohol's effects on pituitary and gonadal hormones has emerged. There now appears to be a disparity between the *acute* and *chronic* effects of alcohol on the hypothalamic-pituitary-gonadal axis and the hypothalamic-pituitary-adrenal axis. Acute alcohol intoxication appears to *stimulate* rather than suppress gonadotropins and ovarian hormones under a variety of experimental conditions (see Mello[66] and Mello *et al.*[73,74] for review). Table 1 summarizes recent data illustrating alcohol's stimulatory effects on pituitary, gonadal and adrenal hormones.

The physiological basis for the alcohol-induced augmentation of LH, prolactin, estradiol and testosterone shown in Table 1 is unclear. It is possible that increased pituitary sensitivity to LHRH stimulation after alcohol administration may reflect a direct alcohol effect on endogenous LHRH or on other hormones, such as estradiol, which are known to modulate pituitary sensitivity to LHRH. Increased estradiol levels after alcohol administration could enhance the LH response to LHRH stimulation, just as the mid-cycle LH surge in normally cycling rhesus females is dependent upon the periovulatory increase in estradiol.[49] The sustained significant elevation in LHRH-stimulated LH after alcohol (165 minutes) in comparison to placebo control (105 minutes)[70] is consistent with earlier studies of estradiol pretreatment in ovariectomized monkeys.[54] There is also evidence that estradiol pretreatment increases pituitary sensitivity to LHRH stimulation in both normal and hypogonadal women[46,56] and in the intact diestrous rat.[2] Consequently, if alcohol administration did increase estradiol levels in ovariectomized monkeys,[70] this could have sensitized the pituitary to produce an augmented LH response to LHRH stimulation. Although ovariectomy reduces circulating estradiol by approximately 60 percent, estrogens are produced in the adrenal and peripheral conversion of androgens to estrogens.[32-34,103] Unfortunately, estradiol was not measured in the ovariectomized females in which alcohol enhanced LHRH stimulated LH.[70] However, estradiol levels were undetectable in our subsequent ongoing studies in ovariectomized females.

The alcohol-related increase in plasma estradiol (E_2) levels after naloxone and naltrexone stimulation[77,80,128] could be accounted for by several mechanisms. It is possible that alcohol may increase estradiol production or decrease estradiol metabolism. We have suggested elsewhere that since intrahepatic ethanol metabolism decreases NAD availability for other coupled oxidative reactions,[17,18,88,89] this might reduce the rate of oxidation of estradiol to estrone and result in elevated estradiol levels.[77,80] Hepatic and gonadal oxidative metabolism of steroids may become rate limiting during alcohol metabo-

Table 1: Alcohol Induced Stimulation of Pituitary, Gonadal and Adrenal Hormones

Subjects	Experimental Conditions	Alcohol Dose	BAL Peak	HORMONES								Sample Frequency	Reference
				Pituitary				Gonadal			Adrenal		
				LH	*FSH*	*PRL*	*ACTH*	*E2*	*Prog*	*Test*	*Cortisol*		
Women n=9 Mid-Luteal	Naloxone 5 mg,i.v.	1 ml/kg	100 ±13 mg/dl	+		++ P<.001		++ P<.004	≠			5-15 minutes	Mendelson et al., 1987
Women n=14 Follicular	Naltrexone 50 mg,P.O.	2.2 ml/kg	123 ±4.3 mg/dl	++ P<.05		+		++ P<.001	[P<.001		+	5-30 minutes	Teoh et al., 1988
Women n=12 Follicular	Basal levels	350 ml	70-75 mg/dl					+ P<.01				5 minutes	Mendelson et al., 1988
Women n=12 Follicular	LHRH 100 mcg,i.v.	0.694 g/kg	113.3 ±7.5 mg/dl					+ P<.0001				5-30 minutes	Mendelson et al., 1989
Women n=12 Mid-Luteal	LHRH 100 mcg,i.v.	0.694 g/kg	121.6 ±11.1 mg/dl					+ P<.01				5-30 minutes	Mendelson et al., 1989
Rhesus Females n=8 Follicular	LHRH 100 mcg,i.v.	2.5-3.5 g/kg	204 ±11 - 338 ±22 mg/dl	+	≠							15-20 minutes	Mello et al., 1986a
Rhesus Females n=5 OVX	LHRH 100 mcg,i.v.	2.5-3.5 g/kg	245 ±26 - 296 ±20 mg/dl	++	+							15-20 minutes	Mello et al., 1986b
Rhesus Males n=5	Naloxone 0.5 mcg/kg,i.v.	2.5-3.5 g/kg	283 ±9 - 373 ±17 mg/dl	+						+		30 minutes	Mello et al., 1985
Men n=6	LHRH 500 mcg	0.695 g/kg	83.1 ±5.5 mg/dl	+	+	++ P<.03				++ P<.001		5 minutes	Phipps et al., 1987

Taken, with permission, from Mello, N.K., Mendelson, J.H., and Teoh, S.K., An overview of the effects of alcohol on neuroencrine function in women, in *Alcohol and Endocrine System*, NIAAA Research Monograph 23, USDHHS, 1992, chapter 7.

lism when relatively low blood alcohol concentrations (45 mg/dl or 10 mmol/l) may saturate human alcohol dehydrogenase isoenzymes and decrease the NAD to NADH ratio. This, in turn, could decrease the rate of oxidation of E_2 to estrone and result in increased E_2 levels.[128]

A similar hypothesis can be advanced to account for the alcohol-induced enhancement of LHRH-stimulated testosterone levels in males.[95] Acute alcohol administration may increase hepatic blood flow,[10,75,122] and ethanol catabolism causes a prompt and dramatic increase in the hepatic NADH-NAD ratio.[29,116] Increased testosterone levels after alcohol and concomitant gonadotropin stimulation may be due in part to increased hepatic and gonadal conversion of precursor steroids such as androstenedione to testosterone as a consequence of increased NADH-NAD ratios during intrahepatic ethanol catabolism. Alternatively, since the LHRH-stimulated increase in LH preceded the increase in testosterone both in human and macaque males,[64,95] it is possible that this elevation in LH levels was sufficient to stimulate testosterone during alcohol intoxication. We interpreted these data to suggest that *acute* alcohol intoxication has minimal effects on hypothalamic-pituitary function.[64]

If alcohol may have either stimulatory or suppressive effects on pituitary and gonadal hormones, depending on the duration of alcohol administration (acute or chronic) and conditions of gonadotropin stimulation (Table 1), this greatly complicates analysis of the mechanisms by which alcohol intoxication induces derangements of the menstrual cycle (see Mello[66] and Mello *et al.*[73,74] for review). It is increasingly apparent that the complex interrelationships between the hypothalamic-pituitary-ovarian axis and the hypothalamic-pituitary-adrenal axis preclude any simplistic conclusion that alcohol acts primarily at one specific target site. It remains to be determined how alcohol may affect the integration and regulation of these systems and how this may be related to alcohol-induced disorders of reproductive function.

SUMMARY AND CONCLUSIONS

A number of disorders of the menstrual cycle are associated with heavy drinking and alcohol abuse. Anovulation, luteal phase dysfunction and irregular menstrual cycles are frequently reported in heavy drinkers and alcohol abusers,[4,47,140] and there appears to be a relationship between the frequency and severity of these disorders and the amount of alcohol consumed.[79,140] Amenorrhea, the most severe disruption of menstrual cycle regularity, has been consistently reported in alcohol-dependent women studied in Europe, the United States and Japan.[41,86,105,111,112,133] Endocrine profiles of amenorrheic alcohol-dependent women are usually characterized by low estrogen levels and high gonadotropin levels, reminiscent of hormonal patterns in early menopause.[41,111,112,133] However, normal estrogen levels are sometimes measured in amenorrheic alcoholic women.[41]

Hyperprolactinemia is another endocrine abnormality frequently observed in alcohol-dependent women[111,113,126,132] as well as in social drinkers and in a primate model of alcoholism.[72,79] Abnormal elevations in prolactin often persist during sobriety,[126,132] but remission of hyperprolactinemia during treatment for alcoholism has also been reported.[111] The contribution of hyperprolactinemia to other endocrine abnormalities associated with alcohol abuse is unclear. Hyperprolactinemia with normal menstrual cycles has been observed in alcohol-dependent women,[126,132] and amenorrhea with normal prolactin levels has been reported in a primate model of alcoholism.[72]

Since a number of other factors (concurrent medical disorders, polydrug abuse, exercise, obesity or anorexia) may also result in amenorrhea,[114] it is not possible to attribute abnormal menstrual cycles in alcoholic women to alcohol alone. However, anovulation and luteal phase dysfunction have also been observed in healthy social drinkers during drinking.[79] These disorders, as well as amenorrhea, have been replicated in a primate model of alcoholism under controlled conditions.[67,72] Thus, the accumulating evidence indicates the generality of clinical observations of reproductive system dysfunction in alcoholic women with liver disease and other medical complications. Valid animal models of alcoholism will be essential for unraveling alcohol's effects on the menstrual cycle.

One consequence of alcohol-related menstrual cycle disorders is that fertility is impaired either by preventing pregnancy or by increasing the risk for spontaneous abortion. Progesterone is essential for the maintenance of early pregnancy and alcohol suppresses progesterone after hCG-stimulation in women.[129] This may account in part for the high incidence of spontaneous abortion associated with alcohol abuse. Yet it is important to recognize that tolerance to the effects of chronic alcohol intoxication occurs in many biologic systems including the reproductive system. Alcohol tolerance is inferred from the fact that alcohol-dependent women conceive and complete pregnancy. Unfortunately, their children may be afflicted with a number of physical and developmental impairments. The dose of alcohol necessary to compromise human fetal development is unknown. However, prospective studies of pregnant women suggest that exposure to relatively low doses of alcohol during the first and second trimester may be associated with morphological abnormalities in the newborn.[20,104] The most severe teratogenic effects of alcohol have been described as the fetal alcohol syndrome.[38,124,125] The fetal alcohol syndrome has also been reproduced in a primate model exposed to weekly episodes of alcohol intoxication under controlled conditions.[15] Data from animal models provides increasingly persuasive evidence that alcohol is a specific teratogen.[15,97] However, the use of other drugs (marijuana, cocaine, opiates) alone or in combination with alcohol is also associated with fetal abnormalities as well as an increased risk for spontaneous abortion.[43,117,127] Alcohol abuse and polydrug abuse may have more adverse effects on fetal growth and development than alcohol abuse alone.[74] As more systematic observations of endocrine status throughout the menstrual cycle are collected in women who abuse alcohol and are alcohol-dependent, a better understanding of the mechanisms underlying alcohol-induced disorders of reproductive function should emerge. We conclude that much remains to be learned about how alcohol influences the regulation and integration of each component of the reproductive system.

ACKNOWLEDGMENTS

Preparation of this review was supported in part by Grants DA 00101 and DA 00064 from the National Institute on Drug Abuse and Grants AA 04368 and AA 06252 from the National Institute on Alcohol Abuse and Alcoholism, ADAMHA. Portions of this review have been adapted from Mello, N.K., Mendelson, J.H., and Teoh, S.K., Neuroendocrine consequences of alcohol abuse in women, in *Prenatal Abuse of Licit and Illicit Drugs*, Hutchings, D.E., Ed., *Annals of the New York Academy of Sciences*, 562, 211, 1989. We are grateful to Lynne G. Wighton for editorial assistance and to Loretta Carvelli for preparation of the manuscript.

REFERENCES

1. Preliminary Estimates from the 1992 National Household Survey on Drug Abuse: Selected Excerpts, National Institute on Drug Abuse, Washington, D.C.: U.S. Government Printing Office, 1993 (June).
2. **Arimura, A., and Schally, V.**, Augmentation of pituitary responsiveness to LH-releasing hormone (LHRH) by estrogen, *Proc. Soc. Exp. Biol. Med.*, 136, 290, 1971.
3. **Balasch, J., and Vanrell, J.**, Corpus luteum insufficiency and fertility: A matter of controversy, *Hum. Reprod.*, 2, 557, 1987.
4. **Becker, U., Tonnesen, H., Kaas-Claesson, N., and Gluud, C.**, Menstrual disturbances and fertility in chronic alcoholic women, *Drug Alcohol Depend.*, 24, 75, 1989.
5. **Berga, S.L., Mortola, J.F., Girton, L., Suh, B., Laughlin, G., Pham, P., and Yen, S.S.C.**, Neuroendocrine aberrations in women with functional hypothalamic amenorrhea, *J. Clin. Endocrinol. Metab.*, 68, 301, 1989.
6. **Boyden, T.W., and Pamenter, R.W.**, Effects of ethanol on the male hypothalamic-pituitary-gonadal axis, *Endocr. Rev.*, 4, 389, 1983.
7. **Brodie, B., and Wentz, A.C.**, An update on the clinical relevance of luteal phase inadequacy, *Semin. Reprod. Endocrinol.*, 7, 138, 1989.
8. **Buchanan, G.C., and Tredway, D.R.**, Hyperprolactinemia and ovulatory dysfunction, in *Human Ovulation*, Hafez, E.S.E., Ed., Elsevier Biomedical Press, Amsterdam, 1979, 255.
9. **Bullen, B.A., Skinnar, G.S., Beitins, I.Z., von Mering, G., Turnbull, B.A., and McArthur, J.W.**, Induction of menstrual disorders by strenuous exercise in untrained women, *N. Engl. J. Med.*, 312, 1349, 1985.
10. **Castenfors, H., Hultman, E., and Josephson, B.**, Effect of intravenous infusions of ethyl alcohol on estimated hepatic blood flow in man, *J. Clin. Invest.*, 39, 776, 1960.
11. **Channing, C.P., Gordon, W.L., Liu, W.-K., and Ward, D.N.**, Physiology and biochemistry of ovarian inhibin, *Proc. Soc. Exp. Biol. Med.*, 178, 339, 1985.
12. **Chiao, Y.-B., and Van Thiel, D.H.**, Biochemical mechanisms that contribute to alcohol-induced hypogonadism in the male, *Alcoholism: Clin. Exp. Res.*, 7, 131, 1983.
13. **Cicero, T.J.**, Common mechanisms underlying the effects of ethanol and the narcotics on neuroendocrine function, in *Advances in Substance Abuse, Behavioral and Biological Research*, Mello, N.K., Ed., JAI Press, Greenwich, CT, 1980, 201.
14. **Cicero, T.J.**, Alcohol-induced deficits in the hypothalamic-pituitary-luteinizing hormone axis in the male, *Alcoholism: Clin. Exp. Res.*, 6, 207, 1982.

15. **Clarren, S.K., Astley, S.J., Bowden, D.M., Lai, H., Milam, A.H., Rudeen, P.K., and Shoemaker, W.J.**, Neuroanatomic and neurochemical abnormalities in non-human primate infants exposed to weekly doses of ethanol during gestation, *Alcoholism: Clin. Exp. Res.*, 14, 674, 1990.
16. **Conn, P.M., and Crowley, W.F.J.**, Gonadotropin-releasing hormone and its analogues, *N. Engl. J. Med.*, 324, 93, 1991.
17. **Cronholm, T., and Sjovall, J.**, Effect of ethanol on the concentrations of solvolyzable plasma steroids, *Biochim. Biophys. Acta*, 152, 233, 1968.
18. **Cronholm, T., Sjovall, J., and Sjovall, K.**, Ethanol induced increase of the ratio between hydroxy- and ketosteroids in human pregnancy plasma, *Steroids*, 13, 671, 1969.
19. **Crowley, W.F., Jr., Filicori, M., Spratt, D.I., and Santoro, N.F.**, The physiology of gonadotropin-releasing hormone (GnRH) secretion in men and women, *Rec. Prog. Horm. Res.*, 41, 473, 1985.
20. **Day, N.L., Jasperse, D., Richardson, G., Robles, N., Sambamoorthi, U., Taylor, P., Scher, M., Stoffer, D., and Cornelius, M.**, Prenatal exposure to alcohol: Effect on infant growth and morphologic characteristics, *Pediatrics*, 84, 536, 1989.
21. **Dierschke, D.J., Hutz, R.J., and Wolf, R.C.**, Induced follicular atresia in rhesus monkeys: Strength-duration relationships of the estrogen stimulus, *Endocrinology*, 117, 1397, 1985.
22. **Dierschke, D.J., Hutz, R.J., and Wolf, R.C.**, Atretogenic action of estrogen in rhesus monkeys: Effects of repeated treatment, *Am. J. Primatol.*, 12, 251, 1987.
23. **diZerega, G.S., and Hodgen, G.D.**, Luteal phase dysfunction infertility: A sequel to aberrant folliculogenesis, *Fertil. Steril.*, 35, 489, 1981.
24. **diZerega, G.S., and Wilks, J.W.**, Inhibition of the primate ovarian cycle by a porcine follicular fluid protein(s), *Fertil. Steril.*, 41, 1094, 1984.
25. **Ellingboe, J., and Varanelli, C.C.**, Ethanol inhibits testosterone biosynthesis by direct action on Leydig cells, *Res. Commun. Chem. Pathol. Pharmacol.*, 24, 87, 1979.
26. **Eskay, R.L., Ryback, R.S., Goldman, M., and Majchrowicz, E.**, Effect of chronic ethanol administration on plasma levels of LH and the estrous cycle in the female rat, *Alcoholism: Clin. Exp. Res.*, 5, 204, 1981.
27. **Filicori, M., Butler, J.P., and Crowley, W.F.**, Neuroendocrine regulation of the corpus luteum in the human: Evidence for pulsatile progesterone secretion., *J. Clin. Invest.*, 73, 1638, 1984.
28. **Filicori, M., Santoro, N., Merriam, G.R., and Crowley, W.F., Jr.**, Characterization of the physiological pattern of episodic gonadotropin secretion throughout the human menstrual cycle, *J. Clin. Endocrinol. Metab.*, 62, 1136, 1986.
29. **Forsander, O., Raiha, N., and Sumalainen, H.**, Alkoholoxydation und bildung von acetoacetat in normaler und glykogenarmer intaker rattenleber, *Hoppe Seyler. Z. Physiol. Chem.*, 312, 243, 1958.
30. **Frisch, R.E.**, Fatness, puberty, menstrual periodicity and fertility, in *Clinical Reproductive Neuroendocrinology*, Vaitukaitis, J.L., Ed., Elsevier Biomedical, New York, 1982, 105.
31. **Frisch, R.E., and McArthur, J.W.**, Menstrual cycles: Fatness as a determinant of minimum weight for height necessary for their maintenance or onset, *Proc. Soc. Exp. Biol. Med.*, 175, 487, 1974.
32. **Gavaler, J.**, Alcohol effects in postmenopausal women: Alcohol and estrogens, in *The Medical Diagnosis and Treatment of Alcoholism, 1st ed.*, Mendelson, J.H., and Mello, N.K., Eds., McGraw Hill, New York, 1992, 623.
33. **Gavaler, J.S.**, Effects of alcohol on endocrine function in postmenopausal women: A review, *J. Stud. Alcohol.*, 46(6), 495, 1985.
34. **Gavaler, J.S.**, Effects of moderate consumption of alcoholic beverages on endocrine function in postmenopausal women: Bases for hypotheses, in *Recent Developments in Alcoholism*, Galanter, M., Begleiter, H., Deitrich, R.A., Goodwin, D.W., Gottheil, E., Paredes, A., Rothchild, A., and Thiel, D.V., Eds., Plenum Press, New York, 1988, 229.

35. **Gavaler, J.S., Van Thiel, D.H., and Lester, R.,** Ethanol: A gonadal toxin in the mature rat of both sexes, *Alcoholism: Clin. Exp. Res.,* 4, 271, 1980.
36. **Goodman, A.L., and Hodgen, G.D.,** The ovarian triad of the primate menstrual cycle, *Rec. Prog. Horm. Res.,* 39, 1, 1983.
37. **Halmesmäki, E., Autti, I., Granström, M.-L., Stenman, U.-H., and Ylikorkala, O.,** Estradiol, progesterone, prolactin, and human chorionic gonadotropin in pregnant women with alcohol abuse, *J. Clin. Endocrinol. Metab.,* 64, 153, 1987.
38. **Hannigan, J.H., Welch, R.A., and Sokol, R.J.,** Recognition of fetal alcohol syndrome and alcohol-related birth defects, in *Medical Diagnosis and Treatment of Alcoholism, 1st ed.,* Mendelson, J.H., and Mello, N.K., Eds., McGraw Hill, New York, 1992, 639.
39. **Hodgen, G.D.,** The dominant ovarian follicle, *Fertil. Steril.,* 38, 281, 1982.
40. **Hollstedt, C., Dahlgren, L., and Rydberg, U.,** Outcome of pregnancy in women treated at an alcohol clinic, *Acta Psychiatr. Scand.,* 67, 236, 1983.
41. **Hugues, J.N., Coste, T., Perret, G., Jayle, M.F., Sebaoun, J., and Modigliani, E.,** Hypothalamo-pituitary ovarian function in thirty-one women with chronic alcoholism, *Clin. Endocrinol.,* 12, 543, 1980.
42. **Hurley, D.M., Brian, R., Outch, J., Stickdale, J., Fry, A., Hackman, C., Clark, I., and Burger, H.G.,** Induction of ovulation and fertility in amenorrheic women by pulsatile low-dose gonadotropin-releasing hormone, *N. Engl. J. Med.,* 310, 1069, 1984.
43. **Hutchings, D.E.,** (Ed.), *Prenatal Abuse of Licit and Illicit Drugs,* New York Academy of Sciences, New York, 1989.
44. **Hutchison, J., Kubik, C., Nelson, P., and Zeleznik, A.,** Estrogen induces premature luteal regression in rhesus monkeys during spontaneous menstrual cycles but not in cycles driven by exogenous gonadotropin-releasing hormone, *Endocrinology,* 121, 466, 1987.
45. **Hutz, R.J., Dierschke, D.J., and Wolf, R.C.,** Role of estradiol in regulating ovarian follicular atresia in rhesus monkeys: A review, *J. Med. Primatol.,* 19, 553, 1990.
46. **Jaffe, R.B., and Keye, W.R.,** Estradiol augmentation of pituitary responsiveness to gonadotropin-releasing hormone in women, *J. Clin. Endocrinol. Metab.,* 39, 850, 1974.
47. **Jones-Saumty, D.J., Fabian, M.S., and Parsons, O.A.,** Medical status and cognitive functioning in alcoholic women, *Alcoholism: Clin. Exp. Res.,* 5, 372, 1981.
48. **Jung, Y., and Russfield, A.B.,** Prolactin cells in the hypophysis of cirrhotic patients, *Arch. Pathol.,* 94, 265, 1972.
49. **Karsch, F.J., Weick, R.F., Butler, W.R., Dierschke, D.J., Krey, L.C., Weiss, G., Hotchkiss, J., Yamaji, T., and Knobil, E.,** Induced LH surges in the rhesus monkey: Strength-duration characteristics of the estrogen stimulus, *Endocrinology,* 92, 1973.
50. **Klopper, A.,** Steroids in pregnancy, in *Clinical Reproductive Endocrinology,* Shearman, R.P., Eds., Churchill Livingstone, Edinburgh, 1985, 209.
51. **Knobil, E.,** On the control of gonadotropin secretion in the rhesus monkey, *Rec. Prog. Horm. Res.,* 30, 1, 1974.
52. **Knobil, E.,** The neuroendocrine control of the menstrual cycle, *Rec. Prog. Horm. Res.,* 36, 53, 1980.
53. **Knobil, E., and Hotchkiss, J.,** The menstrual cycle and its neuroendocrine control, in *The Physiology of Reproduction,* Knobil, E., Neill, J., Ewing, L.L., Greenwald, G.S., Markert, C.L., and Pfaff, D.W., Eds., Raven Press, New York, 1988, 1971.
54. **Krey, L.C., Butler, W.R., Weiss, G., Weick, R.F., Dierschke, D.J., and Knobil, E.,** Influence of endogenous and exogenous gonadal steroids on the action of synthetic RF in the rhesus monkey, *Excerpta Med. Int. Congr. Ser.,* 263, 39, 1973.
55. **Krueger, W.A., Walter, J.B., and Rudeen, P.K.,** Estrous cyclicity in rat-fed ethanol diet for 4 months, *Pharmacol. Biochem. Behav.,* 19, 583, 1983.
56. **Lasley, B.L., Wang, C.F., and Yen, S.S.C.,** The effects of estrogen and progesterone on the functional capacity of the gonadotrophs, *J. Clin. Endocrinol. Metab.,* 1, 820, 1975.
57. **Lex, B.W., Teoh, S.K., Lagomasino, I., Mello, N.K., and Mendelson, J.H.,** Characteristics of women receiving mandated treatment for alcohol or polysubstance dependence in Massachusetts, *Drug Alcohol Depend.,* 25, 13, 1990.

58. **Leyendecker, G., and Wildt, L.,** Control of gonadotropin secretion in women, in *Neuroendocrine Aspects of Reproduction*, Norman, R.L., Ed., Academic Press, New York, 1983, 295.
59. **Marshall, J.C., Case, G.D., Valk, T.W., Corley, K.P., Sauder, S.E., and Kelch, R.P.,** Selective inhibition of follicle-stimulating hormone secretion by estradiol, *J. Clin. Invest.*, 71, 248, 1983.
60. **Martin, J.B., and Reichlin, S.,** Ed., *Clinical Neuroendocrinology, 2nd edition,* F.A. Davis, Philadelphia, 1987.
61. **McArthur, J.W., Bullne, B.A., Beitins, I.Z., Pagano, M., Badger, T.M., and Klibanski, A.,** Hypothalamic amenorrhea in runners of normal body composition, *Endocrinol. Res. Commun.*, 7, 13, 1980.
62. **McLachlan, R.I., Robertson, D.M., Healy, D.L., Burger, H.G., and de Kretser, D.M.,** Circulating immunoreactive inhibin levels during the normal human menstrual cycle, *J. Clin. Endocrinol. Metab.*, 65, 954, 1987.
63. **McNeely, M.J., and Soules, M.R.,** Diagnosis of luteal phase deficiency: A critical review, *Fertil. Steril.*, 50, 1, 1988.
64. **Mello, N., Mendelson, J., Bree, M., Ellingboe, J., and Skupny, A.,** Alcohol effects on luteinizing hormone and testosterone in male macaque monkeys, *J. Pharmacol. Exp. Ther.*, 233, 588, 1985.
65. **Mello, N., Mendelson, J., Palmieri, S., Lex, B., and Teoh, S.,** Operant acquisition of alcohol by women, *J. Pharmacol. Exp. Ther.*, 253, 237, 1990.
66. **Mello, N.K.,** Effects of alcohol abuse on reproductive function in women, in *Recent Developments in Alcoholism*, Galanter, M., Ed., Plenum Press, New York, 1988, 253.
67. **Mello, N.K., Bree, M.P., Mendelson, J.H., Ellingboe, J., King, N.W., and Sehgal, P.K.,** Alcohol self-administration disrupts reproductive function in female macaque monkeys, *Science,* 221, 677, 1983.
68. **Mello, N.K., Ellingboe, J., Bree, M.P., Harvey, K.L., and Mendelson, J.H.,** Alcohol effects on estradiol in female macaque monkey, in *Problems of Drug Dependence 1981*, DHHS (DAM)83-1264, Harris, L.S., Ed., U.S. Government Printing Office, Washington, D.C., 1983, 210.
69. **Mello, N.K., and Mendelson, J.H.,** Drinking patterns during work-contingent and noncontingent alcohol acquisition, *Psychosom. Med.*, 34, 139, 1972.
70. **Mello, N.K., Mendelson, J.H., Bree, M.P., and Skupny, A.S.T.,** Alcohol effects on LHRH stimulated LH and FSH in ovariectomized female rhesus monkeys, *J. Pharmacol. Exp. Ther.*, 239, 693, 1986.
71. **Mello, N.K., Mendelson, J.H., Bree, M.P., and Skupny, A.S.T.,** Alcohol effects on luteinizing hormone-releasing hormone-stimulated luteinizing hormone and follicle-stimulating hormone in female rhesus monkeys, *J. Pharmacol. Exp. Ther.*, 236, 590, 1986.
72. **Mello, N.K., Mendelson, J.H., King, N.W., Bree, M.P., Skupny, A., and Ellingboe, J.,** Alcohol self-administration by female rhesus monkey: A model for study of alcohol dependence, hyperprolactinemia and amenorrhea, *J. Stud. Alcohol,* 49, 551, 1988.
73. **Mello, N.K., Mendelson, J.H., and Teoh, S.K.,** Neuroendocrine consequences of alcohol abuse in women, in *Prenatal Abuse of Licit and Illicit Drugs*, Hutchings, D.E., Ed., Annals of the New York Academy of Sciences, 1989, 211.
74. **Mello, N.K., Mendelson, J.H., and Teoh, S.K.,** Alcohol and neuroendocrine function in women of reproductive age, in *Medical Diagnosis and Treatment of Alcoholism*, Mendelson, J.H., and Mello, N.K., Eds., McGraw-Hill, New York, 1992, 575.
75. **Mendeloff, A.I.,** Effects of intravenous infusions of ethanol upon estimated hepatic blood flow in man, *J. Clin. Invest.*, 33, 1298, 1954.
76. **Mendelson, J.H., Babor, T.F., Mello, N.K., and Pratt, H.,** Alcoholism and prevalence of medical and psychiatric disorders, *J. Stud. Alcohol,* 47, 361, 1986.
77. **Mendelson, J.H., Lukas, S.E., Mello, N.K., Amass, L., Ellingboe, J., and Skupny, A.,** Acute alcohol effects on plasma estradiol levels in women, *Psychopharmacology,* 94, 464, 1988.

78. **Mendelson, J.H., and Mello, N.K.,** Biologic concomitants of alcoholism, *N. Engl. J. Med.,* 301, 912, 1979.
79. **Mendelson, J.H., and Mello, N.K.,** Chronic alcohol effects on anterior pituitary and ovarian hormones in healthy women, *J. Pharmacol. Exp. Ther.,* 245, 407, 1988.
80. **Mendelson, J.H., Mello, N.K., Cristofaro, P., Ellingboe, J., Skupny, A., Palmieri, S.L., Benedikt, R., and Schiff, I.,** Alcohol effects on naloxone-stimulated luteinizing hormone, prolactin and estradiol in women, *J. Stud. Alcohol,* 48, 287, 1987.
81. **Mendelson, J.H., Mello, N.K., Cristofaro, P., Skupny, A., and Ellingboe, J.,** Use of naltrexone as a provocative test for hypothalamic-pituitary hormone function, *Pharmacol. Biochem. Behav.,* 24, 309, 1986.
82. **Mendelson, J.H., Mello, N.K., and Ellingboe, J.,** Acute alcohol intake and pituitary gonadal hormones in normal human females, *J. Pharmacol. Exp. Ther.,* 218, 23, 1981.
83. **Mendelson, J.H., Mello, N.K., Teoh, S.K., and Ellingboe, J.,** Alcohol effects on luteinizing hormone-releasing hormone-stimulated anterior pituitary and gonadal hormones in women, *J. Pharmacol. Exp. Ther.,* 250, 902, 1989.
84. **Mirin, S.M., Mendelson, J.H., Ellingboe, J., and Meyer, R.E.,** Acute effects of heroin and naltrexone on testosterone and gonadotropin secretion: A pilot study, *Psychoneuroendocrinology,* 1, 359, 1976.
85. **Morley, J.E., Baranetsky, N.G., Wingert, T.D., Carlson, H.E., Hershman, J.M., Melmed, S., Levin, S.R., Jamison, K.R., Weitzman, R., Chang, R.J., and Varner, A.A.,** Endocrine effects of naloxone-induced opiate receptor blockade, *J. Clin. Endocrinol. Metab.,* 50, 252, 1980.
86. **Moskovic, S.,** Effect of chronic alcohol intoxication on ovarian dysfunction, *Srp. Arh. Celok.,* 103, 751, 1975.
87. **Murad, F., and Haynes, R.,** Estrogens and progestins, in *The Pharmacological Basis of Therapeutics,* Gilman, A., Goodman, L., Rall, T., and Murad, F., Eds., MacMillan, New York, 1985, 1412.
88. **Murono, E.P., and Fisher-Simpson, V.,** Ethanol directly increases dihydrotestosterone conversion to 5-androstan-3, 17-diol and 5-androstan-3, 17-diol in rat leydig cells, *Biochem. Biophys. Res. Commun.,* 121, 558, 1984.
89. **Murono, E.P., and Fisher-Simpson, V.,** Ethanol directly stimulates dihydro-testosterone conversion to 5-androstan-3, 17-diol and 5 -androstan-3, 17-diol in rat liver, *Life Sci.,* 36, 1117, 1985.
90. **Myers, J.K., Weissman, M.M., Tischler, G.L., Holzer, C.E., Leaf, P.J., Orvaschel, H., Anthony, J.C., Boyd, J.H., Burke, J.D., Krammer, M., and Stoltzman, R.,** Six month prevalence of psychiatric disorders in three communities, *Arch. Gen. Psychiatry,* 41, 959, 1984.
91. **Noth, R.H., and Walter, R.M.J.,** The effects of alcohol on the endocrine system, *Med. Clin. N. Am.,* 68, 133, 1984.
92. **Olster, D.H., and Ferin, N.,** Corticotropin-releasing hormone inhibits gonadotropin secretion in the ovariectomized rhesus monkey, *J. Clin. Endocrinol. Metab.,* 65, 262, 1987.
93. **Ottobre, J., and Stouffer, R.,** Persistent versus transient stimulation of the macaque corpus luteum during prolonged exposure to human chorionic gonadotropin: A function of age of the corpus luteum, *Endocrinology,* 114, 2175, 1984.
94. **Petraglia, F., Sutton, S., Vale, W., and Plotsky, P.,** Corticotropin-releasing factor decreases plasma LH levels in female rats by inhibiting gonadotropin-releasing hormone release into hypophysial-portal circulation, *Endocrinology,* 120, 1083, 1987.
95. **Phipps, W.R., Lukas, S.E., Mendelson, J.H., Ellingboe, J., Palmieri, S.L., and Schiff, I.,** Acute ethanol administration enhances plasma testosterone levels following gonadotropin stimulation in men, *J. Endocrinol.,* 101, 33, 1987.
96. **Prior, J.C., Cox, T.A., Fairholm, D., Kostashuk, E., and Nugent, R.,** Testosterone-related exacerbation of a prolactin-producing macroadenoma: Possible role for estrogen, *J. Clin. Endocrinol. Metab.,* 64, 391, 1987.

97. **Randall, C.L., Ekblad, U., and Anton, R.F.**, Perspectives on the pathophysiology of fetal alcohol syndrome, *Alcoholism: Clin. Exp. Res.*, 14, 807, 1990.
98. **Reame, N.E., Sauder, S.E., Case, G.D., Kelch, R.P., and Marshall, J.C.**, Pulsatile gonadotropin secretion in women with hypothalamic amenorrhea: Evidence that reduced frequency of gonadotropin-releasing hormone secretion is the mechanism of persistent anovulation, *J. Clin. Endocrinol. Metab.*, 61, 851, 1985.
99. **Rebar, R.W.**, Practical evaluation of hormonal status, in *Reproductive Endocrinology*, Yen, S.S.C., and Jaffe, R.B., Eds., W.B. Saunders, Philadelphia, 1986, 683.
100. **Redei, E., Branch, B.J., and Taylor, A.N.**, Direct effect of ethanol on adrenocorticotropin (ACTH) release *in vitro*, *J. Pharmacol. Exp. Ther.*, 237, 59, 1986.
101. **Rivier, C., Rivier, J., and Vale, W.**, Stress-induced inhibition of reproductive functions: Role of endogenous corticotropin-releasing factor, *Science*, 31, 607, 1986.
102. **Rivier, C., and Vale, W.**, Influence of corticotropin-releasing factor (CRF) on reproductive functions in the rat, *Endocrinology*, 114, 914, 1984.
103. **Ross, G.T.**, Disorders of the ovary and female reproductive tract, in *Williams Textbook of Endocrinology, 7th edition*, Wilson, J.D., and Foster, D.W., Eds., W.B. Saunders, Philadelphia, 1985, 206.
104. **Rostand, A., Kaminski, M., Lelong, N., Dehaene, P., Delesrret, I., Klein-Bertrand, C., Querleu, D., and Crepin, G.**, Alcohol use in pregnancy, craniofacial features, and fetal growth, *J. Epidemiol. Commun. Health*, 44, 302, 1990.
105. **Ryback, R.S.**, Chronic alcohol consumption and menstruation, *J. Am. Med. Assoc.*, 238, 2143, 1977.
106. **Sanchis, R., Esquifino, A., and Guerri, C.**, Chronic ethanol intake modifies estrous cyclicity and alters prolactin and LH levels, *Pharmacol. Biochem. Behav.*, 23, 221, 1985.
107. **Santoro, N., Filicori, M., and Crowley, J.**, Hypogonadotropic disorders in men and women: Diagnosis and therapy with pulsatile gonadotropin-releasing hormone, *Endocr. Rev.*, 7, 11, 1986.
108. **Santoro, N., Wierman, M.E., Filicori, M., Waldstreicher, J., and Crowley Jr., W.F.**, Intravenous administration of pulsatile gonadotropin-releasing hormone in hypothalamic amenorrhea: Effects of dosage, *J. Clin. Endocrinol. Metab.*, 62, 109, 1986.
109. **Sauder, S., Frager, M., Case, G., Kelch, R., and Marshall, J.**, Abnormal patterns of pulsatile luteinizing hormone secretion in women with hyperprolactinemia and amenorrhea: Responses to bromocriptine, *J. Clin. Endocrinol. Metab.*, 59, 941, 1984.
110. **Schulz, K.D., Geiger, W., del Pozo, E., and Kunzig, H.J.**, Pattern of sexual steroids, prolactin and gonadotropic hormones during prolactin inhibition in normally cycling women, *Am. J. Obstet. Gynecol.*, 132, 561, 1978.
111. **Seki, M.**, A physiopathological study on ovarian dysfunction in female patients with alcoholism, *Fukuoka Acta Med.*, 79, 738, 1988.
112. **Seki, M., Yoshida, K., and Okamura, Y.**, Hormones in amenorrheic women with alcoholics, *Jpn. J. Fertil. Steril.*, 36, 630, 1991.
113. **Seki, M., Yoshida, K., and Okamura, Y.**, A study on hyperprolactinemia in female patients with alcoholics, *Jpn. J. Alcohol Drug Depend.*, 26, 49, 1991.
114. **Shearman, R.P.**, Secondary amenorrhea, in *Clinical Reproductive Endocrinology*, Shearman, R.P., Ed., Churchill Livingstone, New York, 1985, 493.
115. **Sherman, B.M.**, Hypothalamic control of the menstrual cycle: Implications for the study of anorexia nervosa, in *Neuroendocrinology and Psychiatric Disorders*, Brown, G.M., Koslow, S.H., and Reichlin, S., Eds., Raven Press, New York, 1984, 315.
116. **Slater, T.F., Sawyer, B.C., and Strauli, U.D.**, Changes in liver nucleotide concentrations in experimental liver injury, *Biochem. J.*, 93, 267, 1964.
117. **Smith, C.G., and Smith, M.T.**, Substance abuse and reproduction, in *Seminars in Reproductive Endocrinology*, Thieme Medical Publishers, New York, 1990, 55.
118. **Soules, M.R., Clifton, D.K., Bremner, W.J., and Steiner, R.A.**, Corpus luteum insufficiency induced by a rapid gonadotropin-releasing hormone-induced gonadotropin secretion pattern in the follicular phase, *J. Clin. Endocrinol. Metab.*, 65, 457, 1987.

119. **Soules, M.R., Clifton, D.K., Cohen, N.L., Bremner, W.J., and Steiner, R.A.,** Luteal phase deficiency: Abnormal gonadotropin and progesterone secretion patterns, *J. Clin. Endocrinol. Metab.*, 69, 813, 1989.
120. **Soules, M.R., Clifton, D.K., Steiner, R.A., Cohen, N.L., and Bremner, W.J.,** The corpus luteum: Determinants of progesterone secretion in the normal menstrual cycle, *Obstet. Gynecol.*, 71, 659, 1988.
121. **Soules, M.R., McLachlan, R.I., Marit, E.K., Dahl, K.D., Cohen, N.L., and Bremner, W.J.,** Luteal phase deficiency: Characterization of reproductive hormones over the menstrual cycle, *J. Clin. Endocrinol. Metab.*, 69, 804, 1989.
122. **Stein, S.W., Lieber, C.S., Leery, C.M., Cherrick, G.R., and Abelmann, W.H.,** The effect of ethanol upon systemic and hepatic blood flow in man, *Am. J. Clin. Nutr.*, 13, 68, 1963.
123. **Stouffer, R.L.,** Corpus luteum function and dysfunction, *Clin. Obstet. Gynecol.*, 33, 668, 1990.
124. **Streissguth, A.,** The behavioral teratology of alcohol: Performance, behavioral, and intellectual deficits in prenatally exposed children, in *Alcohol and Brain Development*, West, J., Ed., Oxford University Press, New York, 1986, 3.
125. **Streissguth, A.P., Aase, J.M., Clarren, S.K., Randels, S.P., LaDue, R.A., and Smith, D.F.,** Fetal alcohol syndrome in adolescents and adults, *J. Am. Med. Assoc.*, 265, 1961, 1991.
126. **Teoh, S.K., Lex, B.W., Mendelson, J.H., Mello, N.K., and Cochin, J.,** Hyperprolactinemia and macrocytosis in women with alcohol and polysubstance dependence, *J. Stud. Alcohol*, 53, 176, 1992.
127. **Teoh, S.K., Mello, N.K., and Mendelson, J.H.,** Effects of drugs of abuse on reproductive function in women and pregnancy, in *Addictive Behaviors in Women*, Watson, R., Eds., Humana Press, Clifton, NJ, in press.
128. **Teoh, S.K., Mendelson, J.H., Mello, N.K., and Skupny, A.,** Alcohol effects on naltrexone-induced stimulation of pituitary, adrenal and gonadal hormones during the early follicular phase of the menstrual cycle, *J. Clin. Endocrinol. Metab.*, 66, 1181, 1988.
129. **Teoh, S.K., Mendelson, J.H., Mello, N.K., Skupny, A., and Ellingboe, J.,** Alcohol effects on hCG-stimulated gonadal hormones in women, *J. Pharmacol. Exp. Ther.*, 254, 407, 1990.
130. **Tolis, G.,** Prolactin: Physiology and pathology, in *Neuroendocrinology: Interrelationships of the Body's Two Major Integrative Systems in Normal Physiology and in Clinical Disease*, Krieger, D.T., and Hughes, J.C., Eds., Sinauer Associates, Inc., Sunderland, MA, 1980, 321.
131. **Välimäki, M., Harkonen, M., and Ylikahri, R.,** Acute effects of alcohol on female sex hormones, *Alcoholism: Clin. Exp. Res.*, 7, 289, 1983.
132. **Välimäki, M., Pelkonen, R., Harkonen, M., Tuomala, P., Koistinen, P., Roine, R., and Ylikahri, R.,** Pituitary-gonadal hormones and adrenal androgens in non-cirrhotic female alcoholics after cessation of alcohol intake, *Eur. J. Clin. Invest.*, 20, 177, 1990.
133. **Välimäki, M., Pelkonen, R., Salaspuro, M., Harkonen, M., Hirvonen, E., and Ylikahri, R.,** Sex hormones in amenorrheic women with alcoholic liver disease, *J. Clin. Endocrinol. Metab.*, 59, 133, 1984.
134. **Van Thiel, D.H.,** Ethanol: Its adverse effects upon the hypothalamic-pituitary-gonadal axis, *J. Lab. Clin. Med.*, 101, 21, 1983.
135. **Van Thiel, D.H.,** Ethyl alcohol and gonadal function. Physiology in medicine, *Hosp. Prac.*, 19, 152, 1984.
136. **Van Thiel, D.H., and Gavaler, J.S.,** The adverse effects of ethanol upon hypothalamic-pituitary-gonadal function in males and females compared and contrasted, *Alcoholism: Clin. Exp. Res.*, 6, 179, 1982.
137. **Van Thiel, D.H., Gavaler, J.S., and Lester, R.,** Alcohol-induced ovarian failure in the rat, *J. Clin. Invest.*, 61, 624, 1978.
138. **Wilks, J., and Noble, A.,** Steroidogenic responsiveness of the monkey corpus luteum to exogenous chorionic gonadotropin, *Endocrinology*, 112, 1256, 1983.
139. **Wilks, J.W., Hodgen, G.D., and Ross, G.T.,** Anovulatory menstrual cycles in the rhesus monkeys, the significance of serum, follicle-stimulating hormone/luteinizing hormone ratios, *Fertil. Steril.*, 28, 1094, 1977.

140. **Wilsnack, S.C., Klassen, A.D., and Wilsnack, R.W.,** Drinking and reproductive dysfunction among women in a 1981 national survey, *Alcoholism: Clin. Exp. Res.*, 89, 451, 1984.
141. **Wright, H.I., Gavaler, J.S., Tabasco-Minguillan, J., and Van Thiel, D.H.,** Endocrine effects of alcohol abuse in males, in *Medical Diagnosis and Treatment of Alcoholism*, Mendelson, J.H., and Mello, N.K., Eds., McGraw-Hill, New York, 1992, 341.
142. **Xiao, E., and Ferin, M.,** The inhibitory action of corticotropin-releasing hormone on gonadotropin secretion in the ovariectomized rhesus monkey is not mediated by adrenocorticotropic hormone, *Biol. Reprod.*, 38, 763, 1988.
143. **Yen, S.S.C.,** Clinical applications of gonadotropin-releasing hormone and gonadotropin-releasing hormone analogs, *Fertil. Steril.*, 39, 257, 1983.
144. **Yen, S.S.C., Quigley, M.E., Reid, R.L., Ropert, J.F., and Cetel, N.S.,** Neuroendocrinology of opioid peptides and their role in the control of gonadotropin and prolactin secretion, *Am. J. Obstet. Gynecol.*, 152, 485, 1985.
145. **Zeleznek, A.J.,** Premature elevation of systemic estradiol reduces serum levels of FSH and lengthens the follicular phase of the menstrual cycle in rhesus monkeys, *Endocrinology*, 109, 352, 1981.

Chapter 16

AGING AND ALCOHOL: THE HORMONAL STATUS OF POSTMENOPAUSAL WOMEN

J.S. Gavaler

TABLE OF CONTENTS

0-8493-2451-3/95/$0.00+.50

EFFECTS OF MODERATE ALCOHOL CONSUMPTION ON HORMONE LEVELS IN NORMAL POSTMENOPAUSAL WOMEN

The Study Sample

The effect of moderate alcoholic beverage consumption on postmenopausal hormone levels has been evaluated in study samples of normal postmenopausal women from four countries.[6-9,11] The characteristics of the entire study population of 244 normal postmenopausal women are summarized in Table 1. As may be seen, the study population is composed of relatively young postmenopausal women in whom the duration of menopause is less than a decade. Further, the prevalence of ovariectomy and an acceptable BMI is such that these variables must ultimately be taken into account as determinants of postmenopausal hormonal status.

Simple Hormonal Relationships

The alcohol consumption pattern and relationship of E_2 levels with alcoholic beverage consumption are shown in Table 2. As may be seen, even in a setting where the effects of the known postmenopausal estrogen determinants of ovariectomy and estimated body fat mass are not controlled, E_2 levels are statistically elevated among moderate alcoholic beverage consumers. As may also be seen, the bivariate correlation coefficient of E_2 levels with total weekly drinks, where the abstainers have a value of zero, is statistically significant. This simplistic analytic approach demonstrated no detectable

TABLE 1
Characteristics of the Study Sample of 244 Normal Postmenopausal Women

Characteristic	Value
Age (years)	59.3±0.8
Menopause duration (years)	8.0±0.9
Ovariectomy	19.4%
Body Mass Index (BMI):	
Acceptable (BMI ≤ 24.0)	50.8%
Overweight (BMI > 24 but < 30)	40.3%
Obese (BMI ≥ 30)	8.9%
Use of estrogen replacement therapy (ERT)	0%
Country from which sample drawn:	
U.S.A.	52.2%
Denmark	25.4%
Portugal	13.9%
Spain	8.9%

TABLE 2
The Alcoholic Beverage Consumption Pattern and Relationship to Estradiol (E_2) Levels* in Normal Postmenopausal Women

Characteristic	Value
Alcohol use	76%
Total weekly drinks (among alcohol users)	6.4±0.8
Use of wine	92%
Prevalence of alcohol intake at < 2 drinks/day	87%
Correlation coefficient of E_2 with total weekly drinks	r = 0.36 ($p < 0.001$)
E_2 (pg/ml)	
Abstainers	35.1±3.5
Alcohol users	54.6±4.4 ($p < 0.01$)

*normalizing logarithmic transformations used in analysis

effect of moderate alcoholic beverage consumption on T, the estimate of aromatization of testosterone to estradiol (the E_2:T ratio) nor on levels of the pituitary hormones: LH, FSH and prolactin (PRL).

Several points related to these findings should be emphasized: 1) Daily alcohol consumption of one drink per day has been defined a moderate/acceptable intake level for women by agencies of the United States government.[20] 2) The increases in E_2 occurred in the absence of any change in measures of liver injury or function.[4] 3) The prevalence of alcoholic beverage consumption at levels of less than two drinks per day is 87% in the entire study sample. 4) Although the bivariate correlation coefficient between E_2 levels and total weekly drinks is statistically significant, it must be noted that the actual shape of the regression line is that of a steady increase in E_2 concentrations up to an alcohol intake level of three to six total weekly drinks (or approximately one or fewer drinks per day), above which E_2 concentrations plateau even with higher alcohol intake.

The Findings of Multivariate Analyses

In contrast to the results of the above simplistic analytic approach, the influence of moderate alcoholic beverage consumption on the postmenopausal steroid and pituitary hormone levels evaluated becomes apparent when the following independent variables are controlled for in stepwise multiple linear regression: weight, BMI, ovariectomy (OVEX), age, menopause duration (MENODUR), nationality (NATION), and alcoholic beverage consumption specified as weekly drinks of wine (WINE), total weekly drinks, or the variable distinguishing abstainers from alcohol consumers (DRINK). When

TABLE 3
Hormone Level* Determinants in 244 Normal Postmenopausal Women

Hormone	Adjusted R^2 (F Value) (significance)	Independent Variables Entering the Equation: Variable	β Coefficient	Significance
E_2	$R^2 = 0.344$	NATION	–0.363	$p = 0.000$
	$(F = 30.4)$	WINE	+0.305	$p = 0.000$
	$(p = 0.000)$	OVEX	–0.233	$p = 0.000$
		Weight	+0.169	$p = 0.022$
T	$R^2 = 0.250$	DRINK	+0.406	$p = 0.000$
	$(F = 20.0)$	Age	–0.230	$p = 0.000$
	$(p = 0.000)$	BMI	+0.388	$p = 0.004$
E_2:T	$R^2 = 0.448$	DRINK	+0.482	$p = 0.000$
	$(F = 46.3)$	NATION	–0.300	$p = 0.000$
	$(p = 0.000)$	OVEX	–0.167	$p = 0.001$
		Age	+0.098	$p = 0.076$
LH	$R^2 = 0.164$	BMI	–0.245	$p = 0.000$
	$(F = 18.8)$	OVEX	+0.161	$p = 0.015$
	$(p = 0.000)$	WINE	–0.116	$p = 0.063$
		MENODUR	–0.163	$p = 0.094$
		Age	–0.165	$p = 0.098$
FSH	$R^2 = 0.211$	BMI	–0.288	$p = 0.000$
	$(F = 19.7)$	OVEX	+0.244	$p = 0.000$
	$(p = 0.000)$	NATION	–0.315	$p = 0.000$
		Age	–0.142	$p = 0.035$
		WINE	–0.126	$p = 0.041$
PRL	$R^2 = 0.302$	NATION	+0.531	$p = 0.000$
	$(F = 19.7)$	DRINK	+0.187	$p = 0.001$
	$(p = 0.000)$	Weight	+0.328	$p = 0.009$
		BMI	+0.270	$p = 0.030$
		Age	–0.098	$p = 0.098$

***normalizing logarithmic transformations used in analysis**

the adjusted R^2 values are multiplied by 100 and viewed from the standpoint of the percent of variation explained by the independent variables entering the equation, it must be noted that the R^2 values increase when an alcohol consumption variable is included as an independent variable.[7] As may be seen in Table 3, alcohol consumption enters the equation not only for E_2, but also for T, the E_2:T ratio, LH, FSH and PRL. Based on these findings, it is clear that moderate alcoholic beverage consumption must be added to the list of known hormone determinants in normal postmenopausal women.

HORMONAL RELATIONSHIPS IN POSTMENOPAUSAL WOMEN WITH ALCOHOL-INDUCED CIRRHOSIS

Background

Several studies in postmenopausal women with alcohol-induced cirrhosis and a myriad of studies in alcoholic cirrhotic men have reported an association between altered hormonal status and alcohol abuse sufficient to cause cirrhosis.[1,2,10,11,15-18,21] Frequently E_2, the E_2:T ratio, and PRL have been reported to be increased, and LH and FSH have been reported to be decreased in postmenopausal women with alcohol-induced cirrhosis. In general, the reports in postmenopausal women were based on results obtained in small study samples (n <21) without attention to the established postmenopausal hormone determinants. Further, the relative contribution to alterations in hormonal status made by alcohol abuse and that made by the presence of cirrhosis have not been considered.

The Study Population

Demographic Characteristics and Liver Disease Status

The characteristics of a sample of 66 postmenopausal women with alcohol-induced liver disease are summarized in Table 4. As may be seen, there are no statistical differences in age, weight, BMI or the prevalence of ovariectomy between alcoholic cirrhotic women and normal alcohol-abstaining control women. As may also be seen, these postmenopausal women with alcohol-induced cirrhosis have advanced liver disease; mean levels of biochemical markers of liver function (e.g., albumin, total bilirubin and prothrombin time) and liver injury (e.g., gamma glutamyltranspeptidase [GGTP]) are outside the normal range. The prevalence of major complications of cirrhosis related to portal hypertension is greater than 50% in the entire sample. Child's class is based on levels of albumin, total bilirubin and prolongation of the prothrombin time, as well as on the presence and degree of both ascites and encephalopathy. The sample is composed primarily of women with severe or very severe disease (i.e., Child's class B and C).

Hormonal Status

Comparison to Normal Alcohol-Abstaining Postmenopausal Women

As may be seen in Table 5, the findings of earlier studies have been confirmed and extended by the statistically significant differences in levels of E_2, T, the estimate of aromatization (E_2:T ratio), and all three estrogen-responsive pituitary hormones. Further, the hormonal relationships observed in normal alcohol-abstaining control women are either undetectable (i.e., non-significant correlation coefficients between E_2 and LH, FSH, and BMI) or reversed (i.e., a positive rather than negative significant correlation of age with LH and FSH) in the alcoholic cirrhotic women.

TABLE 4
Characteristics of 66 Postmenopausal Women with Alcohol-Induced Cirrhosis

Characteristic	Value
Age (years)	56.8±1.2*
Body Mass Index (BMI)	24.8±0.6*
Ovariectomy (and hysterectomy)	20.6%*
Natural menopause	67.6%*
Age at natural menopause	45.3±1.0**
Hysterectomy (+ removal of one ovary)	11.8%*
Estrogen replacement therapy	0%*
Albumin (mg/dl)	3.1±0.1***
Prothrombin time (seconds)	14.8±0.4***
Total bilirubin (mg/dl)	5.1±1.0***
GGTP (miu/ml)	136±21***
Ascites	68.2%
Encephalopathy	53.0%
Child's class	
A	24.6%
B	56.1%
C	19.3%

*not significantly different from normal alcohol-abstaining controls
**statistically earlier than in normal postmenopausal women
***outside the normal range

The Effect of Liver Disease Severity on Levels of Hormones

The hormone data summarized in Table 5 clearly demonstrate that the hormonal status of postmenopausal women with alcohol-induced cirrhosis is disrupted in major ways. What is not clear is the relative contribution to alterations in hormonal status made by alcohol abuse and that made by the presence of cirrhosis. The issue of the role of cirrhosis per se can perhaps be obliquely addressed by evaluating whether or not there are changes in levels of hormones associated with varying severities of liver disease. Fortunately, the distribution of the women by Child's class demonstrates that the study sample is composed of women at various stages of cirrhosis severity, a prerequisite for sensibly evaluating relationships between cirrhosis severity and levels of hormones. Based on the direction of differences in postmenopausal hormone levels between normal controls and alcoholic cirrhotic women, if hormone levels are related to cirrhosis severity, the following could be hypothesized: levels of LH, FSH and T might be inversely related, whereas E_2, the E_2:T ratio and PRL might be directly related to severity markers that increase with increasing cirrhosis severity (total bilirubin, prothrombin time, GGTP, and Child's class score), and oppositely correlated with albumin, which diminishes with increased severity.

The results of such analyses are summarized in Table 6. There are several points to be emphasized: 1) Levels of each hormone and the E_2:T ratio are all statistically correlated with at least one measure of cirrhosis severity. 2) These significant correlation coefficients are inverse or direct, consistent with the *a*

TABLE 5
Hormone Levels* and Relationships in 66 Postmenopausal Women with Alcohol-Induced Cirrhosis

Characteristic	**Alcoholic Cirrhotic Woman (n + 66)**	**Normal Alcohol-Abstaining Controls (n = 27)**
E_2 (pg/ml)	62.7±10.4**	27.5±3.3
T (ng/ml)	0.49±0.05**	0.74±0.08
E_2:T ratio	172±27**	44.5±5.8
LH (miu/ml)	8.0±1.4**	24.0±2.8
FSH (miu/ml)	28.0±4.4**	63.3±5.5
PRL (ng/ml)	16.4±2.2**	5.7±0.4
Correlation of age with:		
LH	r = +0.39**	r = –0.51**
FSH	r = +0.37**	r = –0.37**
Correlation of E_2 with:		
LH	r = +0.07	r = –0.32**
FSH	r = +0.08	r = –0.35**
BMI	r = –0.07	r = +0.36**

*normalizing logarithmic transformations used in analysis
**at least $p < 0.05$

TABLE 6
Correlation Coefficients of Hormone Levels* with Markers of Liver Disease Severity in 66 Postmenopausal Women with Alcohol-Induced Cirrhosis

	← *Hormone* →					
	E_2	**T**	**E_2:T**	**LH**	**FSH**	**PRL**
Albumin	ns	ns	ns	+0.260**	+0.273**	–0.213**
Prothrombin time	ns	–0.247**	ns	–0.534**	–0.494**	ns
Total bilirubin	ns	–0.295**	+0.345**	–0.424**	–0.363**	ns
GGTP	+0.354**	ns	+0.369**	ns	ns	+0.283**
Child's score	ns	ns	ns	–0.260**	ns	ns

*normalizing logarithmic transformations used in analysis
**at least $p < 0.05$

priori hypothesis for the direction of a given association. 3) The prevalence of appropriate hormone correlations is higher for the biochemical markers of function and injury than for the composite variable Child's class. Taken together, these findings suggest that the severity of cirrhosis significantly influences levels of hormones; therefore, it must be concluded that, in addition to alcohol abuse, cirrhosis per se makes a major contribution to the observed disruption of postmenopausal hormone levels and interrelationships.

CHANGES IN THE HORMONAL STATUS OF ALCOHOLIC, CIRRHOTIC, POSTMENOPAUSAL WOMEN FOLLOWING LIVER TRANSPLANTATION

Twelve postmenopausal women with alcohol-induced cirrhosis compose the preliminary study group.[12] All were participants both in a study of hormone levels among postmenopausal women with end-stage liver disease as well as in a follow-up study related to whether or not hormone level normalization occurs following liver transplantation for end-stage liver disease. The characteristics of the study sample of alcoholic postmenopausal women are summarized in Table 7. Two points are to be particularly noted: 1) The period of time elapsed following transplantation could be hypothesized to be sufficiently long for normalization of hormonal status to have occurred. 2) Among six women "at risk" for return of menstrual bleeding if the presence of cirrhosis were to have played a role in the induction of premature menopause, one-third experienced reestablishment of cyclic ovarian function.

The data related to changes in postmenopausal hormonal status following liver transplantation are summarized in Table 8 for nine eligible alcoholic women (continued postmenopausal state and no use of ERT). Although the results of this study must be viewed as being preliminary due to the small

TABLE 7
Characteristics of the 12 Alcoholic Postmenopausal Women Following Liver Transplantation

Characteristic	Value
Age at follow-up (years)	54.4±2.2
Time since transplantation (months)	24.4±5.8
BMI	27.3±1.3
Immunosuppression regimen	
Cyclosporine and prednisone	33.3%
FK 506 and prednisone	16.7%
FK alone	50.0%
Use of ERT at time of follow-up	8.3%
Use of alcoholic beverages	25.0%
Return of menstrual bleeding among women "at risk" (age less than 54 and an intact uterus)	2/6

TABLE 8
Changes in Hormone Levels[a] and Hormonal Relationships in 9 Eligible[b] Postmenopausal Women Following Liver Transplantation for Alcohol-Induced Cirrhosis

Hormone	Before Transplantation	After Transplantation
E_2 (pg/ml)	52.5±20.0	64.9±11.8[c]
T (ng/ml)	0.41±0.1	0.16±0.03[c,d]
E_2:T ratio	130±31	525±141[c,d]
LH (miu/ml)	10.6±4.6	31.9±6.2[d]
FSH (miu/ml)	21.1±8.1	61.9±10.9[d]
PRL (ng/ml)	42.4±19.1	16.1±3.3[c]
Estimate of estrogenization of gonadotropin levels		
LH:E_2 ratio (Normals: 87±10)	49±31	65±17[d]
FSH:E_2 ratio (Normals: 230±30)	83±58	128±32[d]
Correlation of E_2 with:		
LH	r = +0.305	r = –0.279
FSH	r = +0.329	r = –0.331
Weight	r = –0.032	r = +0.655[e]
BMI	r = +0.020	r = +0.714[e]

[a]normalizing logarithmic transformation used in analysis
[b]women receiving ERT or experiencing return of menstrual bleeding have been excluded from analysis
[c]significantly different from alcohol-abstaining normal controls
[d]at least $p < 0.05$; paired *t* test for before vs. after transplantation
[e]$p < 0.05$

sample size, the findings are nevertheless tantalizing. Although levels of LH and FSH returned into the normal range, the same is not true for levels of E_2, T, PRL or the E_2:T ratio. Although the correlation of LH and FSH with E_2 has attained the pattern seen in normal controls, the normalization of gonadotropin levels may be more apparent than real: Given that E_2 levels continue to be substantially elevated, the gonadotropin levels may be viewed as being inappropriately high, a view substantiated by the gonadotropin:E_2 ratios in comparison to these ratios in normal controls. The data related to changes and non-changes in postmenopausal hormonal status 2 years following successful liver transplantation raise more questions than they answer: Do these findings reflect the influence of immunosuppression? Do these findings reflect long-term effects of alcohol abuse, of end-stage liver disease, or of the combination of these two factors? Clearly, an immense amount of work remains.

SUMMARY

Data have now been accumulated that demonstrate that moderate alcoholic beverage consumption significantly influences a broad spectrum of hormones in normal postmenopausal women. Data have also been accumulated that demonstrate that the combination of alcohol abuse and alcohol-induced cirrhosis plays a major role in the disruption of postmenopausal hormonal status. Particularly interesting are new data that demonstrate that cirrhosis severity per se influences levels of postmenopausal hormones and their interrelationships. Of major interest are the preliminary findings suggesting that the hormonal status of postmenopausal women with alcohol-induced cirrhosis does not appear to fully normalize following liver transplantation. In summary, then, it is clear that alcoholic beverage consumption, both moderate and abusive, is fully capable of exerting a substantial effect on the hormonal status of postmenopausal women.

ACKNOWLEDGMENTS

This work has been supported by Grant R01 AA06772 from the National Institute on Alcohol Abuse and Alcoholism. The author thanks Marilyn Bonham-Leyba for assistance in preparation of the manuscript.

REFERENCES

1. **Becker, U., Gluud, C., Farholt, S., Bennett, P., Miciv, S., Svenstrup, B., and Hardt, F.,** Menopausal age and sex hormones in postmenopausal women with alcoholic and non-alcoholic liver disease, *J. Hepatol.*, 13, 25, 1991.
2. **Carlstrom, K., Eriksson, S., and Rannevik, G.,** Sex steroids and steroid binding proteins in female alcoholic liver disease, *Acta Endocrinol.*, 111, 75, 1986.
3. **Cobb, C.F., Van Thiel, D.H., Gavaler, J.S., and Lester, R.,** Effects of ethanol and acetaldehyde on the rat adrenal, *Metabolism,* 30, 537, 1981.
4. **Gavaler, J.S.,** Alcohol effects in postmenopausal women: Alcohol and estrogens, in the *Medical Diagnosis and Treatment of Alcoholism,* first edition, Mendelson, J.H., and Mello, N.K., Eds., McGraw Hill, New York, 1992, 623.
5. **Gavaler, J.S.,** Effects of alcohol use and abuse on the endocrine status in expanded study samples of postmenopausal women, *Alcohol Endocr. Syst.*, Zakhari, S., Ed., NIH Pub., 1993, 171.
6. **Gavaler, J.S.,** Alcohol and nutrition in postmenopausal women, *J. Am. Coll. Nutr.*, 12, 347, 1993.
7. **Gavaler, J.S., Deal, S.R., Van Thiel, D.H., Arria, A., and Allan, M.J.,** Alcohol and estrogen levels in postmenopausal women: The spectrum of effect, *Alcoholism: Clin. Exp. Res.*, 17, 786, 1993.
8. **Gavaler, J.S., Love, K., Van Thiel, D.H., Farholt, S., Gluud, C., Monteiro, W., Galvao-Teles, A., Conton-Ortega, T., and Cuervas-Mons, V.,** An international study of the relationship between alcoholic beverage consumption and postmenopausal estradiol levels, *Proc. Fifth ISBRA (International Society for Biomedical Research on Alcoholism) Congress,* Kalant, H., Khanna, J.M., and Israel, Y., Eds., Pergamon Press, New York, 1991, 327.

9. **Gavaler, J.S., and Love, K.,** Detection of the relationship between moderate alcoholic beverage consumption and serum levels of estradiol in normal postmenopausal women: Effects of alcohol consumption quantitation methods and sample size adequacy, *J. Stud. Alcohol,* 53, 389, 1992.
10. **Gavaler, J.S., and Van Thiel, D.H.,** Hormonal status of postmenopausal women with alcohol-induced cirrhosis: Further findings and a review of the literature, *Hepatology,* 16, 312, 1992.
11. **Gavaler, J.S., and Van Thiel, D.H.,** The association between moderate alcoholic beverage consumption and serum estradiol levels in normal postmenopausal women: Relationship to the literature, *Alcoholism: Clin. Exp. Res.,* 16, 87, 1992.
12. **Gavaler, J.S., Van Thiel, D.H., and Deal, S.R.,** The hormonal status of alcoholic postmenopausal women following liver transplantation, *Alcoholism: Clin. Exp. Res.,* 17, 505 (Abstr.), 1993.
13. **Gordon, G.G., Olivo, J., Fereidoon, F., and Southren, A.L.,** Conversion of androgens to estrogens in cirrhosis of the liver, *J. Clin. Endocrinol. Metab.,* 40, 1018, 1975.
14. **Gordon, G.G., Altman, K., Southren, A.L., Rubin, E., and Lieber, C.S.,** Effect of alcohol (ethanol) administration on sex-hormone metabolism in normal men, *N. Engl. J. Med.,* 295, 793, 1976.
15. **Hugues, J.N., Perret, G., Adessi, G., Coste, T., and Modigliani, E.,** Effects of chronic alcoholism on the pituitary-gonadal function of women during menopausal transition and in the postmenopausal period, *Biomedicine,* 29, 279, 1978.
16. **Hugues, J.N., Coste, T., Perret, G., Jayle, M.F., Sebaoun, J., and Modigliani, E.,** Hypothalamo-pituitary ovarian function in thirty-one women with chronic alcoholism, *Clin. Endocrinol.,* 12, 543, 1980.
17. **James, V.H.T., Green, J.R.B., Walker, J.G., Goodall, A., Short, F., Jones, D.L., Noel, C.T., and Reed, M.J.,** The endocrine status of postmenopausal cirrhotic women, *Endocrines and the Liver,* Chiandussi, M., Langer, L., Chopra, I.J., and Martini, L., Eds., Academic Press, San Diego, CA, 1982, 417.
18. **Jasonni, V.M., Bulletti, C., Bolelli, G.F., Franceschetti, F., Bonavia, M., Ciotti, P., and Flamigni, C.,** Estrone sulfate, estrone and estradiol concentrations in normal and cirrhotic postmenopausal women, *Steroids,* 41, 569, 1983.
19. **Longcope, C., Pratt, J.H., Schneider, S., and Fineberg, E.,** Estrogen and androgen dynamics in liver disease, *J. Endocrinol. Invest.,* 7, 629, 1984.
20. **U.S. Dept. of Agriculture and U.S. Dept. of Health and Human Services,** Home and garden Bull. No. 232, *Nutrition and Your Health: Dietary Guidelines for Americans,* Washington, D.C., 1990.
21. **Wright, H.I., Gavaler, J.S., Tabasco-Minguillan, J., and Van Thiel, D.H.,** Endocrine effects of alcohol abuse in males, in *The Medical Diagnosis and Treatment of Alcoholism,* Mendelson, J.H., and Mello, N.K., Eds., McGraw-Hill, New York, 1992, 341.

INDEX

A

D

E

F

G

I

J

K

L

M

N

O

P

R

S

T

V

W

Y

Z